Coral Reefs of the United Kingdom Overseas Territories

Coral Reefs of the World

Volume 4

Series editors: Bernhard M. Riegl, Richard E. Dodge

For further volumes:
http://www.springer.com/series/7539

Charles R.C. Sheppard
Editor

Coral Reefs of the United Kingdom Overseas Territories

Springer

Editor
Charles R.C. Sheppard
School of Life Sciences
University of Warwick
Coventry
United Kingdom

ISSN 2213-719X ISSN 2213-7203 (electronic)
ISBN 978-94-007-5964-0 ISBN 978-94-007-5965-7 (eBook)
DOI 10.1007/978-94-007-5965-7
Springer Dordrecht Heidelberg New York London

Library of Congress Control Number: 2013932342

Cover illustration: Antillogorgia acerosa British Virgin Island. Photo by Charles R.C. Sheppard

Printed on acid-free paper

Springer is part of Springer Science+Business Media (www.springer.com)

Preface

Where and What Are the British Overseas Territories?

The UK's 14 Overseas Territories are highly diverse. They include the world's most remote community (Tristan da Cunha) and one of the richest (Bermuda). They include vast areas of ocean and, in the case of Antarctica, land six times the size of the UK. A UK Government White Paper published in June 2012 gives much useful background on all the Overseas Territories: http://www.fco.gov.uk/resources/en/pdf/publications/overseas-territories-white-paper-0612/ot-wp-0612.

Figure 1 shows the map of the UK Overseas Territories with coral reefs, which are covered in this volume, together with their 200 nautical mile boundaries; this map includes Ascension which has appreciable corals but which does not have coral reefs. Figure 1.1 of Chap. 1 shows the locations of all UK Overseas Territories. Each Overseas Territory has its own relationship with the UK and constitutional relationships continue to evolve. The UK, the Overseas Territories and the Crown Dependencies (the Channels Islands and the Isle of Man) form one undivided Realm. Each Territory has its own constitution, its own government and its own local laws. The constitutions set out the powers and responsibilities of the institutions of government, which for most Territories include a Governor or Commissioner, and elected legislature and ministers. Governors and Commissioners are appointed by the Queen on the advice of ministers in the UK, and in general have responsibility for external affairs, defence, internal security and the appointment, discipline and removal of public officers.

Bermuda is the most populous OT with a population of 66,000, and some OTs, such as British Indian Ocean Territory and British Antarctic Territory, have no permanent populations but do have scientific or military facilities. Territories with resident populations have an office in London.

The Foreign and Commonwealth Office in London is primarily responsible for the Territories, but most government departments also have responsibilities, particularly the Department for Environment, Food and Rural Affairs which assists them in meeting the requirements of various conventions, such as the Convention on International Trade in Endangered Species, Convention on Biological Diversity and Convention on Migratory Species. The Department for International Development has been key for several environmental projects, as has the Department of Energy and Climate Change.

Coral Reefs in the Territories

Eight UK Territories have coral reefs, indeed some are made entirely or mostly from them, and these collectively comprise a significant part of the world's total area of this very highly diverse and productive marine habitat. The marine environment of a few more such as Gibraltar, Ascension and the Cyprus military bases have corals, sometimes in modest profusion, but the corals do not form reefs. The largest reef systems in the Territories include the Pitcairn group in the Pacific, which has four atolls or islands with reefs, and British Indian Ocean Territory which contains five atolls with islands and numerous submerged atolls and banks. Five Territories

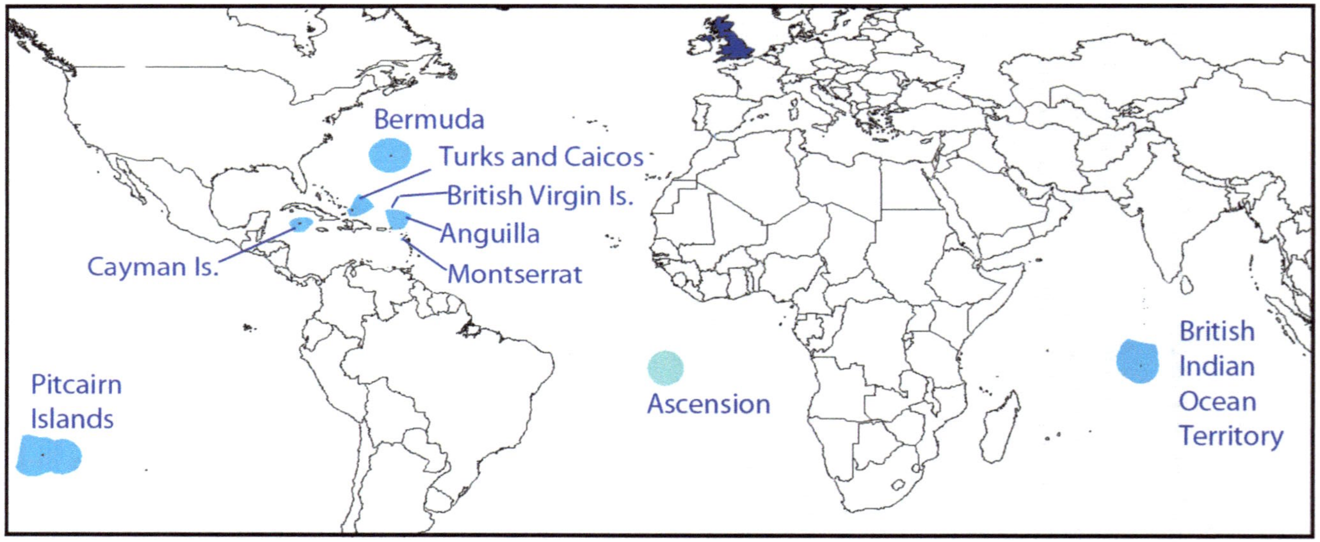

Fig. 1 The UK Overseas Territories which have coral reefs or which have abundant corals, with the extent of their 200 nautical mile territorial boundaries. (Ascension, in paler blue, has corals with no reefs, as do the Mediterranean Territories of Gibraltar and those in Cyprus.)

are in the greater Caribbean region. These have abundant reefs and many islands which are entirely coralline, while Bermuda further north in the Atlantic is an atoll and the most northerly reef system in the Atlantic Ocean. All Territories with reefs have at least one chapter in this volume, as does Ascension which has corals but no reefs.

Environment and Conservation

In the case of multilateral agreements and conservation and wildlife treaties, the UK is responsible for international agreements, though it is the Territory itself that is responsible for implementing them. That the UK is party to a particular conservation agreement does not necessarily mean that it applies to a particular UK Territory, and a territory may not be a signatory to one or other if political, military or logistical concerns are deemed to make that course impossible or impractical. In 2001, Territories signed individual Environment Charters, which are important documents encompassing (mostly very briefly) intentions and responsibilities regarding environmental governance and processes. These are sometimes only aspirational where there is no method of measuring success and limited mechanisms for implementing their terms, and some aspects of the Charters have been taken more seriously than others. The UK Overseas Territories Conservation Forum is an NGO that embraces all Territories, and has the aims of raising public awareness about the wealth of biodiversity in the Territories, warning of potential threats to the environment or to various species, compiling data and facilitating the implementation of conservation conventions, and also of supporting conservation groups and facilitating funding and conservation project management (http://www.ukotcf.org/index.cfm). Most Territories also have their own local NGOs which are engaged, sometimes very actively, in conservation issues, along with or beside government departments. However, government departments in several Territories may be very small and under-resourced, and may be more concerned with, e.g. fisheries than with conservation, or may, indeed, cover both of those commonly conflicting subject areas.

For those tropical Territories that have coral reefs, the amount of scientific or ecological information that is known varies hugely. Bermuda has had a research station, museum and other facilities for many decades, and more is known about its marine environment and habitats than any other. Partly this is due also to its location, where it provided a convenient site for researchers from the UK and the USA, so that much early ecological and descriptive work, and taxonomy, was carried out there. The British Indian Ocean Territory has the largest area of reefs of all of them and has likewise supported a considerable amount of research in recent years. Many Territories in the Caribbean have supported substantial amounts, though Montserrat, famous for the destructive eruptions of its volcano which recently caused extensive damage including obliteration of its Capital, has supported very little investigations of its reefs. Likewise, the several components of Pitcairn in the Pacific have received little study. However, it has been possible to find enough scientific information about all of them for this volume.

As is the case around the world, coral reefs of most Territories have been subject to substantial environmental damage from overuse or extraction of resources. Over-fishing, poor agriculture on adjacent land that causes run-off which adds surplus nutrients and blanketing sediments, input of sewage and marine diseases from effluents, and numerous other forms of pollution have all degraded many reefs of the world, including those of the British UK Territories. Global climate changes have added a further "layer" to the stresses felt by reefs: Warming episodes affected reefs in British Indian Ocean Territories in 1998 and subsequently, and in the Caribbean in 2005 especially. Reefs serve as breakwaters for the land, and mortality of corals from warming has reduced the protection afforded by the reefs. Furthermore, sea level is rising, which brings its own problems to small low islands, and the sea is becoming less alkaline, which causes other stresses, especially to integral components such as calcareous red algae which are critical components of the reef structure. As noted by the FCO (2012) White paper on the Overseas Territories:

> The Intergovernmental Panel on Climate Change has identified the Territories as amongst the 'most vulnerable' and 'virtually certain to experience the most severe impacts' of climate change. This will mean sea level rise; changes in weather patterns, including higher intensity of extreme weather events; coral bleaching; ocean acidification; and sea temperature changes. Other immediate threats include land use change; waste management; invasive species; and threats to habitats from unsustainable development.

The following chapters cover aspects of all those Territories with coral reefs. For some, there is very limited information available, while three have several chapters. These show the Territories to be a remarkable set of places, whose territorial waters greatly exceed than that of the UK and which have a marine diversity far greater than that found around the UK mainland. Chapter 1 maps the reefs of all of them, and notes that the UK is the twelfth biggest coral reef nation in the world. Defining coral reefs precisely is problematic and highly dependent on method, but however a reef is defined (with or without its sandy back-reef area, with or without the seagrass beds that intermingle with them, with or without the biologically connected fish nursery grounds in adjacent mangroves) their biological wealth is incalculable. Their value (or price – not the same at all!) varies according to opinion and method of measurement, and here one Territory has a chapter devoted to trying to assess the value of their reefs, an exercise which may have much meaning to economists but which is anathema to many scientists at the same time! The reefs of the UK Overseas Territories provide huge biodiversity, biological productivity, cultural value and large tourist revenue, and thus are enormously important to the Territories themselves.

Contents

1 **An Appraisal of the Extent and Geomorphological Diversity of the Coral Reefs of the United Kingdom Dependent Territories** 1
S. Hamylton and S. Andréfouët

2 **Coral Reefs of Anguilla** 13
Stuart P. Wynne

3 **Introduction to Reefs and Shorelines of the British Virgin Islands** 23
Shannon Gore

4 **Marine Protected Areas and Management in the British Virgin Islands** 37
Nancy Woodfield Pascoe, Joseph Smith-Abbott, and Shannon Gore

5 **Anegada: An Emergent Pleistocene Reef Island** 47
Shannon Gore

6 **Coral Reefs of the Cayman Islands** 61
Alan Logan

7 **Biology and Ecology of the Coral Reefs of the Cayman Islands** 69
John R. Turner, Croy McCoy, Matt Cottam, Jeremy Olynik, Austin Timothy, Janice Blumenthal, John Bothwell, Fred J. Burton, Phillipe Bush, Paul Chin, Oliver Dubock, Kristian D. Godbeer, James Gibb, Lisa Hurlston, Bradley J. Johnson, Alan Logan, Gene Parsons, and Gina Ebanks-Petrie

8 **Coral Reefs of Montserrat** 89
Andrew Myers

9 **The Reefs of the Turks and Caicos Islands** 97
Alan Logan and Kathleen Sullivan Sealey

10 **Introduction to Bermuda: Geology, Oceanography and Climate** 115
Kathryn A. Coates, James W. Fourqurean, W. Judson Kenworthy, Alan Logan, Sarah A. Manuel, and Struan R. Smith

11 **Biology and Ecology of Corals and Fishes on the Bermuda Platform** 135
Struan R. Smith, Samantha de Putron, Thad J.T. Murdoch, Joanna M. Pitt, and Ivan Nagelkerken

12 **Biogeography, Biodiversity and Connectivity of Bermuda's Coral Reefs** 153
Jan M. Locke, Kathryn A. Coates, Jaret P. Bilewitch, Lyndsey P. Holland, Joanna M. Pitt, Struan R. Smith, and Henry G. Trapido-Rosenthal

13 **Threats to Coral Reefs of Bermuda** 173
Struan R. Smith, Samia Sarkis, Thad J.T. Murdoch, Ernesto Weil, Aldo Croquer, Nicholas R. Bates, Rodney J. Johnson, Samantha de Putron, and Andreas J. Andersson

14 Scleractinia, Octocorallia and Antipatharia of Bermuda's Reefs and Deep-Water Coral Communities: A Taxonomic Perspective Including New Records 189
Jan M. Locke, Jaret P. Bilewitch, and Kathryn A. Coates

15 Total Economic Value of Bermuda's Coral Reefs: A Summary 201
Samia Sarkis, Pieter J.H. van Beukering, Emily McKenzie, Luke Brander, Sebastiaan Hess, Tadzio Bervoets, Lois Looijenstijn-van der Putten, and Mark Roelfsema

16 Ascension Island's Hermatypic but Non-Reef Building Corals 213
Robert Irving

17 British Indian Ocean Territory (the Chagos Archipelago): Setting, Connections and the Marine Protected Area 223
Charles R.C. Sheppard, Brian W. Bowen, Allen C. Chen, Matthew T. Craig, Jeff Eble, Nancy Fitzsimmons, Chai-hsia Gan, Michelle R. Gaither, Matthew Gollock, Shashank Keshavmurthy, Heather Koldewey, Jeanne A. Mortimer, David Obura, Miriam Pfeiffer, Alex D. Rogers, Anne L.S. Sheppard, Catherine Vogler, Gert Wörheide, Ming-Che Yang, and Chris Yesson

18 Coral Reefs of the Chagos Archipelago, Indian Ocean 241
Charles R.C. Sheppard, Mebs Ateweberhan, Allen C. Chen, Alasdair Harris, Rachel Jones, Shashank Keshavmurthy, Carl Lundin, David Obura, Sam Purkis Peter Raines, Bernhard Riegl, Michael H. Schleyer, Anne L.S. Sheppard, Jerker Tamelander, John R. Turner, Shakil Visram, and Sung-Yin Yang

19 The Status of Coral Reef Fish Assemblages in the Chagos Archipelago, with Implications for Protected Area Management and Climate Change 253
Nicholas A.J. Graham, Morgan S. Pratchett, Tim R. McClanahan, and Shaun K. Wilson

20 Coral Islands of the British Indian Ocean Territory (Chagos Archipelago) 271
Peter Carr, Jesse C. Hillman, Mark R.D. Seaward, Scott Vogt, and Charles R.C. Sheppard

21 Contaminants, Pollution and Potential Anthropogenic Impacts in Chagos/BIOT 283
James W. Readman, Francis DeLuna, Ralf Ebinghaus, AntenorNestor Guzman, Andrew R.G. Price, Emily E. Readman, Anne L.S. Sheppard, Victoria A. Sleight, Renate Sturm, Richard C. Thompson, Andrew Tonkin Hendrik Wolschke, Robyn J. Wright, and Charles R.C. Sheppard

22 Coral Reefs of the Pitcairn Islands 299
Robert Irving and Terry Dawson

Index 319

List of Contributors

Andreas J. Andersson Scripps Institute of Oceanography, 9500 Gilman Dr, San Diego, CA 92093, USA, e-mail: aandersson@ucsd.edu

S. Andréfouët Institut de Recherche pour le Développement, U227 Coreus Noumea, Nouméa, New Caledonia, e-mail: serge.andrefouet@gmail.com

Mebs Ateweberhan School of Life Sciences, University of Warwick, Coventry, CV4 7AL, UK, e-mail: mateweberhan@wcs.org

Nicholas R. Bates Bermuda Institute for Ocean Sciences, 17 Biological Lane, St Georges GE01, Bermuda, e-mail: nick.bates@bios.edu

Tadzio Bervoets Institute for Environmental Studies (IVM), Vrije Universiteit University Amsterdam, De Boelelaan 1085, Amsterdam, 1081 HV, The Netherlands, e-mail: manager@naturefoundationsxm.org

Pieter J.H. van Beukering Institute for Environmental Studies (IVM), Vrije Universiteit University Amsterdam, De Boelelaan 1085, Amsterdam, 1081 HV, The Netherlands, e-mail: pieter.van.beukering@vu.nl

Jaret P. Bilewitch School of Biological Sciences, University of Queensland, Brisbane, QLD, 4072, Australia, e-mail: jbilewitch@gmail.com

Janice Blumenthal Department of the Environment, Cayman Islands Government, Grand Cayman, Cayman Islands, e-mail: Janice.Blumenthal@gov.ky

John Bothwell Department of the Environment, Cayman Islands Government, Grand Cayman, Cayman Islands, e-mail: John.Bothwell@gov.ky

Brian W. Bowen Hawai'i Institute of Marine Biology, 1346, Kane'ohe, HI 96744, USA, e-mail: bbowen@hawaii.edu

Luke Brander Institute for Environmental Studies (IVM), Vrije Universiteit University Amsterdam, De Boelelaan 1085, Amsterdam, 1081 HV, The Netherlands, e-mail: lukebrander@gmail.com

Fred J. Burton National Trust for the Cayman Islands, Family Park, 558 South Church Street, Grand Cayman, Cayman Islands, e-mail: fjburton@blueiguana.ky

Philippe Bush Department of the Environment, Cayman Islands Government, Grand Cayman, Cayman Islands, e-mail: Phillippe.bush@gov.ky

Peter Carr School of Life Sciences, University of Warwick, Coventry, CV4 7AL, UK, e-mail: petecarr1@hotmail.com

Allen C. Chen Biodiversity Research Centre, Academia Sinica, 128 Academia Road, Taipei, Nankang, 115, Taiwan, e-mail: acropora.chen@gmail.com

Paul Chin Department of the Environment, Cayman Islands Government, Grand Cayman, Cayman Islands, e-mail: Paul.Chin@gov.ky

Kathryn A. Coates Department of Conservation Services, Marine Ecology Section, 17 North Shore Road, Smith's Parish, FL 03, Bermuda, e-mail: kathryn.coates@dilke.com

Matt Cottam Department of the Environment, Cayman Islands Government, Grand Cayman, Cayman Islands, e-mail: Mat_Cottam@yahoo.com

Matthew T. Craig Department of Marine Sciences and Environmental Studies, University of San Diego, San Diego, CA, USA, e-mail: matthewcraig4@gmail.com

Aldo Croquer Departamento de Estudios Ambientales, Universidad Simon Bolivar, Caracas 1080 A, Venezuela, e-mail: acroquer@usb.ve

Terry Dawson School of the Environment, University of Dundee, Dundee, DD1 4HN, Scotland (UK), e-mail: t.p.dawson@dundee.ac.uk

Francis DeLuna NAVFACFE PWD DG Environmental, PSC 466 Box 5, Gainesville, FPO AP, 96595-0005, USA, e-mail: Francis.DeLuna.RP@fe.navy.mil

Oliver Dubock Department of the Environment, Cayman Islands Government, Grand Cayman, Cayman Islands

Gina Ebanks-Petrie Department of the Environment, Cayman Islands Government, Grand Cayman, Cayman Islands, e-mail: Gina.Ebanks-Petrie@gov.ky

Ralf Ebinghaus Department for Environmental Chemistry, Institute of Coastal Research, Helmholtz-Zentrum Geesthacht, Geesthacht, Germany, e-mail: ralf.ebinghaus@hzg.de

Jeff Eble Center for Insect Science, University of Arizona, 1140 E. South Campus Dr, Tucson, AZ, 85721, USA, e-mail: eble@email.arizona.edu

Nancy Fitzsimmons Faculty of Applied Science, University of Canberra, Canberra, ACT 2601, Australia, e-mail: nancy.Fitzsimmons@canberra.edu.au

James W. Fourqurean Department of Biological Sciences and SERC, Florida International University, 3000 NE 151 Street, North Miami FL33181, USA, e-mail: fourqure@fiu.edu

Chai-hsia Gan Biodiversity Research Centre, Academia Sinica, 128 Academia Road, Taipei, Nankang, 115, Taiwan, e-mail: chgan@gate.sinica.edu.tw

Michelle R. Gaither Ichthyology Section, California Academy of Sciences, San Francisco, CA, 94118, USA, e-mail: michellergaither@gmail.com

James Gibb Department of the Environment, Cayman Islands Government, Grand Cayman, Cayman Islands, e-mail: James.Gibb@gov.ky

Kristian D. Godbeer Department of the Environment, Cayman Islands Government, Grand Cayman, Cayman Islands

Matthew Gollock Zoological Society of London, Regents Park, London, NW1 4RY, UK, e-mail: matthew.gollock@zsl.org

Shannon Gore Conservation and Fisheries Department, Government of the Virgin Islands, 3323, Road Town, Tortola VG1110, British Virgin Islands, e-mail: sd_gore@yahoo.com

Nicholas A.J. Graham ARC Centre of Excellence for Coral Reef Studies, James Cook University, Townsville, QLD, 4811, Australia, e-mail: nick.graham@jcu.edu.au

Antenor Nestor Guzman NAVFACFE PWD DG Environmental, PSC 466 Box 5, Gainesville, FPO AP, 96595-0005, USA, e-mail: AntenorNestor.Guz.RP@fe.navy.mil

S. Hamylton School of Earth and Environmental Sciences, University of Wollongong, Northfields Avenue, Wollongon, NSW 2522, Australia, e-mail: sarah_hamylton@uow.edu.au

Alasdair Harris Omnibus Business Centre, 39-41 North Road, N7 9DP, London, e-mail: al@blueventures.org

Sebastiaan Hess Institute for Environmental Studies (IVM), Vrije Universiteit University Amsterdam, De Boelelaan 1085 Amsterdam1081 HV, The Netherlands, e-mail: sebastiaan.hess@ivm.vu.nl

Jesse C. Hillman Spring Cottage, Bluehill St. Helena Island, South Atlantic, STHL 1ZZ, USA, e-mail: hillman.jesse@gmail.com

Lyndsey P. Holland Biosciences, College of Life and Environmental Sciences, Geoffrey Pope, University of Exeter, Stocker Road, Exeter EX4 4QD, UK, e-mail: lynzholland@yahoo.co.uk

Lisa Hurlston Department of the Environment, Cayman Islands Government, Grand Cayman, Cayman Islands

Robert Irving Sea-Scope Marine Environmental Consultants, Devon, England (UK), e-mail: Robert@sea-scope.co.uk

Bradley J. Johnson Department of the Environment, Cayman Islands Government, Grand Cayman, Cayman Islands, e-mail: Bradley.Johnson@gov.ky

Rodney J. Johnson Bermuda Institute for Ocean Sciences, 17 Biological Lane St Georges GE01, Bermuda, e-mail: rod.johnson@bios.edu

Rachel Jones Zoological Society of London, Regents Park, London, NW1 4RY, UK, e-mail: Rachel.jones@zsl.org

W. Judson Kenworthy 109 Holly Lane, 28516 Beaufort, NC, USA, e-mail: jud.kenworthy@gmail.com

Shashank Keshavmurthy Biodiversity Research Centre, Academia Sinica, 128 Academia Road, Taipei, Nankang, 115, Taiwan, e-mail: shashank@gate.sinica.edu.tw

Heather Koldewey Zoological Society of London, Regents Park, London, NW1 4RY, UK, e-mail: Heather.koldewey@zsl.org

Jan M. Locke Department of Conservation Services, Marine Ecology Section, 17 North Shore Road, Smith's Parish FL 03, Bermuda, e-mail: jml100@gmail.com

Alan Logan Centre for Coastal Studies, University of New Brunswick, 5050, Saint John, E2L 4L5 NB, Canada, e-mail: logan@unbsj.ca

Lois Looijenstijn-van der Putten Institute for Environmental Studies (IVM), Vrije Universiteit University Amsterdam, De Boelelaan 1085 Amsterdam 1081 HV, The Netherlands

Carl Lundin IUCN Marine Programme, Rue Mauverney 28, Gland, 1196, Switzerland, e-mail: carl.lundin@iucn.org

Sarah A. Manuel Department of Conservation Services, Marine Ecology Section, 17 North Shore Road, Smith's Parish, FL 03, Bermuda, e-mail: smanuel@gov.bm

Croy McCoy Department of the Environment, Cayman Islands Government, Grand Cayman, Cayman Islands, e-mail: Croy.McCoy@gov.ky

Tim R. McClanahan Wildlife Conservation Society, Marine Programs, Bronx, NY, 10460, USA

Emily McKenzie Joint Nature Conservation Committee, Peterborough, UK, e-mail: Emily.McKenzie@wwfus.org

Jeanne A. Mortimer Department of Biology, University of Florida, Gainesville, FL, USA, e-mail: Mortimer@ufl.edu

Thad J.T. Murdoch Bermuda Zoological Society, 40 North Shore Road, Smith's Parish FL04, Bermuda, e-mail: tjmurdoch@gov.bm

Andrew Myers Montserrat Reef Project, 289, Brades, Montserrat, Eastern Caribbean or British West Indies, e-mail: montserrat.reef.project@gmail.com

Ivan Nagelkerken School of Earth and Environmental Sciences, University of Adelaide, Adelaide SA 5005, Australia, e-mail: ivan.nagelkerken@adelaide.edu.au

David Obura CORDIO East Africa, #9 Kibaki Flats, Kenyatta Beach, Bamburi Beach 10135, Mombasa, 80101, Kenya, e-mail: dobura@cordioea.org

Jeremy Olynik Department of the Environment, Cayman Islands Government, Grand Cayman, Cayman Islands, e-mail: Jeremy.Olynik@gov.ky

Gene Parsons Department of the Environment, Cayman Islands Government, Grand Cayman, Cayman Islands, e-mail: Gene.Parsons@gov.ky

Miriam Pfeiffer RWTH Aachen University, Templergraben 55, Aachen, 52056, Germany, e-mail: pfeiffer@geol.rwth-aachen.de

Joanna M. Pitt Department of Environmental Protection, Marine Resources Section, 3 Coney Island Road, St. Georges CR04, Bermuda, e-mail: jpitt@gov.bm

Morgan S. Pratchett ARC Centre of Excellence for Coral Reef Studies, James Cook University, Townsville, QLD 4811, Australia, e-mail: Morgan.pratchett@jcu.edu.au

Andrew R.G. Price Environment Department, York University, YO1 5DD, York, UK

School of Life Sciences, University of Warwick, Gibbet Hill Campus, CV4 7AL, Coventry, UK, e-mail: andrew.price@warwick.ac.uk

Sam Purkis National Coral Reef Institute, Nova Southeastern University, Oceanographic Center, 8000 North Ocean Drive, Dania Beach, FL, 33004, USA, e-mail: purkis@nova.edu

Samantha de Putron Bermuda Institute for Ocean Sciences, 17 Biological Lane, St Georges GE01, Bermuda, e-mail: samantha.deputron@bios.edu

Peter Raines 71 Badminton Road, London, SW12 8BL, UK, e-mail: rainespeter@gmail.com

James W. Readman Plymouth Marine Laboratory, Plymouth, PL1 3DH, UK, e-mail: jwre@pml.ac.uk

Emily E. Readman Plymouth Marine Laboratory, Plymouth, PL1 3DH, UK, e-mail: jwre@pml.ac.uk

Bernhard Riegl National Coral Reef Institute, Nova Southeastern University, Oceanographic Center, 8000 North Ocean Drive, Dania Beach, FL, 33004, USA, e-mail: rieglb@nova.edu

Mark Roelfsema Institute for Environmental Studies (IVM), Vrije Universiteit University Amsterdam, De Boelelaan 1085 Amsterdam 1081 HV, The Netherlands

Alex D. Rogers Department of Zoology, University of Oxford, The Tinbergen Building, South Parks Road, Oxford, OX1 3PS, UK, e-mail: alex.rogers@zoo.ox.ac.uk

Samia Sarkis Department of Conservation Services, Marine Ecology Section, 17 North Shore Road, Smith's Parish FL 03, Bermuda, e-mail: scsarkis@gov.bm

Michael Schleyer Oceanographic Research Institute, Marine Parade, 10712, Durban, 4056, South Africa, e-mail: schleyer@ori.org.za

Mark R.D. Seaward School of Life Sciences, University of Bradford, Bradford, BD7 1DP, UK, e-mail: M.R.D.Seaward@Bradford.ac.uk

Anne L.S. Sheppard School of Life Sciences, University of Warwick, Coventry, CV4 7AL, UK, e-mail: a.sheppard@warwick.ac.uk

Charles R.C. Sheppard School of Life Sciences, University of Warwick, Coventry, CV4 7AL, UK, e-mail: charles.sheppard@warwick.ac.uk

Victoria A. Sleight School of Marine Science and Engineering, Plymouth University, Plymouth, PL4 8AA, UK

Joseph Smith-Abbott National Parks Trust of the Virgin Islands, 57 Main Street, Road Town, Tortola, British Virgin Islands, e-mail: director@bvinpt.org

Struan R. Smith Natural History Museum, Bermuda Aquarium, Museum and Zoo, 40 North Shore Road, Smith's Parish, FL04, Bermuda, e-mail: srsmith@gov.bm

Renate Sturm Department for Environmental Chemistry, Institute of Coastal Research, Helmholtz-Zentrum Geesthacht, Geesthacht, Germany, e-mail: Renate.Sturm@hzg.de

Kathleen Sullivan Sealey Department of Biology, University of Miami, Coral Gables, FL, USA, e-mail: ksealey56@gmail.com

Jerker Tamelander UNEP Division of Environmental Policy Implementation, UN, Rajdamnern Nok Av, Bangkok, 10200, Thailand, e-mail: tamelander@un.org

Richard C. Thompson School of Marine Science and Engineering, Plymouth University, Plymouth, PL4 8AA, UK, e-mail: R.C.Thompson@plymouth.ac.uk

Austin Timothy Department of the Environment, Cayman Islands Government, Grand Cayman, Cayman Islands, e-mail: Timothy.Austin@gov.ky

Andrew Tonkin School of Marine Science and Engineering, Plymouth University, Plymouth, PL4 8AA, UK

Henry G. Trapido-Rosenthal Department of Biology, Division of Natural Sciences and Mathematics, Chaminade University of Honolulu, Honolulu, HI 96816, USA, e-mail: henry.trapido-rosenthal@chaminade.edu

John R. Turner School of Ocean Sciences, Bangor University, Wales, UK, e-mail: j.turner@bangor.ac.uk

Shakil Visram Biodiversity Research Centre, Academia Sinica, 128 Academia Road Taipei, Nankang, 115, Taiwan, e-mail: shakilvisram@yahoo.co.uk

Catherine Vogler Department für Geo- und Umweltwissenschaften Paläontologie & Geobiologie, Ludwig-Maximilians-Universität, Richard-Wagner-Str. 10, 80333, Munich, Germany, e-mail: catherine.vogler@gmail.com

Scott Vogt Naval Facilities Engineering, Far East Command Yokosuka, Yokosuka, Japan, e-mail: Scott.Vogt@fe.navy.mil

Ernesto Weil Department of Marine Sciences, University of Puerto Rico, 9000, Mayagues, PR, 00680, USA, e-mail: ernesto.weil@upr.edu

Shaun K. Wilson Department of Environment and Conservation, Marine Science Program, Kensington, WA, 6151, Australia

Hendrik Wolschke Department for Environmental Chemistry, Institute of Coastal Research, Helmholtz-Zentrum Geesthacht, Geesthacht, Germany, e-mail: hendrik.wolschke@hzg.de

Nancy Woodfield Pascoe National Parks Trust of the Virgin Islands, 57 Main Street, Road Town, Tortola, British Virgin Islands, e-mail: planning@bvinpt.org

Gert Wörheide Department für Geo- und Umweltwissenschaften Paläontologie & Geobiologie, Ludwig-Maximilians-Universität, Richard-Wagner-Str. 10, Munich 80333, Germany, e-mail: woerheide@lmu.de

Robyn J. Wright School of Marine Science and Engineering, Plymouth University, Devon PL4 8AA, UK

Stuart P. Wynne Department of Fisheries and Marine Resources, Government of Anguilla, The valley, Anguilla, e-mail: stuart@stuartwynne.co.uk

Ming-Che Yang Biodiversity Research Centre, Academia Sinica, 128 Academia Road Taipei, Nankang, 115, Taiwan, e-mail: jaymcyang@gmail.com

Chris Yesson Zoological Society of London, Regents Park, London, NW1 4RY, UK, e-mail: Chris.Yesson@ioz.ac.uk

Sung-Yin Yang Biodiversity Research Centre, Academia Sinica, 128 Academia Road, Taipei, Nankang, 115, Taiwan, e-mail: likefh@gmail.com

An Appraisal of the Extent and Geomorphological Diversity of the Coral Reefs of the United Kingdom Dependent Territories

1

S. Hamylton and S. Andréfouët

Introduction

The 14 Dependent Territories governed by the United Kingdom (UK) Foreign and Commonwealth Office include in alphabetic order Anguilla, Ascension Island and Tristan da Cunha, British Antarctic Territory, Bermuda, British Indian Ocean Territory, British Virgin Islands, Cayman Islands, Falkland Islands, Gibraltar, Montserrat, Pitcairn Island, St Helena, South Georgia and South Sandwich Islands, Sovereign Base Areas on Cyprus and the Turks and Caicos Islands (Fig. 1.1). UK governance responsibilities for these territories include the strengthening of democracy, environmental protection, improvement of public services and law enforcement (Oldfield and Sheppard 1997)

Seven of these territories incorporate substantial reef systems, including Anguilla, Bermuda, the British Indian Ocean Territory (also known as the Chagos Islands), the British Virgin Islands, the Cayman Islands, the Pitcairn Islands and the Turks and Caicos Islands. These territories are composed of small remote islands that support a disproportionately large area of reefs, lagoons and associated marine biodiversity. The total reef area inside the Dependent Territories mapped by the Millennium Mapping Project is 4,712 km^2, which makes the UK approximately the twelfth reef nation of the World.

At the global scale, current estimates of national and regional reef areas derive from a variety of sources including marine charts and maps derived from remote sensing satellite images. They provide estimates that vary widely in accuracy. Basic information (reef or non-reef) on the location, extent and geomorphological nature of reefs systems has been compiled in the *World Atlas of Coral Reefs* published by the United Nations Environment Programme World Conservation Monitoring Centre (UNEP-WCMC, Spalding et al. 2001). This atlas provided a global overview of reef distribution, however, the level of detail was largely determined by the variable availability and specification (scale, accuracy, precision) of marine charts across different reef regions. As a consequence, the inventory was often inconsistent from one area to another (Wabnitz et al. 2010). Other areas have since benefited from specific high resolution mapping projects that have provided accurate estimates of reef areas, itemized in different categories that follow a predefined typology of reef types and habitats (for instance, for Hawaii see Rohmann et al. 2005).

Starting in 2004, a global inventory of reef geomorphological diversity and units has produced consistent maps of reef areas worldwide. The data source is made of Landsat 7 Enhanced Thematic Mapper Plus (ETM+) images, completed by occasional Landsat 5 and Aster satellite images. The project, named Millennium Coral Reef Mapping Project (MCRMP) has produced GIS files for many different parts of the world, as well as selected electronic atlases produced for non-GIS users. These regional atlases include Islands of the Central and Western Indian Ocean (Andréfouët et al. 2009b), Papua New Guinea (Andréfouët et al. 2006b) and the French Territories (Andréfouët et al. 2008). Other projects have distributed degraded (both in spatial and thematic resolutions) versions of the MCRMP products, for the Caribbean (Burke et al. 2004) and globally (Burke et al. 2010). The principles used to design the MCRMP typology, as well as the main hierarchical structure are described elsewhere (Andréfouët et al. 2006a; Andréfouët 2011). The typology and products have proved to be relevant in various contexts: for instance, we can cite geological appraisals (Andréfouët et al. 2009a), conservation planning (Green et al. 2009; Dalleau et al. 2010; Allnutt et al. 2012), fisheries and food security (Bell et al. 2009) and enhancement of the inventories of

S. Hamylton (✉)
School of Earth and Environmental Sciences, University of Wollongong, Northfields Avenue, Wollongong, NSW 2522, Australia
e-mail: sarah_hamylton@uow.edu.au

S. Andréfouët
Institut de Recherche pour le Développement, U227 Coreus Noumea, Nouméa, New Caledonia
e-mail: serge.andrefouet@ird.fr

C.R.C. Sheppard (ed.), *Coral Reefs of the United Kingdom Overseas Territories*, Coral Reefs of the World 4,
DOI 10.1007/978-94-007-5965-7_1, © Springer Science+Business Media Dordrecht 2013

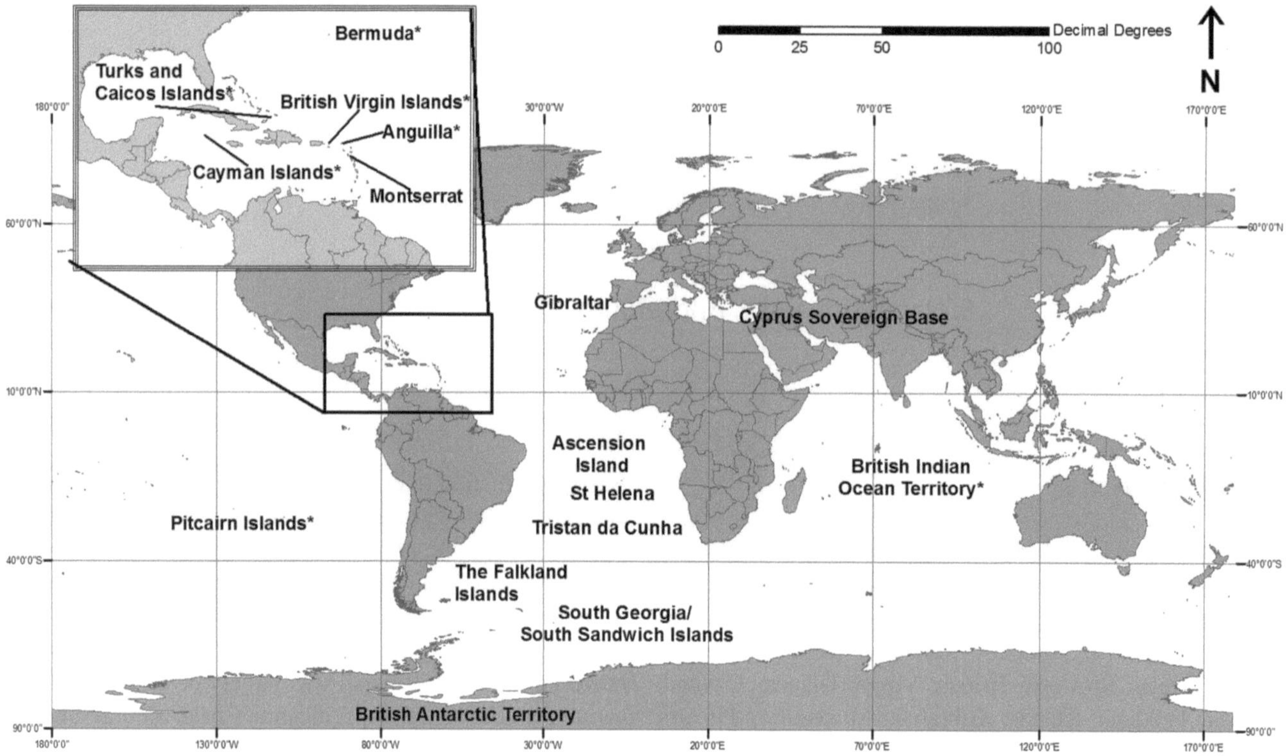

Fig. 1.1 Location of the UK Dependent Territories. *Asterisk* denotes reef territories

specific habitats such a seagrass beds (Wabnitz et al. 2008) or seamounts (Allain et al. 2008). Ongoing applications include connectivity and vulnerability to climate change assessments.

This chapter uses MCRMP products to conduct a geomorphological appraisal of the reef systems of the UK Dependent Territories with the following three objectives:

1. To identify where the coral reefs of the Dependent Territories are,
2. To generate a consistent and accurate measurement of reef areas, and
3. To delineate the morphological zones associated with these reef systems.

Methods

From a geomorphic perspective, coral reefs are three-dimensional structures that have evolved over geological timescales according to local sea level variations, subsidence, tectonics, hydrodynamic and climate forcing, and dominant living community types. They range in area from 1 to 100 km^2 in extent (Hopley 2011). As a result, reefs display a myriad of shapes and structures at a scale that can be resolved by high spatial resolution (1–30 m) optical spaceborne sensors down to a water depth of about 40 m in very clear waters. Typical depth penetration limit is around 20–30 m.

The UK Dependent Territories assessment was carried out using GIS layers generated as part of the Millennium Coral Reef Mapping Project (MCRCP). High resolution Landsat 7 ETM+ satellite images of coral reefs were interpreted using segmentation and photo-interpretation techniques to delineate regions belonging to different morphometric groups within a globally applicable typology of 800 classes (Andréfouët et al. 2006a). A reef "typology" refers to the definition of categories of reef objects according to a series of characteristics relevant for a given purpose (Andréfouët 2011).

The MCRCP hierarchical typology employs five hierarchical levels (Andréfouët et al. 2009a):

- Level 1: discriminates between oceanic and continental reefs;
- Level 2: discriminates the main reef complexes. Atolls, banks, uplifted atolls and islands can be either oceanic or continental. Then, the continental patch reefs, barrier reefs, fringing reefs and marginal structures are defined.

- Level 3: discriminates further details within each of the Level 2 blocks that are too numerous to cite here, but include for instance barrier, fringing and patch reefs of islands (either oceanic or continental), as well as different categories of these types: outer barrier, coastal barrier, multiple barrier, faro barrier, etc.; or lagoon exposed fringing, ocean-exposed fringing, etc.
- Level 4: defines the geomorphological units discernable on Landsat imagery within each of the previous blocks, including forereef, reef flat, pass, sedimentary terraces, enclosed lagoon, reef island, etc.
- Level 5: combines categories for Levels 1–4 to provide a final typology of 800 classes worldwide, although any single reef complex is likely to include between 1 and 20 classes at most. A level 5 label is thus the concatenation of the Levels 1–4. For instance a "Oceanic"/"Island"/"Coastal Barrier Reef"/"Reef Flat" makes a Level 5 description. Each combination is unique.

This appraisal was primarily conducted at Level 3 of the typological hierarchy, which provided an optimal simple, yet detailed level to facilitate comparison between the different territories.

In addition to the various geomorphologic attributes, each of the MCRCP polygon shapefiles has an associated "Reef" attribute that denoted whether the geomorphic unit supports significant coral communities (with a 1 value assigned to reefs and 0 assigned to non-reefs). For instance, the Level 4 classes "forereef", "reef flat", "subtidal reef flat", "pass", "pinnacle", etc. are considered as Reef, whereas terrace (i.e., sedimentary areas), lagoon, etc. are not considered part of the coral reef *per se*. To calculate the overall reef areas in the present assessment, we considered all polygons with a Reef attribute of value 1, for which the geometry calculator was employed to calculate the area, followed by the summary statistics tool to sum the areas of all the reef polygons.

Reef extent will clearly depend on the definition used: the "Reef" definition used here was largely consistent and compatible with a classical definition of coral reefs. For instance, it was compatible with the definition that state that coral reefs are "physical structure which has been built up and continues to grow over decadal time scales, as a result of the accumulation of calcium carbonate laid down by hermatypic corals and other organisms" (Spalding et al. 2001). Other definitions as provided by *The Encyclopedia of Modern Coral Reefs* elaborate further on the geomorphological components of coral reefs, including coral tracts (large areas of indefinite extent) and massive structures (in basal area and thickness and wave resistance) (Done 2011). These definitions were also consistent with the MCRMP definition.

Results

The reef systems of the Dependent Territories included 19 Level 3 reef classes, covering a total reef area of 4,712 km^2 (this figure relates to the areas that were identified as Reef only, as explained above). These included reef areas from atolls, barrier, banks, fringing and patch categories (Table 1.1).

If we compare the MCRCP estimates with Spalding et al. (2001), it can be seen from Fig. 1.3 that in many cases (5 out of 7 nations), the UNEP-WCMC study delineated more extensive reef areas than the MCRMP. The differences are significant in several instances (reaching 139% in the case of the British Virgin Islands). Similar discrepancies have been reported elsewhere (Andréfouët et al. 2006b; Wabnitz et al. 2010), illustrating the inherent variability associated with the different mapping approaches.

Discussion: Characteristics of Each Territory

For each Territory, we provide the Level 1 and Level 2 MCRMP label, and some information from the literature combined with the new inventory at Level 3 (Fig. 1.2).

Turks and Caicos (*Oceanic/Island*)

The limestone islands of the Turks and Caicos Islands group stretch across the northern extent of the relatively small Turks Bank and the much larger Caicos Bank (area 3,933 km^2). The margins of these banks slope down to a deeper shelf structure at a depth of 20–30 m that descends into oceanic water exceeding 4 km depth (Sullivan et al. 1994). Across the northern shore of the Caicos Islands is a coastal barrier reef complex with exposed fringing reefs on the oceanward aspect. Geologically, the subaerial islands that have developed in the Turks and Caicos Islands (area 924 km^2) consist of oolithic limestone sediments, with eolianite hills that have developed on the windward shores reaching up to 75 m above sea-level and karst limestone cliffs (Wanless and Dravis 1989). Of the overall typology, the Turks and Caicos Islands consisted of 11 different classes, with the shelf terrace occupying 60% of the mapped shallow platform.

Anguilla (*Oceanic/Island*)

Anguilla is a low coralline island (area 192 km^2), which has developed on a volcanic base, as part of the Lesser Antilles arc, which stretches 800 km across the eastern margin of

Table 1.1 A morphometric summary of the reef areas of the Dependent Territories (N.B. The right hand side column quotes % cover of the five most dominant geomorphic classes and therefore does not sum to 100%)

Territory	Area mapped/km²	Reef area/km²	# classes	Five dominant level 3 classes
Anguilla	2,537	43	7	*Shelf slope* (89%) *Main land* (8%) *Ocean exposed fringing* (2%) *Shelf terrace* (1%) *Bank lagoon* (0.4%)
Bermuda	733	340	7	*Island lagoon* (34%) *Outer Barrier complex* (32%) *Ocean exposed fringing* (13%) *Intra-lagoon patch reef complex* (11%) *Main land* (8%)
British Indian Ocean Territory	15,639	2,859	10	*Drowned atoll* (77%) *Atoll lagoon* (15%) *Atoll rim* (4%) *Drowned Bank* (3%) *Bank lagoon* (1%)
British Virgin Islands	4,500	138	9	*Shelf slope* (91%) *Lagoon exposed fringing* (3.5%) *Outer barrier reef complex* (2.8%) *Intra-lagoon patch reef complex* (1%) *Shelf patch reef complex* (0.8%)
Cayman Islands	471	471	9	*Main Land* (56%) *Island lagoon* (16%) *Coastal Barrier reef complex* (7%) *Ocean exposed fringing* (6%) *Shelf terrace* (5%)
Turks and Caicos Islands	6,885	822	11	*Shelf terrace* (60%) *Main land* (14%) *Shelf slope* (12%) *Shelf structure* (8%) *Coastal barrier reef complex* (3%)
Pitcairn Islands	89	39	9	*Main land* (52%) *Shelf slope* (20%) *Bank barrier* (10%) *Atoll rim* (7%) *Ocean exposed fringing* (7%)

the Caribbean Sea (Stein et al. 1982). The total area of reef systems mapped for Anguilla was 2,537 km² comprised of submerged banks and terraces, a shelf slope and fringing reefs. The fringing reefs have developed along the north and south coast of the island together with a number of offshore cays that support smaller reef platforms. The dominant morphological feature mapped for Anguilla was the submarine shelf shared with St Martin to the southeast, which occupied 89% of the area of the reef system. Along the northern extent of this Bank are exposed linear segments of fringing reef along the shelf edge. This 17 km-long reef along the southeast coast is considered to be one of the most important unbroken reefs in the eastern Caribbean (Putney 1982).

British Virgin Islands (*Oceanic/ Island*)

The Virgin Islands constitute the eastern extremity of the Greater Antilles arc and, in administrative terms, the shallow shelf on which the Virgin Islands sit can be subdivided into the US Virgin islands in the lower south western portion of the shelf and the British Virgin Islands on the upper north eastern portion. There are 40 uplifted volcanic islands, small cays,

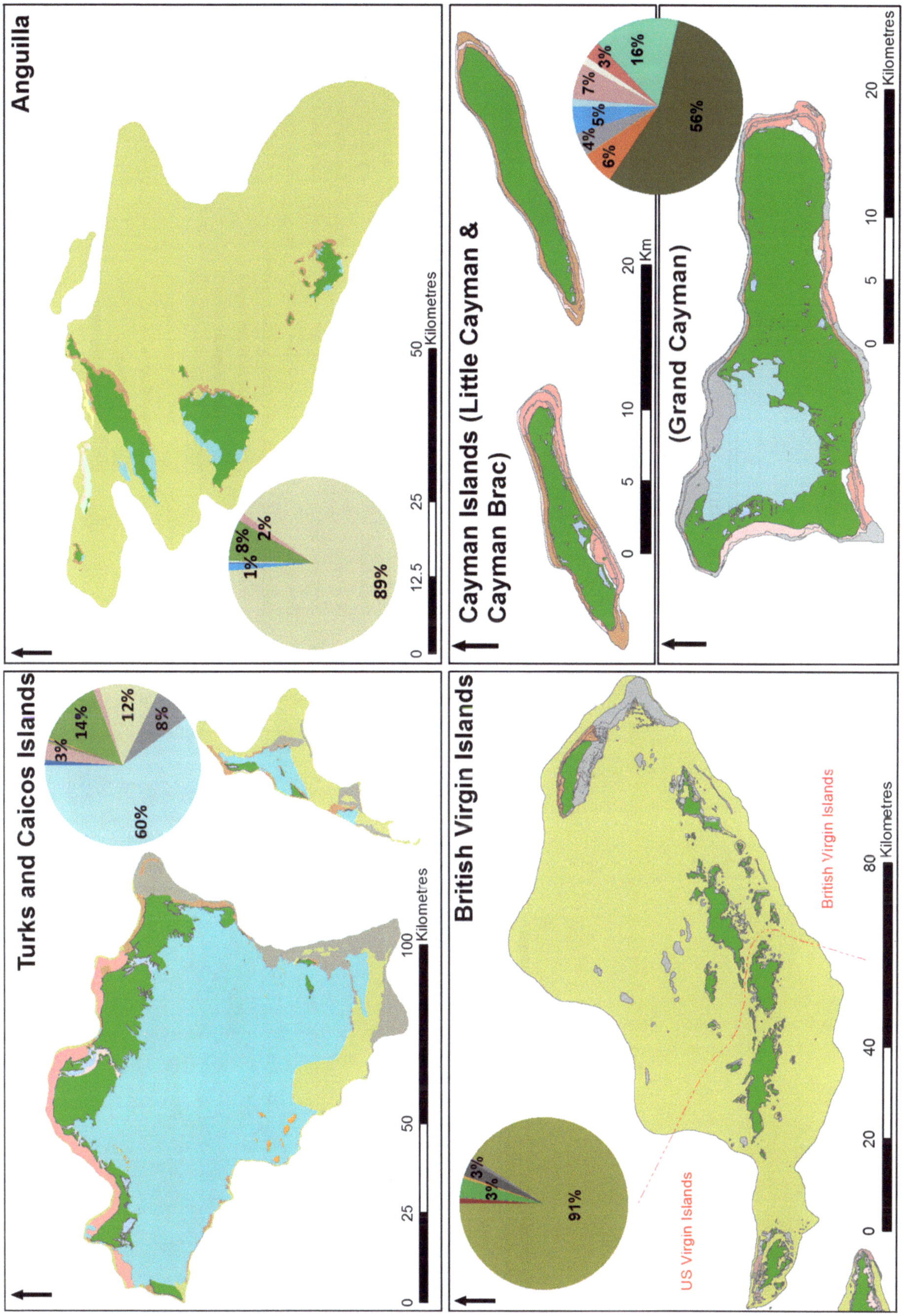

Fig. 1.2 Geomorphological maps of the reef systems of the UK Dependent Territories. Pie charts indicate % coverage of the geomorphic units for each territory (see key for units)

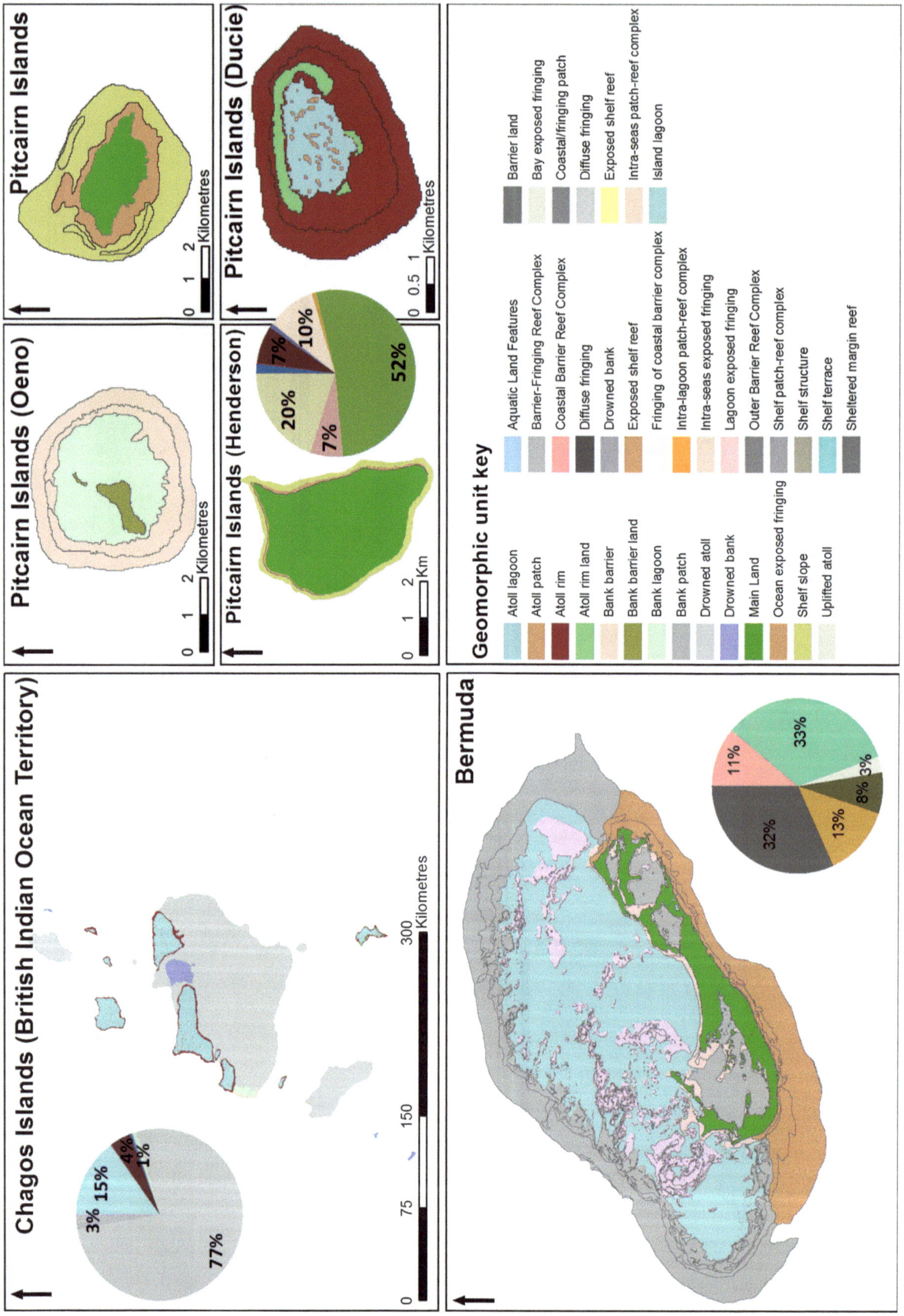

Fig. 1.2 (continued)

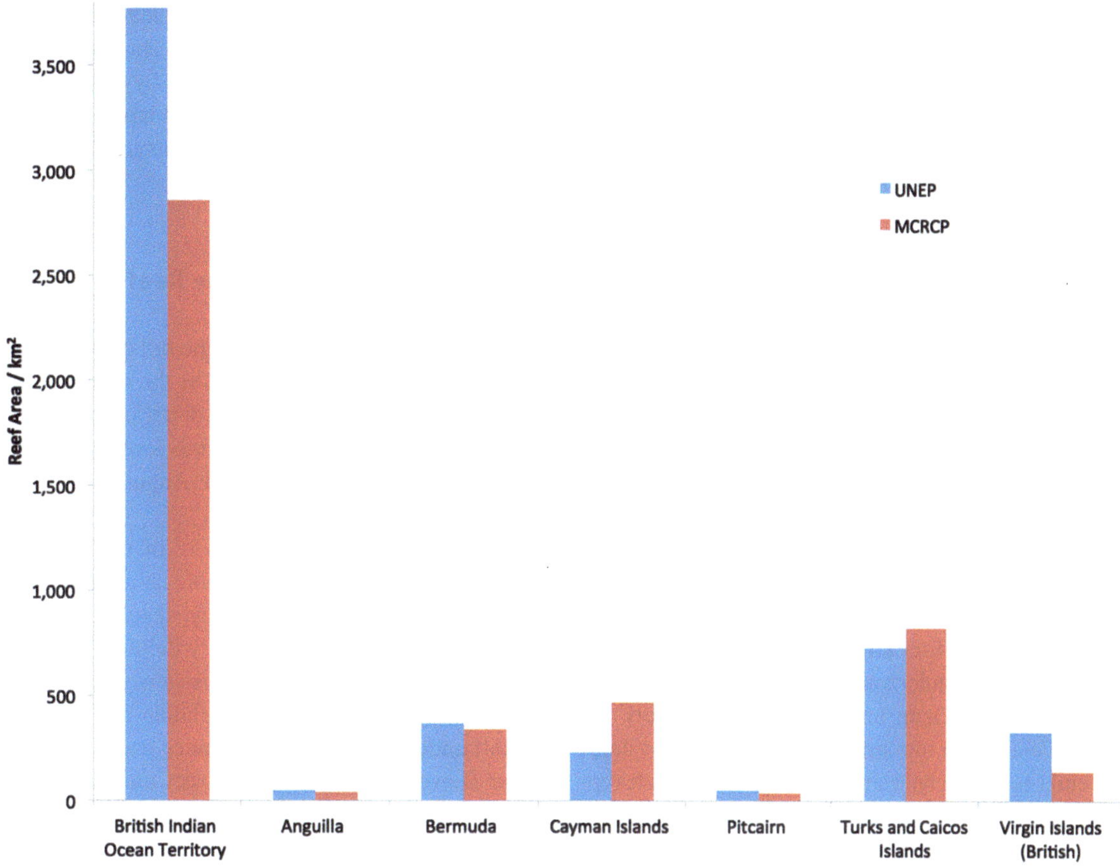

Fig. 1.3 Reef areas for each of the Dependent Territories compared between the previous reference (World atlas of coral reefs, Spalding et al. 2001, in *blue*) and the millennium coral reef Mapping project (*red*)

reef platforms and rocks in the group, the largest of which is Tortola (54 km^2). These rise from the Puerto Rican shelf, which sits at 65 m below sea level. The small shelf patch reef complex has largely developed within a matrix of volcanic uplift around the larger islands across the central shelf area. On the eastern windward side an outer barrier reef complex has developed in association with Anegada, a relatively flat emergent coral limestone platform (altitude 8 m) (Oldfield et al. 1999).

Cayman Islands (*Oceanic/Island*)

The flat, low lying Cayman Islands consist of three islands: Grand Cayman, Cayman Brac and Little Cayman, which sit at the western end of the Greater Antilles group. The islands sit along the Cayman Ridge, which forms the northern margin of the east–west aligned Oriente Transform Fault (Brunt and Davies 1994). The Cayman Islands have a collective land area of 261 km^2, which supports a series of terraced fringing and barrier reefs upon which spur and groove formations have developed that display considerable variability in structural form in relation to local wave power dynamics around the islands (Roberts 1974). A total reef area of 126 km^2 is supported by these islands in the form of an outer and coastal barrier reef complex, exposed fringing reef and shelf terrace.

British Indian Ocean Territory (*Oceanic/Island*)

The British Indian Ocean Territory lies at the southernmost extension of the north–south aligned Chagos-Laccadive Ridge and is composed of a limestone cap several 100 m thick that has developed over the hotspot that now lies under Reunion (Sheppard and Wells 1988; Parson and Evans 2005). The reef systems of the British Indian Ocean Territories are comprised of 2,859 km^2 reef area, which fall into ten classes related to the major morphological units (Andréfouët et al. 2009b). These include five atolls (The Great Chagos Bank (the largest atoll structure in the world at 9,210 km^2), Diego Garcia, Egmont, Peros Banhos and Salomon). There is also an atoll whose islands disappeared and which became awash in the past 250 years (Blenheim), and many drowned banks of which Speakers Bank, Pitt Bank, Victory Bank and Centurion Bank are perhaps the best known. Each of the atolls has substantial lagoons, ranging in size from 11 to 940 km^2 with carbonate rims of varying degrees of subaerial exposure around their perimeter. All atolls and submerged banks appear to be actively growing reefs (Sheppard and Wells 1988).

Pitcairn Islands (*Oceanic/ Island*)

The Pitcairn group is comprised of four widely spaced atolls and islands in the South Pacific Ocean that fall along two geological structural lineations associated with hotspot activity of the clockwise-spreading Pacific plate (Spencer 1995). These four structures are Pitcairn Island (a volcanic island), Henderson Island (an uplifted atoll) and two small atolls, Oeno and Ducie. Collectively the islands can be classified into nine geomorphic units, dominated by emergent volcanic and reef islands, which represent 52% of the area mapped. Pitcairn is a volcanic island that rises 3.5 km from the seafloor with a peak that stands 347 m above sea level with continuous narrow fringing reef around it (Benton and Spencer 1995). The atoll of Henderson Island is a reef-capped volcano that was uplifted as a result of crustal loading by the adjacent Pitcairn volcano (Fosberg et al. 1983; Wells and Jenkins 1988), giving rise to several unique biodiversity characteristics for which Henderson has been designated a UNESCO World Heritage Site. Oeno atoll has a marked outer reef rim perimeter, with an island of area 0.7 km^2 that has developed at the centre of the lagoon. Ducie atoll (6 km^2) is the most easterly atoll of the Indo-Pacific reef province and the southern most atoll of the world, thought to be the surface expression of a field of seamounts (Spencer 1995).

Bermuda (*Oceanic/Island*)

Bermuda is comprised of 150 isolated coral limestone islands in the Sargasso Sea, western Atlantic Ocean, that have formed along the rim of an extinct submarine volcano approximately 1,000 km east of the North Carolina on the continental USA coastline. The extinct volcano sits on top of the Bermuda Platform, a topographic high of the Bermuda Pedestal, a basement that lies in water depths around 75 m (Vacher and Rowe 1997). The land area (56 km^2) is predominantly comprised of a network of ten main islands that are joined by causeways. The extinct volcano rim surrounds a substantial island lagoon (246 km^2). The Bermuda reef system (total area 677 km^2) is formed by the most northerly coral reefs in the world, which form a large outer barrier reef structure that encompasses an island lagoon of area 246 km^2.

Further Refinements to the Reef Inventory

Figures 1.4 and 1.5 demonstrate the full detail (Level 5) contained in the MCRCP products for the Chagos Islands and Anguila. The additional level of detail is apparent from the number of classes represented at this level, which are 24 and 71 for the Chagos islands and Anguila respectively (as opposed to 10 and 22 at Level 3).

The identification of 4,712 km^2 of reef within the UK Dependent Territories was possible using a remote sensing dataset of the requisite accuracy, resolution, consistency and completeness for consistently delineating shallow reef morphological units. Such consistency is important for regions that span the Atlantic, Indian and Pacific Oceans yet fall under a common governance framework. All of the major (Level 2) morphological reef units identified in the MCRCP global typology including atolls, banks, uplifted atolls, islands, patch reefs barrier reefs, fringing reefs and marginal structures are represented in the UK Dependent Territories.

As satellite remote sensing images have become increasingly available at a resolution commensurate with reef landform morphological variability (1–100 km^2), morphometrics derived from them represent an important source of information for managing both global environmental change and anthropogenic influences on reefs. Assessment of reefs according to morphogenetic phenomena, such as tectonic activity, sea level rise, sediment and hydrodynamics, provides a fundamental basis on which ecological dynamics and the impacts of human activities can be superimposed and understood. To this end, this geomorphological assessment, generated for the first time from consistent images of the UK Dependent territories, presents a useful foundation for the incorporation of morphodynamic information into marine environmental management decisions and policies.

While this study draws on the best available data for the time being, this assessment could be further improved by combining higher specification remote sensing data and targeted ground referencing field campaigns in a coordinated manner across the UK Dependent Territories. This additional effort will allow mapping in greater detail fine geomorphological structures (e.g., spur and grooves, different types of reef flats) as well as biological assemblages and benthic cover.

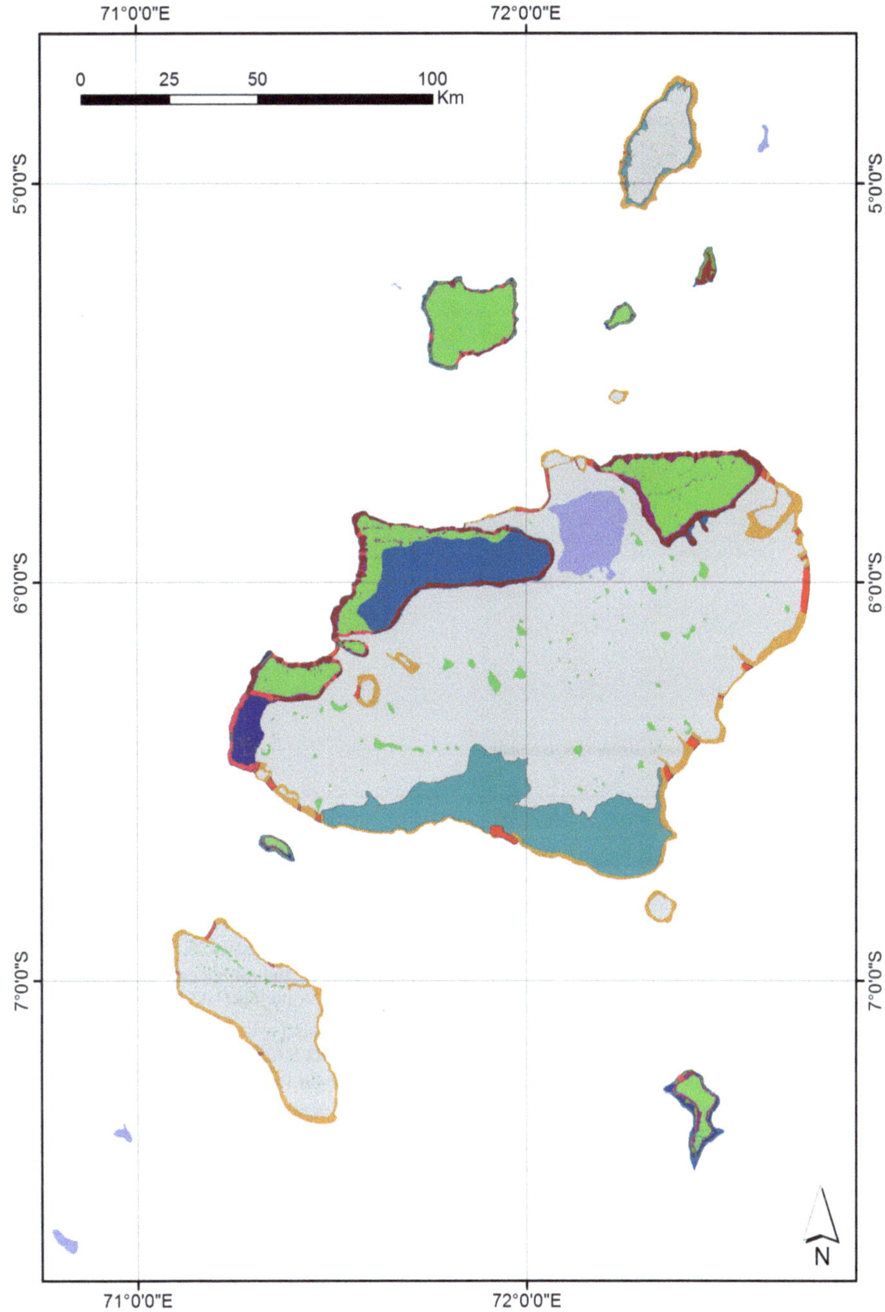

Fig. 1.4 The Chagos Islands (or British Indian Ocean Territory), illustrating the detail included at Level 5 in the MCRCP geomorphological map

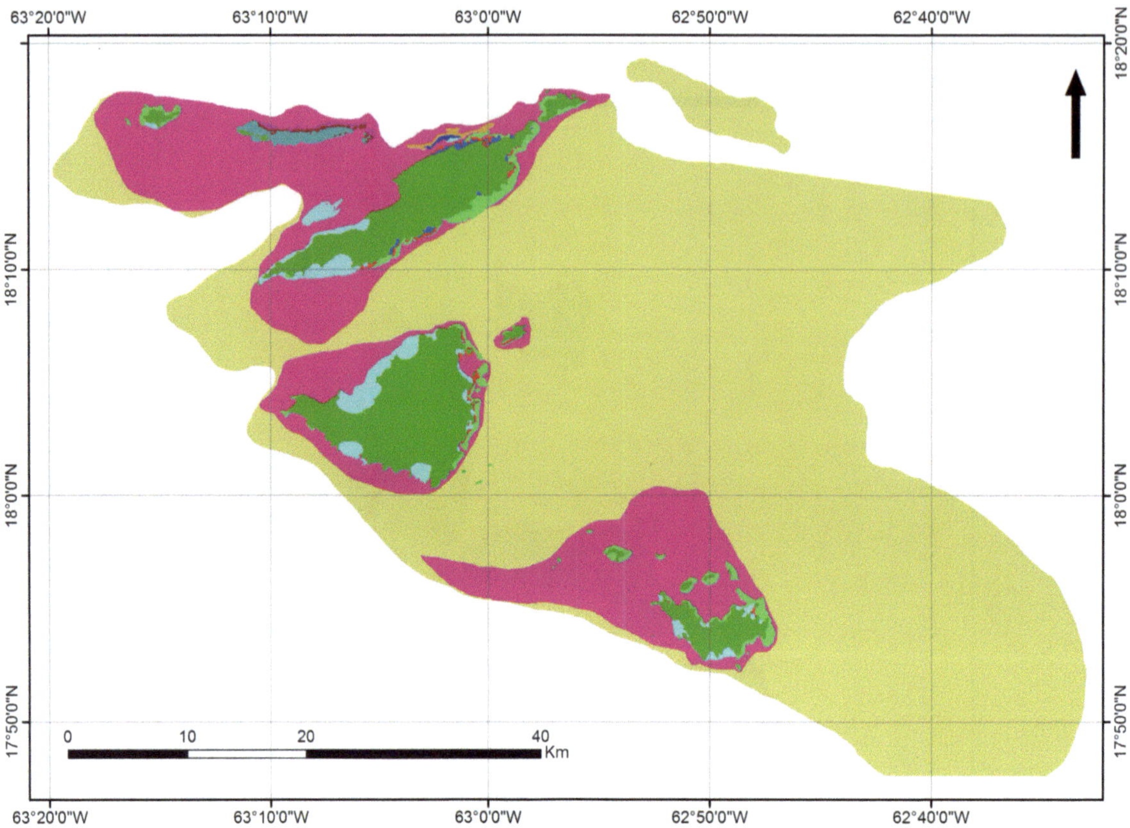

Fig. 1.5 Anguila, illustrating the detail included at Level 5 of the MCRCP Geomorphological map

References

Allain V, Kerandel J, Andréfouët S, Magron F, Clark M, Kirky D, Muller-Karger FE (2008) Enhanced seamount location database for the western and central Pacific Ocean: screening and cross-checking of 20 existing datasets. Deep Sea Res I 55:1035–1047

Allnutt TF, McClanahan TR, Andréfouët S, Baker M, Lagabrielle E, McClennen C, Rakotomanjaka A, Tianarisoa T, Watson R, Kremen C. (2012) Comparison of marine spatial planning methods in Madagascar demonstrates value of alternative approaches. PlosOne 7(2): e28969. doi: 10.1371/journal.pone.0028969

Andréfouët S (2011) Reef typology. In: Hopley D (ed) Encyclopedia of modern coral reefs. Springer, Dordrecht, p 906

Andréfouët S, Muller-Karger FE, Robinson JA, Kranenburg CJ, Torres-Pulliza D, Spraggins SA, Murch B (2006a) Global assessment of modern coral reef extent and diversity for regional science and management applications: a view from space. In: Proceedings of 10th international coral reef symposium, Okinawa, pp 1732–1745

Andréfouët S, Chagnaud N, Chauvin C, Kranenburg C (2006b) Atlas of the Papua new Guinea coral reefs, Centre IRD de Nouméa, Nouméa,Nouvelle-Calédonie,IRD/CRISP/IMaRS/NASA/Worldfish, Nouméa, 30 p, + 10 maps

Andréfouët S, Chagnaud N, Chauvin C, Kranenburg C (2008) Atlas of the French overseas coral reefs, Centre IRD de Nouméa, Nouméa, Nouvelle-Calédonie, CD-ROM

Andréfouët S, Cabioch G, Flamand B, Pelletier B (2009a) A reappraisal of the diversity of geomorphological and genetic processes of New Caledonian coral reefs: a synthesis from optical remote sensing, coring and acoustic multibeam observations. Coral Reefs 28:691–707

Andréfouët S, Chagnaud N, Kranenburg C (2009b) Atlas of western and central Indian Ocean coral reefs, Centre IRD de Nouméa, Nouméa, Nouvelle-Calédonie, CD-ROM

Bell JD, Kronen M, Vunisea A, Nash W, Keeble G, Demmke A, Andréfouët S (2009) Planning the use of fish for food security in the Pacific. Mar Policy 33:64–76

Benton T, Spencer T (1995) The Pitcairn Islands: biogeography, ecology and prehistory. Academic, London

Brunt MA, Davies JE (1994) The Cayman Islands: natural history and biogeography. Kluwer Academic, Dordrecht

Burke L, Maidens J, Spalding M, Kramer P, Green E, Greenhalgh S, Nobles H, Kool J (2004) Reefs at risk in the Caribbean. World Resources Institute, Washington, DC

Burke L, Reytar K, Spalding M, Perry A (2010) Reefs at risk revisited. World Resources Institute, Washington, DC

Dalleau M, Andréfouët S, Wabnitz CCC, Payri C, Wantiez L, Pichon M, Friedman K, Vigliola L, Benzoni F (2010) Use of habitats as surrogates of biodiversity for efficient coral reef conservation planning in Pacific Ocean Islands. Conserv Biol 24:541–552

Done T (2011) Coral reef, definition. In: Hopley D (ed) Encyclopedia of modern coral reefs. Springer, Dordrecht, pp 261–266

Fosberg FR, Sachet MH, Stoddart DR (1983) Henderson Island (southeastern Polynesia): summary of current knowledge. Atoll Res Bull 329:1–18

Green A, Smith SE, Lipsett-Moore G, Groves C, Peterson N, Sheppard S, Lokani P, Hamilton R, Almany J, Aitsi J, Bualia L (2009) Designing a resilient network of marine protected areas for Kimbe Bay, Papua New Guinea. Oryx 43:488–498

Hopley D (ed) (2011) Encyclopedia of modern coral reefs. Springer, Dordrecht

Oldfield S, Sheppard S (1997) Conservation of biodiversity and research needs in the UK Dependent Territories. J Appl Ecol 34:1111–1121

Oldfield S, Procter D, Fleming LV (1999) Biodiversity: the UK overseas territories, report published by the Joint Nature Conservation Committee. ISBN: 1 86107 502 2

Parson L, Evans A (2005) Seafloor topography and tectonic elements of the Western Indian Ocean. Philos Trans R Soc 363(1826):15–24

Putney AD (1982) Survey of conservation priorities in the Lesser Antilles. Final report, Caribbean Environment Technical Report, Caribbean Conservation Association, Barbados

Roberts HH (1974) Variability of reefs with regard to changes in wave power around an island. In: Proceedings of the second international coral reef symposium, vol 2. Brisbane, pp 497–512

Rohmann SO, Hayes JJ, Newhall RC, Monaco ME, Grigg RW (2005) The area of potential shallow water tropical and sub-tropical coral ecosystems in the United States. Coral Reefs 24:370–383

Sheppard C, Wells S (1988) Coral reefs of the world volume 2: Indian Ocean, Red Sea and Gulf. UNEP regional Seas directories and bibliographies. IUCN, Gland/Cambridge, UK, pp 1–389

Spalding M, Ravilious C, Green E (2001) World atlas of coral reefs. University of California Press, Berkeley

Spencer T (1995) The Pitcairn Islands, South Pacific Ocean: plate tectonics and climatic contexts. In: Benton T, Spencer T (eds) The Pitcairn Islands: biogeography, ecology and prehistory. Academic, London, pp 13–32

Stein S, Engela J, Weins DA (1982) Subduction seismicity and Tectonics in the Lesser Antilles Arc. J Geophys Res 87:8642–8664

Sullivan KM, Chiappone M, Lott C (1994) Abundance patterns of stony corals on platform margin reefs of the Caicos Bank. Bahamas J Sci 1:2–12

Vacher HL, Rowe MP (1997) Geology and hydrogeology of Bermuda. In: Vacher HL, Quinn TM (eds) Geology and hydrogeology of carbonate Islands. Elsevier, New York, pp 35–91

Wabnitz CC, Andréfouët S, Torres-Pulliza D, Müller-Karger FE, Kramer PA (2008) Regional-scale seagrass habitat mapping in the Wider Caribbean Region using Landsat sensors: applications to conservation and ecology. Remote Sens Environ 112:3455–3467

Wabnitz CCC, Andréfouët S, Muller-Karger FE (2010) Measuring progress toward global marine conservation targets. Front Ecol Environ 8:124–129

Wanless HR, Dravis JJ (1989) Carbonate environments and sequences of the Caicos platform: field trip guidebook T374. American Geophysical Union, Washington, DC

Wells SA, Jenkins M (1988) Coral reefs of the world volume 3: central and Western Pacific. UNEP regional seas directories and bibliographies. IUCN, Gland/Cambridge, UK, pp 1–329

Coral Reefs of Anguilla

2

Stuart P. Wynne

Introduction

Anguilla (18°12.80N and 63°03.00W) is the most northerly of the Caribbean Leeward Islands. It is a low lying coralline Iandmass approximately 24 km long. Its central portion, at its widest point, has a width of approximately 5 km, tapering off rapidly towards easterly and westerly ends (Fig. 2.1). Most of it is a limestone platform (Fig. 2.2), with Crocus Hill, at an altitude of 65 m, being the highest point above sea level. The surrounding shallow platform, the Anguilla bank, is the most northern formation of the Lesser Antilles, separated from the Greater Antilles by the Anegada Gap. These islands and shallow substrates are based on a much larger limestone platform which is contiguous with several adjacent islands, most notably St Martins (Hubbard 1989). Anguilla has only a very few small extrusive basalts and tuffs, and thus contrasts strongly with adjacent islands which include substantial volcanic structures which emerge through the limestone to several 100 m high. A long chain of submerged shallow water off the northern side of Anguilla provides much substrate for reefs and for huge expanses of sandy substrate. This covers a total of about 14,600 ha of sublittoral substrate less than 20 m deep.

Aside from mainland Anguilla there are a number of offshore cays and rocky outcroppings, the most significant being Sombrero Island, Dog Island, Scrub Island, Prickly Pear Cays, Sandy Island, Anguillita and Scilly Cay (Fig. 2.3). Various protected areas are present in Anguillian waters whose purposes vary from preservation of historical sites to conservation of reef and seagrass areas. Table 2.1 gives details of these protected areas, and baseline surveys of the 'ecologically based' marine parks (see Fig. 2.1) were conducted in 2007 where 30 sites were surveyed as part of a project funded by the British Government through the Overseas Territories Environment Programme to increase the efficacy of the management of these areas(Wynne 2007a). In general the majority of Anguilla's underwater habitat consists of coral reefs, seagrass beds, sand and/or algal flats. A detailed atlas of these habitats was produced in 1994 by the Natural Resources Institute of the United Kingdom (NRI 1995; Sheppard et al. 1995).

S.P. Wynne (✉)
Department of Fisheries and Marine Resources,
Government of Anguilla, The valley, Anguilla
e-mail: stuart@stuartwynne.co.uk

Economy and Services

The Valley is Anguilla's capital and the only true 'town' on the island with a population of 1169 (2001 Census). Other settlements on the island overlap onto each other and as such their names relate more to geographical locations rather than 'villages'. All main services are based in The Valley, for example post office, Government buildings, hospital, library, as are most shops, banks, schools and telecommunications. Anguilla's economy relies heavily on tourism, a sector that began expanding rapidly during the 1980s. Prior to this only a few 1,000 foreigners visited the island annually, but by the end of the following decade visitor numbers were estimated to be up to 80,000 per annum (Gell and Watson 2000). The other important economic sectors are financial services, construction and Government services. In terms of dollar value other sectors such as agriculture and fishing are not of major importance. Aside from local demand the fishing industry is driven by the increasing tourist sector and concerns exist regarding fisheries sustainability and the ecological impact of such activities on the extensive coral reefs that surround the island.

Fishing Industry

Thirty years ago, Olsen and Ogden (1981) estimated the maximum sustainable yield (MSY) for demersal stocks to be 2,740 t for finfish and 230 t of lobster and 230 t of conch

C.R.C. Sheppard (ed.), *Coral Reefs of the United Kingdom Overseas Territories*, Coral Reefs of the World 4,
DOI 10.1007/978-94-007-5965-7_2, © Springer Science+Business Media Dordrecht 2013

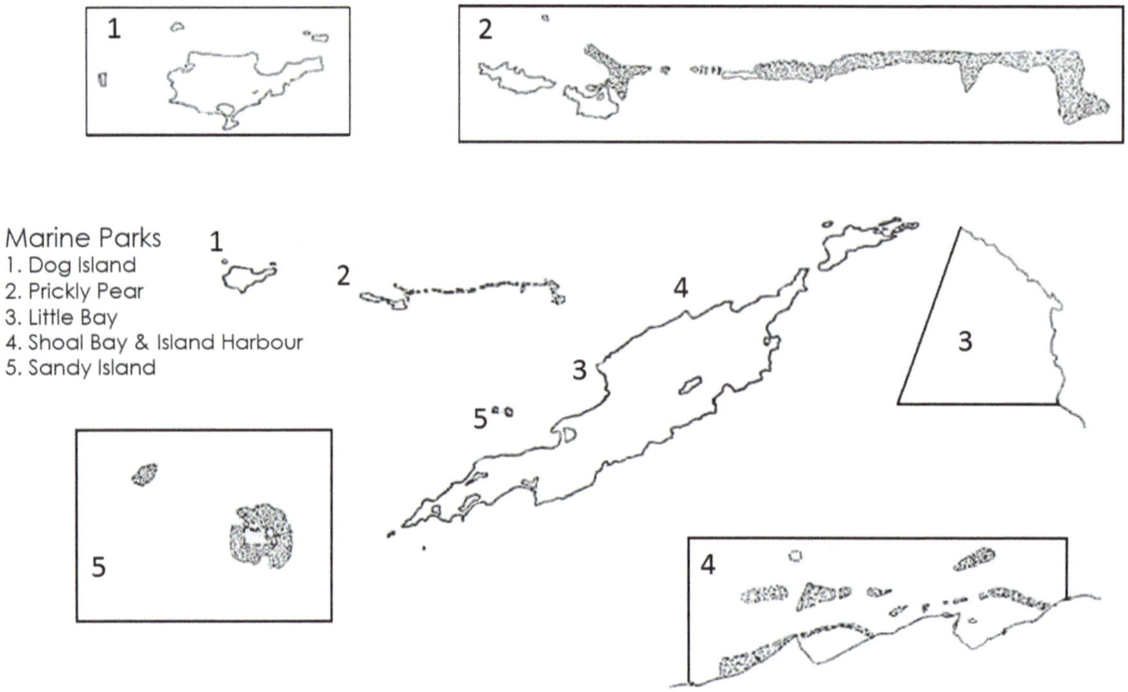

Fig. 2.1 Map of Anguilla. The main 'ecologically based' marine protected areas. Three other areas afforded protection are not illustrated: Sombrero Land/Sea Reserve, Junks Hole Marine Park, and Rendezvous Bay (Photograph S. Wynne)

Fig. 2.2 Limestone cliffs along the northern coast of Anguilla (Photograph S. Wynne)

and other shellfish, with 80% of the production potential coming from the 22% of the shelf area containing coral reef communities. Nearshore resources by 2000 had declined under increasing pressure (Gell and Watson 2000). Export duties allowed an estimate of lobster exports in 1979 of at least 28,364 kg, though the true value was probably much higher. In 1980, fishers were discarding scarids and snappers up to 2.3 kg as trash fish and fish pots used *Epinephelus striatus*, now commercially extinct in many parts of the Caribbean, as lobster bait (Salm 1980). However, by 1987, signs of overfishing were apparent for lobster and potfish (Stephenson 1992).

Fig. 2.3 Typical small islands around the main island of Anguilla: Scilly Cay in the north (Photograph S. Wynne)

Table 2.1 Details of Anguilla's marine protected areas. It should be noted that at the time of writing legislation for these areas is limited although this situation is poised to change in the next few years

Name	Status	Area (km^2)	Purpose and Description
Dog Island	Marine park	c.10	Ecological (reef) marine park surrounding Dog Island and three cays. Popular fishing area (traps). Turtle nesting on beaches
Prickly Pear and Seal Island	Marine park	c.33	Ecological (reef) marine park surrounding Prickly Pear and entire Seal Island Reef. Popular fishing area (traps). High levels of tourism (yachting) around Prickly Pear East
Sandy Island	Marine park	c.5	Ecological (reef and seagrass) marine park surrounding Sandy Island. Popular fishing area (traps). High levels of tourism (yachting). Foraging juvenile turtles
Shoal Bay-Island Harbour	Marine park	c.19	Ecological (reef) marine park stretching from Fountain Beach to Island Ridge. Popular fishing area (traps and spear). High levels of tourism (beach goers). Foraging and nesting turtles
Little Bay	Marine park	c.1	Ecological (seagrass) marine park triangular area originating from Pelican and Flat Cap points. Popular fishing area (hand net, hook and line). High levels of tourism (yachting). Important foraging turtle area
Junks Hole	Marine park	<1	Historical site whose boundary is that within a radius of 500 yards from the ship wrecked Spanish Galleon El Buen Consejo
Sombrero Island	Land-Sea Reserve	c.1	Historical site whose boundary is that within 2000 yards of the islands centre. Important for bird populations. Marine portion is mixed reef and deep water habitat much of which is popular with fishers
Rendezvous Bay	No Anchoring Zone	c.2	Site who is given special protection from anchoring under the Cruising Permit Act. Extensive seagrass beds exist in the area

Today, Anguilla's fishing industry is largely artisanal with the majority of fishers owning one small open vessel powered by outboard engines (average length 7.4 m) and employing only one other crew member. There are two main target groups, lobster and finfish, which are geographically separated and serve distinct markets. There is also a small conch fishery present. The finfish fishery, in terms of total weight, accounts for approximately 67% of landings, and although most is sold locally some fishers do sell their catch in St Martin. Hotels & restaurants in Anguilla generally purchase deep demersal and offshore pelagic species along with the majority of lobster landings, whereas demand for home consumption is largely for reef species. This essentially means pressure on Anguilla's coral reef systems is driven mainly by

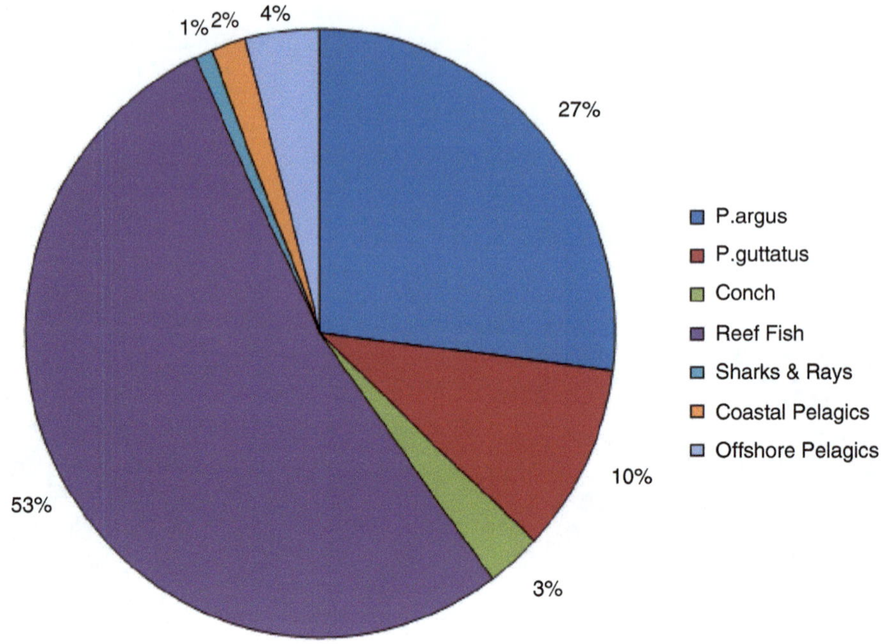

Fig. 2.4 Composition of landings made by Anguilla's fishing industry split into the six main target groups (Murry 2008) (Photograph S. Wynne)

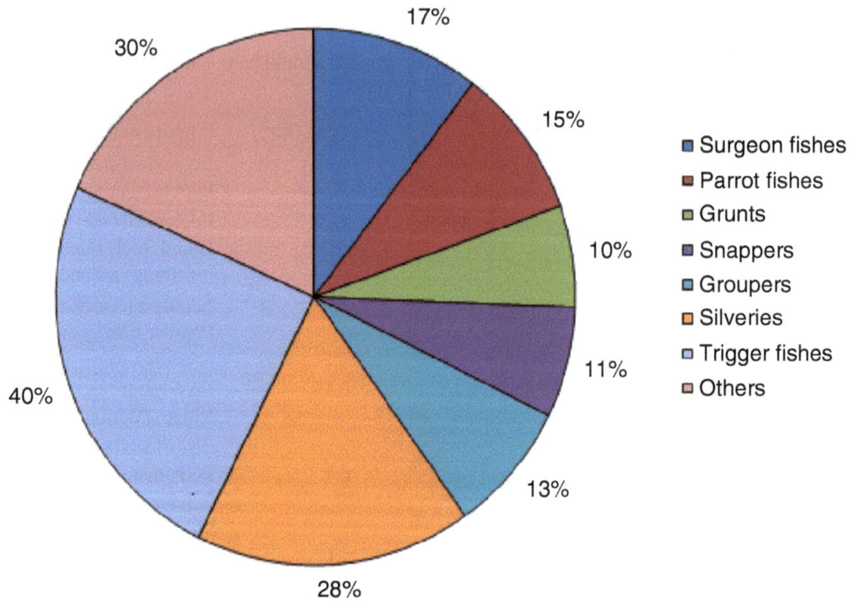

Fig. 2.5 Family composition of reef fish caught by local fishers. Data collected via interviews during 2009 (provided by Professor Nancy Carder). These results have been used rather than those presented by Wynne (2010) as they represent what the fishers caught rather than what they landed. Parrotfishes for example are often used as bait and so, aside from rarer large individuals, do not get brought ashore (Photograph S. Wynne)

local demand for food, not tourism. An exception to this is the small 'crayfish' fishery (*Panulirus guttatus*, a species closely related to *Panulirus argus*, the main lobster species targeted) that is harvested in shallow reef areas. In 1999 it was estimated that c.4.5% of the population was employed in the fishing industry, an industry that in 2006 provided only 1.4% of the islands GDP (XCD 6.11 million). There are plans to encourage fishers to move towards large (offshore) pelagic species, which at present are mainly the target of the sports fishing industry. This would relieve pressure on reef resources and also satisfy the growing tourist market which generally favours the bone-free fillets that these species provide. Reef fish comprise an estimated 53% of landed catch (Fig. 2.4), with typical composition as detailed in Fig. 2.5. Data collected as part of the 2007 fisheries census suggested that there are currently around 300 fishers operating in Anguilla (Murray 2008). However, recreational fishing in Anguilla is a popular past-time with most people partaking in it at some point in their daily lives. Spear fishing (by locals) is currently permitted in all parts of Anguilla's marine environment and is a popular past-time for many islanders. It is likely though that due to its highly efficient nature, management measures will be introduced in the near future to limit its impact.

Fig. 2.6 Skeletal *Acropora* reef on the south coast of Anguilla (Photograph S. Wynne)

Coral Reefs

Much of the shallow sublittoral in Anguilla supports seagrass beds and expanses of sand and algae, as well as reefs. As far as its reefs are concerned, Anguilla is surrounded by a mixture of fringing, barrier and patch reefs interspersed by the sand/algal flats and seagrass beds. Shallow, sheltered habitats support large areas of seagrasses, mostly of extensive shallow beds of *Thalassium testudinum* (turtle grass) with some deeper beds of *Syringodium filiforme* (manatee grass), and coral reefs intersperse with these (Gell and Watson 2000). On the whole the reefs exist in shallower regions although there are areas of extensive deeper reefs beyond recreational dive limits. In the 1970s coral communities cover approximately 22% of the entire Anguillan shelf area (Olsen and Ogden 1981). In general the reefs around Anguilla can be split into two geographical regions: the northern coastal region and the southern coastal region.

Southern Coast

The southern coastal region, at least historically, can be categorised as an *Acropora sp.* dominated fringing reef system that is ecologically and structurally influenced by its exposed nature. Although historical data are lacking, this system appears to have degraded severely since the early 1980s when it was reportedly still in a relatively pristine condition. Exact reasons for this demise are unclear but hurricane damage during the 1980s (and later in the 1990s) had a severe effect, and White Band Disease that is known to have been affecting *Acropora sp.* throughout the region at the time, had also severely affected Anguilla by the 1990s (Sheppard et al. 1995; Bythell and Buchan 1996). The reefs simply did not recover after these events and in most shallow reef areas which exhibited classical elkhorn reef structures, the coral cover was essentially dead (Sheppard et al. 1995) (Fig. 2.6). Currently the south coast reef areas are characterised by *Acropora palmata* rubble with very low benthic diversity or low relief rocky pavement areas with scattered soft corals, macroalgae, sponges and small hard corals (for example *Montastraea sp., Diploria sp., Siderastrea sp. & Agaricia sp.*). Some relatively extensive seagrass beds also exist along the south coast.

Northern Coast

The northern coastal region is dominated by patch, barrier and fringing reef systems interspersed by sand and/or algal flats, with a combined estimated area of 14,600 ha (Sheppard et al. 1995). In a similar way to the south coast it appears that historically there were extensive *Acropora sp.* reefs that exist today mainly as vast rubble fields. Having said this, a few patches of intact *Acropora palmata* stands can still be found along the Seal Island reef system, much of which appears to be new growth, and large expanses can be found of Porites also (Fig. 2.7).

Although much reduced today, vast areas of *Montastraea spp.* were also known to have been present historically in northern coastal regions. This genera is still the most dominant in the area but of the existing colonies many are exhibiting signs of

Fig. 2.7 Extensive colony of *Porites* off Sandy Island, North coast (Photograph S. Wynne)

Fig. 2.8 *Montastraea annularis* infected by Yellow Blotch Disease in Shoal Bay East, Anguilla (Photograph taken by S. Wynne in June 2007)

disease (mainly Yellow Blotch Disease) which is probably the cause of its overall decline (see Fig. 2.8). Although northern reef areas appear to be slowly degrading, it does not appear to be as severely affected as the south coast, with extensive areas of relatively high diversity still present (albeit with high levels of macroalgae). This diversity still extends to the hard coals with dominant genera being *Montastraea sp., Diploria sp., Siderastrea sp. & Porites sp.* As with south coast reef systems, reasons for the degradation of the northern reef system are likely a mixture of stressors that probably interact with each other in complex ways. For example, climatic factors such as hurricanes and bleaching physically damage corals that then find it difficult to recover because (for example) eutrophication encourages disease, and macroalgae is more able to dominate in their place.

Grazing species become less able to keep the macroalgae under control, a problem that is exacerbated by reduced densities of *Diadema antillarum* since the mass mortality event in the 1980s (Tuya et al. 2004) and by overfishing of herbivorous fish species. All of this ultimately leads to habitat degradation.

In 1994, overall 435 ha of dead elkhorn reef was mapped in Anguilla, much of it located across the entrances to bays and off several sheltered beaches (Sheppard et al. 1995). It appears that other than some renewed growth in some areas, most remains as areas of rubble.

Temporal Changes

Most reports suggest Anguilla's reef systems have been in decline since before the 1980s although data to back this up are sparse. The extensive mapping study in 1994 showed that large expanses of the shallow elkhorn (*Acropora palmata*) was dead, most likely due to White Band Disease (Sheppard et al. 1995). A study investigating the impacts of hurricane Luis in 1995 (Bythell and Buchan 1996) compared Anguilla to the British Virgin Islands and concludes that its reefs had 'undergone extensive mortality of the dominant coral *Acropora palmata*' since they were first affected by White Band Disease in 1976. The earliest marine ecological survey work conducted in Anguilla was undertaken by the Bellairs Institute (Barbados) who conducted reasonably extensive survey work in 1990 (Oxenford and Hunte 1990). Since 2007 annual monitoring began (see following section), and in 2009 a dedicated study was also conducted that attempted to revisit the sites used by the Bellairs Institute in 1990 (Wynne 2010). Although these sites were established pre-GPS a number of them were relocated and surveys replicated. The comparative study revealed that over the last 20 years both hard corals and reef fish populations have declined (although there is some variability between sites). Combining all the sites, coral cover has dropped from an average of 14% in 1990 to 4% in 2009. This is a 70% decline in less than 20 years. At one site close to Forest Bay the decline was as high as 98%. Overall, reef fish populations have dropped by approximately a third. The 1990 study did not record fish size so biomass estimates are not possible. This situation is of significant concern because, even though White Band Disease reportedly began to affect Anguilla's reef systems in 1976, the Bellairs study concluded that in 1990 there were a 'variety of diverse and attractive marine habitats (which are) in relatively good condition, with little apparent impact from human activities'. This certainly does not seem to be the case today. On a positive note, macroalgae levels were more variable since 1990 with some sites showing an increase while others showed a decrease. Reasons for this remain unclear but it is likely that it is at least in part thanks to a partial recovery of *Diadema antillarum* since the mass mortality event in the 1980s. Although some areas of reef still have very limited numbers present others appear to be recovering and have very high densities.

Present Monitoring

In 2007 the Department of Fisheries and Marine Resources initiated the Anguillian Marine Monitoring Programme (AMMP), a long-term project annually assessing key sites within Anguilla's shallow marine environment. The surveys at these sites were in areas that did not exceed 15 m in depth. In the first year five sites were surveyed as part of a pilot study (Wynne 2008a) that tested methodology and served to train research staff. Over the following years more sites were added to the programme (Wynne 2008b), so that by 2010 the full complement of 15 sites had been reached. Ten of these sites were located on coral reef areas and five on seagrass beds. At the ten coral reef sites full benthic surveys are conducted annually using SCUBA equipment, together with fish counts that assess total species diversity and size class/abundance of commercially and ecologically important families. Full survey methodologies can be found in Wynne (2007b). At the ten sites surveyed hard coral cover ranges from almost 15% to less than 1% and macroalgae cover ranges from almost 30% to less than 1% (Table 2.2). Coral diseases at the monitoring sites are sometimes relatively common and can affect more than 50% of colonies present (Table 2.3). The most prevalent diseases are currently yellow blotch disease infecting *Montasraea sp.* (see Fig. 2.8) and cyanobacterial overgrowths/infections affecting a variety of species but most noticeably *Siderastrea siderea*. White Band Disease is not often recorded but this is likely due to the current paucity of *Acropora palmata*.

Across the ten reef sites surveyed the five most abundant fish species were *Thalassoma bifasciatum* (1,240 ha^{-1}), *Acanthurus coeruleus* (867 ha^{-1}), *Acanthurus bahianus* (840 ha^{-1}), *Scarus iserti* (787 ha^{-1}) & *Chromis multilineata* (413 ha^{-1}). The abundance measure is mean number of individuals per hectare of available habitat, i.e. that representative of the study sites. Of the commercially and ecologically important fish families surgeonfishes (*Acanthuras sp.*) are the most prolific followed by parrotfishes (*Scarus sp.* & *Sparisoma sp.*). Mean results for all families included are in Table 2.4, and the mean size class of these fish families is 10–15 cm (Fig. 2.9) in terms of relative biomass. This measure is a simplified method of calculating biomass based purely on length of fish.

The current monitoring effort has concluded that Anguilla's shallow (<15 m) coral reef systems are in a relatively poor state of health with low hard coral cover, high macro algae cover and smaller than expected mean fish size classes. The corals that are present are threatened by numerous factors including (but not limited to) disease, climate change,

Table 2.2 Percentage covers of main benthic characteristics of the ten coral reef sites monitored as part of AMMP. Results are for all surveys combined between 2007 and 2009. For individual results see Wynne 2008a, b and Wynne 2010. Other benthic characteristics not detailed include (but are not limited to) sand, calcareous algae and mobile invertebrates

Site Name	Bare rock[a]	Macroalgae	Hard coral	Soft coral	Sponges
Anguillita	81.7	0.7	4.3	3.8	2.4
Sandy Island	54.0	5.5	14.5	1.4	3.3
Long Reef	66.6	3.1	11.1	2.1	1.1
Limestone Bay	57.6	27.3	5.1	3.5	4.2
Shoal Bay East	65.2	19.1	10.3	2.4	0.3
Island Harbour	61.1	27.3	1.0	1.6	0.7
Scrub Island	65.7	18.2	3.8	1.3	0.7
Forest Bay	84.1	8.2	3.0	0.5	0
Little Harbour	83.0	10.9	3.1	1.0	2.0
Sile Bay	66.3	13.2	0.9	0	0
MEAN	**68.5**	**13.4**	**5.7**	**1.8**	**1.5**

[a]Bare rock refers to rock or dead coral skeletons covered with turf algae, coralline algae, and/or light sediment

Table 2.3 Coral health results across the ten coral reef sites monitored as part of AMMP. It should be noted that the sites that have the highest percentage of corals 100% healthy also have the lowest percentage cover of hard corals (Table 2.2)

Site Name	Coral tissue healthy	Colonies 100% healthy	Coral tissue diseased	Colonies with disease	Coral tissue recently dead	Coral tissue long dead	Colonies exhibiting mortality
Anguillita	97.3	92.8	0.3	1.9	0.3	2.2	7.2
Sandy Island	81.1	45.5	0.1	1.3	2.6	16.2	54.5
Long Reef	92.1	77.6	0.0	0.0	1.0	6.9	22.4
Limestone Bay	98.6	94.8	0.0	0.0	0.0	1.4	5.2
Shoal Bay East	87.7	75.0	0.4	8.8	1.7	10.2	25.0
Island Harbour	97.0	76.4	0.0	0.0	0.9	2.1	23.6
Scrub Island	96.2	90.7	0.0	0.0	0.2	3.6	9.3
Forest Bay	100	100	0.0	0.0	0.0	0.0	0.0
Little Harbour	94.9	79.3	0.0	0.0	0.9	4.2	20.7
Sile Bay	100	100	0.0	0.0	0.0	0.0	0.0
MEAN	**94.5**	**83.2**	**0.1**	**1.2**	**0.8**	**4.7**	**16.8**

Table 2.4 Percentage of total number of individuals within each size class across all surveyed reef sites split into commercially and ecologically important fish families

	<5 cm	5–10 cm	10–15 cm	15–20 cm	20–25 cm	25–30 cm	30–35 cm	35–40 cm	40–45 cm	45–50 cm	>50 cm	Total
Surgeonfishes	7.38	14.57	22.39	6.51	0.25	0.00	0.00	0.00	0.00	0.00	0.00	**51.10**
Parrotfishes	8.56	12.61	4.67	3.12	2.13	0.98	0.80	0.38	0.05	0.00	0.00	**33.31**
Grunts	3.14	0.06	0.69	1.11	0.49	0.04	0.03	0.02	0.00	0.00	0.00	**5.57**
Snappers	0.02	0.02	0.15	0.35	0.00	0.02	0.00	0.00	0.00	0.00	0.00	**0.54**
Groupers	0.02	0.06	0.20	0.46	0.07	0.03	0.02	0.03	0.00	0.00	0.00	**0.88**
Silveries	0.03	0.20	0.08	0.19	0.30	0.08	0.18	0.02	0.00	0.04	0.03	**1.14**
Triggerfishes	0.00	0.00	0.15	0.50	0.53	0.13	0.11	0.02	0.00	0.00	0.00	**1.43**
Goatfishes	0.00	0.04	0.92	0.86	0.03	0.03	0.00	0.00	0.00	0.00	0.00	**1.87**
Angelfishes	0.00	0.00	0.02	0.04	0.03	0.04	0.04	0.00	0.00	0.00	0.00	**0.17**
Butterflyfishes	0.06	0.53	0.02	0.00	0.00	0.00	0.00	0.00	0.00	0.00	0.00	**0.60**
Squirrelfishes	1.30	0.07	0.16	0.08	0.02	0.00	0.00	0.00	0.00	0.00	0.00	**1.63**
Large Wrasses	0.09	0.09	0.02	0.05	0.07	0.05	0.03	0.00	0.00	0.00	0.00	**0.39**
Other	0.30	0.54	0.10	0.05	0.11	0.09	0.07	0.06	0.02	0.02	0.02	**1.37**
Total	**20.88**	**28.78**	**29.56**	**13.30**	**4.03**	**1.49**	**1.28**	**0.52**	**0.07**	**0.06**	**0.05**	**100**

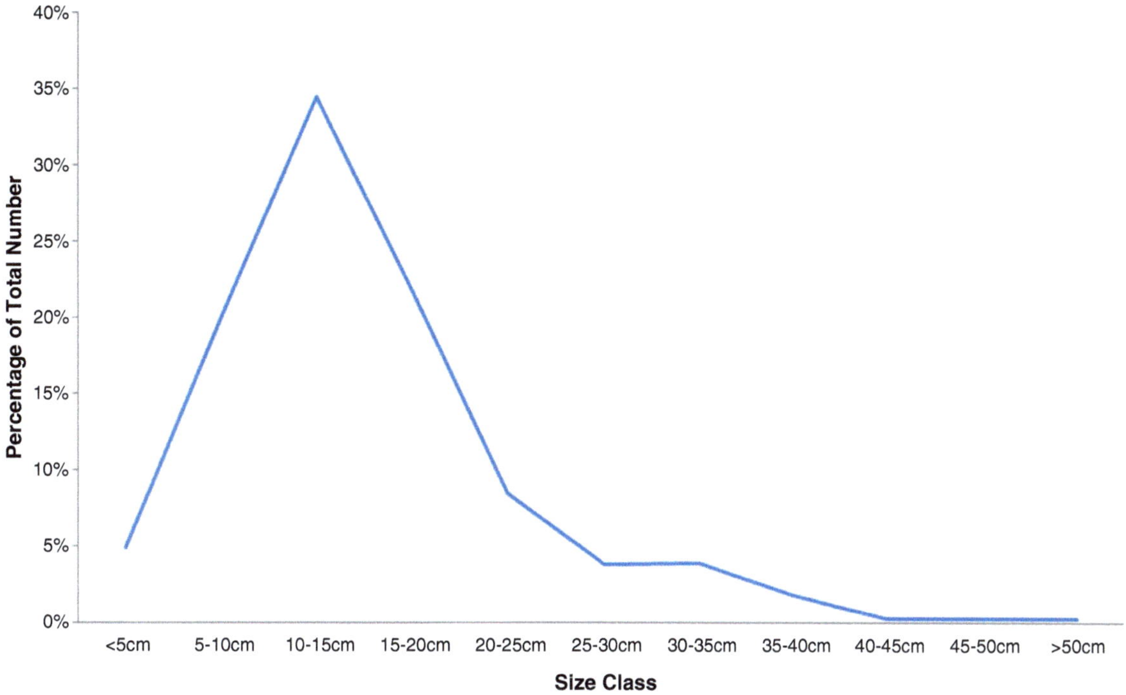

Fig. 2.9 Percentage of relative biomass that each size class of all commercially and ecologically important fish families accounts for across all surveyed reef sites (Photograph S. Wynne)

overfishing of various fish species and organic pollution. These threats are difficult to separate as they are all interlinked, and difficult to manage as their sources can be numerous and often outside management areas. For example, although organic pollution can be derived from numerous local sources (yachts illegally dumping grey water; leeching from poorly maintained septic tanks; and salt ponds, a natural sink for organic nutrients, being connected to the sea – either anthropogenically by pipes or naturally via storm breaches) it is likely that the pollution is predominantly influenced by regional factors, for example heavy nutrient loads introduced into the Caribbean Sea via the Orinoco outflow. Despite this, local management is essential where possible as it will only be through all nations doing whatever they are able that the regional influences may be mitigated.

Lionfish

During the latter part of 2009 the Department of Fisheries and Marine Resources lead a public awareness initiative to inform people of the pending arrival of the invasive Lionfish (*Pterois volitans*), and also laid out a targeted eradication response plan (Wynne 2009) that would be initiated once their arrival had been confirmed. This confirmation came on the 16th August 2010 when a local dive operator spotted an individual close to Anguillita Cay. A few days later the specimen was captured by Fisheries Officers, photographed, and brought ashore for positive identification. In the months that followed sighting frequency began to increase and by the end of the year 'permanent arrival' had become a reality. Government effort is being targeted at bathing beaches and important reef areas, with future monitoring and reports by the public tracking their spread and impact. It is likely that this invasive species will unfortunately become another threat to Anguilla's coral reef ecosystem and as such new management measures are needed to mitigate its impact. These measures will be all the more urgent as with no known predators in the Caribbean this species can decimate juvenile fish populations and negatively affect local livelihoods. The Department of Fisheries and Marine Resources currently issues ELF capture devices to dive operators to help safely target this venomous species.

The Future

Changes to the management of Anguilla's marine environment are long overdue, and results from current monitoring combined with the arrival of the Lionfish mean these changes are now more urgent than ever. Indeed, plans are afoot to bring in a number of new management measures that are designed to promote the sustainability of Anguilla's marine resources, including its coral reefs. Plans include (but are not limited to) restricting spearfishing, the introduction of closed seasons, and enhanced management of the Marine Parks to include areas closed to fishing. The mention of such plans has already come up against public resistance, and until this matter is addressed any management strategy will likely fail. With this in mind the present focus is on public outreach, and a programme of educating stake holders so that they understand why management measures are being introduced and the benefits that they will receive in the long term. Plans are also in place to increase surveillance and enforcement capabilities of Governmental Departments through officer training workshops and the overhaul of the present Fisheries Department vessel.

References

Bythell JC, Buchan KC (1996) Impact of hurricane Luis on the coastal and marine resources of Anguilla. Marine Ecological Survey produced for the British Development Division of the Overseas Development Administration by the University of Newcastle, United Kingdom

Gell F, Watson M (2000) Chapter 39: UK overseas territories in the Northeast Caribbean. In: Sheppard CRC (ed) Seas at the millennium: an environmental evaluation, vol 1. Elsevier Science, Anguilla/Montserrat, pp 615–626

Hubbard DK (ed) (1989) Terrestrial and marine geology of St Croix. In: Proceedings of 12th Caribbean geological conference, US Virgin Islands, vol 2, 13pp

Murry PA (2008) Strategic plan for the management of the fisheries sector of Anguilla 2008–2012. Unpublished report prepared for the Department of Fisheries and Marine Resources by the OECS Environment and Sustainable Development Unit

NRI (1995) Habitat atlas and database of Anguilla, Natural Resources Institute and Overseas Development Administration, Large format, 100 cm × 50 cm, 18 pp

Olsen DA, Ogden JC (1981) Management planning for Anguilla's fishing industry. Draft report prepared for Eastern Caribbean natural area management program, US Virgin Islands

Oxenford HA, Hunte W (1990) A survey of marine habitats around Anguilla, with baseline community descriptors for coral reefs and seagrass beds. Report for the Department of Agriculture and Fisheries, Government of Anguilla, by the Bellairs Institute, McGill University, St.James, Barbados

Salm RV (1980) Anguilla. Coral Reefs and Marine Parks Potential. Consultancy report for selection and design of marine parks and reserves. Eastern Caribbean natural area management program, unpubl rep. Available from Island Resources Foundation

Sheppard CRC, Matheson K, Bythell JC, Murphy P, Blair MC, Blake B (1995) Habitat mapping in the Caribbean for management and conservation: use and assessment of aerial photography. Aquat Conserv: Mar Freshw Ecosyst 5:277–298

Stephenson A (1992) Anguilla. country profile 1991. Anguilla Government

Tuya F, Boyra A, Sanchez-Jerez P, Barbera C, Haroun R (2004) Can one species determine the structure of the benthic community on a temperate rocky reef? The case of the long-spined seaurchin Diadema antillarum in the eastern Atlantic. Hydrobiologia 519:211–214

Wynne S (2007a) Ecological baseline survey of Anguilla's five marine parks. Produced by the Department of Fisheries and Marine Resources for the Anguillian National Trust as part of an Overseas Territories Environmental Programme funded project. Copies can be obtained by contacting fisheriesmr@gov.ai

Wynne S (2007b) Anguilla marine monitoring programme survey protocol. Produced by the Department of Fisheries and Marine Resources for the Government of Anguilla. Copies can be obtained by contacting fisheriesmr@gov.ai

Wynne S (2008a) Anguilla marine monitoring programme pilot study. Produced by the Department of Fisheries and Marine Resources for the Government of Anguilla. Copies can be obtained by contacting fisheriesmr@gov.ai

Wynne S (2008b) Anguilla marine monitoring programme 2008 Annual report, Produced by the Department of Fisheries and Marine Resources for the Government of Anguilla. Copies can be obtained by contacting fisheriesmr@gov.ai

Wynne S (2009) Lionfish response plan – Anguilla BWI. Produced by the Department of Fisheries and Marine Resources for the Government of Anguilla. Copies can be obtained by contacting fisheriesmr@gov.ai

Wynne S (2010) Status of Anguilla's marine resources 2010. 2009 AMMP report, Produced by the Department of Fisheries and Marine Resources for the Government of Anguilla. Copies can be obtained by contacting fisheriesmr@gov.ai

3 Introduction to Reefs and Shorelines of the British Virgin Islands

Shannon Gore

Introduction

The British Virgin Islands (BVI) have long been a popular tourist destination because of the islands' idyllic sandy white beaches, crystal blue waters and dramatic scenic vistas. These natural resources have allowed tourism to become a significant driver of the contemporary economy along with tax and asset protection for offshore financial businesses. The country's low population of 25,383 (Central Intelligence Agency 2011) and limited large scale developments have helped maintain most of these natural resources. However, like many Caribbean islands, future economic growth and development threaten the very reason why people visit the islands.

Oceanographic and Geologic Setting

The British Virgin Islands (BVI) are a small archipelago located 95 km east of Puerto Rico in the north-eastern Caribbean. The Atlantic Ocean is to the north of the BVI and the Caribbean Sea to the south (Fig. 3.1). Geologically, these islands form the eastern edge of the Greater Antilles island chain and with Puerto Rico (PR) and the United States Virgin Islands (USVI), excluding St. Croix; collectively rise from the Puerto Rican/Virgin Island (PR/VI) platform. This bank is a carbonate microplate deposited over an inactive and subsiding island arc during the Oligocene to Holocene between the North American and Caribbean Plates and has undergone extensive submarine and subaerial erosion (Byrne et al. 1985; Masson and Scanlon 1991; Van Gestel et al. 1999). The PR/VI platform is bounded by the Puerto Rican Trench 125 km to the north (depth 8,395 m) and the Anegada Passage to the southeast (depth 1,915 m). The southern edge of the platform is marked by the Virgin Islands Basin (depth 4,091 m), the south-western edge by the Muertos Trough (depth 5,550 m) and the western edge by the Mona Passage (depth 4,000 m).

The BVI has 60+ high volcanic islands (max. elevation of 522 m) composed of Cretaceous volcanic rock and their sedimentary products (Helsley 1960). These strata generally trend east–west and slope moderately to steeply towards the north. The oldest rocks (the Water Island Formation, developed during the Lower Cretaceous) are exposed on the south-western islands of the BVI (Pelican Island, the Indians, the majority of Norman Island, Peter Island and Dead Chest as well as the southern portions of Salt and Cooper Islands). The youngest exposed rock (the Necker Formation (Middle Eocene)) comprises the smaller islands to the northeast of Tortola and Virgin Gorda (the majority of Guana Island, the northern side of Great Camanoe, and northern portions of the Dog Islands, all of the Seal Dog Islands, Moskito, Prickly Pear and Necker Islands) (Figs. 3.2, 3.3, and 3.4).

Towards the end of the volcanic period that occurred approximately 35–40 million years ago, volcanic rocks were intruded by granite-like magma which formed the coarse-grained igneous rocks of the Virgin Island Batholith (Figs. 3.5 and 3.6). This incorporates the eastern end of Tortola, southern end of Great Camanoe, all of Little Camanoe, Beef Island and Scrub Island, as well as the majority of the Dog Islands and Virgin Gorda (Smith et al. 1998).

The islands are generally separated into two island chains by the Sir Francis Drake channel, which is approximately 6 km wide channel and with a maximum depth of 50 m. The southern chain includes all the islands between Norman and Necker Islands and the northern chain includes Little Tobago (just west of Jost Van Dyke) to the Dog Islands.

Anegada, in contrast to the high volcanic islands of the BVI, is a carbonate emergent reef island with a maximum elevation of 8 m (Martin-Kaye 1959) (Fig. 3.7 and see Chap. 3). The island flanks the far north-eastern margin of the PR/VI platform and

S. Gore (✉)
Conservation and Fisheries Department,
Government of the Virgin Islands, 3323, Road Town,
Tortola VG1110, British Virgin Islands
e-mail: sd_gore@yahoo.com

C.R.C. Sheppard (ed.), *Coral Reefs of the United Kingdom Overseas Territories*, Coral Reefs of the World 4,
DOI 10.1007/978-94-007-5965-7_3, © Springer Science+Business Media Dordrecht 2013

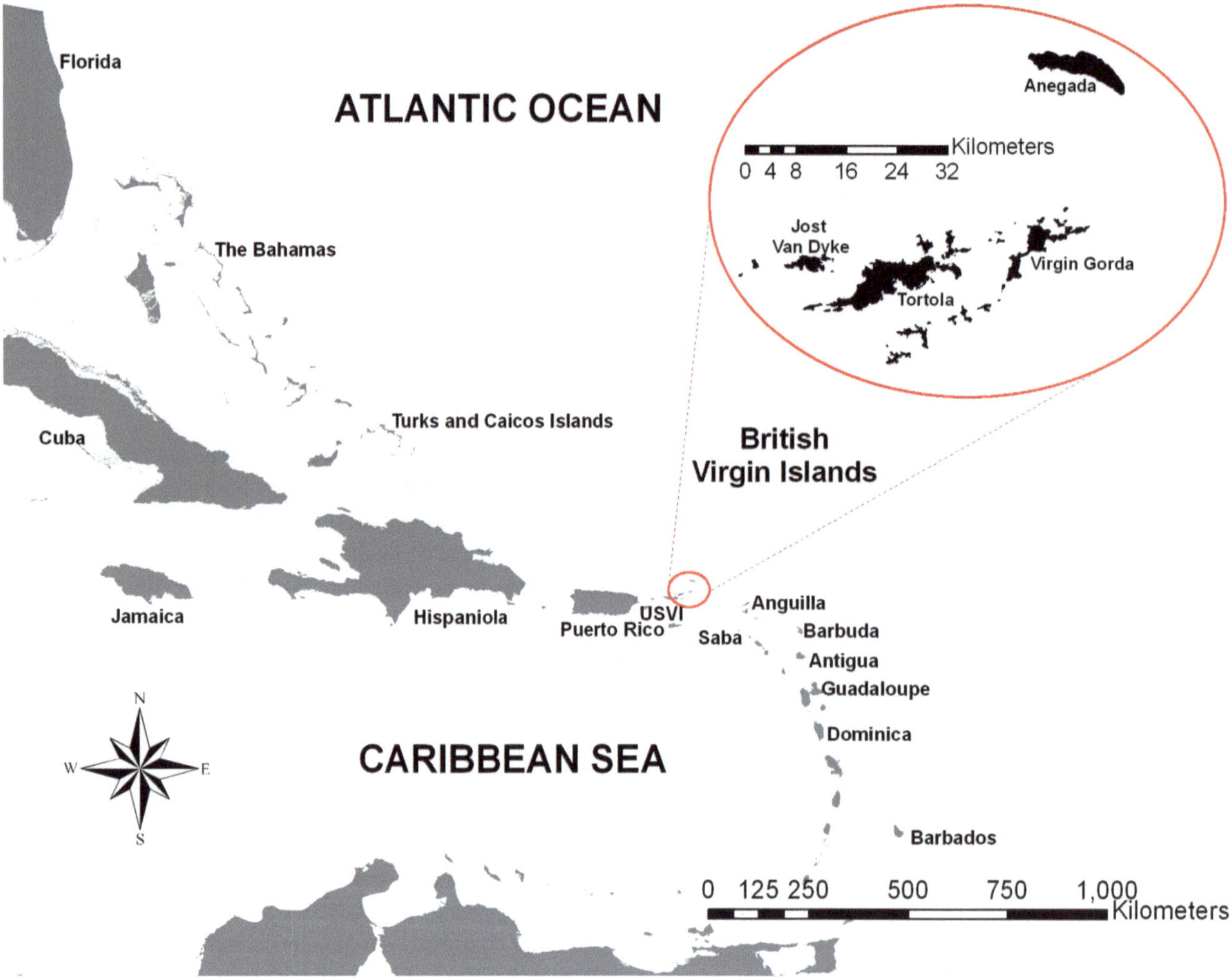

Fig. 3.1 Location map of the British Virgin Islands

Fig. 3.2 Underwater reef mosaic along the coast of Dead Chest (Credit A. Jenik)

Fig. 3.3 Dead Chest Island is one of the oldest islands in the BVI and is part of the Water Island Formation (Credit S. Gore)

Fig. 3.4 The Indians are geologically part of the Water Island Formation (Credit S. Gore)

is approximately 25 km northeast of the BVI island chains. It is the only reef island on the platform and is comprised of two main elements adjacent to each other, a Pleistocene limestone flat and limestone ridge (Anegada Limestone Formation) on and around which contemporary coastal sedimentary environments have developed (Anegada Ridge Plain Formation).

Fig. 3.5 Large granite boulders of the Virgin Gorda Batholith along the southern coast of Fallen Jerusalem. Aerial view (Credit S. Gore)

Fig. 3.6 The Baths (Credit S. Gore)

Climate

The BVI lie within the Northeast Trade Wind Zone where prevailing wind and waves approach from the east-northeast to southeast. Average wind direction varies throughout the year but generally, the predominant winds are from the east-northeast during the winter (December to February); east during the spring (March to May) and east to southeast during the summer and fall (June to November).

Fig. 3.7 Anegada and its surrounding reef (Credit S. Gore)

With the exception of tropical storms and hurricanes, the highest wind speeds occur from December to February and June to July. Winds generally blow between 6 and 9 m per second (ms^{-1}) but can be as low as 3 ms^{-1} during the fall and during the winter, gusts may reach over 15 ms^{-1}.

The hurricane season extends from June to November with August and September being the most active months (Hubbard 1989). The earliest recorded hurricane in the BVI was in 1713 (Pickering 1983). Since 1851, the earliest data set documented by the National Weather Service (HURDAT data from the US National Hurricane Centre), 35 hurricanes have passed over or within 100 km of the BVI, 17 of which were major hurricanes (category 3 or above) (NOAA 2011a). Hurricanes and storms generally pass the BVI from the southeast to the northwest but high-energy waves may approach the islands from any direction.

From October to April, large south running swells, locally called "groundseas", come from the north and are generated by storms in the North Atlantic Ocean. These long-period, high energy waves commonly have wave heights greater than 1 m with peak heights of up to 5 m. However, extreme swell events other than from tropical storms or hurricanes are rare and are considered as 20-year events. The last two documented events for the BVI occurred in October 1991 (NWS 2008) and March 2008 (Cooper et al. 2008). In 2008, a low-pressure system moved off the mid–Atlantic states and generated extreme wave growth with very long period swells (NWS 2008) (Fig. 3.8). These propagated across the Atlantic to the northeast coast of South America and as far as the West African coast and Antarctica. At the same time, a NOAA data buoy (#41043) located approximately 270 km north of the US Virgin Islands recorded waves greater than 4.5 m with a period of 17 s (NOAA 2011b). Extreme wave heights of 6–9 m were common around Puerto Rico with waves reaching 9–12 m across the outer reefs. In Anegada, the fringing reef significantly attenuated incoming waves and despite some localized erosion, waves did not breach any of the ridges along the north shore (Atwater et al. 2012).

The BVI lies in the path of the Equatorial Current that sets westward with an average speed of 0.3 m/s (Hydrographer of the Navy 1978). Tides are mixed semi-diurnal, and the tidal ranges around Tortola are typically between 0.29 and 0.65 m (National Oceanography Centre 2011).

Temperature in the BVI varies little throughout the year, and average daytime temperatures range from 25°C to 29°C and drop about 6°C at night.

Seawater temperatures around the Virgin Islands generally range from 26°C to 29°C but long-term water temperature records from nearby St. John, USVI, indicate an increase in mean monthly seawater temperatures by 0.6°C/decade over the past 20 years (Edmunds 2004). The inflow of North Atlantic waters through the nearby Anegada Passage (Johns et al. 2002) may allow for mixing of cooler waters from upwelling but no long term monitoring data exists for waters specifically around the BVI.

Rainfall amounts to approximately 1,150 mm per year collectively in the BVI, but Anegada experiences the least amount, approximately 890–1,016 mm per year (Downs 1997). The heaviest periods of rainfall occur during October and November. In November 2003, for example, an average

Fig. 3.8 Extreme swell event in March 2008 (Credit S. Gore)

of 508 mm (20 in.) of rain fell over a 5 day period in the BVI (DDM 2003). Record breaking rainfall occurred in 2010, with 634 mm (24.98 in.) falling over the BVI within 3 days. This caused extensive flooding throughout the BVI, and it was considered a 50-year flood event by the BVI Department of Disaster Management (DDM). On Anegada, however, flooding and consequent damage was minimal as floodwaters are distributed and absorbed through Anegada's extensive wetland ecosystems.

Coastal Features and Reefs

The Territory has a total land area of 153 km^2 and approximately 420 km of coastline with settings that vary significantly throughout the Territory. This variability is based on differences in the islands' underlying geology, directional orientation to seasonal wind and waves, sedimentary characteristics (i.e. grain size and composition), and morphologic diversity based on planforms, adjacent seaward features (coral reefs, marine vegetation, carbonate platforms) and adjacent terrestrial environments (watersheds and associated features). Coral reef flats may be broad, as in Virgin Gorda (Fig. 3.9) and a reef flat which connects Jost Van Dyke with Little Jost Van Dyke (Fig. 3.10), and there are small sandy cays such as the aptly named Sandy Cay which is a National Park (Fig. 3.11).

The high islands of the BVI are dominated by steep sloping hills and a diverse range of coastal features including moist and dry coastal woodlands, freshwater drainage pathways locally called 'ghuts', salt ponds, wetlands, mangroves (Fig. 3.12), lagoons, sandy and coralline rubble beaches, and cliffs. A number of low-lying mangrove cays are developing over coral reef/rubble flats nearshore along the southern coast of Tortola and the south-central section of neighbouring Beef Island. Additionally, one vegetated sand cay exists as part of an incipient tombolo (Sandy Spit, Fig. 3.13) and an unvegetated sand cay is forming over a narrow elongated reef flat ~340 m off the southern coast of Necker Island.

Marine habitats around the volcanic islands include seagrass meadows, muds, carbonate banks and coral reefs. The majority of reefs are fringing reefs which have developed nearshore on a narrow shelf which averages 500 m in width between the low tide shoreline and the 20 m bathymetric depth contour. Cross-sectional reef morphology shows a strong correlation between reef ecology and degree of exposure to waves (Woodroffe 2002; Geister 1977) but due to variable wind and wave exposure and the close proximity of neighbouring islands within the BVI, there is rarely a clear distinction between windward and leeward reefs around the high islands. Additionally, many of the fringing reefs lack an identifiable reef crest, partially due to white-band disease that affected *Acroporids* in the late 1970s (Aronson and Precht 2001).

The contemporary coastline of Anegada has a near continuous carbonate beach (fine to coarse-grained sands) with local occurrences of beachrock and red mangroves (*Rhizophora mangle*). The western side of Anegada, also

Fig. 3.9 Fringing reef in Taylor's Bay, Virgin Gorda (Credit S. Gore)

Fig. 3.10 Reef flat between Jost Van Dyke and Little Jost Van Dyke (Credit JAG. Cooper)

known as the Anegada Ridge Plain Formation, consists of shore-parallel dune ridges and shallow hyper-saline wetlands. The eastern side of the island, the Anegada Limestone Formation, is an emergent Pleistocene limestone flat on and around which contemporary sedimentary environments have developed. Subtidal marine environments around Anegada include carbonate sands and muds, seagrasses, algal plains and coral rock.

Fig. 3.11 Sandy Cay National Park (Credit S. Gore)

Fig. 3.12 Red mangroves in Hans Creek, Beef Island (Credit S. Gore)

Fig. 3.13 Sandy Spit (Credit S. Gore)

Anegada's surrounding Horseshoe Reef covers 133 km^2 and is one of the largest contiguous coral reefs in the Caribbean. It is composed of two distinctive facies, a high energy reef front on the northeast windward side of the island, and a series of patch reefs orientated and aligned systematically to prevailing currents and waves along the island's southern leeward side (Brown and Dunne 1980). These particular reefs display a high species diversity and greater coral abundance than those reefs found on the northern side of the island (Dunne and Brown 1979).

The fringing reef on the north side of Anegada extends beyond the eastern end of the island to the southeast for 14 km and at an area called the "White Horses" turns 90° to the southwest for another 8.5 km. The section of the reef extending beyond the eastern tip of the island is considered a barrier reef (http://www.reefbase.org) despite the lagoon area being an open bank (Hubbard et al. 2008). This reef system is the primary source of carbonate sediments found between Anegada and Tortola (the Anegada Bank) and provides carbonate rich sediments to the Atlantic facing beaches along the northern island chain.

Threats to Marine Resources

Prior to the 1960s, the economy of the British Virgin Islands was based on subsistence fishing and farming (Encontre 1989). However, with the natural environment of the BVI regarded as an economic asset coupled with the increase in international travel, the development of a tourism industry marked a historical change in economic thinking. The first cruise ship to visit the Territory came in 1960 followed by the opening of Laurence Rockefeller's Little Dix Bay Resort on Virgin Gorda in 1964. In 1969, the first yacht charter company opened; marking the beginning of what would later result in the greatest concentration of charter yachts in the world (Spalding et al. 2007). The reefs remained a prime attraction (Fig. 3.14) with a modest recovery of staghorn corals in some places (Fig. 3.15).

Since the onset of a tourism-based economy, a paradigm shift occurred from preserving natural resources for local subsistence to the exploitation of natural resources for economic gain. Early legislation reflected foresight to protect natural resources at the time but as the connection between people and the environment diminished, legislation fell short of protecting the assets for which people would visit the BVI. Additionally, the environmental legislation that did exist did not keep up with the advancement of scientific knowledge.

Updating environmental legislation has since become a well-recognized need in the BVI, with some success including the VI Fisheries Act, 1997; the VI Fisheries Regulations, 2003; the VI Planning Act, 2004; and the BVI Systems Plan 2007–2017 which proposes the protection of 30% of each marine habitat type. However, despite some legislation protecting environmental resources, there are still gaps. For

Fig. 3.14 Reefs of the BVI are one main attraction for tourism (Credit S. Gore)

Fig. 3.15 Staghorn corals have made a small recovery in some parts of the BVI (Credit S. Gore)

example, enforcement of legislation often lacks the resources (personnel, patrol vessels, vehicles) needed to implement penalties. Even the penalties are often not sufficient to deter people from violating the laws in the first place.

Other gaps in legislation include the lack of specific regulations, such as water quality standards to control sedimentation caused by unpaved roads and other types of coastal developments. Additionally, management plans such as those for watersheds and beaches are limited to ensure marine resource sustainability. One consequence of lack of regulations or adherence to them is garbage dumped on reefs, and heavy sedimentation after rains (Fig. 3.16).

Fig. 3.16 Heavy sedimentation due to unpaved roads cut into the hillside along the northern shore of Tortola (Credit L. Jarecki)

Fig. 3.17 Bleaching from 2005 (Credit A. Jenik)

Lack of (or in some cases limited) legislation and management plans are not the only problems threatening marine resources in the BVI. Coral reefs in the BVI, like many other Caribbean reefs, experienced the epidemic of white-band disease that affected *Acroporids* in the late 1970s (Aronson and Precht 2001) as well as the region-wide disease-induced mass mortality of *Diadema antillarum* in 1983–1984 (Carpenter 1988). Additionally, hurricanes and tropical storms passing near or through the BVI have further caused deterioration of reefs around the BVI. In 2005, the average water temperatures in the Caribbean region were the warmest in over 150 years which led to the most severe bleaching event ever recorded in the basin (Eakin et al. 2010) (Fig. 3.17). Water temperatures remained higher than the long-term average for over 10 weeks in the Virgin Islands, and along with the consequent outbreak of white plague disease, live coral cover was reduced by 60% in the BVI (Hime 2008) and neighbouring USVI (Miller et al. 2009).

Fig. 3.18 Grounding at the RMS Rhone Marine Park, Salt Island (Credit A. Jenik)

Other threats to the marine environment include groundings (Fig. 3.18) and anchor damage from yachts, ghost fishing from the loss of fishing pots and the invasion of lionfish (*Pterois volitans*) which first appeared in March 2010. Despite these threats upon marine resources in BVI, the Territory's population remains low, and large scale developments remain minimal but are, however, on the rise.

References

Aronson RB, Precht WF (2001) White-band disease and the changing face of Caribbean coral reefs. Hydrobiologia 460:25–38

Atwater BF, ten Brink US, Buckley M, Halley RS, Jaffe BE, Lòpez-Venagas AM, Reinhardt EG, Tuttle MP, Watt S, Wei Y (2012) Geomorphic and stratigraphic evidence for an unusual tsunami or storm a few centuries ago at Anegada, British Virgin Islands. Nat Hazard 63(1):51–84

Brown BE, Dunne RP (1980) Environmental controls of patch-reef growth and development. Mar Biol 56(1):85–96

Byrne DB, Suarez G, McCann WR (1985) Muertos trough subduction—microplate tectonics in the northern Caribbean. Nature 317:420–421

Carpenter RC (1988) Mass mortality of a Caribbean sea urchin: immediate effects on community metabolism and other herbivores. Proc Natl Acad Sci USA 85:511–514

Central Intelligence Agency (CIA) (2011) The CIA world fact book. https://www.cia.gov/library/publications/the-world-factbook/geos/vi.html. Accessed 2 July 2012

Cooper JAG, Jackson DWT, McKenna J, Gore S (2008) Morphological impact of a high energy swell event on Caribbean Island shorelines. In: Joint Meeting of the Geological Society of America, Soil Science Society of America, American Society of Agronomy, Crop Science Society of America, Gulf Coast Association of Geological Societies with the Gulf Coast Section of SEPM (eds) Geological Society of America Abstracts with Programs, vol 40. Houston. p 91

Department of Disaster Management (2003) Annual report – 2003. Department of Disaster Management, Tortola, 39p. http://www.bviddm.com/index.php?action=doc&category_id=5. Accessed 2 July 2012

Downs MA (1997) Anegada sea turtle recovery project community assessment, 54p. Available from the Island Resource Foundation Website http://www.irf.org/mission/planning/turtles.doc. Accessed 9 May 2012

Dunne RP, Brown BE (1979) Some aspects of the ecology of reefs surrounding Anegada, British Virgin Islands. Smithsonian Institution, Washington, DC

Eakin CM, Morgan JA, Smith TB, Liu G, Alvarez-Filip L, Baca B, Bouchon C, Brandt M, Bruckner A, Cameron A, Carr L, Chiappone M, James M, Crabbe C, Day O, de la Guardia-Llanso E, DiResta D, Gilliam D, Ginsburg R, Gore S, Guzmán H, Hernández-Delgado EA, Husain E, Jeffrey C, Jones R, Jordán-Dahlgren E, Kramer P, Lang J, Lirman D, Mallela J, Manfrino C, Maréchal JP, Mihaly J, Miller J, Mueller E, Muller E, Noordeloos M, Oxenford H, Ponce-Taylor D, Quinn N, Ritchie K, Rodríguez S, Rodríguez-Ramírez A, Romano S, Samhouri J, Schmahl G, Steiner S, Taylor M, Walsh S, Weil E, Williams E (2010) Caribbean corals in crisis: record thermal stress, bleaching and mortality in 2005. PLoS One 5(11):e13969

Edmunds PJ (2004) Juvenile coral population dynamics track rising seawater temperature on a Caribbean reef. Mar Ecol Prog Ser 269:111–119

Encontre P (1989) Why does the tourism dollar count? An Introduction to the economics of tourism in the British Virgin Islands. British Virgin Islands Tourist Board, Road Town, 139pp

Geister J (1977) The influence of wave exposure on the ecological zonation of Caribbean coral reefs. In: Proceedings of third international coral reef symposium,vol 2, Geology, Rosenstiel School of Marine and Atmospheric, Miami, pp 23–29

Helsley CE (1960) Geology of the British Virgin Islands. Unpublished Ph.D. thesis, Princeton University, Princeton, 219pp

Hime S (2008) Effect of marine based tourism in the British Virgin Islands. Unpublished Ph.D. thesis, University of East Anglia, Norwich, 213pp

Hubbard DK (1989) Modern carbonate environments of St. Croix and the Caribbean: a general overview. In: Hubbard DK (ed) Terrestrial and marine geology of St. Croix, US Virgin Islands, vol 9. West Indies Laboratory Special Publication, St. Croix, pp 85–94

Hubbard D, Burke R, Gill I, Ramirez W, Sherman C (2008) Coral reef geology: Puerto Rico and the US Virgin Islands. In: Riegl BM, Dodge RE (eds) Coral reefs of the USA, vol 1. Springer, Dordrecht, pp 263–302

Hydrographer of the Navy (1978) West Indies pilot, 2nd edn. Hydrographer of the Navy, London, 385pp

Johns WE, Townsend TL, Fratantoni DM, Wilson WD (2002) On the Atlantic inflow to the Caribbean sea. Deep-Sea Res I 49:211–243

Martin-Kaye PHA (1959) Reports on the geology of the Leeward and British Virgin Islands. Voice, Castries, 117pp

Masson DG, Scanlon KM (1991) The neotectonic setting of Puerto Rico. Geol Soc Am Bull 103:144–154

Miller J, Muller E, Rogers C, Waara R, Atkinson A (2009) Coral disease following massive bleaching in 2005 causes 60% decline in coral cover on reefs in the US Virgin Islands. Coral Reefs 28(4):925–937

National Oceanographic and Atmospheric Administration (2011a) National Hurricane Center Data Archive, "Best-track" database (HURDAT). http://www.nhc.noaa.gov/pastall.shtml. Accessed 12 June 2012

National Oceanographic and Atmospheric Administration (2011b) National Data Buoy Center. http://www.ndbc.noaa.gov/news_archive.shtml. Accessed 12 June 2012

National Oceanography Centre (2011) British overseas territories (British Virgin Islands). http://www.pol.ac.uk/ntslf/tidalp.html. Accessed 10 June 2011

National Weather Service Weather Forecast Office (NOAA) (2008) Extreme atlantic swell event of march 2008. http://www.srh.noaa.gov/sju/?n=2008_swell01. Accessed 2 July 2012

Pickering VW (1983) Early history of the British Virgin Islands: from Columbus to emancipation. Falcon Publications International, British Virgin Islands, 248pp

Smith AL, Schellekens JH, Diaz AL (1998). Batholiths as markers of tectonic change in the north-eastern Caribbean. In: Lidiak EG, Larue DK (eds) Tectonics and geochemistry of the northeastern Caribbean. Geologic society of America special paper, 322, Boulder, pp 99–122

Spalding MD, Fox HE, Allen GR, Davidson N, Ferdana ZA, Finlayson M, Halpern BS, Jorge MA, Lombana A, Lourie SA, Martin KD, McManus E, Molnar J, Recchia CA, Robertson J (2007) Marine ecoregions of the world: a bioregionalization of coast and shelf areas. Bioscience 57:573–583

Van Gestel JP, Mann P, Grindlay NR, Dolan JF (1999) Three-phase tectonic evolution of the northern margin of Puerto Rico as inferred from an integration of seismic reflection, well, and outcrop data. Mar Geol 161(2–4):257–286

Woodroffe CD (2002) Coasts: form, process and evolution. Cambridge University Press, Cambridge, 623pp

Marine Protected Areas and Management in the British Virgin Islands

4

Nancy Woodfield Pascoe, Joseph Smith-Abbott, and Shannon Gore

Introduction

The British Virgin Islands (BVI) has 14 declared protected areas in the marine environment, including one marine park and 13 fisheries protected areas, managed by the National Parks Trust of the Virgin Islands (NPTVI) and the Conservation and Fisheries Department (CFD) respectively. In addition to this, the '*British Virgin Islands Protected Areas System Plan 2007–2017*' (Gardner et al. 2007), approved by Cabinet in 2008, identifies another 40 areas for inclusion in the marine protected area (MPA) network, with designation in progress and pending completion in 2013 (Fig. 4.1). The health of the natural resources contained in these sites varies, with some sites having healthy ecosystems while the habitats and resources in others are severely stressed and have been included in the MPA network because they are adjacent to terrestrial national parks and require management as part of an ecosystem approach. An example of the latter is The Baths National Park.

Within the BVI, there are several different management objectives for the protected areas, which led to the adoption of the management categories developed by IUCN-The World Conservation Union (IUCN 1994) as the official framework for management of these areas in the BVI. The overall goals of protected areas in the BVI include conservation, sustainable resource use, recreation, economic development, education and community involvement.

The NPTVI is one of the oldest national trusts in the Caribbean region, created in 1961 under the National Parks Ordinance (1961) with an ability to declare national parks, and later, marine parks under the Marine Parks and Protected Areas Ordinance (1979). The National Parks Ordinance was recently revised and is now known as the National Parks Act (2006). Protected areas can also be declared under the Fisheries Act (1997) with a primary focus on protecting areas that are important for the life cycle of fisheries, such as spawning areas and juvenile habitats.

The BVI Marine Environment and the MPA

There are many different habitats throughout the BVI, including coral reefs, mangroves, seagrass meadows, and sandy beaches (utilised for turtle nesting). There are four main species of mangrove in the BVI, including: red mangrove (*Rhizopora mangle*), black mangrove *(Avicennia germinans)*, white mangrove *(Laguncularia racemosa)* and buttonwood *(Conocarpus erectus)*. There are six different species of seagrass, including turtle grass *(Thalassia testudinum)*, manatee grass *(Syringodium filiforme)*, shoal grass *(Halodule wrightii)*, paddle grass *(Halophila decipiens)*, widgeon grass *(Ruppia maritime)* and *Halophila baillonis* (Anderson et al. 1986).

Every island in the BVI is surrounded by coral reefs of varying size, health and composition. The Anegada Horseshoe Reef is the third largest contiguous reef in the Eastern Caribbean at 63 km (39 miles) long, containing both patch reefs and barrier reefs. There are 63 popular dive sites, which include 57 coral reef sites and 6 artificial reefs that have been created by shipwrecks.

Coral Reefs of the BVI

Most reefs in the BVI other than Anegada are considered fringing reefs under the classic definition; they grow nearshore and lack a sizeable lagoon separating them from shore. However, reef morphology varies throughout the Territory because of variations in exposure to wind and waves as well

N. Woodfield Pascoe (✉) • J. Smith-Abbott
National Parks Trust of the Virgin Islands, 57 Main Street, P.O. Box 860, Road Town, Tortola, British Virgin Islands
e-mail: planning@bvinpt.org; director@bvinpt.org

S. Gore
Conservation and Fisheries Department, Government of the Virgin Islands, P.O. Box 3323, Road Town, Tortola, VG1110, British Virgin Islands
e-mail: sd_gore@yahoo.com

C.R.C. Sheppard (ed.), *Coral Reefs of the United Kingdom Overseas Territories*, Coral Reefs of the World 4, DOI 10.1007/978-94-007-5965-7_4, © Springer Science+Business Media Dordrecht 2013

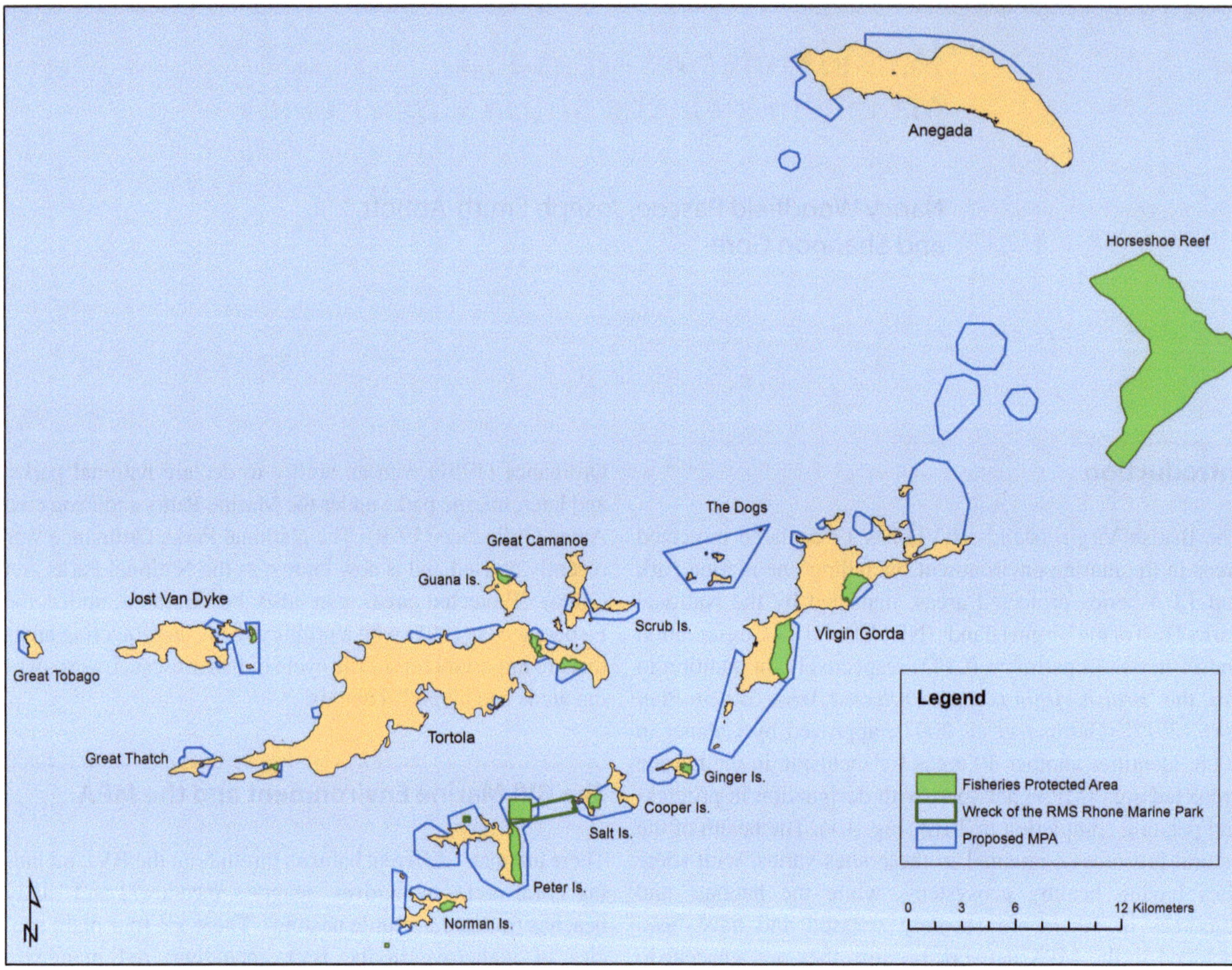

Fig. 4.1 Proposed marine protected area network for the BVI (Created by NPTVI)

as to proximity to neighbouring islands. Because incoming waves are significantly attenuated by coral reefs (Young 1989) due to the change in water depth and friction (Roberts et al. 1975), Gore (2011) recognized this was an important factor on beach form and classified these fringing reefs into three categories, based on their morphology. Each type of fringing reef is briefly described below.

'Headland attached linear reefs' are parallel to the coast, usually not more than 500 m offshore but are connected closely to land in some way, usually a headland (Fig. 4.2). This particular type of fringing reef in the BVI has a distinct reef crest, usually composed of *Acropora palmata* but since the outbreak of white-band disease throughout the Caribbean region in the late 1970s (Aronson and Precht 2001), *Montastrea annularis* has become the dominant reef species in the BVI. The forereef often has the highest coral coverage since it is based furthest from land-based pollutants and is exposed to higher wave energy which increases flushing.

A 'terrace flat' is a low gradient, broad lagoon (less than 500 m) composed of unconsolidated sediments, seagrasses, algae, small patch reefs and most commonly coral rubble (Fig. 4.3). Strong wave energy dissipation occurs over the initial fore reef, and subsequent wave reduction occurs through the shoaling process of the shallow flats.

'Graded reefs' are continuous reef slopes that differ from reef flats in that the profile from shore deepens much quicker, often less than 500 m from the shoreline so waves do not shoal (Fig. 4.4). These types of reefs have a higher relief, and coral coverage on them is often greater than those found on reef flats. Graded reefs usually extend from the shore seaward uniformly or may have spur and groove formations.

The conservation of these different reef habitats, along with others including mangroves, has been the goal of a programme to create a network of marine protected areas across the BVI. The NPTVI began its marine conservation programme in 1991 due to the rapidly increasing number of pleasure boats cruising BVI waters as the charter tourism industry grew steadily within the Territory.

The NPTVI has established a network of mooring buoys in order to eliminate anchoring on the reefs. A marine conservation

Fig. 4.2 Headland attached linear reef at Oil Nut Bay, Virgin Gorda (Photo by S. Gore)

Fig. 4.3 Terrace flat reef at Palm Bay, Scrub Island (Photo by S. Gore)

permit is required by law to use the NPTVI mooring buoys and these can be purchased from all BVI charter and dive companies and also at all marine ports of entry. Across this network there are over 160 mooring buoys that have been installed at coral reefs and sensitive marine habitats to prevent anchor damage and are patrolled and maintained daily by a team of NPTVI Marine Wardens. GIS and GPS technology is used to map the mooring buoy locations and report maintenance and patrolling activities on a weekly basis. Anchoring is illegal at the RMS Rhone Marine Park. The NPTVI monitors the fleet size of the local charter industry with frequent surveys, with over 600 vessels recorded in 2010. However,

Fig. 4.4 Graded reef at Beef Island (Photo by S. Gore)

Fig. 4.5 The Baths National Park, the most visited tourist destination in the BVI (Photo by F. Peters)

this figure does not include private vessels based in the BVI, or day-sail and term charter vessels that originate from the US Virgin Islands (USVI), or the many transient private vessels travelling through the Caribbean that may visit the BVI. In order to manage the demand for its mooring sites, NPTVI have a 90 minute time limit for mooring use, with no overnight use as the moorings are frequently at locations exposed to swell. Issues concerning carrying capacity for these dive and snorkel sites will be the subject of an upcoming study as part of an overall carrying capacity report for the NPTVI protected area system, which will include both marine and terrestrial areas (Fig. 4.5).

Future activities also include the completion of a management plan for the network of MPAs with a focus on zoning, permitted activities, education and sustainable use. The NPTVI will re-engage stakeholders consulted in the MPA network design process and broaden its consultation process to include more land based businesses, Government Departments and those indirect stakeholders whose activities may be impacting MPAs directly or indirectly through improper land based activities.

The United States Coral Reef Task Force recognized the NPTVI in 2006 when it was awarded a Coral Reef Task Force award for its efforts to improve the management of marine resources in the BVI through the development of a network of marine protected areas with use of sound science and stakeholder input, the development of strong partnerships, and a commitment to multilateral environmental agreements.

Designing the MPA Network

The network of marine protected areas in the BVI was designed by the NPTVI, in collaboration with CFD and other partners as part of an Overseas Territories Environment Programme (OTEP) funded project entitled, '*Assessment and Improved Management of New and Existing Marine Protected Areas in the British Virgin Islands*' (2004–2006). The overall goal of the NPTVI was to create a network of marine protected areas that met the following criteria (Woodfield Pascoe and Smith Abbott 2010):

- To create a Marine Protected Area (MPA) network that reflects the major marine and coastal habitats of the BVI;
- To protect 30% of the important biological habitats across the BVI. (e.g. hard corals, soft corals, seagrasses, mangroves, turtle nesting beaches, fishery habitats);
- To cluster protected areas together so that they can be easily managed; and,
- To ensure that there are marine protected areas distributed across the BVI in order to ensure 'resilience' within the network.

The first project goal was to ground-truth near-shore marine habitats in order to update a 1991 GIS dataset of the coastal resources of the BVI (Blair Myers et al. 1992), which had been the basis for the initial MPA network design. The updated GIS dataset was then used with the decision making program MARXAN, and in this programme the NPTVI collaborated with The Nature Conservancy who provided expertise and training.

The BVI was divided into three geographic units to build resilience into the network through even distribution of MPAs in each unit. This process eliminated the potential to place heavy reliance on an extensive reef system around the island of Anegada to the detriment of other areas (Fig. 4.6).

Certain areas were 'locked' into the MPA network. These included areas that NPTVI and CFD had identified as important due to their biodiversity, their importance as fish nurseries or fish breeding habitats. This information came from local fishermen and from previous projects determining the locations of spawning aggregations conducted by CFD in collaboration with partners in the USVI. CFD also have a long term sea turtle monitoring programme that provided information on important nesting beaches and foraging areas.

Additionally the software was programmed to select at least 30% of each biological habitat type (i.e. soft corals, hard corals, algae, seagrasses, mangroves) from the coastal resource GIS dataset (Fig. 4.7).

Four potential MPA network scenarios were created with varying levels of clustering of MPA areas and locked-in areas. These maps were then taken to stakeholders for feedback, including fishermen, dive operators, the charter boat industry and relevant government departments. To ensure participation in the stakeholder review process, meetings were organized on the four main islands in the BVI (Tortola, Anegada, Virgin Gorda and Jost Van Dyke) using existing organizations such as fisheries associations, the Charter Yacht Society, the Dive Operators Association, and the Marine Association.

Because the NPTVI has a long established relationship with the charter and dive industry which has come from over 25 years of managing mooring buoys in sensitive reef sites, it was relatively easy to ensure the participation of this sector of the marine industry, and the fisheries extension officers at CFD were critical in assisting with this process with regard to fishermen. Separate meetings were held for fishermen and the dive/charter industry due to potential conflicts of interests, and to make participants feel more comfortable when providing feedback. The main outcomes of the meetings was to have stakeholders draw on large printed maps that displayed the four MPA network models, indicating areas they currently use for fishing, diving, and anchoring, in addition to making suggestions of areas that should be protected.

Stakeholders were also asked to select the MPA network model they preferred the most. The MPA network model eventually selected was the one with the highest level of clustering and locked-in areas. Some amendments were then made to this MPA network based upon the stakeholder feedback. Very small isolated areas were also removed to leave larger MPAs more evenly distributed across the BVI.

The final stakeholder selection was also the preferred choice of NPTVI and CFD because the high level of clustering and number of locked-in areas meant that the MPA network included all the important biodiversity areas that the two organisations wished to protect and because the individual MPAs were large enough areas that entire reef systems of all types were well represented, on both northerly and southerly coastal exposures (see Fig. 4.1).

This final map was included in the proposed *British Virgin Islands Protected Areas System Plan 2007–2017* (Gardner et al. 2007). The NPTVI then collaborated with the Survey

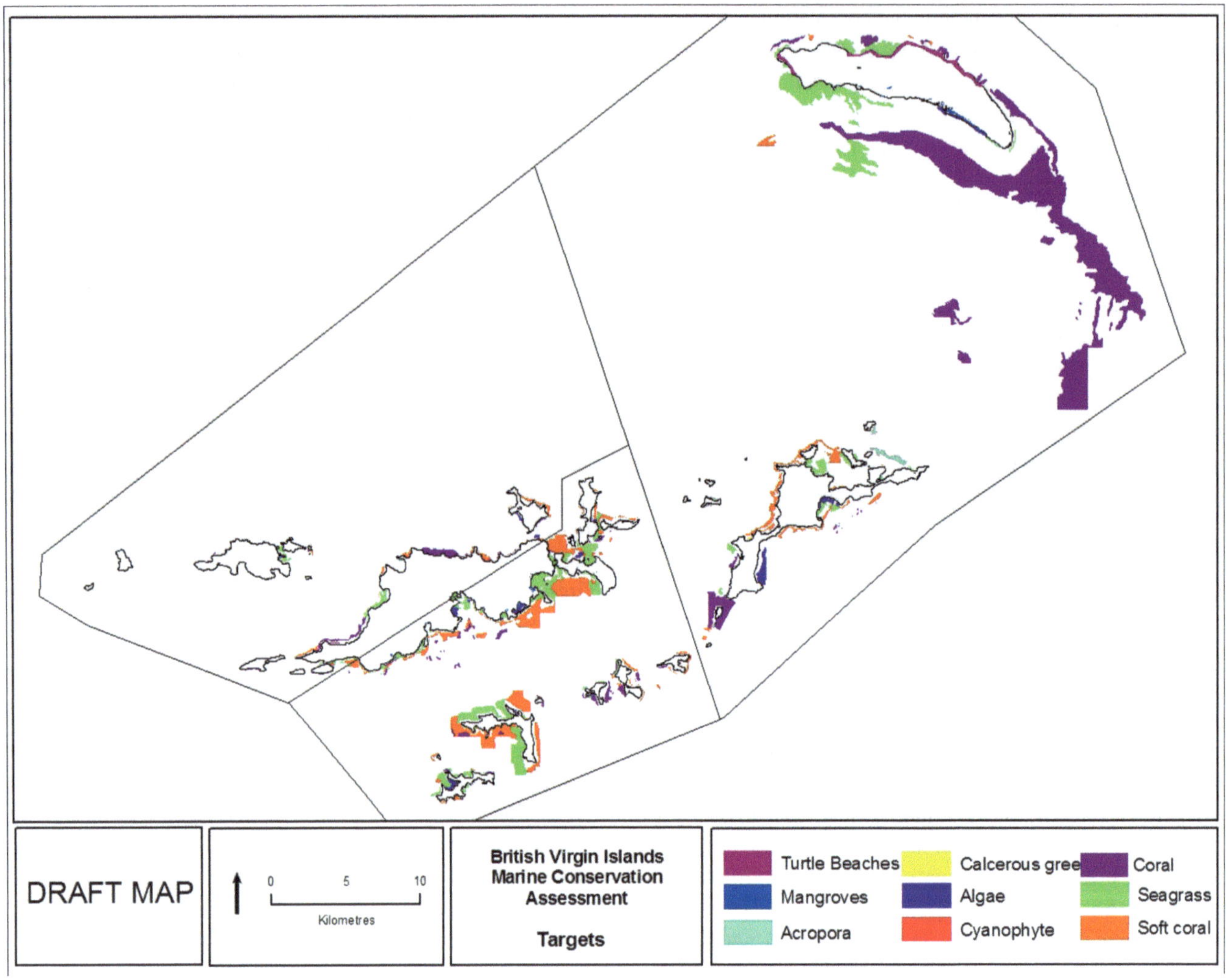

Fig. 4.6 The geographic regions of the BVI that were utilized in the MARXAN analysis (Created by A. Huggins)

Department to create the legal maps for these areas so that they can be officially designated as MPAs. The management categories developed by IUCN - The World Conservation Union have been adopted in the System Plan to define all marine and terrestrial areas. Each area within the MPA network has been assigned a management category using the information gathered from the stakeholders and from the assessment of conservation value attached to the area, e.g. Strict Nature Reserve, Protected Landscape/Seascape, Marine Park, Habitat Management Area, Managed Resource Area, and Species Management Area (Table 4.1). These maps will then be used to consultatively create the zoning plan for the MPAs where permitted activities can occur, as part of the management planning exercise, in accordance with the requirements of the National Parks Act 2006 (Fig. 4.8).

In addition to this, the boundaries of the MPAs and zones of permitted activities within them will be identified using marker buoys. As the MPA network is very extensive, a public relations campaign will be run to inform all stakeholders of the different management categories, the prohibited activities and the zones in the relevant MPAs. Promulgation will include all media sources, such as internet, newspapers, publications and brochures in the BVI, the US Virgin Islands and Puerto Rico.

Lessons Learned in Designing an MPA Network

1. In designing the MPA network, NPTVI first identified areas that were rich in biodiversity and important for fisheries and locked these areas into the software so that they would be included in every MPA network scenario. The MARXAN software also requires in-depth information about the potential human stressors to the marine environment and their threat range. The greatest human impacts identified included anchorages, mooring buoy fields, landfill sites, marinas, hazardous material locations, hotels, sedimentation sources, marine dumps and urban

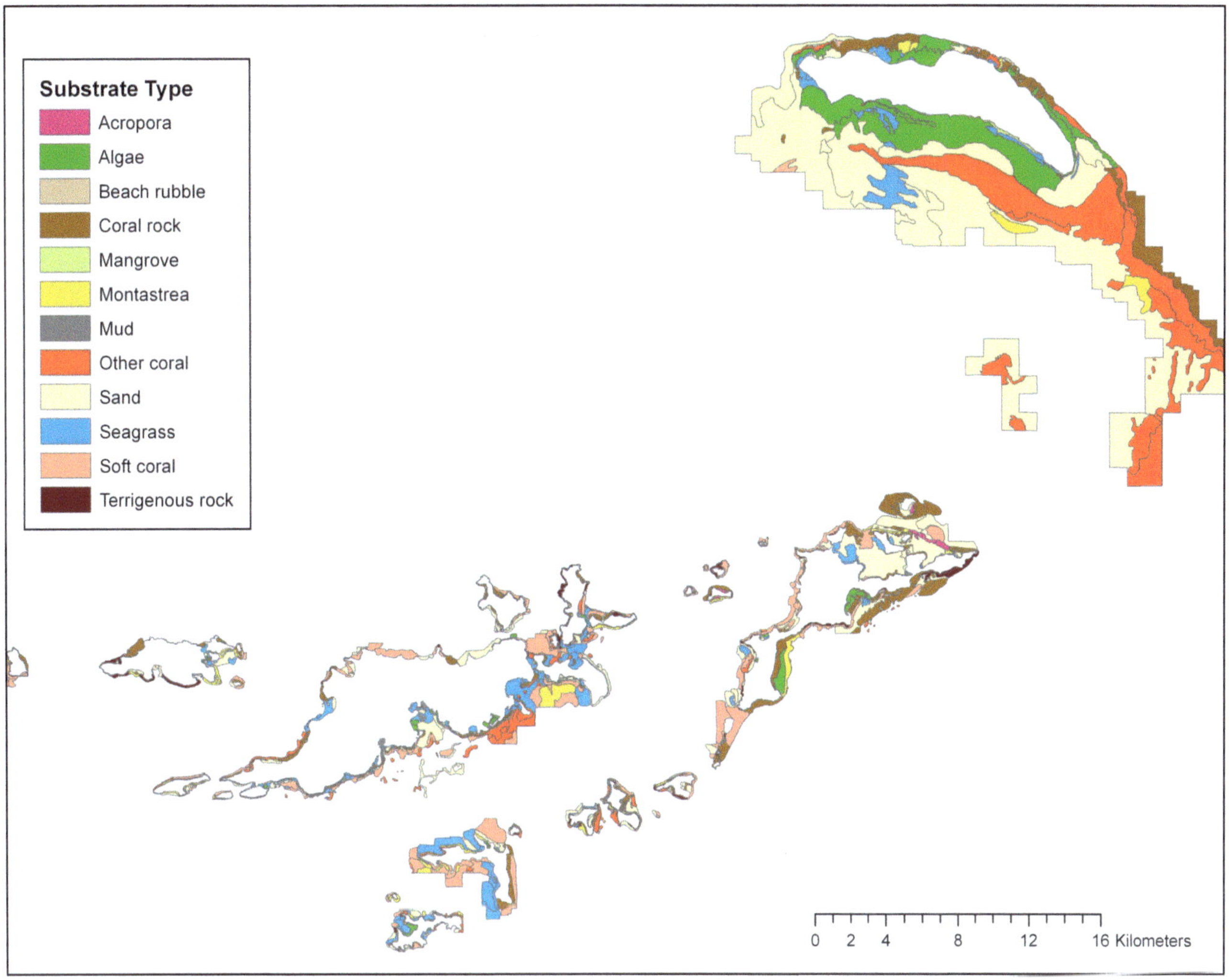

Fig. 4.7 The BVI Coastal resource atlas that was produced in 2006 (Created by NPTVI)

Table 4.1 MPA network management categories

Strict Nature Reserve – The conservation and management objectives are to preserve habitats, ecosystems and wildlife species in as undisturbed state as possible while maintaining biodiversity and established ecological processes, and to secure examples of the natural environment primarily for scientific study and research, environmental monitoring, and science education
Protected Landscape/Seascape – an area of land and/or sea where the interaction of people and nature over time has produced a distinctive character with significant aesthetic, ecological or cultural heritage value and often with high biological diversity, where management aims to maintain and support this harmonious interaction of people and nature
Marine Park – a natural area of sea containing a representative sample of major natural regions, features, or scenery, or where wildlife, habitats, or surface features are of special significance and are not materially altered by current human occupation or exploitation, to be maintained in perpetuity with conservation and management objectives: to protect natural and scenic areas of national and international significance for spiritual, scientific, educational, recreational, or tourism purposes
Managed Resource Area – an area of land and/or sea large enough to absorb sustainable resource uses without detriment to its long-term natural values, to be managed to ensure long-term protection and maintenance of biological diversity while providing a sustainable flow of natural products and services to meet community needs

areas. Therefore the potential MPA scenarios generated were a reflection of the special biodiversity areas, the high level of clustering of MPAs that was a requirement, and the minimum predetermined permitted distance of each from human stressors. These criteria meant that some of the areas selected for the MPA network were naturally protected due to their geographic location as being areas with the least human influence and greatest biodiversity and many were also located around some of the uninhabited southern cays.

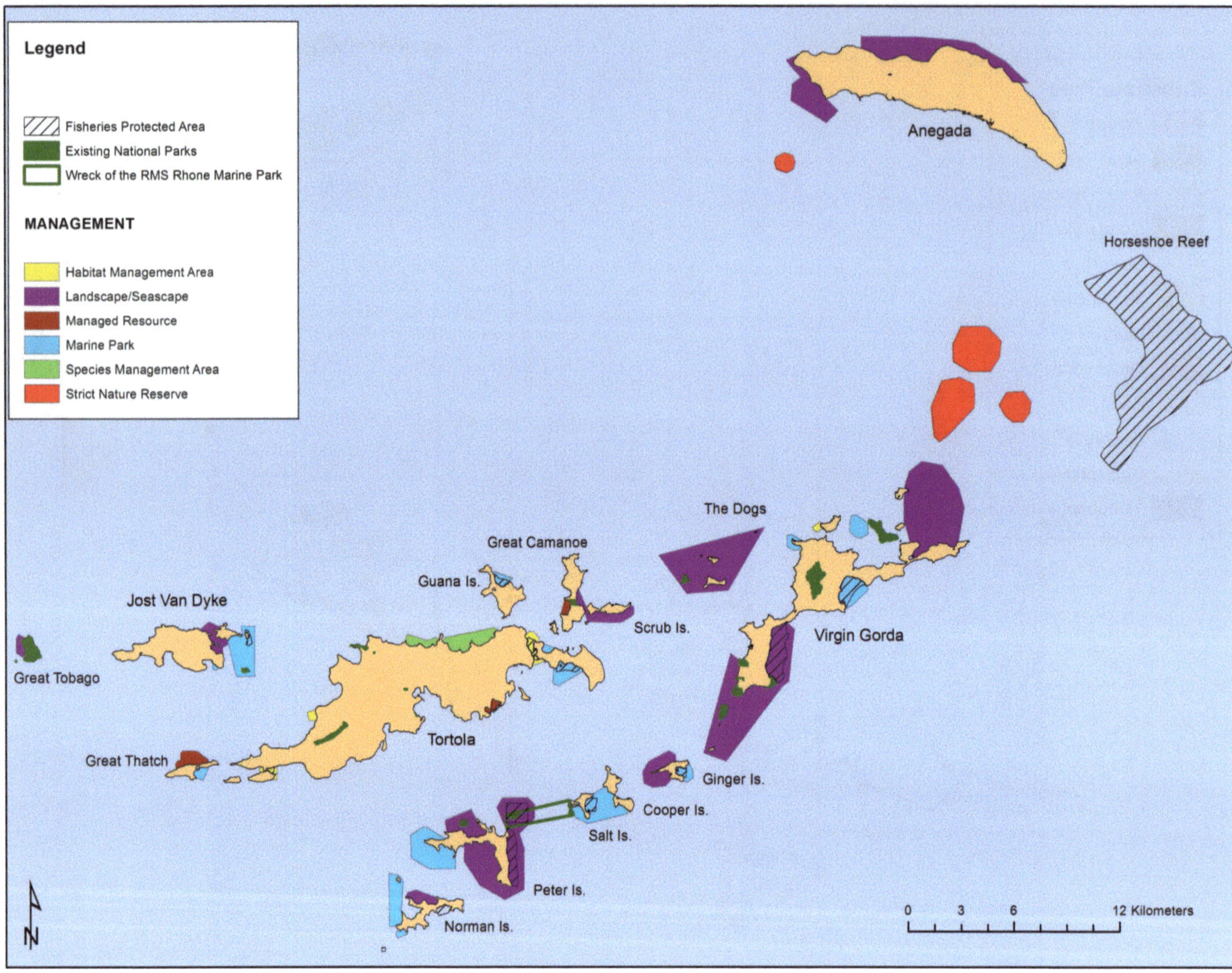

Fig. 4.8 Management categories for the MPA network, based upon the IUCN framework for protected area management (Created by NPTVI)

2. Stakeholder meetings and government involvement throughout the planning process ensured that everyone was aware of the MPA goals (the 30% goal became very well known in the BVI and regionally.)
3. It is critical to plan field work in a strategic manner that ensures optimal field assessment and representation of geographic units. The BVI was divided into three geographic units for the MARXAN analysis to ensure equal ground truthing of polygons to maintain accuracy in the selection of 30% of each habitat type for inclusion in the MPA network.
4. A greater understanding of the stakeholder groups is important. As the NPTVI did not traditionally have a relationship with fishermen it was not always possible to engage fisherman. Therefore, working with the CFD fisheries extension officers was essential and helped improve these relationships.
5. There are broad issues of fisherman incomes directly conflicting with conservation. Therefore, future activities should aim to create alternative sources of income. This dilemma is ongoing and NPTVI will utilise the stakeholder meetings for the MPA management planning process to generate ideas about alternatives in the future, but overall sustainable use of the marine environment by all sectors is the only certain way forward.
6. The way in which information is presented to stakeholders can affect how much feedback is provided. For example, using large paper maps was most successful, and people were also more inclined to attend meetings when they saw that their opinions were being recorded and taken into consideration.
7. The building of trust between government and community entailed continued engagement of stakeholders throughout the MPA planning process, particularly when zoning areas. In the present case, the final scenario that was selected was agreed upon in its entirety by all stakeholders.
8. Many small island nations do not have access to university experts or scientific researchers so field work can be limited by capacity issues and resources. Therefore, scientific, management, and monitoring training is an important part

Fig. 4.9 Taylors Bay Fisheries protected area, on the southern coast of Virgin Gorda (Photo by A. Dickerson)

of the long term project goals. Finding the right people to undergo training is equally important to ensure that capacity is retained within an organization.

9. Building resilience into the system by using both geographic distribution across an area and by using natural features can reduce conflict between stakeholders and the requirements of conservation. Some areas included in the MPA network, both fulfilled the strict criteria for inclusion and are located on the north or south sides of islands that are naturally too exposed, deep or rough to be utilized by stakeholders. Thus conflict was avoided in protecting the area and the 30% goal of habitat protection was still achieved, an example being Taylor's Bay Fisheries Protected Area (Fig. 4.9). The MPA network includes an array of reef types due to the large areas that have been selected for protection.

Monitoring Change Within MPAs

Long term monitoring is required to measure the effectiveness of the MPA network and ten monitoring sites were established across the BVI as part of the OTEP funded project, with locations in all represented habitats.
However the NPTVI has insufficient capacity to conduct regular monitoring of these transects and so continued collaboration with CFD is important to assess the health and status of the MPAs. NPTVI Marine Wardens are primarily responsible for the maintenance of the mooring buoys leaving little to no time for research activities. In the long term, a marine biologist and research assistant would be based at the NPTVI and would work in collaboration with the CFD on monitoring transects. Collaboration between these two departments is critical in order to share resources, technical skills and maximize efficiency of effort. The marine biologists at CFD assisted in the establishment of the OTEP project monitoring sites and also conduct periodic monitoring across the Territory, particularly during extreme events, e.g. mass bleaching or coral disease.

Threats to the MPA Network

The BVI has been threatened by both natural disasters and anthropogenic impacts. Hurricanes have frequently impacted the area over time, whilst flooding from torrential rains has resulted in landslides which subsequently harm the marine resources due to increased sedimentation. Most recently, the bleaching event of 2005 has had devastating impacts, resulting in almost 90% of the BVI reefs being bleached at that time and live coral cover reduced by 60% (Hime 2008).

Human impacts on the BVI are vast and include: anchor damage from charter and private vessels, increased numbers of mega-yachts and mini cruise ships; coastal development of marinas, hotels, and villas; over harvesting of conch (Fig. 4.10), spiny lobster and whelks; unregulated fishing practices that include fish pots and spears; increased sedimentation due to development on adjacent steep hillsides, the creation of unpaved roads and improper erosion control; sewage discharge from charter and private vessels and ocean outfall disposal of terrestrial waste; and from too many vessels

Fig. 4.10 Overharvesting of conch has led to the creation of enormous conch middens along Anegada's southeastern coast (Photo by N. Woodfield Pascoe)

that originate from the BVI, USVI, Puerto Rico and throughout the Caribbean.

However, despite these threats, the creation of a national network of carefully regulated MPAs will go some way to mitigating the threats, and provides a mechanism for management of the reefs and other marine resources of the BVI.

References

Anderson M, Lund H, Gladfelter E, Davis M (1986) Ecological community type maps and biological community descriptions for Buck Island reef national monument and proposed marine sites in the British Virgin Islands. Virgin Islands resource management cooperative, Biosphere reserve research report no. 4, West Indies Laboratory, Fairleigh Dickinson University, St. Croix

Aronson RB, Precht WF (2001) White-band disease and the changing face of Caribbean coral reefs. Hydrobiologia 460:25–38

Blair Myers CN, Sheppard CRC, Bythell JC (1992) A coastal resource atlas of the British Virgin Islands. Natural Resource Institute, Chatham, 4pp and 22 maps

Gardner L, Smith Abbott J, Woodfield-Pascoe N (2007) British Virgin Islands protected areas system plan 2007–2017. BVI National Parks Trust, Tortola

Gore S (2011) Beach geomorphology and management in the British Virgin Islands. Ph.D. thesis, University of Ulster, Coleraine, 243p

Hime S (2008) Effect of marine based tourism in the British Virgin Islands. Unpublished Ph.D. thesis, University of East Anglia, Norwich, 213pp

IUCN Commission on Parks and Protected Areas (1994) Guidelines for protected area management categories. IUCN, Gland

Roberts HH, Murray SP, Suhayda JN (1975) Physical processes in a fringing reef system. J Mar Res 33:233–260

Woodfield Pascoe N, Smith Abbott J (2010) British Virgin Islands Case Study. Coral reefs: a reef resilience toolkit module, the nature conservancy. http://www.reefresilience.org/Toolkit_Coral/C8_BVI.html

Young IR (1989) Wave transformation over coral reefs. J Geophys Res 94:9779–9789

5 Anegada: An Emergent Pleistocene Reef Island

Shannon Gore

Introduction

Anegada was named after the Spanish word meaning "flooded" or "drowned," and the island, when first found, may have looked a lot different than it does today. Anegada's extensive salt pond and wetland systems, particularly on the western side of the island, create the impression that the Spanish translation of the island's name still holds true today (Fig. 5.1). The future of this low-lying island, with a maximum elevation of 8 m (Martin-Kaye 1959) is uncertain due to the potential impacts from climate change such as erosion and subsequent flooding that may result from sea level rise (Mimura et al. 2007) and the increased frequency and intensity of storms (Bender et al. 2010). However, Anegada's geologic history and the oceanographic processes that shape the island reveal a different story. Since the island lacks extensive real estate and commercial development and has a low population (288 in 2006, (DGO 2007)), the island's natural ecosystems, processes and past geologic signatures of extreme events have been maintained and preserved. Even though limited research has been carried out on the island, what is known reveals a resilient island; contradicting the name it was originally given.

Anegada's Physical Setting

The island of Anegada is 38 km^2 in area and extends 17 km along a west to southeast arc (Dunne and Brown 1979) along the north eastern most part of the Puerto Rico/Virgin Islands platform (Fig. 5.2). A wide coral reef fringes the northern coast and extends to the southeast of the island for approximately 17 km. The island's orientation exposes it to the northeast trade winds, Atlantic swell waves and to occasional tropical cyclones. Its location 125 km south of the Puerto Rico trench makes the island susceptible to seismic activity (ten Brink et al. 2004). Anegada's low elevation also makes the island vulnerable to potential impacts from climate change such as erosion and island inundation (Mimura et al. 2007) resulting from the increased frequency and intensity of storms (Bender et al. 2010) and sea-level rise.

Geologic Genesis

Anegada is entirely different from the rest of the British Virgin Islands (BVI), United States Virgin Islands (USVI), and Puerto Rico all of which make up the Puerto Rican/Virgin Islands (PR/VI) platform (with the exception of St. Croix) (Fig. 5.3). This platform is part of the earth's crust that is actually a tectonic microplate which fits like a jigsaw piece between the North American and Caribbean plates (Van Gestel et al. 1999; Masson and Scanlon 1991). All the islands on this microplate, except for Anegada, are volcanic in origin and those islands that make up the BVI formed roughly 80 million years ago (Helsley 1960).

Anegada, on the other hand, formed as part of a massive coral reef system during the last Interglacial highstand (Marine Isotope Stage 5e) between 130,000 ± 2k and 119,000 ± 2k years before present (ybp) (Chen et al. 1991; Hearty et al. 2007). During this period, sea levels were higher than today, somewhere in the range of 2.5–15 m higher in the Caribbean (Hearty et al. 2007). Reef accretion along the windward (north-eastern to eastern) edge of the platform, like other windward facing Caribbean reefs, grew under similar conditions found today (Geister 1977).

Over the 100,000 years following the last interglacial highstand, the climate became cooler, and sea levels lowered to a maximum of 120 m below current levels (Fairbanks 1989; Bard et al. 1990). Ice volumes in North America, high latitude Europe and in the Antarctic ice sheets reached maximum levels ~30,000years ybp which led to the last glacial

S. Gore (✉)
Conservation and Fisheries Department, Government of the Virgin Islands, 3323, Road Town, Tortola VG1110, British Virgin Islands
e-mail: sd_gore@yahoo.com

C.R.C. Sheppard (ed.), *Coral Reefs of the United Kingdom Overseas Territories*, Coral Reefs of the World 4,
DOI 10.1007/978-94-007-5965-7_5, © Springer Science+Business Media Dordrecht 2013

Fig. 5.1 Wetlands of Western Anegada (Credit: S. Gore)

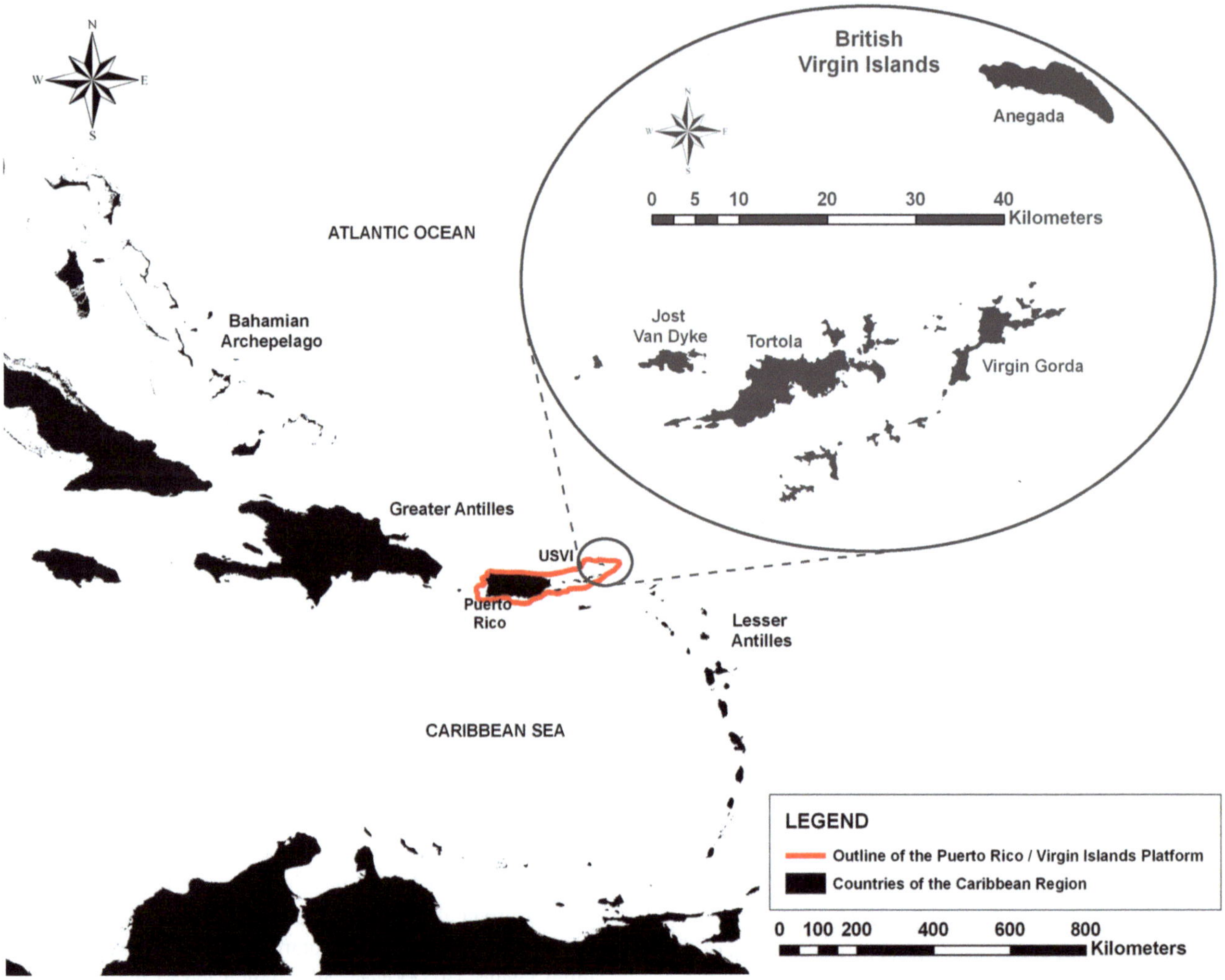

Fig. 5.2 Location of the BVI in the wider Caribbean region

Fig. 5.3 Satellite image of Anegada (Credit: Image Science and Analysis Laboratory)

Fig. 5.4 Anegada Limestone Formation (Credit S. Gore)

maximum (LGM) over the following 10,000 years (Lambeck et al. 2002). During this time (~20,000–30,000 ybp), the entire PR/VI platform was exposed (Bush et al. 1995) and was a time in which one would have been able to walk to Puerto Rico from Anegada. The coral reef that formed the north-eastern part of the PR/VI platform had died off and began transforming into a solid slab of limestone called the Anegada Limestone Formation from years of exposure to sun, wind, rain and the many other elements of nature (Gore 2012) (Figs. 5.4 and 5.5).

Fig. 5.5 Exposed corals seen in the Anegada Limestone Formation (Credit S. Gore)

About 20,000 years ago, following the last glacial maximum, sea levels in the Caribbean started rising. Reconstruction of past sea levels has been modelled through the use of corals, particularly *Acropora palmata* in the Caribbean (Lightly et al. 1982) since this species grows specifically within a narrow depth range and is exclusively confined to the reef crest. Dating these corals provides for a widely used means to identify past sea levels (Blanchon and Shaw 1995; Toscano and Macintyre 2003). Using this method, three periods of postglacial rapid sea level rise, termed melt-water pulse (MWP) occurred ~14,000 ybp, ~11,000 ybp and ~8,000 ybp (Blanchon and Shaw 1995; Toscano and Macintyre 2003). During these periods, rates of sea level rise were in the range of 35 mm year^{-1} up to 60 mm year^{-1} (Blanchon 2011). During the third period, sea levels rose 6.5 m and reefs back-stepped in an attempt to "keep up" with rapid sea level rise (Neumann and Macintyre 1985). This response is seen in many Caribbean reefs (Hubbard 1997) and is suggested to exist around Anegada (Dunne and Brown 1979). Additionally, beachrock found offshore up to 150 m from the modern coast indicates a former shoreline, possibly just before the last melt-water pulse 8,000 ybp (Fig. 5.6).

The island of Anegada today is divided into two distinct geologic formations that geographically split Anegada in half (Fig. 5.7). The eastern side of the island, the Anegada Limestone Formation is distinguished by the indurated coral reef limestone with modified karst topography including solution pits and sinkholes, locally called slobs (Howard 1970). This formation forms an arcuate ridge along the north central to north-eastern side of the island and gently slopes southward. Coral sampled at ~2–3 m above sea level, from a quarry located at the eastern end of Anegada, was dated using U/Th232 to be ~121,096 ybp (US ten Brink, personal communication, 4 April 2011), thus confirming its formation during the Pleistocene.

The western side of the island, as well as a narrow strip of unconsolidated sediments parallel to the northern edge of the now lithified emergent reef (Anegada Limestone Formation) is comprised of dune and beach ridge complexes, lagoons and mangroves. This part of the island is named the Anegada Ridge Plain Formation (Gore 2012). Since little is known about the controls on sediment supply in island formation (Woodroffe 2002), as well as a lack of dating investigations on Anegada's ridge plain, the exact temporal formation sequence for this portion of the island is unknown. However, it is speculated as sea levels began to lower following the last interglacial highstand, a period of intense storms possibly occurred in the Caribbean (Hearty 1997; Hearty et al. 2007) which would have helped form the inner wave-built ridges. While this area was sub-aerially exposed during the last glacial maximum, an eolian sand veneer may have formed over these basal ridges. As sea level started to rise again, ridges continued to develop, most likely during high energy events, from sediments produced by the frontal reef.

Although no other island on the PR/VI platform resembles Anegada's limestone and ridge-plain characteristics, Anegada has a few similarities with the limestone islands of the Bahamian Archipelago to the northwest (including the Bahamas, Turks and Caicos Islands, the Mouchoir, Silver and Navidad Banks), as well as the islands to the southeast of Anegada along the outer Lesser Antilles arc, also known as

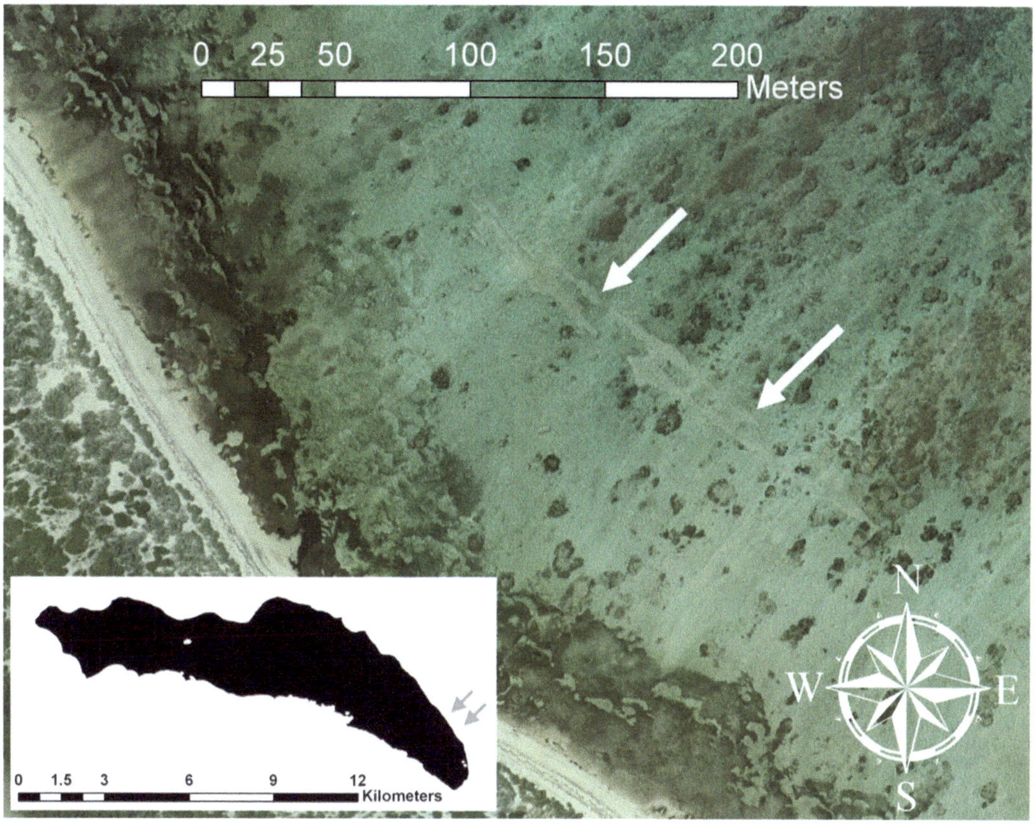

Fig. 5.6 Location of beachrock and former shoreline (Credit: BVI Survey Dept.)

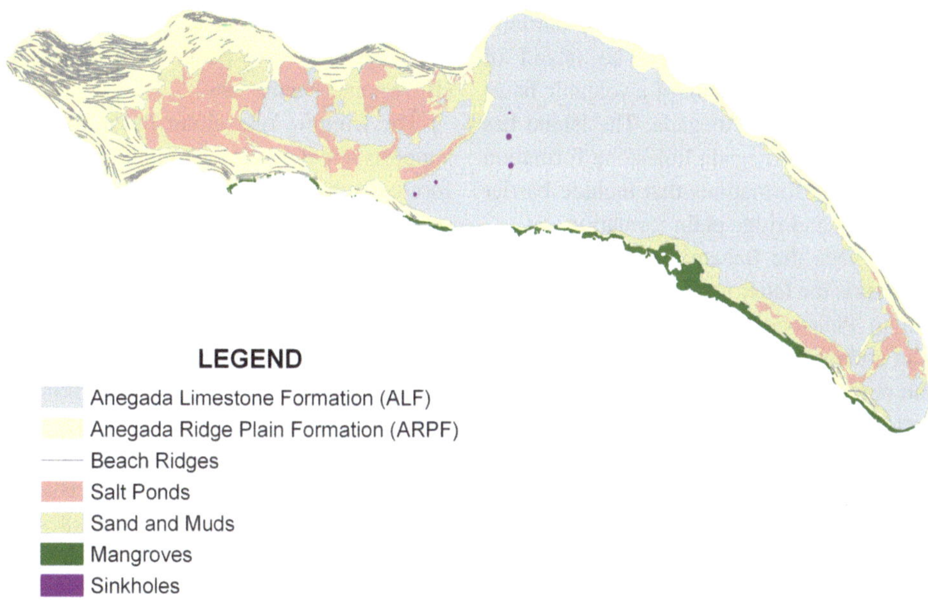

Fig. 5.7 Geology of Anegada

the limestone Caribbees (Anguilla, Dog, Sombero, St. Martin/St. Maarten, St. Barthélemy, Antigua, Barbuda, Guadeloupe, Marie-Galante and Grande Terre).

Similar to the Anegada limestone formation, most of the major islands of the Bahamian archipelago are exposed Pleistocene reefs (Cant 1977) with elevations 5 m above the present sea level (Hearty 1998). Some of these islands also display multiple Holocene beach-accretion ridges that formed in the low lying areas (Enos 2011), similar to those found in the Anegada ridge plain formation.

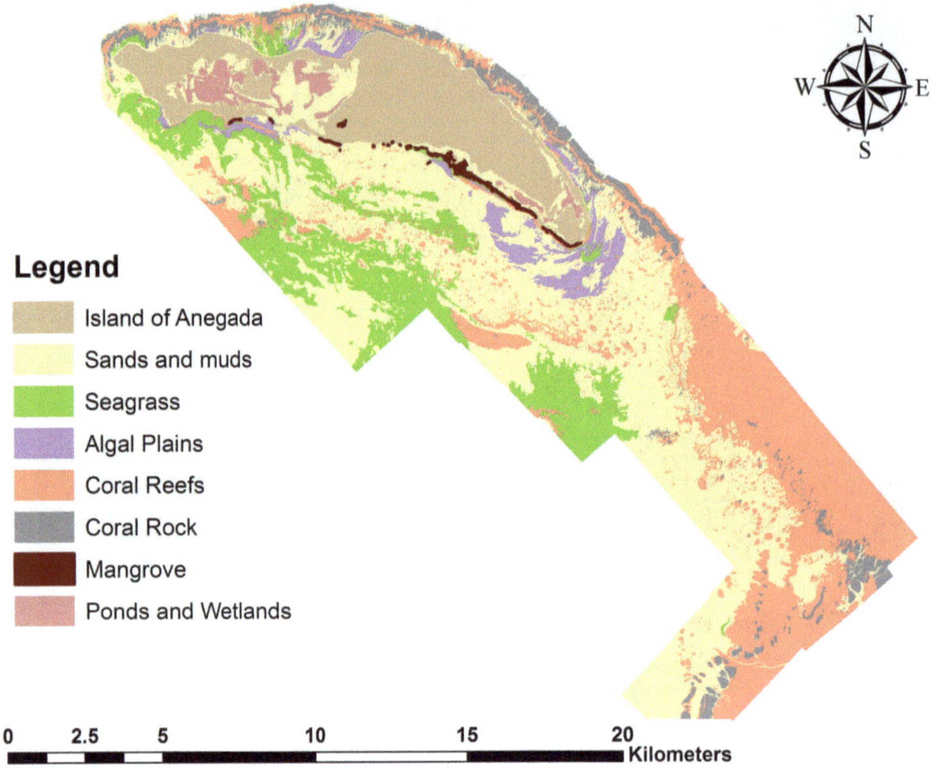

Fig. 5.8 Marine and Coastal habitats of Anegada

The limestone Caribbees also have exposed Eocene to Pleistocene limestones which are built entirely or partially over the volcanics (Martin-Kaye 1959). The island of Barbuda in particular lacks any exposure of a volcanic basement and most closely resembles Anegada. The island has karstic features similar to the Anegada limestone formation as well as similar Holocene formations that include barrier reefs, lagoons, mangroves and ridge plain formations.

The differences between the Bahamian archipelago and the limestone Caribbees is the latter form a shallow veneer of limestone (Eocene to Pleistocene in age) over volcanics while the Bahamian archipelago limestones reach depths up to 10 km (Sheridan et al. 1988). The depth of limestone and the underlying geology of Anegada are currently unknown. Additionally, Anegada is separated from the Bahamas by the Puerto Rican trench (depth 8,395 m) to the north and the Anegada Passage (depth 1,915 m) separates the island from the limestone Caribbees to the southeast.

Modern Coastal Systems of Anegada

Coral Reefs

Anegada's Horseshoe Reef collectively covers approximately 133 km^2 (Fig. 5.8) and is composed of two distinctive facies, a high energy reef front along the north-eastern windward side of the island, and a series of patch reefs with a marked northwest/southeast orientation aligned to prevailing wind and wave patterns on the southern leeward side (Brown and Dunne 1980).

The fringing reef along the northern shore of Anegada supports a reef system with lagoon, back reef, reef crest, and forereef sub-environments. This section of the reef is broken only by a few narrow channels. The width between the northern Anegada shoreline and reef crest (the lagoon area) show spatial variability ranging between 5 m and up to 3 km. The lagoon floor comprises primarily carbonate sands, seagrasses and/or algae. The north-western lagoon has well-developed patch reefs while further east, patch reefs are less established (Brown and Dunne 1980).

A barrier reef extends approximately 14 km southeastwardly beyond the eastern tip of the island to an area charted as "The White Horses" where the horseshoe shape of the reef turns almost 90°. From the White Horses, deep patch reefs (30 m depth) run in a south westerly direction for 8.5 km, to approximately 9.5 km northeast of Pajero's Point, Virgin Gorda.

During the mid-1980s, a healthy reef ecosystem existed on Anegada with live coral coverage reaching 100% in a few areas (Anderson et al. 1986). However, throughout the Caribbean, an epidemic of white-band disease affected Acroporids during the late 1970s (Aronson and Precht 2001) and was followed by a disease-induced mass mortality of

Fig. 5.9 Aerial image of the eastern side of Anegada and current channels running parallel to the coast (Source BVI Survey Dept.)

Diadema antillarum in 1983–1984 (Carpenter 1988). Additionally a number of hurricanes which passed the BVI added to the shift of reef-building corals from Acroporids to *Montastraea annularis* and a reduced the live coral coverage to 8–19.5% by the late 1990s (Nemeth et al. 2003).

Seagrass Beds

Seagrass beds cover approximately 28 km^2 of the shallow (<10 m depth) marine area surrounding Anegada (Blair Myers et al. 1992) and include *Thalassia testudinum, Syringodium filiforme, Halodule wrightii*, and *Halophila baillonis* (Anderson et al. 1986). The densest seagrass beds are located to the south and west of the island. Large patches of seagrasses occur off the north shore in Windless and Bones Bight, where wave energy is attenuated by the fringing reef and off the eastern tip of the island, seagrasses, algae and sand form distinctive channels running parallel with strong current flows (Fig. 5.9).

The accumulations of algae and seagrasses (also called beach cast) can be seen along most of the entire coastline year-round (Fig. 5.10). Large accumulations that form conspicuous wedge-shaped blocks called banquettes (Boudouresque and Meinesz 1982) often form between Pomato Point and Setting Point, particularly following hurricanes.

Mangroves

The south central to south-eastern shoreline of Anegada is composed of mangrove forest rather than sandy beaches (Fig. 5.11). Two types of mangrove forest habitat are represented. Several small communities of mangroves (*Rhizophora mangle*) cluster along the central shoreline and are not directly linked to any inland bodies of water. The second type of mangroves are continuous "mangrove wetlands', described by Jarecki (2004) as an ecosystem which includes a hypersaline aquatic habitat (i.e. a salt ponds), the pond's shoreline and its fringing mangroves. The south eastern mangrove wetlands of Anegada include three salt ponds and four species of mangroves (*Rhizophora mangle, Avicennia germinans, Laguncularia racemosa* and *Conocarpus erectus*). The existence of mangrove forests along this section of Anegada's south eastern coastline is indicative of a stable or prograding coastline.

The same four species of mangroves that are found on the eastern side of the island are also found in the western interior of Anegada. Mangroves fringe seven interconnected salt ponds which collectively comprises one of the largest remaining mangrove wetland systems in the Lesser Antilles (Scott and Carbonell 1986; Jarecki 2004). In November 1999, this area was awarded recognition of international importance as a Ramsar site (Ramsar Site#983). (http://ramsar.wetlands.org/Database/Searchforsites/tabid/765/language/en-US/Default.aspx).

Fig. 5.10 Banquettes of seagrass along the southern coast of Anegada (Source S. Gore)

Fig. 5.11 Walkway to the fishermen's dock on the southern coast (Credit: Conservation and Fisheries Department)

Salt Ponds

The interior ponds of Anegada were once marine bays connected directly to the sea through inlets on the north shore of the island but were altered to a hypersaline state due to an extreme sedimentary event, such as a tsunami, that blocked marine waters from entering the bay (Atwater et al. 2012). A thick sheet of sand found as far as 2 km inland containing grains of the foraminiferan *Homotrema rubrum*, reducing in concentration landwards, is indicative of a marine overwash

Fig. 5.12 Flamingo Pond, Western Anegada (Credit S. Gore)

origin (Pilarczyk and Reinhardt 2012; Mackenzie et al. 1965). This layer of foraminiferan rich sediments lies directly over the limestone bedrock or marine pond mud and below a laminated cyanobacterial mat (Atwater et al. 2012). This stratigraphic change from marine pond to hypersaline pond indicates marine waters had to have been blocked from entering the bay. This change is believed to have been caused by a transatlantic tsunami that formed after the Lisbon earthquake of 1755 (Atwater et al. 2012).

Sedimentary features of ponds in Anegada include laminated cyanobacterial mats, organic muds and sand. Water depths within the western ponds respond to seasonal changes in the mean sea level rather than rainfall and evaporation cycles (Jarecki 2004).

Within Anegada's western salt ponds (Fig. 5.12), a number of fetch-limited barrier islands (described by Cooper et al. 2007) have formed and reflect local variable wind directions (Fig. 5.13). It is unknown if these islands formed before or after the ponds turned hypersaline. Although shoreline processes of these particular islands are unknown, the vegetation has changed. Schomburgk (1832) mentions these small islands as having more "vigorous" vegetation than the surrounding banks during his visit in 1831 but today, vegetation is sparse because of years of grazing by feral animals (Downs 1997).

Beaches

The near continuous sandy coastline around Anegada varies spatially due to their orientation to prevailing wind and waves, planforms and sedimentary characteristics. Additionally, they differ temporally and can be divided into four physiographic regions.

The first region is located on the south central to south eastern Anegada coastline, fringing the Anegada Limestone Formation and consists of a mangrove forest described previously (Fig. 5.14). This region has prograded a maximum of 60 m between 1953 and 2002 (Gore 2012). The entire southern coastline exhibits finer grained sands than the northern shore beaches; however, in contrast to the southeastern coastline, the southern coastline of the Ridge Plain Formation consists of characteristic cuspate forelands developed by a sequence of low ridges

Fig. 5.13 Fetch limited barrier islands found within the salt ponds in Anegada (Credit S. Gore)

Fig. 5.14 Mangroves along the south eastern coast of Anegada (Credit BVI Survey Dept.)

where waves approach from two different directions (Fig. 5.15). The series of forelands along this section of Anegada form distinctive windward (facing southeast) and leeward (facing southwest) beaches and demonstrate varying degrees of accretion and erosion.

The third distinct region of coastline consists of beaches along the northern Ridge Plain Formation which are generally exposed to the prevailing wind and waves and create a series of broad crenulated bays separated by wide sand bluffs (Fig. 5.16). Historic changes along this region

Fig. 5.15 Cuspate forelands on the south western side of Anegada (Credit: BVI Survey Dept.)

Fig. 5.16 Anegada's north western coastline (Credit J. Scheiner)

Fig. 5.17 1953 Aerial photo of the north western coast of Anegada (Credit BVI Survey Dept.)

alternating periods of accretion and erosion by up to ±90 m (Gore 2012).

In general, beaches along the Ridge Plain Formation have had the most dramatic temporal changes. The far western side of the island has shown up to 285 m of erosion between 1861 and 2009 while to the south, up to 135 m has accreted (Gore 2012). This suggests this margin of the island has morphologically adjusted to prevailing conditions by shifting in a counter clockwise movement.

Beaches along the northern Anegada Limestone Formation, the fourth region, also face prevailing wind and waves but are relatively straight compared to the north western shoreline (Fig. 5.17). Crenulated bays along this section of the coast only exist where narrow channels break in the fringing reef (Dunne and Brown 1979).

Unlike the other coastlines around Anegada, this region has exposed outcrops of both Pleistocene bedrock and beachrock which limits shoreline erosion (Fig. 5.18). As a result, the temporal changes on the western side of the island, compared to the dramatic temporal changes on the western side of the island, show more relative stability with no more than ±30 m of accretion/erosion between 1953 and 2002.

Dune Ridges

From a bird's-eye view, multiple foredune ridges found in the Anegada Ridge Plain Formation (Fig. 5.19) run parallel

Fig. 5.18 Exposed beachrock near Loblolly Bay (Credit S. Gore)

Fig. 5.19 The Anegada Ridge Plain formation named after the shore parallel ridges (Credit: BVI Survey Dept.)

to shore and reflect an abundant sand supply, strong onshore winds and the local vegetation's ability to trap sand (Sanderson et al. 1998; Woodroffe 2002). However, since beaches and dunes accustomed to high-energy regimes require extreme storm activity to cause significant morphological impact (Cooper et al. 2004), the existence of the north coast fringing reef that induces significant wave attenuation and energy dissipation offshore, renders the exact temporal formation sequence of the Ridge Plain Formation unknown.

Ridges found along the north ridge plain rise up to 4 m (Fig. 5.20) and are higher than those along the southern shore of the island (Atwater et al. 2012). However, sand deposits thicken towards the west (~10 m) and to the south (Insular Environments 1973) which may be indicative of antecedent topography sloping in a southerly direction, similar to the Limestone Formation to the east.

Conclusion

Jumping to the conclusion that Anegada is eventually going to flood and disappear because of climate change is merely a misunderstanding of this dynamic reef island's ability to aggressively adjust to change. The trans-Atlantic tsunami and over 30 hurricanes passed Anegada within 150 km since 1713. Over this period, at least seven dune ridges continued to form along the north shore from high energy waves capable of moving sand landward. Mangroves have also continued to prograde along the southern coast. This shows the island's resiliency is based on Anegada's current ability to adjust to both extreme events and prevailing conditions. Although the coastline may display alternating areas of erosion and accumulation of sand, in some cases nearly 300 m of change, the long-term result has been a counter-clockwise rotation of the western side of the island.

Fig. 5.20 High dunes found along the north western shores of Anegada (Credit: S. Gore)

A better understanding of the formation of Horseshoe reef in relation to Anegada's development of the Ridge Plain Formation may provide greater insight for those changes resulting from sea level rise. Additionally, gaining a better understanding of how the loss of live coral coverage may be morphologically altering Horseshoe Reef and what those implications may have on the shoreline is still needed. However, this type of research may become more difficult over time with extensive coastal development imminent.

Without taking into consideration Anegada's coastal dynamics, coastal developments could cause the island to lose its resiliency to natural events. This in turn could threaten the island's biodiversity that currently sustains globally and regionally important species (McGowan et al. 2006a) of breeding seabirds (McGowan et al. 2006b), plants (Clubbe et al. 2004), populations of sea turtles (McGowan et al. 2008) and the critically endangered Anegada Rock Iguana (*Cyclura pinguis*) (Bradley and Gerber 2008). Depending on how development proceeds, the fragile nature of the island could forever be changed. Only then can we blame ourselves for allowing such an amazing island to drown and live up to its name.

References

Anderson M, Lund H, Gladfelter E, Davis M (1986) Ecological community type maps and biological community descriptions for Buck Island reef national monument and proposed marine sites in the British Virgin Islands. Virgin Islands resource management cooperative report, 4, 236pp

Aronson RB, Precht WF (2001) White-band disease and the changing face of Caribbean coral reefs. Hydrobiologia 460:25–38

Atwater BF, ten Brink US, Buckley M, Halley RS, Jaffe BE, Lòpez-Venagas AM, Reinhardt EG, Tuttle MP, Watt S, Wei Y (2012) Geomorphic and stratigraphic evidence for an unusual tsunami or storm a few centuries ago at Anegada, British Virgin Islands. Nat Hazard 63(1):51–84

Bard E, Hamelin B, Fairbanks RG (1990) U–Th ages obtained by mass spectrometry in corals from Barbados: sea level during the past 130,000 years. Nature 346:456–458

Bender MA, Knutson TR, Tuleya RE, Sirutis JJ, Vecchi GA, Barner ST, Held IM (2010) Modeled impact of anthropogenic warning on the frequency of intense Atlantic hurricanes. Science 327:454–458

Blair Myers CN, Sheppard CRC, Bythell JC (1992) A coastal resource Atlas of the British Virgin Islands. Natural Resource Institute, Chatham, 4pp and 22 maps

Blanchon P (2011) Meltwater pulses. In: Hopley D (ed) Encyclopedia of modern coral reefs: structure form and process. Springer, New York, pp 683–690

Blanchon P, Shaw J (1995) Reef drowning during the last deglaciation: evidence for catastrophic sea-level rise and ice-sheet collapse. Geology 23:4–8

Boudouresque CF, Meinesz A (1982) Découverte de l'herbier de Posidonie, vol 4. Cahiers Parc National de Port-Cros, Hyeres, 79pp

Bradley KA, Gerber GP (2008) Conservation of the Anegada Iguana (*Cyclura pinguis*). Iguana 12:79–85

Brown BE, Dunne RP (1980) Environmental controls of patch-reef growth and development. Mar Biol 56(1):85–96

Bush DM, Webb RMT, González Liboy J, Hyman L, Neal WJ (1995) Living with the Puerto Rico Shore. Duke University Press, Durham, p 193

Cant R (1977) The role of coral deposits in building the margins of the Bahama Bank. In: Taylor DL (ed) Proceedings of the 3rd international coral reef symposium. Geology, vol 2, Rosensteil School of Marine and Atmospheric Science, Miami, pp 9–13

Carpenter RC (1988) Mass mortality of a Caribbean sea urchin: immediate effects on community metabolism and other herbivores. Proc Natl Acad Sci USA 85:511–514

Chen JH, Curran HA, White B, Wasserburg GJ (1991) Precise chronology of the last interglacial period. 234U/230Th data from fossil coral reefs in the Bahamas. Geol Soc Am Bull 103:82–97

Clubbe C, Gilman M, Acevedo-Rodriguez P, Walker R (2004) Abundance, distribution and conservation of regionally endemic plant species on Anegada, British Virgin Islands. Oryx 36:342–346

Cooper JAG, Jackson DWT, Navas F, McKenna J, Malvarez G (2004) Identifying storm impacts on an embayed, high-energy coastline: examples from western Ireland. Mar Geol 210(1–4):261–280

Cooper JAG, Lewis DA, Pilkey OH (2007) Fetch limited barrier islands: overlooked coastal landforms. GSA Today 17:4–9

Deputy Governor's Office (2007) Sister Island profile of Anegada. Deputy Governor's Office, Road Town, Tortola. http://www.dgo.gov.vg/index.php?mpageid=160. Accessed 26 April 2011

Downs MA (1997) Anegada sea turtle recovery project community assessment report, Island Resources Foundation, Washington, DC, 47p. http://www.irf.org/turtle.htm. Accessed 9 May 2012

Dunne RP, Brown BE (1979) Some aspects of the ecology of reefs surrounding Anegada, British Virgin Islands. Atoll Res Bull 236:1–85

Enos P (2011) Bahamas. In: David H (ed) Encyclopedia of modern coral reefs: structure form and process, Springer-Verlag earth science series. Springer, New York, pp 683–690

Fairbanks RG (1989) A 17,000-year glacio-eustatic sea level record: influence of glacial melting dates on the Younger Dryas event and deep ocean circulation. Nature 342:637–642

Geister J (1977) The influence of wave exposure on the ecological zonation of Caribbean coral reefs. In: Proceedings of third international coral reef symposium. Geology, vol 2, Rosenstiel School of Marine and Atmospheric, Miami, pp 23–29

Gore S (2012) Beach geomorphology and management in the British Virgin Islands. Ph.D. thesis, University of Ulster, Coleraine, 234pp

Gore S, Cooper JAG, Jackson DWT, Jarecki L (2012) Spatial variability of beach geomorphology in the British Virgin Islands. Submitted to Zeitschrift für Geomorphologie

Hearty PJ (1997) Boulder deposits from large waves during the last Interglaciation on north Eleuthera Island, Bahamas. Quat Res 48:326–338

Hearty PJ (1998) The geology of Eleuthera Island, Bahamas: a Rosetta stone of Quaternary stratigraphy and sea level history. Quat Sci Rev 17(4–5):333–355

Hearty PJ, Hollin JT, Neumann AC, O'Leary MJ, McCulloch M (2007) Global sea-level fluctuations during the last interglaciation (MIS 5e). Quat Sci Rev 26((17–18):2090–2112

Helsley CE (1960) Geology of the British Virgin Islands. Unpublished Ph.D. thesis, Princeton University, Princeton, 219pp

Howard J (1970) Reconnaissance geology of Anegada Island. The Caribbean Research Institute, University of the Virgin Islands, St. Thomas, 18pp

Hubbard DK (1997) Reefs as dynamic systems. In: Birkland C (ed) Life and death of coral reefs. Chapman & Hall, New York, pp 43–67

Insular Environments, LTD (1973) Anegada, British Virgin Islands: environmental considerations for development. Prepared for Tillyard and Partners, Charlotte Amalie/St. Thomas, Unpublished, 89pp

Jarecki L (2004) Salt ponds of the British Virgin Islands: investigations in an unexplored ecosystem. Unpublished Ph.D. thesis, University of Kent, Canterbury, 183pp

Lambeck K, Yokoyama Y, Purcell T (2002) Into and out of the Last Glacial Maximum: sea level change during Oxygen Isotope stages 3 and 2. Quat Sci Rev 21:343–360

Lightly RG, Macintyre IG, Stuckenrath R (1982) *Acropora palmata* reef framework: a reliable indicator of sea level in the western Atlantic for the past 10,000 years. Coral Reefs 1(2):125–130

Mackenzie FT, Kulm LD, Cooley RL, Barnhart JT (1965) *Homotrema rubrum* (Lamarck), a sediment transport indicator. J Sediment Petrol 35:265–272

Martin-Kaye PHA (1959) Reports on the geology of the Leeward and British Virgin Islands. Voice, Castries, 117pp

Masson DG, Scanlon KM (1991) The neotectonic setting of Puerto Rico. Geol Soc Am Bull 103:144–154

McGowan A, Broderick AC, Clubbe C, Gore S, Godley BJ, Hamilton M, Lettsome B, Smith-Abbott J, Woodfield NK (2006a) Darwin initiative action plan for the coastal biodiversity of Anegada, British Virgin Islands, 13pp. http://www.seaturtle.org/mtrg/projects/anegada/. Accessed 12 April 2010

McGowan A, Broderick AC, Gore S, Hilton G, Woodfield NK, Godley BJ (2006b) Breeding seabirds in the British Virgin Islands. Endanger Spec Res 2:15–20

McGowan A, Broderick AC, Frett G, Gore S, Hastings M, Pickering A, Wheatley D, White J, Witt MJ, Godley BJ (2008) Down but not out: marine turtles of the British Virgin Islands. Anim Conserv 11:92–103

Mimura N, Nurse L, McLean RF, Agard J, Briguglio L, Lefale P, Payet R, Sem G (2007) Small Islands. In: Parry ML, Canziani OF, Palutikof JP, van der Linden PJ, Hanson CE (eds) Climate change 2007: impacts, adaptation and vulnerability. Contribution of working group II to the fourth assessment report of the intergovernmental panel on climate change. Cambridge University Press, Cambridge UK, pp 687–716

Nemeth RS, Quandt A, Requa L, Rothenberger JP, Taylor MG (2003) A rapid assessment of coral reefs in the Virgin Islands (part 1: stony corals and algae). In: Lang JC (ed) Status of coral reefs in the Western Atlantic: results of initial surveys, Atlantic and Gulf Rapid Reef Assessment (AGRRA) program. Smithsonian Press, Washington, DC, pp 544–566

Neumann AC, Macintyre IG (1985) Reef response to sea level rise: keep-up, catch-up or give up. In: Proceedings of the fifth international coral reef congress, vol 3, Tahiti/French Polynesia, pp 105–110

Pilarczyk JE, Reinhardt EG (2012) *Homotrema rubrum* (Lamark) taphonomy as an overwash indicator in marine ponds from Anegada, British Virgin Islands. Nat Hazard. 63(1):85–100

Sanderson PG, Eliot I, Fuller M (1998) Historical development of a foredune plain at Desperate Bay, Western Australia. J Coastal Res 14(4):1187–1201

Schomburgk RH (1832) Remarks on Anegada. J R Geogr Soc Lond 2:152–170

Scott DA, Carbonell M (1986) A directory of neotropical wetlands. International Union for Conservation of Nature and Natural Resources/International Waterfowl Research Bureau, Cambridge/Slimbridge, 684pp

Sheridan RE, Mullins HT, Austin JA Jr, Ball MN, Ladd JW (1988) Geology and geophysics of the Bahamas. In: Sheridan RE, Grow JA (eds) The Atlantic continental margin, vol 1–2. Geological Society of America, Boulder, pp 329–364

ten Brink US, Danforth WW, Polloni CF, Andrews B, Llanes P, Smith S, Parker E, Uozumi T (2004) New seafloor map of the Puerto Rico trench helps assess earthquake and tsunami hazards. EOS Trans, Am Geophys Union 85:349–360

Toscano MA, Macintyre IG (2003) Corrected western Atlantic sea-level curve for the last 11,000 years based on calibrated 14C dates from *Acropora palmata* framework and intertidal mangrove peat. Coral Reefs 22:257–270

Van Gestel JP, Mann P, Grindlay NR, Dolan JF (1999) Three-phase tectonic evolution of the northern margin of Puerto Rico as inferred from an integration of seismic reflection, well, and outcrop data. Mar Geol 161(2–4):257–286

Woodroffe CD (2002) Coasts: form, process and evolution. Cambridge University Press, Cambridge, 623pp

Coral Reefs of the Cayman Islands

6

Alan Logan

Introduction

The British Overseas Territory of the Cayman Islands lies in the north-western part of the Caribbean Sea between 19°15′–19°45′N and 79°44′–81°27′W and consists of three islands, Grand Cayman, Cayman Brac and Little Cayman, the last two known as the Sister Islands. All are low-lying and are prominences on the submerged Cayman Ridge which is an extension of the east–west trending Sierra Maestra mountain range of south-eastern Cuba (Roberts 1977) (Fig. 6.1). The ridge is bordered to the south by the Cayman Trench where depths exceed 5,500 m and the Yucatan Basin to the north-west with depths around 4,500 m (Wells 1988; Spalding et al. 2001). Grand Cayman, lying approximately 250 km south of Cuba and 280 km north-northwest of Jamaica, is the largest of the three islands at almost 200 km^2, while Cayman Brac and Little Cayman are smaller and arranged *en echelon* at about 120 km east-north-east of the main island. The prevailing trade winds are generally from the north-west in winter and the south-east in summer and the tidal range is less than 1 m (Wells 1988).

This chapter discusses the submarine topography of the nearshore areas, coral reef geomorphology and zonation, biotic communities and substrate types of the reef complex as a whole. Then, the next chapter discusses a variety of related reef topics such as the ongoing programme of reef assessment as related to anthropogenic and other threats, coral bleaching and diseases, and iconic reef fish such as groupers and lionfish. This chapter concludes with a short overview of the regulatory structure and reef management methods designed to protect the reefs, mechanisms which assume special importance where their accessibility and pristine condition are the basis for a thriving diver-tourism industry.

A. Logan (✉)
Centre for Coastal Studies, University of New Brunswick,
5050, Saint John, NB E2L 4L5, Canada
e-mail: logan@unbsj.ca

Submarine Topography

The geology of the Cayman Islands is now well known (Matley 1926; Brunt et al. 1973; Roberts 1977; Jones 1994). Jones (1994) has discussed the tectonic setting of the islands, in particular the active spreading centre of the Mid-Cayman Rise and its relation to transform fault movement in the region (Fig. 6.1). The nearshore submarine topography acts as a base for reef development, in that a narrow submerged fore-reef shelf about 0.5–2.0 km (average 500 m) in width surrounds all three islands and has two well-marked submarine terraces at 8–10 m and 20 m depth (Roberts 1977; Logan 1981) which are remarkably similar for all three islands. This suggests regional stability and contemporaneous reef and lagoonal development that allow the recognition of similar substrates and associated communities from all three islands (see substrate-community maps in Roberts 1988, 1994; Logan 1988, 1994). These terraces are described for Grand Cayman by Rigby and Roberts (1976) and Roberts (1977, 1994), and for Cayman Brac and Little Cayman by Logan (1994). The shallow (or upper) terrace slopes gradually from either the shoreline or fringing reef to a depth of between 8 and 10 m where there is a former sea cliff, now heavily colonized by reef growth, sloping down to about 15 m depth. The shallow terrace often shows spur-and-groove reef development. The deep (or lower) terrace has a depth of between 15 and 20 m and consists of a sand plain with scattered patch reefs and remnants of spur-and-groove. The seaward edge of the sand plain terminates at a slight ridge, beyond which is a steep drop-off down the fore-reef slope into deep water. Occasionally, as at Bloody Bay on Little Cayman, the deep terrace is absent and the drop-off occurs at the edge of the shallow terrace as an almost vertical wall. Typical profiles showing the submarine terraces and their reef zones are shown in Fig. 6.2a, b. The fore-reef slope beyond 40 m depth is virtually unexplored but fathometer profiles from Grand Cayman by Rigby and Roberts (1976) show a very steep wall to about 150 m, beyond which is a slight reduction in slope which probably represents a fan of proximal reef sediments stacked up against the base of the old reef wall.

C.R.C. Sheppard (ed.), *Coral Reefs of the United Kingdom Overseas Territories*, Coral Reefs of the World 4,
DOI 10.1007/978-94-007-5965-7_6, © Springer Science+Business Media Dordrecht 2013

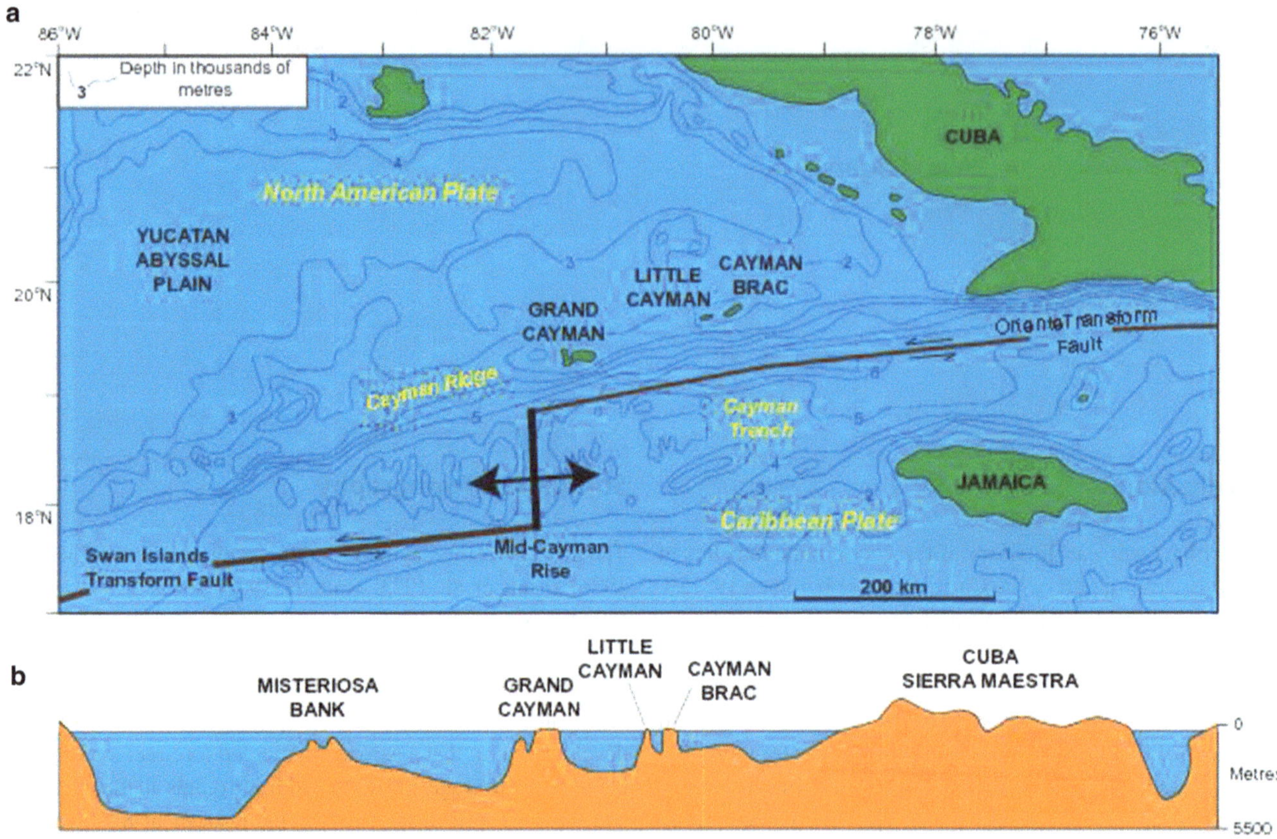

Fig. 6.1 Map and section of the central Caribbean to show the Cayman Islands as ridge pinnacles and the postulated relative movement of plates in the region (After Jones 1994)

The Reef Complex and Its Facies

The dominant influence on shallow marine substrates and their communities in the Cayman Island is the prolific growth of reef-building corals and coralline algae which help establish the biogenically-constructed limestone coral reefs. Initial reef growth on a shallow shelf under optimum conditions eventually leads to a complex of reef and reef-dependent environments termed the *reef complex* (Henson 1950). Here, over time, there is an interplay of constructional and destructional processes which results in reef growth on the one hand and the formation of reef-derived sediments on the other. The presence of a fringing reef near sea level and spur-and-groove structure on the deeper reefs reduces the effects of waves and provides quiet-water conditions in the lee of the reefs, where lagoon and shore communities can develop.

Lagoonal substrates are dominated by sediments varying in grain size from fine sands to coarse rubble. Sandy areas are often inhabited by sparse algae such as species of *Halimeda*, *Penicillus*, *Avrainvillia* and *Udotea*, while shells of the infaunal bivalve *Codakia orbicularis* are scattered over the surface of the sand. Where gravel or rubble occurs, attached green algae and *Sargassum* are present, as well as the brown algae *Padina* and *Turbinaria* and scattered coral heads of *Agaricia agaricites,* along with sea urchins belonging to *Echinometra* and *Diadema.* Marine grasses are patchy in distribution and dominated by *Thalassia testudinum*, with *Syringodium filiforme* and species of *Halodule* sometimes forming mixed stands (Logan 1994, Fig. 6.4). The sponge *Tedania ignis* is common in these grass beds, along with calcareous green algae and echinoids, while small coral colonies belonging to *Porites* occur between the blades. *Callianassa* and/or *Arenicola* sand mounds are present throughout the grass beds. Patch reefs dominated by species of *Montastrea* (Logan 1994, Fig. 6.6) are commonly seen in lagoonal areas, with single stands of *Acropora palmata* (Logan 1994, Fig. 6.5) in the lee of the fringing reef. Other corals encountered are *Porites astreoides*, *P. porites*, *Diploria strigosa*, *D. clivosa*, *Colpophyllia natans*, *Siderastrea siderea* and *Agaricia agaricites*, the latter often occurring as a prominent understorey species. The hydrozoan *Millepora complanata* in bladed growth form is also locally present (Logan 1994, Fig. 6.9).

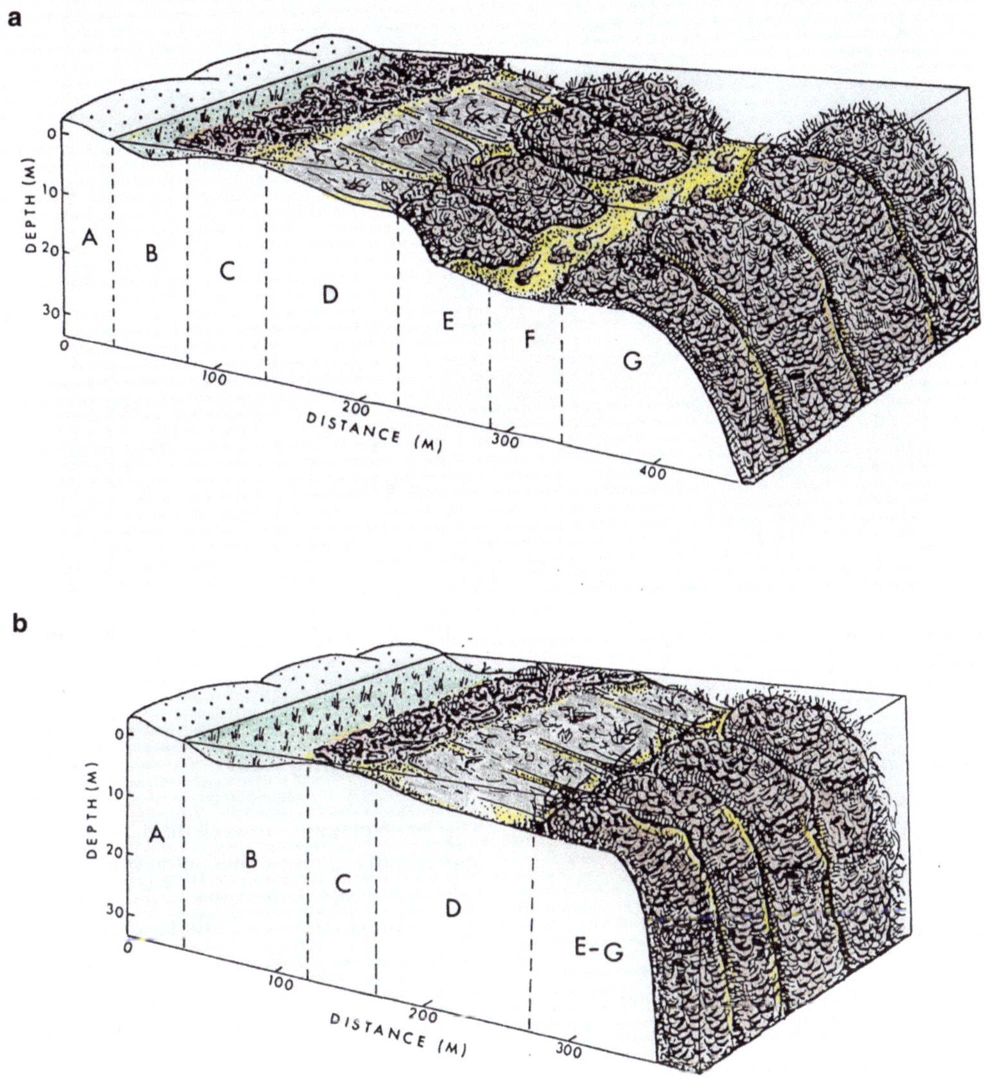

Fig. 6.2 (**a**) Idealised three-dimensional offshore profile typical of all three Cayman Islands where shallow and deep terraces are present. (**b**) Profile off Bloody Bay, Little Cayman with no deep terrace and sand plain. Key to substrates: *A* shoreline, *B* lagoon, *C* active fringing reef, with rubble zone behind, *D* barren rock pavement, *E* shallow terrace reefs with spur-and-groove, *F* sand plain, *G* deep terrace reefs (Modified after Logan 1994)

Coral Reef Zonation

Coral reefs, consisting mainly of a consortium of scleractinian corals and calcareous algae, are found at three depth levels in the Cayman Islands (Rigby and Roberts 1976; Roberts 1977, 1994; Logan 1994). The shallowest reefs (not including lagoonal patch reefs) form a linear wave-resistant crest at sea level responsible for lagoonal development in its lee around most of Grand Cayman and Little Cayman but only rarely on Cayman Brac. This reef is essentially a narrow fringing reef built on the apex of two oppositely-sloping surfaces and thrives in high energy wave conditions. The main reef-builders are massive wave-resistant corals, although even they may suffer extensive damage during hurricanes, resulting in the formation of a rubble flat zone on its landward side. A barren rock pavement separates the active fringing reef from a second reef development on the shallow terrace while a sand plain separates this reef from a third reef development at 15–20 m depth which extends down the fore-reef slope to the limit of coral growth at about 70 m. The zonation pattern of the reefs is remarkably similar for all three islands. Where submerged Pleistocene Ironshore Formation outcrops across bays, it usually acts as a locus for developing fringing reefs at or near sea level, although Blanchon and Jones (1997) have shown that storm-derived rubble may also control the location and architecture of some fringing reefs around Grand Cayman. Landwards of

Fig. 6.3 Waves breaking over nearshore fringing reef and associated rubble flat with storm deposits of boulder ramparts on the shore, near Brac Reef Hotel, Cayman Brac

these reefs are the lagoonal facies of grass beds, patch reefs and sand deposits, all comprising an integral part of the reef complex (Logan 1994).

Fringing Reef Rubble Flat

Shorewards of the active fringing reef is a zone of rubble derived from the mechanical breakdown of the reef by storms. This zone is extensively developed along the south coast of Little Cayman from The Flats to Sandy Point near the east end of the island (Logan 1994, Fig. 6.7) and is gradational with typical lagoonal substrates at depths close to sea level. Here wave energy is high and limestone blocks made up of dead fronds of *Acropora palmata* and zoned blades of *Millepora complanata* have been tossed shorewards into the rubble zone. There is a general dearth of living corals, only low-relief robust forms such as *Porites astreoides*, *P. porites*, *Diploria clivosa*, *Siderastrea siderea*, and *Agaricia agaricites* can survive here. Crustose coralline algae are common but calcareous green algae such as *Halimeda* are rare. Active bioerosion by parrot fish, sea urchins and the sponge *Cliona* contribute to the barren nature of this zone. Furthermore, much of the rubble flat is exposed at low tide (Logan 1994, Fig. 6.8) and bears the brunt of wave activity, thus providing a constant source of biologically and physically derived sediments for the lagoonal and shore areas and even forming boulder ramparts on the shore during the most violent of storms (Fig. 6.3).

Active Fringing Reef

The breaker zone around the Cayman Islands is narrow and usually defines the outer limit of bays and sounds (Fig. 6.4). This is the zone of the reef crest, dominated by an *Acropora-Millepora* thicket, and spans the *Palythoa-Millepora* and *Acropora palmata-Diploria strigosa* zones described for Caribbean fringing reefs by Geister (1977, 2011). According to this scheme *Acropora cervicornis* should occur shorewards of the fringing reef but this species is now rare almost everywhere in the Caymans, although it survives as isolated stands in the barren rock pavement zone seawards of the fringing reef and occasionally near the drop-off. *Acropora palmata* occurs as large arborecent colonies robust enough to withstand the constant surf, the upwardly-inclined fronds preferentially orientated towards the open ocean. Gaps between colonies are often filled by the hydrozoan *Millepora complanata* which forms a low hedge of vertically-inclined, upwardly-flaring blades, the flat faces facing the direction of wave advance (Logan 1994, Fig. 6.9). Understorey species include the zoanthid *Palythoa caribaeorum* and low relief colonies of species of the corals *Porites*, *Diploria*, *Agaricia* and *Montastrea*, with the foraminiferan *Homotrema rubrum* common in interstices. Fenner (1993) listed ten coral species from the reef crest in Cayman Brac. Although the reef crest acts as a protective barrier for the enclosed sounds and bays, channels do occur, allowing lagoonal water to periodically drain back to the open sea (see Fig. 6.4).

Fig. 6.4 Development of fringing reefs at Frank Sound and Gun Bay, east end of Grand Cayman resulting in lagoons with sands and grass beds (*dark patches*). Note banded reef zones on seaward side of fringing reefs. (Google Earth)

Barren Rock Pavement

Seawards of the fringing reef a shallow terrace forms the upper part of the shelf and comprises a barren rock pavement to a depth of about 8 m depth, merging seawards into the well-developed coral reefs of the shallow terrace. The barren rock pavement zone occurs around all three islands and comprises a rock surface gently dipping seawards from either the fringing reef or the shoreline (where a fringing reef is absent) to near the seaward edge of the shallow terrace. Subparallel grooves, named radial grooves by Rigby and Roberts (1976), extend seawards and may join up with sand-filled grooves of the spur-and-groove zone of the shallow terrace reefs. Typically the grooves are 3–4 m in width, about 1 m in depth relative to the adjacent rock surface, and have a U-shaped profile. Although they appear to be erosional the origin of these groves is unclear, and their possible influence on spur-and-groove development on the shallow terrace is not known. The rock pavement between the grooves is sparsely colonised by a variety of organisms capable of withstanding the relatively high wave energy environment of this zone. These include isolated heads of the corals *Acropora palmata*, *Porites astreoides*, *Diploria clivosa*, *D. strigosa* and species of *Montastrea*. Common algae include *Stypopodium*, *Padina*, *Galaxaura*, *Amphiroa*, *Dictyota*, *Turbinaria* and *Sargassum*. Stands of gorgonians are also common, particularly the sea fans *Gorgonia flabellum* and *G. ventalina* which may show alignment of the plane of the colony normal to the direction of wave advance for stability (Logan 1994, Fig. 6.11). Nevertheless, in the most violent of storms many sea fans are uprooted and end up on the nearby beaches.

Shallow Terrace Reefs

The barren rock pavement zone grades seawards at depths of 6–8 m into a fully-fledged, highly-diverse coral reef colonising the seaward edge of the shallow terrace and, in places, draped over remnants of the old sea cliff (Fig. 6.2a). This is the zone of spur-and-groove, a feature seen in reefs worldwide (Shinn 2011) where coral spurs and intervening grooves are developed in response to wave energy conditions. This zone is present around almost all of Grand Cayman and Little Cayman, but less so around Cayman Brac (see substrate maps in Roberts 1988, 1994 for Grand Cayman and Logan 1988, 1994 for Cayman Brac and Little Cayman). Relief between spur tops and adjacent grooves averages 3–4 m. Spur tops are colonised mainly by branching or domal coral growth forms but the flanks are dominated by platy growth forms that form overlapping shingle-like structures. Coalescence of the sides of adjacent spurs often result in roofing over to form caves and tunnels (Logan 1981) which support shade-loving communities (coelobites), with a coralline algae community at the tunnel entrance, grading into a demosponge-bryozoan community further into the tunnel and a sclerosponge-brachiopod community in dimly-lit areas with less than 1%

Fig. 6.5 Beginning of sand plain at junction with shallow terrace reef off Seaview Hotel, Grand Cayman. Note sticks from dead branching corals on sand. Pipe fish is about 0.5 m long

surface illumination, where biotic coverage is much reduced (Logan 1981, Fig. 4; Logan 1994, Figs. 6.13–6.14). The dominant coral on the spurs is *Montastrea annularis* and its sibling species which adopt a variety of growth forms from columnar-lobate to massive on the open reef to platy on the flanks. Species of *Diploria* and *Porites* are common, representing the *Diploria-Montastrea-Porites* coral community found throughout the Caribbean and in Bermuda. Bifacial *Agaricia agaricites* and *A. tenuifola* are also common, both box-like in growth form in open areas, but platy on the reef flanks. *Acropora cervicornis* was formerly common on the spur tops but is now rare. Other corals include *Colpophyllia natans, Siderastrea siderea, Montastrea cavernosa, Porites astreoides, Eusmilia fastigiata, Dichocoenia stokesi, Manicina areolata, Mussa angulosa, Meandrina meandrites* and species of *Diploria, Scolymia* and *Mycetophyllia.* Other groups represented include gorgonians, green algae and demosponges, plus the shade-loving sclerosponges *Goreauiella auriculata* and *Ceratopora nicholsoni.*

Sands

The grooves between the spurs are floored by sediments ranging in texture from coarse coral rubble to fine sands. The coarse sediments comprise fragments of the branching corals *Acropora cervicornis* and *Porites porites*, the fine sands *Halimeda* segments and the red foraminiferan *Homotrema rubrum.* These sands continue onto a sand plain of upto 350 m (Fig. 6.5) separating the spur-and-groove zone of the shallow terrace from the reefs of the deep terrace at the edge of the drop-off and down the fore-reef slope. The sand plain is a barren zone, with only a few isolated lens-shaped patch reefs, but forms a consistent and easily mappable feature on aerial photographs (Roberts 1988; Logan 1988). However, in Bloody Bay, Little Cayman, for a distance of about 2 km, the deep terrace and its associated sand plain are inexplicably absent, the shallow and deep terrace reefs merging in water as shallow as 7 m to produce a very shallow drop-off known as the Little Cayman Wall, popular with divers. This spectacular area is part of the Bloody Bay to Jackson's Point Marine Park (Wells 1988) and is illustrated in Logan (1994, Figs. 6.15–6.19) and in Fig. 6.6.

Deep terrace and Fore-Reef Slope Reefs

Sand plain sediments at their seaward edge are banked up against a prominent lip of coral reef at about 15 m depth, the reef edge forming a dam with occasional gaps that allow sediments to be funnelled down the steep fore-reef slope into deep water (Roberts 1983) (Fig. 6.2a, b). Initially the seaward slope of the reef is gradual but increases rapidly below about 20 m. Remnants of spur-and-groove are represented by massive buttresses which overhang the steep slope and harbour cryptic habitats not yet studied. These reefs show high coral coverage and diversity (Fenner 1993). Many coral colonies show platy growth form in response to diminished

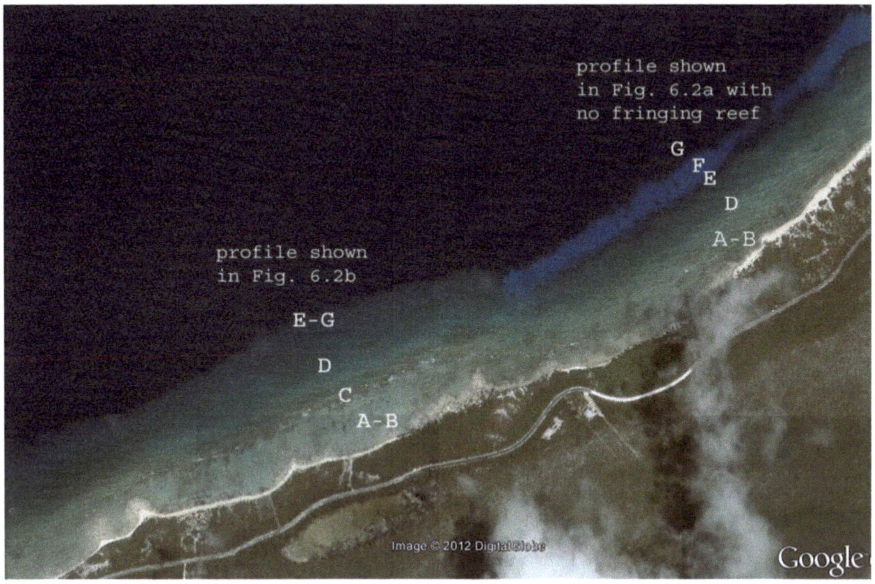

Fig. 6.6 Zonation of reefs in Bloody Bay, Little Cayman. Upper profile as shown in Fig. 6.2a, lower profile as in 6.2b

Fig. 6.7 Deep terrace reef, Little Cayman Wall, off Jackson's Point, 30 m depth, showing *Montastrea cavernosa* (*left*), red sponge *Haliclona rubescens* and alcyonarians

light. Enormous plates of *Agaricia* are attached precariously to the steep slope by their narrow bases and become unstable at the slightest disturbance. Massive hemispherical mounds of *Montastrea cavernosa* (Fig. 6.7) and large sheets of *Mycetophyllia ferox* occur. Multi-coloured sponges exhibit a wide variety of growth forms, from encrusting to whip-like to tubular to vasiform, with large barrel sponges belonging to *Xestospongia* in evidence. Gorgonians are abundant, as well as a host of other invertebrates such as crinoids, bryozoans, molluscs and ascidians.

References

Blanchon P, Jones B (1997) Anatomy of a fringing reef around Grand Cayman: storm rubble, not coral framework. J Sediment Res 67:1–16

Brunt MA, Giglioli MEC, Mather JD, Piper DJW, Richards HG (1973) The Pleistocene rocks of the Cayman Islands. Geol Mag 110(3):209–221

Fenner DP (1993) Some reefs and corals of Roatan (Honduras), Cayman Brac, and Little Cayman. Atoll Res Bull 388:30

Geister J (1977) The influence of wave exposure on the ecological zonation of Caribbean coral reefs. In: Proceedings of the third international coral reef symposium, vol 1, Miami, pp 23–29

Geister J (2011) West Indian coral reef classification. In: Hopley D (ed) Encyclopedia of modern coral reefs, structure, form and process. Springer Science, New York, 6p

Henson RRS (1950) Cretaceous and Tertiary reef formations and associated sediments in the Middle East. Bull Am Assoc Petrol Geol 34:215–238

Jones B (1994) Geology of the Cayman Islands. In: Brunt MA, Davies JE (eds) The Cayman Islands: natural history and biogeography. Kluwer Academic, Dordrecht, pp 13–49

Logan A (1981) Sessile invertebrate coelobite communities from shallow reef tunnels, Grand Cayman, B.W.I. In: Proceedings of the fourth international coral reef symposium, vol 2, Manila, pp 735–744

Logan A (1988) Lesser Caymans. Swamps and shallow marine substrates 3, Cayman Islands 1: 25,000 (map), Overseas Development Natural Resources Institute, UK

Logan A (1994) Reefs and lagoons of Cayman Brac and Little Cayman. In: Brunt MA, Davies JE (eds) The Cayman Islands: natural history and biogeography. Kluwer Academic, Dordrecht, pp 105–124

Matley CA (1926) The geology of the Cayman Islands (British West Indies) and their relation to the Bartlett Trough. Quart J Geol Soc Lond 82:352–387

Rigby JK, Roberts HH (1976) Grand Cayman Island: geology, sediments and marine communities. Brigham Young Univ Geol Stud 4:97–122

Roberts HH (1977) Field guidebook to the reefs and geology of Grand Cayman Island, B.W.I. Field guidebooks. In: Proceedings of third international coral reef symposium, Miami, 41p

Roberts HH (1983) Shelf margin reef morphology: a clue to major off-shelf sediment transport routes, Grand Cayman Island, West Indies. Atoll Res Bull 263:22

Roberts HH (1988) Grand Cayman. Swamps and shallow marine substrates 1 and 2, Cayman Islands 1 : 25,000 (map), Overseas Development Natural Resources Institute, UK

Roberts HH (1994) Reefs and lagoons of Grand Cayman. In: Brunt MA, Davies JE (eds) The Cayman Islands: natural history and biogeography. Kluwer Academic, Dordrecht, pp 75–104

Shinn EA (2011) Spurs and grooves. In: Hopley D (ed) Encyclopedia of modern coral reefs, structure, form and process. Springer Science, New York, 4p

Spalding MD, Ravilious C, Green EP (2001) World atlas of coral reefs. Cayman Islands, University of California Press, Berkeley, USA, pp 141–146

Wells SM (ed) (1988) Coral reefs of the world, vol 1. Cayman Islands, 85–95. UNEP Regional Seas Directories and Bibliographies. IUCN Gland, Switzerland and Cambridge, U.K, 373p

Biology and Ecology of the Coral Reefs of the Cayman Islands

7

John R. Turner, Croy McCoy, Matt Cottam, Jeremy Olynik, Austin Timothy, Janice Blumenthal, John Bothwell, Fred J. Burton, Phillippe Bush, Paul Chin, Oliver Dubock, Kristian D. Godbeer, James Gibb, Lisa Hurlston, Bradley J. Johnson, Alan Logan, Gene Parsons, and Gina Ebanks-Petrie

Introduction

In this chapter a variety of topics is elaborated upon, including climate, ocean currents, reef and associated habitats, the ongoing programme of reef assessment and protection as related to anthropogenic and other threats, coral bleaching and disease, and the increasing problem of the invasive lionfish. Also included is information on the regulatory structure and reef management methods designed to protect the reefs in the Cayman Islands, which assume special importance where their accessibility and pristine condition are the basis for a thriving diver-tourism industry.

General

The Cayman Islands UK Overseas Territory is made up of three small low-lying subtropical islands in the NW Caribbean. The islands are tips of an underwater mountain chain and as such are characterized by a very narrow coastal shelf, usually of less than 1 km in width, with considerable reef development upon it, which falls steeply to very deep water close to shore. With a total land mass of approximately 260 km², the majority of the islands 56,000 population live on the largest and most developed island of Grand Cayman. Little Cayman and Cayman Brac are located about 100 km to the North East and are considerably less developed. Immense economic, social, and environmental change has come very rapidly to the Cayman Islands. These remote western Caribbean Islands were undiscovered until 1503, remained unsettled until about 1700 (Craton 2003), and for hundreds of years stayed nearly unchanged—such that they came to the attention of the outside world in the 1950s as the "islands time forgot". It was not until the 1960s that unprecedented tourism and financial booms began, with expanding human population and development which catapulted the islands into the twenty-first century and bringing attendant threats to biodiversity (Ebanks and Bush 1990). In recent years, changes due to local stressors have accelerated, and it has become clear that even isolated oceanic islands will have to face upcoming global challenges such as climate change.

Tourism and Finance

Scheduled air service to the islands began in 1948 (Giglioli 1994) and due to their tropical climate, friendliness, safety, and pristine waters, tourism in the Cayman Islands began to boom, growing from 3,440 visitors in 1966 to 76,600 in 1975 (Ebanks and Bush 1990). Cruise ships first visited in the late 1960s (Craton 2003) and in 2010 tourist arrivals to the Cayman Islands reached 1,597,800 by sea and 288,300 by air (Cayman Islands Government 2011b). In addition to a growing tourism industry, by the 1970s the Cayman Islands had become a major centre for international finance (Craton 2003). Through immigration, the population of the Cayman Islands expanded rapidly, more than doubling between the 1960s and 1980s (from 8,511 in 1960 to 25,900 in 1988) and

J.R. Turner (✉)
School of Ocean Sciences, Bangor University, Wales, UK
e-mail: j.turner@bangor.ac.uk

C. McCoy • M. Cottam • J. Olynik • A. Timothy • J. Blumenthal
J. Bothwell • P. Bush • P. Chin • O. Dubock • K.D. Godbeer
J. Gibb • L. Hurlston • B.J. Johnson • G. Parsons • G.E.-Petrie
Department of the Environment, Cayman Islands Government, Grand Cayman, Cayman Islands
e-mail: Croy.McCoy@gov.ky; Mat_Cottam@yahoo.com; Jeremy.Olynik@gov.ky; Timothy.Austin@gov.ky; Janice.Blumenthal@gov.ky; John.Bothwell@gov.ky; Phillippe.bush@gov.ky; Paul.Chin@gov.ky; James.Gibb@gov.ky; Bradley.Johnson@gov.ky; Gene.Parsons@gov.ky; Gina.Ebanks-Petrie@gov.ky

F.J. Burton
National Trust of the Cayman Islands, Family Park, 558 South Church Street, Grand Cayman, Cayman Islands
e-mail: fjburton@blueiguana.ky

A. Logan
Centre for Coastal Studies, University of New Brunswick, 5050, Saint John E2L 4L5, NB, Canada
e-mail: logan@unbsj.ca

C.R.C. Sheppard (ed.), *Coral Reefs of the United Kingdom Overseas Territories*, Coral Reefs of the World 4,
DOI 10.1007/978-94-007-5965-7_7, © Springer Science+Business Media Dordrecht 2013

nearly doubling again (to 54,878) by 2010 (Cayman Islands Government 2011a). Due to the rise in population and tourism, development also boomed, with clearing of land, dredging of canals and other major works beginning in the 1960s (Ebanks and Bush 1990). With the fast pace of development came an urgent need for natural resource management and conservation legislation (Ebanks and Bush 1990). In 1978, a Marine Conservation Law was enacted to protect coral reefs (a major source of tourism dollars through the diving industry) and important fishery species such as turtles, conch, and lobster (Cayman Islands Government 1978). Protections were expanded in 1986, when a system of Marine Protected Areas was put in place to protect fish populations and other species (Cayman Islands Government 1986). On land, introduction of conservation legislation has lagged behind: there is still a need for a National Conservation Law to designate terrestrial protected areas and implement conservation plans for key habitats and species (DaCosta-Cottam et al. 2009), but the Bill has stalled.

Natural History

Knowledge of the biodiversity of the Cayman Islands has grown greatly since their discovery. The logs of early explorers provided the original records of turtle and crocodile populations around the Cayman Islands (Smith 2001). The first scientific expeditions to the Cayman Islands began with natural history collections in the late 1800s and early 1900s and the Oxford Expedition in 1938 documented much of the flora and fauna in the islands (Davies and Brunt 1994). Long-term local studies started with the foundation of the Mosquito Research and Control Unit (MRCU) in 1965. Established to control rampant mosquito populations, its founder Dr. Marco Giglioli became increasingly involved in environmental research and monitoring of development. Under his leadership, the MRCU conducted early studies such as the Natural Resources Study of 1974–1975, which provided baseline data on the marine environment and highlighted the need for environmental management (Davies and Brunt 1994). This led to the formation of a Natural Resources Unit within MRCU, and in the mid-1990s, the formation of the Cayman Islands Department of Environment (DoE). The DoE is now the main government agency responsible for the management of natural resources in the Cayman Islands. To this end, the department monitors habitats such as coral reefs, seagrass beds and mangroves, and conducts research and monitoring programs on various key species, including grouper, conch, lobster, and turtles. Non-governmental organizations have also played a key role in environmental research and conservation education: these include the National Trust for the Cayman Islands (founded in 1987) and more recently the Central Caribbean Marine Institute (CCMI) and the Guy Harvey Research Institute. The Cayman Islands have also participated in regional initiatives such as the Caribbean Coastal Marine Productivity Program (CARICOMP) and the DoE and local NGOs have collaborated with overseas organizations on research studies (e.g. Godley et al. 2004; DaCosta-Cottam et al. 2009). Most recently, a new UK Government Department of Environment and Rural Affairs (DEFRA) Darwin Initiative project with Bangor University UK and The Nature Conservancy USA has been instituted to assess the effectiveness of marine parks and plans to meet future challenges.

While immense changes have occurred since the Cayman Islands were discovered, particularly since the 1960s (Ebanks and Bush 1990), the environment in the Cayman Islands now faces even greater challenges, including invasive species such as lionfish, coral bleaching, and ocean acidification, to name only a few. Changes brought about by these emerging threats may rapidly rival the challenges of the past 500 years, necessitating current conservation action, foresight, and planning.

Geography

The Cayman Islands consist of three small low-lying islands situated in the middle of the Caribbean Sea southwest of Cuba, and are the peaks of a submerged ridge that runs westward from the Cuban Sierra Maestra mountain range formed entirely from calcareous marine. The largest of the three islands, Grand Cayman, now has a population of 53,100 whilst Little Cayman and Cayman Brac (collectively known as the Sister Islands) are sparsely populated with only 2,300 inhabitants between them (Economics and Statistics Office, Government of Cayman Islands).

Oceanography

The Islands are geographically situated on the northern boundary of the Caribbean main stream current, which is a spin-off of the North Atlantic Gyre, composed of the North Equatorial Current and the Guyana Current, which enters the Caribbean through passages in the Lesser Antilles (Kinder 1983). According to Stoddard (1980), ocean currents in the Cayman Islands move predominantly in a northwesterly direction, with velocities recorded on the largest island, Grand Cayman averaging 30 cms^{-1} and exceeding 35 cms^{-1} for nearly 20% of time during which it was monitored (Darbyshire et al. 1976). Data from the Department of Environment static current drifters released in the winter at Little Cayman and Cayman Brac show a strong component moving to the southeast, looping back towards the islands before reaching the island of Jamaica. Thereafter, current movement is seemingly westward towards Grand Cayman before entering the loop current between Yucatan and into the Gulf of Mexico. Drifter data for Grand Cayman also show a strong southeast-bound current; however at about 18°S, it picks up the Caribbean main stream current, travelling westward, passing by the Misteriosa Banks then entering the loop current between the Yucatan and into the Gulf of Mexico. Currents around the Cayman Islands are complex in terms of small scale spatial diversity (including stratification), duration, intensity, and velocity. This complexity may be attributed to their location and the physiographic properties of the Caribbean basin, over which the generally westward moving Caribbean main-stream current flows. Starting from about 75°W, the seabed topography of the northwest Caribbean is disrupted by such features as the Nicaraguan Rise, the Cayman Trench, the Cayman Ridge, and the Cayman Rise. In general, the Cayman Islands, due to their small size and large-scale oceanic regimes, seem to be the focal point of a somewhat confused "crossroads" of the main stream Caribbean current, exhibiting inconsistent patterns and unexplained loop currents as it is forced northwest through the trough between Jamaica and the Nicaraguan Rise, then encounters the deep (>6,000 m) Cayman Trench, then forced up and over the shallower Cayman Ridge as it is funnelled towards the Yucatan Channel and into the Gulf of Mexico.

Climate

The climate of the Cayman Islands is influenced by their location in the western Caribbean Basin. The islands have a tropical marine climate with two distinct seasons: a wet season from May to November and a relatively dry season from December to April. Air temperatures (recorded on Grand Cayman) range from a low of 11.20°C to a high of 36.50°C, but the mean monthly air temperatures only range from 24.75° C (February) to 28.40°C (July). North-easterly trade winds predominate for most of the year, with hurricanes occurring mainly between August and November. Direct hits from hurricanes strike the islands about once every 10 years, however the islands are brushed by tropical storms or hurricanes every 2.21 years. Tompkins (2005) estimates that between 1887 and 1987 a tropical cyclone passed within 100 miles of Grand Cayman once every 2.7 years, and passed directly over it once every 12.5 years. Hurricanes *Gilbert* (1989) and *Ivan* (2004) were particularly significant in recent years, with *Gilbert* decimating *Acropora* populations, and *Ivan* stripping sand and soft corals from shallower reefs, most notably around Grand Cayman (Croy McCoy pers. comm.) Tropical storms are common in both the wet (summer) and dry (winter) seasons (Blanchon and Jones 1997). However, while hurricanes often strike in late summer, winter storms are often associated with cold northerly fronts. High frequency and severity of storms led Blanchon and Jones (1997) to suggest that severe storms are the primary physical agent impacting the marine environment in the Cayman Islands.

Marine Habitats

In the Cayman Islands, coral reefs, lagoons, seagrass beds and mangroves constitute the major coastal interface habitats, but beaches, maritime cliffs and ironshore are also of significance. Here we elaborate on aspects of coral reefs, lagoons, seagrass beds, mangroves and beaches.

Coral Reefs

Coral reefs have been outlined in the preceding chapter. Here, specific features such as spur-and-groove which varies depending upon the exposure of the coastline, are expanded upon (McCoy et al. 2010). The orientation of the islands in relation to the winter storms from the north to northeast and the summer storms and hurricanes from the south to southeast result in distinct energy zones. Grand Cayman has three types; the south and east coasts of the island are the high-energy exposed windward aspect, the north is the moderate energy protected-windward aspect and the west is the low energy leeward aspect (McCoy et al. 2010). The northeast to southwest orientation of the Sister Islands results in only two energy zone; high-energy exposed windward aspects in the south and moderate energy protected-windward aspects in the north (Dromard et al. 2011). Development of spur-and-groove on all three islands depends upon the exposure of the coastline to winter storms from the north and north-east and summer storms and hurricanes from the south to southeast. The south and east coasts of all three islands have the highest energy zones overall and therefore the greatest development of spur-and-groove structure (McCoy et al. 2010; Dromard et al. 2011). Coral cover values for the Cayman Islands as a whole were 25% in 1997 and declined to 18% by 1999. Mean coral cover was stable between 1999 and 2004 but declined to 14% in 2006, there being negligible change in 2008, followed by an unusual bleaching event in deepwater around Grand Cayman only in 2009 (Fig. 7.1). This has resulted in coral cover in 2011 of about 11%.

The major habitats and their coverage on Cayman coral reefs are shown in Table 7.1, and are mapped in Fig. 7.2a–c. Table 7.1 defines these as (1) *aggregate reef* where hard coral cover (alive and dead) exceeds 70% substrate coverage and soft corals and sponges are also present. (2) *Spur and groove* is usually associated with the seaward edge of the reef crest, and with the edge of the fore reef, near the escarpment, orientated perpendicular to shore and escarpment and typically composed of hard coral cover (alive and dead), exhibiting a high vertical relief relative to the surrounding pavement and

Fig. 7.1 Deep reef at Grand Cayman deep reef from 20 m to about 60 m, showing bleached corals in September 2009 (Photo: Patrick Weir)

sand channels. The spurs are usually formed by accreting hard corals, and the grooves usually comprise sand or hardbottom. (3) *Individual patch reefs* are isolated coral formations, dominated by hard corals, although some soft corals and sponges may be present. (4) *Aggregated patch reefs* are aggregated coral colonies, where colonies (alive and dead) exhibit > 70% substrate coverage and hard corals generally dominate, although some soft corals and sponges may be present. Confined areas of bare sand or hardbottom are present within the matrix of the reef aggregation. (5) *Reef rubble* consists of unstable coral rubble, usually found on the back reef portions of the reef crest. Reef rubble is often colonised with filamentous or other macroalgae. (6) *Reef crest* is semi-emergent or emergent coral reef. (7) *Sand plains* consist of expanses of uncolonised sediment (ranging from coarse sand to silt) located between the shallow and deep terrace reefs. (8) *Colonised hardbottom* exhibits coral cover within the range of 10–70% of the substrate. Dominant features are low-relief pavement or rubble, or low-relief rock and sand grooves, colonised by algae, soft corals, and sparse hard corals, which are dense enough to partially obscure the underlying rock. (9) *Uncolonised hardbottom* is pavement, often dominated by algae but exhibiting a hard coral, soft coral, and sponge cover of <10%. (10) The *Wall* is a near-vertical or vertical slope extending from the shelf-margin to great depths and characterised by abundant coral and sponge colonisation from the drop-off to 120 m depth. (11) *Beachrock* is cemented sand derived from calcite precipitating out of seawater, resulting in the formation a flat rock-like substrate.

Marine Protected Areas in the Cayman Islands include Marine Parks, Replenishment Zones, Environmental Zones, No Dive Zones and Wildlife Interaction Zones: totalling 19,311 acres for Grand Cayman, 2,281 acres for Little Cayman, and 914 acres for Cayman Brac. Total for the Cayman Islands: 22,506 acres (91.08 km^2). Although coral reefs surround the three islands, good examples of reefs are at Seven Mile Beach marine park, Southwest Point (Sand Cay High Heads), and North Side Reefs in Grand Cayman; White Bay marine park in Cayman Brac, and Bloody Bay marine park in Little Cayman. Coral reefs throughout Cayman are protected through the Marine Conservation Law and Regulations and coral reefs within Marine Parks have an additional level of protection. However, the Marine Park Regulations do not afford outright protection against activities related to construction and development.

The coral reefs of Cayman are affected by environmental factors, some of which operate on a global scale or regional scale such as changes in Earth's atmosphere, ocean acidification, and dust storms. Other factors, such as marine pollution and nutrification, are generally more localized, and linked to issues of coastal management.

Coral Bleaching

In the Cayman Islands, the first observance of a coral bleaching event occurred in 1983. Bleaching also occurred in 1987, 1991, 1994, 1998, 2003 and 2005. The acute global coral bleaching event of 1998 resulted in the highest mortality measured to date. During the 1997–1998 El Niño weather event, local sea surface temperatures rose above 30°C for 25 days, from August 9th through to September 3rd. Surveys indicated approximately 90% of all corals in Grand Cayman were affected, and an observed mortality of 10% at two 9 m-deep reef sites on the North Side of the island were recorded (Bush and McCoy *unpub.*). A near 1°C increase in SSTs in the tropics has been observed over the past century, and some corals are already living at or very near to their upper thermal tolerances. A regional temperature increase of +1°C relative to the 1961–1990 mean is expected by the end of 2100 which will likely result in coral bleaching across the entire Caribbean (McWilliams et al. 2005). More alarming projections of more than 1°C by 2015 and even about 2°C have been made for the 2050s for the Caribbean Sea (Sear et al. 2001; Nurse et al. 2001; Sheppard and Rioja 2005). The 2005 bleaching event, which affected coral reefs throughout the Caribbean, also affected coral reefs in Cayman Islands, but with little mortality, and an event in 2009 caused

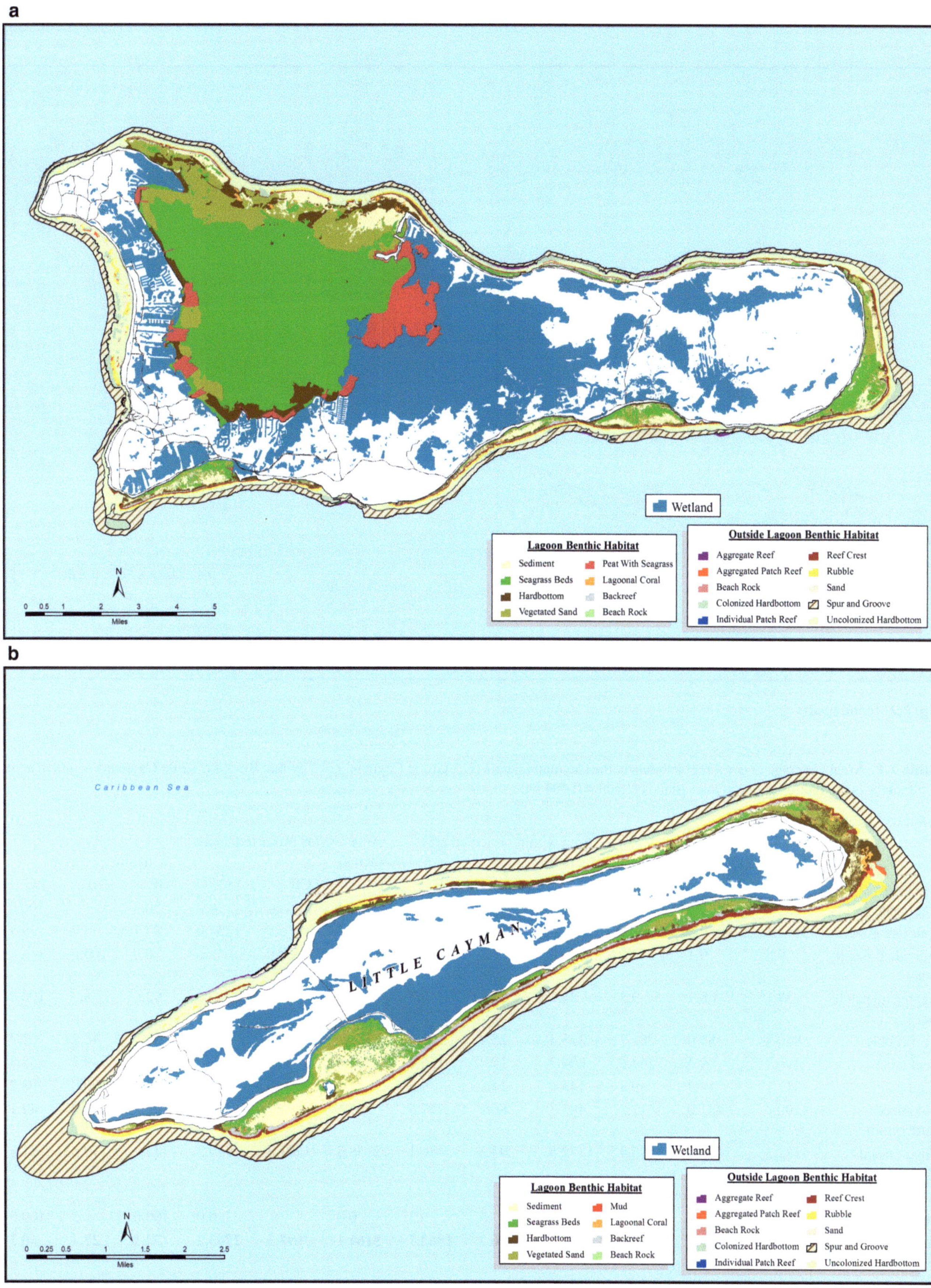

Fig. 7.2 (**a**) Map of Grand Cayman illustrating coral reef and lagoon habitats. (**b**) Map of Cayman Brac illustrating coral reef and lagoon habitats. (**c**) Map of Little Cayman illustrating coral reef and lagoon habitats

c

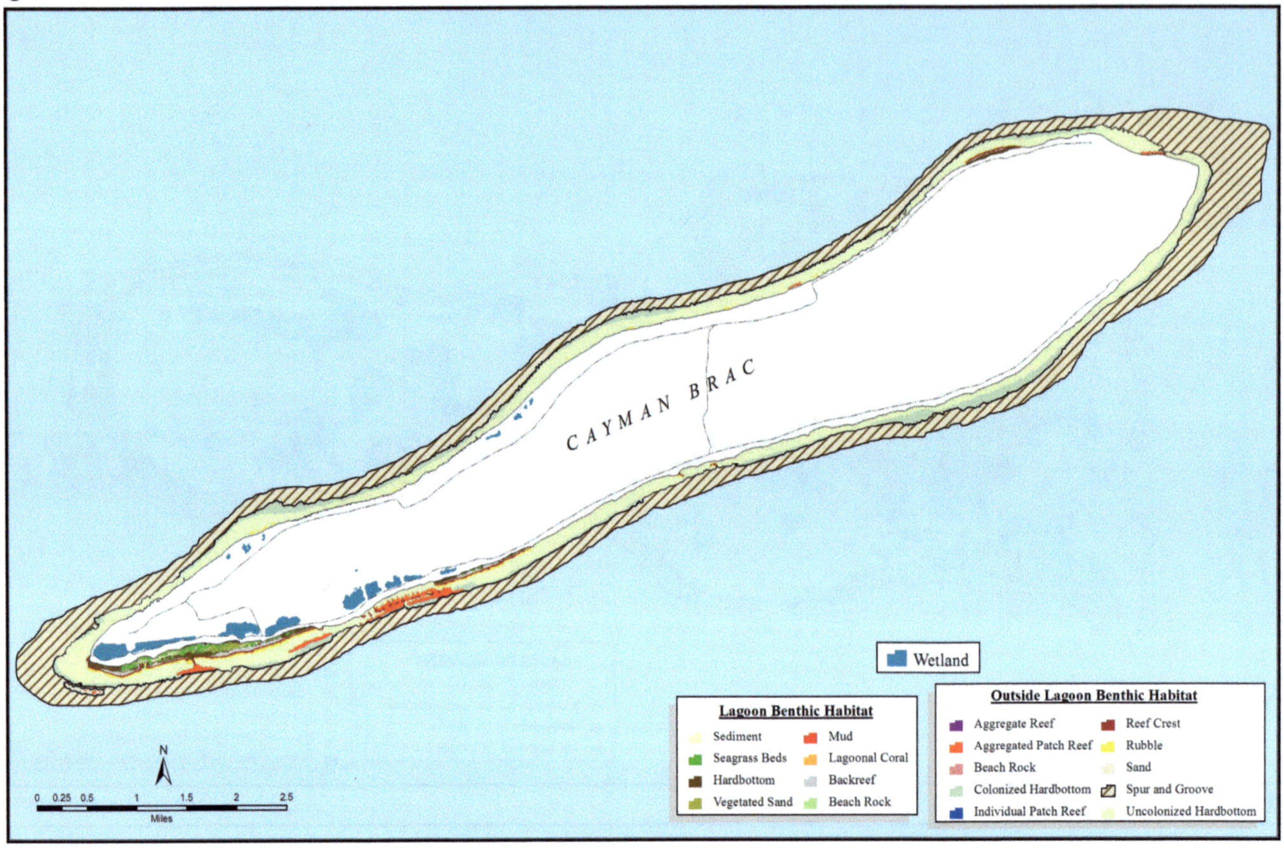

Fig. 7.2 (continued)

Table 7.1 Areal coverage of coral reef habitats in the Cayman Islands (*GC* Grand Cayman, *CB* Cayman Brac, *LC* Little Cayman). Units in acres in all tables is the unit used in Cayman Islands (1 acre = 0.404 ha)

Coral reef habitats												
	Total area (acres)			Area within protected areas (acres)			Area outside protected areas (acres)			% habitat protected		
Category	GC	CB	LC	GC	CB	LC	GC	CB	LC	GC	CB	LC
Aggregate reef	122.9	0.0	7.8	61.8	x	7.8	61.1	x	0.0	50.3	x	100.0
Spur and groove	5153.1	2940.3	2045.1	1397.4	903.5	811.1	3755.7	2036.9	1234.0	27.1	30.7	39.7
Individual patch reef	0.8	0.1	0.0	0.0	0.1	x	0.8	0.0	x	0.0	100.0	x
Aggregate patch reef	37.1	68.7	9.8	34.1	53.6	9.8	3.0	15.1	0.0	92.0	78.0	100.0
Reef rubble	840.3	55.0	281.7	325.4	25.4	175.2	514.9	29.6	106.6	38.7	46.2	62.2
Reef crest	496.7	39.4	264.6	204.2	19.4	155.6	292.4	20.0	109.0	41.1	49.2	58.8
Sand plain	217.1	14.8	9.4	145.4	14.8	9.3	71.7	0.0	0.2	67.0	100.0	98.2
Colonised hardbottom	1190.9	481.0	611.7	485.3	86.2	257.5	705.6	394.7	354.2	40.7	17.9	42.1
Uncolonised h'bottom	4131.3	1494.5	1395.4	1372.5	293.3	436.1	2758.8	1201.2	959.3	33.2	19.6	31.3
Wall	*No measurements—vertical feature*											
Beachrock	6.4	0.0	0.3	6.4	x	0.3	0.0	x	0.0	100.0	x	100.0
Total	**12196.7**	**5093.7**	**4625.9**	**4032.6**	**1396.2**	**1862.7**	**8164.1**	**3697.5**	**2763.2**	**33.1**	**27.4**	**40.3**

bleaching on the deep terrace around Grand Cayman. The mass bleaching event of summer 2009 was the response to a hot deep water gyre that stretched all the way from the surface down to 460 m, with temperature loggers at the time recording elevated sea temperatures of over 30°C around Grand Cayman. This was accompanied by a period of calm weather and absence of cloudcover (Croy McCoy pers. comm.) meaning that solar radiation was additionally high. The mass bleaching of coral colonies that followed was particularly severe on Grand Cayman where the hot water gyre was centred. Bleaching occurred but to a less severe degree on Little Cayman, which was on the periphery of the gyre. Surprisingly, being situated just a short distance away, Cayman Brac almost escaped the coral bleaching event. From observations made at the time, it is clear that bleaching was much more severe on deeper reefs than on shallower ones. Colonies at depth rapidly expelled all zooxanthellae, turned bright white and remained so for several months. In contrast, many coral colonies situated much closer to the surface appeared to regain at least some degree of pigmentation much more rapidly (Croy McCoy pers. comm.) thus indicating re-colonisation by zooxanthellae. Figure 7.1 shows bleached colonies at depth, and bleaching occurred between 20 and 60 m , making this bleaching event highly atypical when compared to most solar induced bleaching events.

Coral Diseases

The reefs of Grand Cayman have similar or lower levels of prevalence of coral diseases at the community level when compared with other countries around the Caribbean basin (Weil, Cróquer & McCoy, *unpub.*). The most common and perhaps most severe diseases affecting reefs of Cayman Islands are yellow band, white plague and black band diseases. Yellow band disease affects the three species of *Montastraea* (*M. faveolata, M. franksi* and *M. annularis*) and normally persists, producing extensive mortality on individual colonies. White plague affects a wide range of hosts, including the major reef builders, and may kill extensive areas of living tissue relatively quickly. This disease tends to be seasonal but not persistent, with higher levels of prevalence during warmer months. Black band disease presents as a dark red or black microbial mat, which migrates over the surface of the coral, resulting in tissue degradation, and exposure of the coral skeleton. A study in 2011 (Hillyer, unpublished MSc thesis) found that disease prevalence in hard corals was very patchy at small scales, and varied according to island and aspect. Prevalence was highest on Cayman Brac and Little Cayman (combined disease 13.10% ± 1.19% SE, and 10.91% ±1.53% SE) on southern coasts, explained in part by the density of sensitive hosts.

Physical Destruction of Coral Reefs

Physical destruction of coral reefs due to anchor damage from commercial and recreational boating has contributed significantly to the degradation of reefs in the Cayman Islands. Coral reefs in areas exposed to heavy marine traffic, including cruise ships, for example, in the vicinity of George Town harbour, have been almost completely destroyed. Further, coastal development has caused nutrient and sediment-rich runoff from the terrestrial environment, exacerbated by land clearance, roads construction and emplacement of impervious surfaces, combined with the removal of buffering coastal vegetation. A major cause of impact has been dredging and channelling for fill and access, which impacts coral reefs within the footprint of activity, and beyond through increased sediment loading.

Sounds (Lagoons)

The majority of the inshore coast of the Cayman Islands comprises reef-protected shallow saltwater lagoons. Good examples on Grand Cayman are North Sound, Frank Sound, East End, Bodden Town lagoon and Pease Bay Lagoon, and on Cayman Brac: Dick Sessinger's Bay and North East Bay, and on Little Cayman: South Hole Sound (Fig. 7.3), Mary's Bay, Charles Bight, Preston Bay and Point of Sand. In the Cayman Islands, the term "Sound" is most commonly used to describe lagoon areas. The largest sound in the Cayman Islands is North Sound on Grand Cayman, the second-largest semi-enclosed lagoon in the Caribbean. The key habitats in the sounds are: backreef areas of dead, unstable coral rubble and rocks located on the landward side of the fringing reef, often colonised with filamentous or other macroalgae; lagoon corals; hardbottoms of low-relief pavement or rubble, often colonised by algae; seagrass beds; unvegetated mud and sand sediments; vegetated sediments; mud; and beachrock (Table 7.2).

The varied habitats and substrates support a high diversity of marine invertebrates. In addition to echinoderms such as urchins and starfish, the complement of species includes commercially significant species, including Queen conch *Strombus gigas* and Spiny lobster *Panulirus argus*. Whelks *Cittarium pica* constitute fisheries in the Cayman Islands, of sufficient size to impact natural populations to the extent that they have required interventory management. Key species, such as Queen conch and Spiny lobster, though subject to regulatory management, remain at population levels well below historic norms. Lagoon areas include "no-take" areas (Replenishment Zones) for Queen conch and Spiny lobster. No marine life may be taken by diving or removed by excavation without prior written approval of the Governor-in-Cabinet.

Though local waters are actively policed by Marine Enforcers of the Department of Environment, poaching remains an issue. Affected species include Queen conch *Strombus gigas*, Spiny lobster *Panulirus argus,* and Green turtle *Chelonia mydas.* The Sandbar in Grand Cayman's North Sound supports large numbers of Southern stingrays *Dasyatis americana* (Fig. 7.4). This quasi-natural tourist attraction is one of Grand Cayman's most popular features with cruise-ship visitors. The sheltered, productive waters of the sounds provide habitat to adult marine turtles, which feed on

Fig. 7.3 Reef, lagoon and mangroves at South Hole Sound, Little Cayman

Table 7.2 Areal coverage of lagoonal habitats in the Cayman Islands (1 acre = 0.404 ha)

Lagoon habitats												
	Total area (acres)			Area within protected areas (acres)			Area outside protected areas (acres)			% habitat protected		
Category	GC	CB	LC	GC	CB	LC	GC	CB	LC	GC	CB	LC
Seagrass	15808.8	36.3	452.4	7803.4	2.8	283.3	8005.3	33.6	169.1	49.4	7.6	62.6
Sediment	2109.7	13.9	540.8	1339.8	0.6	343.5	769.9	13.3	197.4	63.5	4.0	63.5
Hardbottom	2097.2	60.4	311.5	1099.0	18.1	204.1	998.2	42.3	107.4	52.4	30.0	65.5
Vegetated sand	4021.0	11.7	336.4	1751.5	0.1	252.6	2269.5	11.6	83.8	43.6	0.5	75.1
Mud	2472.5	0.0	0.0	1969.8	x	x	502.7	x	x	79.7	x	x
Lagoonal coral	143.9	0.0	25.4	68.8	x	23.0	75.1	x	2.3	47.8	x	90.8
Backreef	319.9	34.9	154.1	146.2	8.3	93.9	173.7	26.6	60.2	45.7	23.7	61.0
Beachrock	11.2	0.8	7.6	2.6	0.0	4.6	8.7	0.8	3.0	22.9	5.9	60.3
Total	**26984.1**	**158.0**	**1828.2**	**14181.1**	**29.8**	**1205.0**	**12803.0**	**128.2**	**623.2**	**52.6**	**18.8**	**65.9**

coral reefs and seagrass beds, and the shallows and intertidal areas support a variety of shorebirds and waders. Least tern (Egg bird) *Sterna antillarum,* and Bridled tern *Sterna anaethetus* dive for food in nearshore waters. Common migratory (non-breeding) seabirds include the Royal tern (Sprat bird, Old Tom) *Sterna maxima.* The combination of sheltered and productive waters, aesthetic appeal, and proximity to population centres, contribute to significant recreational pressure on lagoons. Swimming, snorkelling, scuba diving, recreational fishing, and pleasure boating and sailing, are the main activities, enjoyed by visitors and residents alike. Recreational pressure on lagoons is generally increasing. Diversification of recreational activities, such as the recent growth in the popularity of personal watercraft (jet skis), has added to pressures on popular areas.

Seagrass Beds

In the Cayman Islands, seagrasses, along with coral reefs and mangroves, constitute one of the three major coastal interface communities. As highly productive habitats, seagrasses provide a nursery for the juvenile stages of many marine organisms, and contribute to sediment stability and water clarity. The Cayman Islands support extensive seagrass beds in shallow lagoonal back reef areas, where fringing coral reefs shelter them from wave-action and storms. Significant seagrass beds can be found in North Sound, Frank Sound, and at the East End of Grand Cayman, and in Dick Sessinger's Bay on Cayman Brac, and in South Hole Sound and Mary's Bay on Little Cayman. The total coverage of seagrasses locally is approximately 7,000 ha, with the most extensive beds located in North Sound,

Fig. 7.4 Stingray, *Dasyatis americana* at Sandbar, North Sound, Grand Cayman

Table 7.3 Areal coverage of seagrass habitats in the Cayman Islands (1 acre = 0.404 ha)

Seagrass habitats	Total area (acres)			Area within protected areas (acres)			Area outside protected areas (acres)			% habitat protected		
Category	GC	CB	LC	GC	CB	LC	GC	CB	LC	GC	CB	LC
Seagrass beds	15808.8	36.3	452.4	7803.4	2.8	283.3	8005.3	33.6	169.1	49.4	7.6	62.6

Grand Cayman (Table 7.3). Several different species of seagrasses are found in the Cayman Islands, although Turtle grass *Thalassia testudinum* is the dominant species. Green turtles *Chelonia mydas* and sea urchins are among the few animals which are able to digest cellulose and feed directly on living seagrasses. Seagrasses, however, provide a substrate and a source of organic matter which supports a diversity of epiflora and fauna, including diatoms, algae, sponges, amphipods, polychaete worms and echinoderms. This diversity of infauna attracts larger predatory species, including fish. As such, seagrass beds underpin numerous food webs, and include culturally, commercially and trophically important species. Seagrasses are highly sensitive to changes in water quality, including clarity and salinity. Since the late-1960s, local seagrass beds have been severely impacted by extensive dredging of shallow lagoons to facilitate access, and dredging for fill, using (often unscreened) cutter-head hydraulic and mechanical dredges. In 2001, the DoE resurveyed the original 1976 Wickstead Report (see reference in Wells 1988) sites and found local seagrass beds to be significantly impacted by dredging activity, both directly, through the removal of substrate and physical modification of the environment, and indirectly, through the introduction of particulate matter into the water column. A policy on no further commercial dredging in the North Sound was established in 1997; however, navigational channels and other projects deemed "minor" continue to receive approval. The unwritten nature of this moratorium, and the fact that it does not extend to areas of coral reef, seagrasses and lagoons outside of North Sound, make it weaker protection than these key ecosystems need. The environmental impacts of recent and previous development activities remain evident to this day.

Mangroves

In the Cayman Islands, "mangrove trees" comprise four species: Black mangrove *Avicennia germinans*, White mangrove *Laguncularia racemosa*, Red mangrove *Rhizophora mangle*, and Buttonwood *Conocarpus erectus*. A tolerance for wet and salty conditions is a common feature of all four;

Fig. 7.5 Mangrove, Grand Cayman mangrove wetland

Table 7.4 Areal coverage of mangrove habitats in the Cayman Islands (1 acre=0.404 ha)

Mangrove habitat												
	Total area (acres)			Area within protected areas/buffers (acres)			Area outside protected areas/buffers (acres)			% habitat protected		
Category	GC	CB	LC	GC	CB	LC	GC	CB	LC	GC	CB	LC
Seasonally flooded mangrove shrubland/woodland	697.3	19.3	700.0	65.5	0.1	26.0	631.8	19.3	674.0	9.4	0.3	3.7
Seasonally flooded mangrove forest/woodland	12138.6	23.4	464.2	1467.2	0.1	9.6	10671.4	23.4	454.7	12.1	0.3	2.1
Tidally flooded mangrove shrubland/woodland	477.6	0.0	0.0	394.5	x	x	83.1	x	x	82.6	x	x
Tidally flooded mangrove forest/woodland	2802.0	0.0	19.0	1046.6	x	0.0	1755.4	x	19.0	37.4	x	0.0
Total	**16115.4**	**42.7**	**1183.2**	**2973.8**	**0.1**	**35.6**	**13141.7**	**42.6**	**1147.7**	**18.5**	**0.3**	**3.0**

however, their specific tolerances are markedly different. Red mangrove is the pioneering species, and typically constitutes the entirety of the seaward fringe of mangrove forest. Buttonwood, by comparison, occupies the opposite extreme of this range, preferring the driest and least saline environments of the four mangrove species.

Good examples of mangroves can be found in the Central Mangrove Wetland (ca. 8,500 acres), North Sound Mangrove Buffer and Barkers Mangrove on Grand Cayman (Fig. 7.5); Westerly ponds on Cayman Brac; and the Crown Wetlands (especially Tarpon Lake and surrounds), Booby Pond (northern mangrove fringe), and South Hole Sound in Little Cayman. While all species of mangrove are highly tolerant of root submersion in water, this tolerance remains within critical boundaries. Normally, oxygen concentrations decline in the pneumatophores (aerial roots) during high tide, and recover quickly during low tide, when the roots are once more exposed to the air. Immersion of the pneumatophores for more than a few days, however, results in a sharp decline in oxygen stored within the roots, effectively "drowning" the trees, and resulting in the mass mortality of submerged forest. Hence, mangroves are highly intolerant of elevated levels of standing water, and susceptible to interruptions to natural drainage. Large scale die-offs of mangrove result in canopy loss and decay of the underground root system, resulting in the exposure and oxidation of the peat layer below. This results in subsidence of the peat layer, often resulting in the formation of permanent pools.

Habitat and plant assemblages associated with the Black mangrove *Avicennia germinans*, White mangrove *Laguncularia racemosa*, Red mangrove *Rhizophora mangle*, and Buttonwood *Conocarpus erectus*, incorporate the following vegetation formations, as described by Burton (2008): (1) Seasonally flooded evergreen sclerophyllous forest; (2) Tidally flooded mangrove forest; (3) Seasonally flooded/saturated sclerophyllous evergreen woodland; (4) Tidally flooded evergreen woodland; (5) Seasonally flooded/saturated evergreen shrubland; (6) Saturated sclerophyllous evergreen shrubland and (7) Tidally flooded evergreen shrubland (Table 7.4).

Terrestrial protected areas including mangroves in the Cayman Islands are limited to Animal Sanctuaries, National Trust property, and the mangrove fringe associated with the North Sound Environmental Zone. The Animal Sanctuaries established under the Animals Law (1976), incorporate four significant inland pools, ponds and mangrove lagoons (two in Grand Cayman, one in Cayman Brac, one in Little Cayman), extending to a total of 341 acres. As of Jan. 2009, National Trust owned/shared ownership properties, protected under the National Trust for the Cayman Islands Law (1987), extended to a total of ca. 3,109 acres.

Mangrove constitutes one of the Cayman Islands' most undervalued and severely impacted habitats. Coastal mangrove contributes to biodiversity through provision of a secure nursery area. Protected from large predators within the matrix of the mangrove root system, the larvae and juvenile forms of many reef and open sea species grow in mangrove, before moving seaward as they mature. Spiny lobster *Panulirus argus* spends up to 2 years maturing in mangrove roots. Mangrove provides habitat to a variety of crabs, including *Eurytium limosum* and *Aratus pisonii,* the grapsid crab *Sesarma angustipes,* the fiddler crab *Uca speciosa,* and land crabs including *Gecarcinus lateralis* and the White Land crab *Cardisoma guanhumi.* Habitat loss and busy coastal roads inflict a heavy toll on land crabs, which of necessity undertake periodic mass-migration to the sea to lay their eggs. Many fish typically associated with coral reefs are obligate mangrove dwellers in their juvenile stages. Mangrove is also most significant from a terrestrial perspective, with respect to its complement of birdlife and an important roost for several species of local significance, including West Indian Whistling-duck *Dendrocygna arborea* and Greater Antillean grackle *Quiscalus niger*. Black mangrove *Avicennia germinans* provides nesting habitat for a significant proportion of the islands' Grand Cayman parrot *Amazona leucocephala caymanensis* and the White-crowned pigeon *Patagioenas leucocephala*. Mangrove is also of particular value to resident and migratory waders, such as the Snowy egret *Egretta thula*. While the floral diversity of mangrove is predominately restricted to the four mangrove species, dry keys within the mangrove complex contribute to the floral diversity of the system, with species such as Mahogany *Swietenia mahagoni,* Red birch *Bursera simaruba* and Manchineel *Hippomane mancinella*. The endemic and critically endangered herb *Agalinis kingsii* also occurs locally with in mangrove shrubland in the Cayman Mangrove Wetland.

Traditionally, mangrove has been regarded as worthless land, and a breeding ground for mosquitoes. In 1965 the *Mosquito Research and Control Unit*, MRCU, was established, and rapidly implemented a systematic dyking and canalisation programme for the mangroves, in combination with ground-based fogging, and aerial application of larvicide. The effect was to radically reduce the population of mosquitoes throughout the islands, however, local attitudes to mangrove or "swamp" improved little. With the concurrent economic boom associated with the advent of the banking and tourism industry, mangrove was targeted for profitable residential and canal development. Besides important roles in nutrient regulation, carbon sink, and coastal protection, the mangrove has an important role in rainfall production. Saturated air derived from the moist understory, and transpiration from the leaf surface, rises above the Central Mangrove Wetland and develops into localised cloud. The clouds are carried westward by the prevailing wind, contributing to the rainfall of central and western Grand Cayman. Rainfall in these areas is some 40% higher than in districts on the windward side of the Central Mangrove Wetland. In addition, the mangrove provides important sources of freshwater. The hydrological influences and ironpan formation associated with large mangrove areas contribute to elevation of the freshwater table in land peripheral to the wetland, resulting in the formation of some of the island's most fertile farm and grassland. Canalisation and development disrupt this function, causing salinisation of freshwater lenses, and depleting terrestrial freshwater availability.

Beaches

In the Cayman Islands, sandy beach and cobble originates as an erosional product of calcareous algae and the coral reefs which surround much of the islands, carried to the shore by coastal currents and storm events. Due to its unconsolidated and mobile nature, sandy beach and cobble is susceptible to lateral movement under the influence of local currents and storms, resulting in the migration of the beach and associated communities along the shoreline. The mobile nature of unconsolidated sandy beach and cobble also extends landward from the active shoreline to incorporate the beach ridge. Though naturally vegetated with coastal shrubland, the beach ridge remains an active component of the beach, prone to recover its dynamic nature in the event of erosion or lateral migration of the foreshore. The coral sand which forms the sandy beaches of the Cayman Islands is typically "fine sand". The *Seven Mile Beach Nourishment Project*, implemented by the Department of Environment (2004), determined the mean grain size on Seven Mile beach to be 0.4 mm, with an 88% shell content. This property contributes greatly to the aesthetics of the beach. The natural dimensions of fine sands limit desirable options for artificial beach restoration, excluding coarser sand sources from consideration. As such, beach sand should be regarded as a valuable and largely irreplaceable national resource, and should be maintained and managed as such. Cobble beaches and boulders ramparts comprise large reef fragments. On some beaches, cobble dominates the entirety of the shoreline. In others, cobble exists in conjunction with fine sand and a variety of intermediates, most often forming an extant cobble ridge at the top of the beach, abutting, and subsumed beneath, the permanent vegetation lines. The Crown owns and is legally responsible for areas of the seabed and beach, up to and including the mean high water mark. As the set-back for developments in coastal areas is generally measured in relation to the mean high water mark, consideration of the predictions of climate change, including increase in storm severity and sea-level rise, requires reconsideration of revision of survey baselines. Key beach habitats are: Seven Mile Beach, East End Beach, Smith's Barcadere, Barkers Beach and Rum Point (artificial) in Grand Cayman; Public Beach on Cayman Brac, and Spott Bay, Point of Sand, Preston Bay Beach, South Hole Sound and Owen Island on Little Cayman.

Cayman's beaches are important for nesting turtles. Historically, Cayman was considered the largest rookery for Green turtles *Chelonia mydas* in the Caribbean. Estimates place the original rookery at over one million individuals. Sandy beach provides the sole nesting habitat for the Cayman Island's remnant population of marine turtles. During a typical nesting season, from May to October, *DoE* staff find on average 43 nests in Grand Cayman, 12 nests in Cayman Brac, and 11 nests in Little Cayman. Typically, a female will lay 3–6 nests per season. Currently, annual nesting is credited to less then 20 individuals per species. Nesting turtles are notoriously site-specific. Localised degradation of critical beaches may disproportionately impact a small population. Beaches support a variety of birdlife, especially herons, waders and other shorebirds, with shallows, strandline flotsam and maritime invertebrates providing important food sources. The Antillean nighthawk (Rickery-Dick) *Chordeiles gundlachii* nests in bare areas of beach ridge. The fast-draining, shifting, salt-exposed environment is too extreme to support significant plant diversity; however a few highly tolerant species survive, such as Juniper *Suriana maritima*, Lavender *Argusia gnaphalodes*, Bay vine *Ipomoea pes-caprae,* Inkberry *Scaevola plumieri*, Cocoplum *Chrysobalanus icaco,* and the endemic Tea banker *Pectis caymanensis robusta*. Ghost crabs *Ocypode quadrata* are common along beaches, where they excavate burrows in the sand, and the Grand Cayman Curly-tailed lizard (Lion lizard) *Leicephalus carinatus varius* remains a common sight on sandy beaches, though it is probably in severe decline, with many large colonies lost to beach-front development. Sister Isles Rock Iguanas *Cyclura nubila caymanensis* also nest in this habitat, preferring the beach ridge and associated shrubland. Historically, the American crocodile *Crocodylus acutus* was a beach nester, though Cayman's population has long since been extirpated.

In several places in the Cayman Islands, highly organic and sometimes eutrophic ponds, pools and mangrove lagoons are separated from adjacent clear-water coastal lagoons by the filtering barrier of the beach ridge. Beach ridge integrity is a significant factor in natural storm defence. Structured in large part by successive storm events, the natural beach ridge provides an effective barrier to storm surge. Levelling, excavation and devegetation compromises the structural integrity of the beach ridge, weakening its function as a physical barrier, and facilitating the ingress of storm surge. In many cases, the natural forces of erosion may rapidly come to bear on damaged areas, exploiting weakness in a positive feedback loop, widening fissures and channels, deepening holes, uprooting vegetation and exacerbating loss of sandy beach and cobble.

The natural environment is the mainstay of the Cayman Islands' tourism product, and sandy beaches are integral to both the landscape and seascapes of the Islands, and contribute disproportionately to the natural aesthetic, benefiting quality of life for residents through the provision of scenic vistas, and contributing a financial premium to commercial undertakings. Inappropriate coastal development may thus be expected to have a very significant impact on the perception of the Cayman Islands as being "spoilt" or "unspoilt" by residents and holiday makers.

Cayman's Ironshore and Maritime Cliffs

The majority of the rocky coastline of the Cayman Islands comprises ironshore: white limestone, which weathers to a grey colour. The "Ironshore formation" was named by Matley (1924), referring the hard calcrete crust or caliche that typically develops on the weathered surface of the rock (Jones 2000). Though locally generally sharp and jagged, on a larger scale, ironshore constitutes a mostly flat or gently sloping topography.

Maritime cliffs are formed at the junction between land and sea, as a result of erosion. Notable examples on Grand Cayman include the vertical maritime cliffs of Pedro St James (max. 17 m) and High Rock (max. 12 m). The lack of any protective fringing reef in these areas results in the cliff-tops being exposed to heavy wave action during severe storms.

On Cayman Brac, the Bluff (max. 46 m) represents the Cayman Island's most spectacular maritime cliff feature (Fig. 7.6). The Bluff is most extant in the eastern portion of the Brac, where the cliff falls vertically to the sea surface. However, a low-lying coastal platform bounds the majority

Fig. 7.6 Ironshore and the Bluff, on Cayman Brac

of the Bluff, separating the majority of the relict maritime cliff from the sea. Wave-cut notches in the cliff provide clear indication of previously elevated sea levels. This partial separation of the cliff face from the immediate stresses of the marine environment creates a unique habitat, and gives rise to species found nowhere else in the islands. The coastal platform also supports the majority of the population of Cayman Brac; however, since the advent of Hurricane *Ivan*, development of the high land on the top of the Bluff has accelerated. Landward, maritime cliffs and ironshore support a sparse but intriguing vegetation, which often adopts a dwarf, prostrate, sometimes almost encrusting form, in response to the intense environmental conditions of soil and freshwater scarcity, wind exposure and salt-spray. Under these conditions, trees such as Buttonwood *Conocarpus erectus* suffer a natural bonsai effect: their diminutive size belying their age.

Seawards, maritime cliffs and ironshore generally become increasingly denuded of vegetation, but support increasing populations of marine invertebrates, such as chitons (*Polyplacophora* sp.). Due in part to the porous nature of limestone, permanent rock pools and associated communities are not a feature generally associated with ironshore. Maritime cliffs and ironshore are currently critically under-represented within the protected areas of the Cayman Islands.

Geologically, the Ironshore Formation comprises some of the youngest rock in the Cayman Islands. Ironshore is composed of soft fossiliferous white limestone, deposited through numerous transgressions of the sea during the Pleistocene period. Cores into the ironshore have dated the surface layers to 129,000 years old, with deeper layers up to almost 500,000 years old (Vézina et al. 1999; Jones 2000). The majority of high land, bluff and maritime cliffs are older Cayman Formation dolostone, probably dating to the Lower / Middle Miocene period, ca. 5–15 million years old. The maritime cliffs of Pedro Bluff comprise the Pedro Castle Formation dolostone and limestone. This younger rock, dating from the Pliocene (ca. 2 million years old) overlies the Cayman Formation in areas limited to the region of Pedro Castle, Grand Cayman, and the West end quarry, Cayman Brac. The oldest rocks in the Cayman Islands are the Brac Formation limestone or sucrosic dolostone. Forming the lower parts of the maritime cliffs at the eastern end of Cayman Brac, Brac Formation rock dates from the Lower / Upper Oligocene period, some 28 million years ago.

Beach rock represents a dynamic feature of the shoreline, constantly forming and eroding. This developing limestone generally forms smooth, seaward-dipping sheets, and is found in areas along the edge of the shoreline (Moore 1973). These accretions are ongoing, as is evidenced by the numerous modern artefacts to be found cemented into the matrix of the rock (Jones and Goodbody 1982). Beach rock is also highly prone to erosion and physical damage during high seas (Jones and Goodbody 1984).

By virtue of their inaccessibility, maritime cliffs and ironshore have important landscape value, representing some of the most natural/least modified terrestrial environment.

However, the majority of the habitats constitute bare rock, and as such, much of this area is of little biodiversity significance. However, where conditions permit, highly specialized conditions promote the establishment of niche-adapted flora and fauna. One such example is the colourful Rock crab *Grapsus grapsus*. Another is the herb *Verbesina caymanensis*. This critically endangered endemic plant is known only from a small cluster of individuals, growing on the relict marine cliffs below Peter's Cave, Cayman Brac. The maritime cliffs do provide habitat to some of the Cayman Islands most spectacular seabirds, most notably Brown booby *Sula leucogaster*, and White-tailed tropic bird *Phaethon lepturus*. In the Cayman Islands, Brown boobies are known only from a disjunct nesting colony ranged along the maritime cliffs constituting the eastern point of Cayman Brac. White-tailed tropic birds also nest along the Bluff, most notably in the vicinity of Peter's Cave. A small breeding colony of White-tailed tropic birds nested along the maritime cliffs at Pedro St James Bluff, ca. 17–20 birds, 1996–1997. No birds have been reported at this site since the advent of Hurricane *Ivan*, 2004. Vidal Key, a small (<0.1 ha) ironshore cay off the Barkers peninsula, West Bay, supports the Cayman Islands only known colony of Bridled terns *Sterna anaethetus* (ca.20 pairs). Least tern (Egg bird) *Sterna antillarum* occasionally nests on ironshore; however, this species has been forced onto man-made habitats such as exposed marl, in the advent of disturbance of much of its natural habitat. The Antillean nighthawk (Rickery-Dick) *Chordeiles gundlachii* nests on ironshore and sandy beaches, which has similarly needed to adapt to nesting in cleared areas of marl. Some maritime cliffs, especially those on Cayman Brac, incorporate significant caves. In addition to their geological interest, some support biological interest including breeding colonies of birds, and bats, and Peter's Cave and Rebecca's Cave, Cayman Brac, are of cultural significance.

Management of Coastal Resources

Management of reefs, sounds, seagrasses, mangroves, beaches and rocky shore areas includes protection from a variety of non-ecologically sound processes. Here coral reefs are selected as an example, but many of these potential problems apply to the other habitats.

Management of Coral Reefs

Sediments: the active dredging of nearshore lagoons for fill and access generates particulate matter. The fine nature of this residue enables it to remain suspended in the water column for long periods, during which time it may be carried some distance from the activity area. In suspension, this residue contributes to increased turbidity, reduced clarity and increased light attenuation through the water column, compromising the productive capacity of photosynthetic organisms, and associations including seagrasses and corals. On settlement, these sediments have the capacity to smother seagrass beds and damage the fragile feeding mechanisms of coral polyps. Due to their fine nature, these sediments are prone to resuspension as a result of modest wave action, or currents generated by passing water craft. Due to the confined, reef-protected nature of much dredged seabed, residue has a tendency to remain captive, repeatedly cycling between suspension and sedimentation, migrating within, and impacting, the nearshore system.

Coastal development: nutrient and sediment-rich runoff from the terrestrial environment, exacerbated by land clearance, roads construction and emplacement of impervious surfaces, combined with the removal of buffering coastal vegetation, contribute to an increased incidence of stressors on coral reefs, in combination with an erosion of natural filtration mechanisms.

Dredging/channelling: for fill and access, this activity impacts coral reefs within the footprint of activity, and beyond through increased sediment loading.

Antiquation of legislation: though progressive in their time, the current protective measures for coral reefs (outside of Marine Park areas) in the Cayman Islands have remained unchanged since 1978. Since the Marine Parks Regulations were established in 1986, the population of the Cayman Islands has risen from 21,545 to 56,729 in 2011 (Government of Cayman Islands Economics and Statistics Office 2012), and annual visitor numbers to the Islands have increased to nearly two million; placing coral reefs under increasing pressure. Protection measures for coral reefs should be dynamic and responsive to change if they are to effectively address the evolving complement of pressures to which they are subjected, and the expectations of a tourism industry increasingly knowledgeable with regard to environmental provision. With other destinations setting the current benchmark by protecting their entire coral reef system, the Cayman Islands can no longer be regarded as ecologically forward-thinking with less than 10% of its coral reefs included within marine protected areas.

Anchor damage: physical destruction of coral reefs due to commercial and recreational boating has contributed significantly to the degradation of reefs in the Cayman Islands. Coral reefs in areas exposed to heavy marine traffic, for example, in the vicinity of George Town harbour, have been almost completely destroyed.

Nutrification (*eutrophication*): addition of nutrients into coastal waters promotes a shift in the natural complement of reef species, encouraging the rapid growth of algae, which has the potential to smother corals and compete for space. Sewage effluent and fertilizer runoff are the most common sources of coastal nutrification. In the Cayman Islands, examples of nutrification and algal domination of degraded reefs may be seen adjacent to the effluent outlet for Boatswain's Beach (previously Cayman Turtle Farm).

Ocean acidification: Ocean acidification is of major long-term concern for coral reefs as it reduces the capacity of corals to calcify. Ocean surface pH is estimated to have decreased from approximately 8.25 to 8.14 since the beginning of the industrial era, (Jacobson 2005) and it is estimated that it will drop by a further 0.3–0.4 units by 2100 as the ocean absorbs more anthropogenic CO_2 (Orr et al. 2005). As ocean pH falls, so does the concentration of calcium carbonate, which is normally present at supersaturating concentrations. When carbonate becomes under-saturated, structures made of calcium carbonate are vulnerable to dissolution. Research has already found that corals experience reduced calcification or enhanced dissolution when exposed to elevated CO_2 (Gattuso et al. 1998).

Spearfishing: this popular recreational sport targets top-predators and removes large adult breeding stock from the reef environment.

Diver damage: the living tissue of coral polyps is sensitive to physical pressure and abrasion. Permanent death of polyps can result from divers touching the surface of corals, either deliberately with their hands, or accidental through trailing equipment or fins. The modern diving community is generally better educated regarding the sensitivity of coral reefs, resulting in a much reduced impact from individual divers; however, this reduction in impact is largely offset by the extent of historical degradation and the current high levels of usage of remnant coral reefs.

Invasive species: Red lionfish *Pterois volitans* were first reported on Cayman reefs in 2008 (see later section).

Accidental grounding: grounding events of recreational and commercial vessels are a regular occurrence. Grounding may result from boats slipping moorings during severe weather, or poor navigation.

Storm damage: there is evidence that modest storm action may be beneficial to reef health, cleaning corals surfaces of excessive epifauna and epiflora, especially marine algae, and removing dead or dying skeletons. However, wave action, strong currents, abrasion and scour arising from severe storms can significantly impact reef structure. Studies throughout the Caribbean show that hurricanes on average cause a 17% decline in coral cover in the year following the storm (Gardner et al. 2005). Climate change predictions are for more intense storms.

Pollution: poor water quality encourages the establishment and the spread of infectious diseases among corals. Industrial pollutants, such as copper, can impact the natural development of corals. There is also evidence that tributyltin (TBT), a key ingredient in anti-fowling paints, significantly impacts coral growth and recruitment. Vessel grounding events thus have the potential not only to inflict immediate physically damage on coral heads, but also impact recovery through TBT contamination (Negri et al. 2002; Smith et al. 2003). The impact of these factors on Cayman coral reefs is largely unstudied.

Oil-spill: pollution events arising from large scale wrecking and small-scale accidental and deliberate release of oil into the marine environment have significant potential to impact coral reefs and associated flora and fauna.

Dust: global air currents link the Caribbean with Africa, carrying dust to the Cayman Islands from the Sahara. This process is most pronounced during positive phases of the North Atlantic Oscillation (Prospero and Nees 1986), and over time is believed to have supplied the Cayman Islands with the characteristic "red mold" earth which supports some of the best farm and grassland. Dust events have been linked to a decline in the health of coral reefs across the Caribbean and Florida, primarily since the 1970s (*U. S. Geological Survey*).

Marine litter: while coral skeletons appear rocky and fairly impervious to physical damage, the living polyps are delicate, and highly sensitive to even slight physical pressure or abrasion.

Artificial installations: a significant decline in the quality of Cayman's coral reefs in recent years has resulted in increasing attempts to diversify the "attractiveness" of diving sites through augmentation of natural features with artificial installations. Examples include the *Shipwreck City* project, which aims to deliberately wreck ships as a dive feature off Grand Cayman, and a concrete *"Lost City of Atlantis"* off Cayman Brac.

To better understand the dynamics of the Cayman Islands reefs, the *Department of Environment* has commenced a long-term *Coral Reef Monitoring Programme*, encompassing all three islands, towards determining the scale of the problems facing coral reefs, and to provide conservation authorities with data to foster better management practice of local reefs.

In 1988, the *Marine Conservation Regulations* made it illegal to damage coral by anchoring in the Cayman Islands. The *Department of Environment* maintains a network of over 300 permanent moorings around the islands, towards minimizing anchor damage. The maximum fine imposed to-date is CI$150,000 for 130 m^2 of damaged reef. In 2007, the Cayman Islands Government strengthened protective regulations on the legal harvest of marine turtles.

The establishment of the proposed Barkers National Park would represent the first protected area in Cayman to incorporate a full continuum of habitats, from coastal shrubland and mangrove, to beach, lagoon and coral reef. Dive tour operators work in cooperation with the *Department of Environment* are educating divers towards minimising impact on coral reefs, including participation in the annual *Reef Watch* survey.

Since the mid 1980s, under the *Regulations of the Marine Conservation Law* "whoever anchors any vessel exceeding 20 feet in length or a commercial vessel ... in such a manner that damage is caused to the coral by an anchor, chain or any similar contrivance, is guilty of an offence."

Under the *Marine Conservation Law* "Whoever, unless licensed ... intentionally cuts, carves, injures, mutilates, removes, displaces or breaks any underwater coral or plant growth or formation in Cayman waters is guilty of an offence." Also under the *Marine Conservation Law* "any person who, while equipped with any kind of underwater breathing apparatus, takes any marine life in Cayman waters is guilty of an offence."

While these clauses effectively protect all corals in the Cayman Islands from a variety of threats, this protection is incomplete. Unforeseen events such as boat groundings, land-based stresses, people walking on shallow coral reefs, etc. remain unaddressed. In 1986 *Marine Conservation Regulations* created strict licensing procedures for spear fishing. Importation of new spear guns and parts into the country is illegal under these regulations. Though this was envisioned as slowly phasing out this activity, it has been only partially successful.

Coral reef monitoring began being undertaken by *DoE* biannually, as part of CARICOMP, and is now expanded and undertaken annually at 62 permanent sites, by DoE in collaboration with Bangor University and The Nature Conservancy through the DEFRA Darwin Initiative project to Enhance an Established Marine Protected Area System. The project aims to fully protect between 30 and 50% of the shelf of the Cayman Islands.

Key Marine Species in Cayman Waters

In addition to research programmes focused on habitats and ecosystems, recent scientific research in the Cayman Islands has focused on key species, including major commercial species such as turtles, conch, and lobster, and species important to tourism such as reef fish (Nassau groupers) and southern stingrays. Here two such species are focussed upon.

Nassau Grouper *Epinephelus Striatus*

The Nassau grouper *Epinephelus striatus* (Fig. 7.7) is a large sea bass—a prominent member of the 12 species of groupers reported for the Cayman Islands. It previously comprised one of the most economically important spawning aggregation based artisanal fisheries in the Caribbean, however, it is now severely over-fished. It ranges from inshore to about 100 m, and is to be found in coral reef, mangrove, seagrass and estuarine habitats, from southern Brazil, throughout the Caribbean, western Yucatan, Bahamas, Florida and Bermuda.

Fig. 7.7 Nassau grouper *Epinephelus striatus*

Western Atlantic: Bermuda, Florida, Bahamas, Yucatan Peninsula and throughout the Caribbean to Brazil. Eastern Gulf of Mexico: only off the coast of Yucatan, at Tortugas and off Key West.

Endangered throughout its range, and locally extinct in many locations, the Nassau grouper is listed as Endangered A2ad on the IUCN Red List. Of the areas in the Cayman Islands known to have hosted spawning aggregations (SPAGs), only one, Little Cayman West End, still hosts (albeit reduced) reproductively viable numbers. Two sites host remnant aggregations, the reproductive viability of which remains unknown (Cayman Brac East End and Twelve Mile Bank NE End). Three other sites are considered non-functional (Little Cayman East End and Grand Cayman East End & South West Point). Two other areas of similar geomorphological and hydrological characteristics are anecdotally reported as historical spawning aggregation sites.

Epinephelus striatus is protected under the Marine Conservation Law (2003 Revision) Section 6(7) & 9, and the Marine Conservation (Grouper Spawning Areas) Directives 2003. Under the current Marine Conservation Law, fishing for Nassau grouper is seasonally prohibited in eight Designated Grouper Spawning Areas DGSA. Nassau grouper may not be taken by spear, or in any Marine Park area. The legal minimum size for take is 12 in. (30.5 cm). Pending legislation, Nassau grouper would be protected under the National Conservation Law (Schedule II). The Department of Environment and the Marine Conservation Board are the lead bodies for local protection.

As top level predators on the reefs, Nassau grouper lead a predominantly solitary existence for most of the year. With the onset of the full moons in January and February, individuals migrate to specific locations where they form brief (ca. 10 day) broadcast spawning aggregations, spawning *en masse*. Located off deep shelf promontories, these aggregations represent 100% of the species annual reproductive output.

Spawning gives rise to pelagic larvae, which settle out on reefs within 40 days. Nassau grouper life history characteristics include large size, long-lived, slow-growth, and contracted reproductive period, coupled with high spawning site fidelity. These factors combine to makes this species especially vulnerable to over-exploitation. At 7 or 8 years of age, most will recruit to spawning aggregations as reproductive adults. The oldest recorded individual was 29 years of age. Anecdotal evidence suggests they may live much longer. Populations have declined an estimated 60% during the past three decades. In the Cayman Islands, spawning aggregations have been fished down to a level such that natural replenishment is inhibited.

Historical fishing pressure: fishing pressure on spawning aggregations is overwhelmingly responsible for the endangered status of this once prolific species. Historically, thousands of Nassau grouper were taken annually from SPAG areas by a small but effective local artisanal fishing fleet. Prior to the 2001 discovery of the Little Cayman West End SPAG, catch from these spawning sites had dwindled steadily, becoming insignificant due to overexploitation. The estimated spawning population of the "new" Little Cayman West End site was more than halved within two seasons, following discovery. This "boom and bust" dynamic of spawning aggregation fisheries is typified throughout the tropics. In 2004, an 8-year ban on fishing Nassau grouper in designated SPAGs was legislated in an attempt to maintain the viability of the remaining Little Cayman SPAG, and towards facilitating the potential recovery of remnant SPAGs and associated local stocks. This ban was successfully extended for another 8 years in 2012.

Poaching: as is the case with other protected species, it is likely that some background poaching occurs. Compliance with regulations has, however, generally been good during the first 5 years of the ban (at least for the 5 'nearshore' designated SPAGs), affording an opportunity for reestablishment and replenishment. There is, however, suspected fishing at the 12-Mile Bank site.

Current fishing pressure: Nassau grouper is a species of significant commercial and recreational interest. Individuals are taken primarily by hand-line, fish traps, and spear gun. Cessation of SPAG fishing has afforded the most immediate protection from the decimation of already depleted local stocks, however, the very limited shelf area of the Cayman Islands may mean that grouper populations remain sensitive to the continued lower level harvest of adults outside the spawning season. An attempt to lessen this potential impact was implemented recently, in the form of a ban on the spearing of this species, addressing a method of take regarded as responsible for the majority of take of adults during the non-spawning months.

Aesthetic value: Nassau grouper are considered an iconic species of tropical Atlantic coral reefs. As such, they are an important benchmark for reefs, in attracting dive tourism in the Caribbean. During 1987–2001, the *DoE* undertook annual monitoring of the Cayman Island's SPAG fishery. Catch, size of fish, and sex were recorded. Results indicated a strongly significant and steady decrease in both overall catch and individual size over this period. A 1995 a proposal to reduce fishing by 50% went without political support. Subsequent to the 2001 SPAG discovery, and the subsequent unregulated take of fish (approximately 4,000 fish during the spawning seasons of 2001 and 2002), wide-spread public outcry engendered political support for protection. In 2002 legislation for an alternate year fishing strategy dictated that 2003 be a non-fishing year. Calculations showed that with the resumption of fishing in 2004, even with a catch quota of 12 grouper per boat per day, what remained of this SPAG would be decimated. A bag limit of one grouper per person per day was introduced between May and October and a closed season from November to April. A further 8 year ban introduced in 2012 was made more palatable by an increased bag limit of four grouper per person per day outside of the closed season.

Invasive Red Lionfish *Pterois Volitans*

Red lionfish *Pterois volitans* favour coral reefs and rocky outcrops, with a wide distribution in the central and eastern Indo-Pacific. It is now spread through much of the Caribbean. *Pterois volitans* (Fig. 7.8a) is found in depths ranging from 10 to 80 m (and has been reported fished from much deeper depths), and reaches 40 cm long in the Caymans. It is a predator, showed that its main diet was a wide range of fish, and crustaceans. The first individual was reported locally in Feb 2008, in Little Cayman, and then another was caught in Cayman Brac in Oct. 2008, and five more caught locally in Jan. 2009. By 2012, this invasive fish can be found on reefs around all three islands at all depths (Fig. 7.8b), and although some juveniles can be found in the peat escarpments in seagrass beds, they do not seem to have penetrated mangrove systems. Lionfish are voracious predators on reef fish, and a recent study by *DoE* of stomach contents of lionfish from the Cayman islands by *DoE* has revealed prey to include wrasses, damselfish, parrotfish, surgeonfish, goatfish, squirrelfish, blennies, gobies, basslets, trumpetfish, lizardfish, hawkfish, mantis shrimp, Pederson shrimp, crabs, seagrass, and even lionfish. The impact on reef communities is likely to be significant, and is the subject of collaborative research between *DoE* and Bangor University.

How lionfish became established in the Atlantic remains unknown, however, it is accepted that several fish were introduced into the marine waters of Biscayne Bay, Florida, as a result of Hurricane *Andrew* in 1992. Lionfish have now been documented along the US east coast, from Florida through Massachusetts, east to Bermuda, and south throughout the Bahamas, and the Caribbean, including Turks and Caicos, Jamaica and Cuba. The red lionfish *Pterois volitans* is invasive in Cayman waters, and requires implementation of active control to prevent its spread. Lionfish represent a serious

a

b

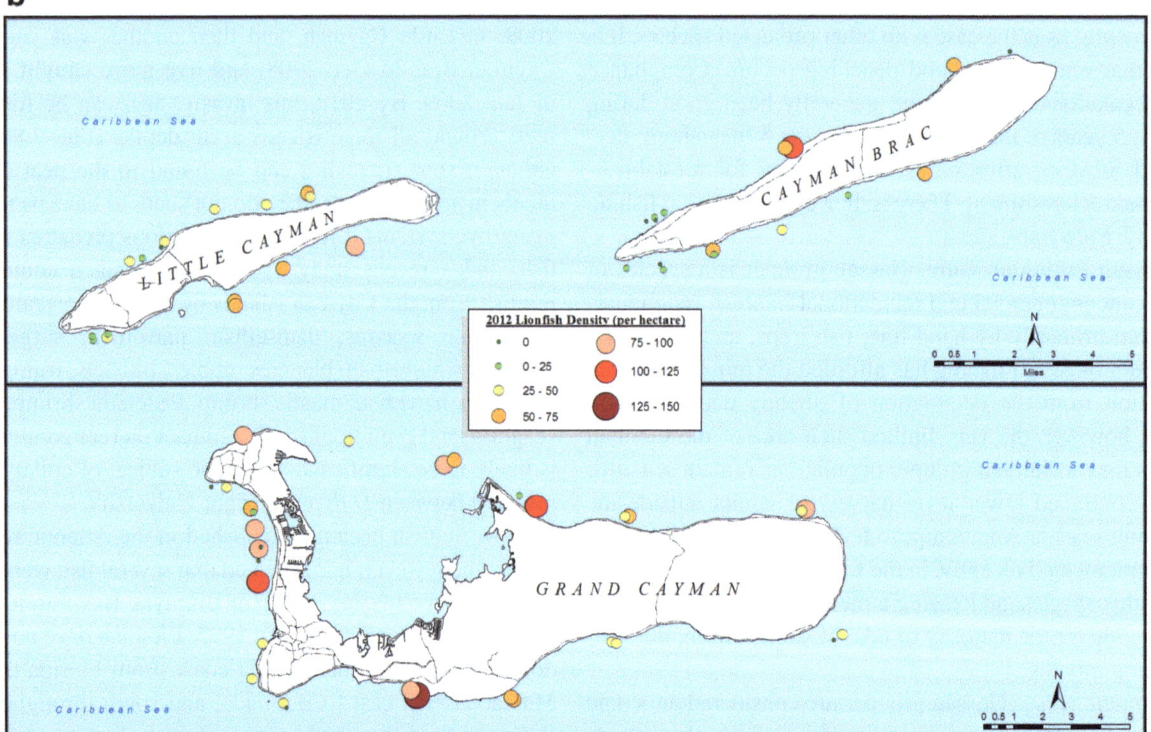

Fig. 7.8 (**a**) The invasive lionfish, *Pterois volitans*. (**b**) Densities of lionfish in the three Cayman Islands per hectare, in 2012

threat to both native marine life and, because it is extremely venomous, human health and safety. Red Lionfish have been subject to an intensive control programme in the Cayman Islands since 2008. Following training from REEF, the Cayman Islands DoE embarked on an extensive circuit of training courses, enabling professional divers to become trained and licensed Lionfish cullers. It may be that it will be found to be preyed upon by one of the top level predators on the reefs, the Nassau grouper.

References

Blanchon P, Jones B (1997) Hurricanes control on shelf-edge-reef architecture around Grand Cayman. Sedimentology 44:479–506

Burton FJ (2008) Vegetation classification for the Cayman Islands. In: Threatened plants of the Cayman Islands: the red list. Royal Botanic Gardens Kew, Richmond

Cayman Islands Government (1978) Marine Conservation Law, 1978. Gazette no. 9, 1995 (Suppl no. 2), 12pp

Cayman Islands Government (1986) Marine Conservation (Marine Parks) Regulations, 1986. Extraordinary Gazette of 21 Feb 1986 (Suppl no. 1), 1 pp

Cayman Islands Government (2011a) The Cayman Islands 2010 population and housing census: preliminary report, The Economics and Statistics Office, George Town

Cayman Islands Government (2011b) The Cayman Islands' compendium of statistics 2010. The Economic and Statistics Office, George Town

Craton M (2003) Founded up the seas: a history of the Cayman Islands and their people. Ian Randle Publishers, Kingston

DaCosta-Cottam M, Olynik J, Blumenthal J, Godbeer KD, Gibb J, Bothwell J, Burton FJ, Bradley PE, Band A, Austin T, Bush P, Johnson BJ, Hurlston L, Bishop L, McCoy C, Parsons G, Kirkconnell J, Halford S, Ebanks-Petrie G (2009) Cayman Islands National Biodiversity Action Plan 2009. Cayman Islands Government, Department of Environment, Grand Cayman

Darbyshire J, Bellamy I, Jones B (1976) The Cayman Islands natural resources study. Part III. Results of the investigations into the physical oceanography. Ministry of Overseas Development, London

Davies JE, Brunt MA (1994) In: Brunt MA, Davies JE (eds) The Cayman Islands: natural history and biogeography. Kluwer Academic, Dordrecht, pp 1–12

Dromard D, McCoy C, Turner JR (2011) Measuring the performance of marine protected areas: the case of Little Cayman and Cayman Brac, Cayman Islands. Proc Gulf Caribb Fish Inst 63:247–253

Ebanks GC and PG Bush (1990). The Cayman Islands: a case study for the establishment of marine conservation legislation in small island countries. In: Proceedings of the 1990 congress on coastal and Marine tourism—a symposium and workshop on balancing conservation and economic development, vol 1, Newport, pp 197–200

Gardner TA, Cote IM, Gill JA, Grant A, Watkinson AR (2005) Hurricanes and Caribbean coral reefs: impacts, recovery patterns, and role in long-term decline. Ecology 86:174–184

Gattuso JP, Frankignoulle M, Bourge I, Romaine S, Buddemeier RW (1998) Effect of calcium carbonate saturation of seawater on coral calcification. Global Planet Change 18:37–46

Giglioli MEC (1994) The boom years in Grand Cayman; environmental deterioration and conservation. In: Brunt MA, Davies JE (eds) The Cayman Islands: natural history and biogeography. Kluwer Academic, Dordrecht, pp 509–526

Godley BJ, Broderick AC, Campbell LM, Ranger S, Richardson PB (2004) An assessment of the status and exploitation of marine turtles in the UK overseas territories in the Wider Caribbean. Final project report to the Department of Environment, Food and Rural Affairs and the Foreign and Commonwealth Office, 253pp

Jacobson MZ (2005) Studying ocean acidification with conservative, stable numerical schemes for nonequilibrium air-ocean exchange and ocean equilibrium chemistry. J Geophys Res Atmos 110:D07302

Jones B (2000) Geology of the Cayman Islands. University of Alberta, Edmonton, Alberta, Canada (unpublished data)

Jones B, Goodbody QH (1982) The geological significance of endolithic algae in glass. Can J Earth Sci 19:671–678

Jones B, Goodbody QH (1984) Biological alteration of beachrock on Grand Cayman Island, British West Indies. Bull Can Petrol Geol 32:201–215

Kinder TH (1983) Shallow currents in the Caribbean Sea and Gulf of Mexico as observed with satellite-tracked drifters. Bull Mar Sci 33:239–246

Matley CA (1924) Geological survey of the Cayman Islands. Annual general report, Jamaica, pp 41–45

McCoy CM, Dromard CR, Turner JR (2010) An evaluation of Grand Cayman MPA performance: a comparative study of coral reef fish communities. Proc Gulf Caribb Fish Inst 62:345–353

McWilliams JP, Côté IM, Gill JA, Sutherland WJ, Watkinson AR (2005) Accelerating impacts of temperature-induced coral bleaching in the Caribbean. Ecology 86:2055–2060

Moore CH (1973) Intertidal carbonate cementation, Grand Cayman, West Indies. J Sediment Petrol 43:591–602

Negri AP, Smith LD, Webster NS, Heyward AJ (2002) Understanding ship-grounding impacts on a coral reef: potential effects of antifoulant paint contamination on coral recruitment. Mar Pollut Bull 44:111–117

Nurse L, Sem G, Hay JE, Suarez AG, Wong PP, Briguglio B, Ragoonaden S (2001) Small Island states. In: McCarthy JJ, Canziani OF, Leary NA, Dokken DJ, White KS (eds) Climate change 2001: impacts, adaptation and vulnerability. Contribution of working group II to the third assessment report of the intergovernmental panel on climate change. Cambridge University Press, Cambridge, UK, pp 843–875

Orr JC, Fabry VJ, Aumont O, Bopp L, Doney SC, Feely RA, Gnanadesikan A, Gruber N, Ishida A, Joos F, Key RM, Lindsay K, Maier-Reimer E, Matear R, Monfray P, Mouchet A, Najjar RG, Plattner GK, Rodgers KB, Sabine CL, Sarmiento JL, Schlitzer R, Slater RD, Totterdell IJ, Weirig MF, Yamanaka Y, Yool A (2005) Anthropogenic ocean acidification over the twenty-first century and its impact on calcifying organisms. Nature 437:681–686

Prospero JM, Nees RT (1986) Impact of the North African drought and El Niño on mineral dust in the Barbados trade winds. Nature 320:735–738

Sear C, Hulme M, Adger N, Brown K (2001) The impact of global climate change on the UK overseas territories: issues and recommendations. Natural Resources Institute, University of Greenwich/Tyndall Centre for Climate Change Research, University of East Anglia, Kent/Norwich

Seven Mile Beach Nourishment Project, George Town, Cayman Islands, Bahamian Sand Source Analysis (2004) Report, Department of Environment, Cayman Islands Government

Sheppard CRC, Rioja NR (2005) Sea surface temperature 1871–2099 in 38 cells in the Caribbean region. Mar Environ Res 60:389–396

Smith RC (2001) The maritime heritage of the Cayman Islands (new perspectives on maritime history and nautical archaeology). University Press of Florida, Gainesville

Smith LD, Negri AP, Philipp E, Webster NS, Heyward AJ (2003) The effects of antifoulant-paint-contaminated sediments on coral recruits and branchlets. Mar Biol 143(4):651–657, ISSN 0025-3162

Stoddard DR (1980) Little Cayman: ecology and significance. Atoll Res Bull 241:1–180

Tompkins EL (2005) Planning for climate change in Small Islands: insights from national hurricane preparedness in the Cayman Islands. Global Environ Change Part A 15:139–149

Vézina J, Jones B, Ford D (1999) Sea-level highstands over the last 500,000 years: evidence from the ironshore formation on Grand Cayman, British West Indies. J Sediment Res 69:317–327

Wells SM (ed) (1988) Coral reefs of the world, vol 1. UNEP Regional Seas Directories and Bibliographies/IUCN, Gland/Cambridge, UK, pp 85–95, Cayman Islands, 373p

Coral Reefs of Montserrat

8

Andrew Myers

Island Overview

Montserrat is located within the Leeward Islands of the Eastern Caribbean at approximately 16° 45′ 0" N, 062° 12′ 0" W. The neighbouring islands are Antigua 45 km to the northeast; Redonda, a small uninhabited island 25 km to the northwest; Nevis 55 km to the northwest, and Guadeloupe 60 km to the southeast (Fig. 8.1).

The island is volcanic in nature with multiple peaks. Montserrat is divided into three distinct sections correlated to three volcanic ranges. Those ranges are, from oldest to newest: the Silver Hills in the north (403 m); Center Hills, site of the island's nature reserve (741 m); and the Soufriere Hills, the youngest, tallest and currently active volcano (1,070 m) (Le Friant et al. 2009). Montserrat is an island of steep mountains, dense forest, deep valleys and ghauts, multiple fresh water springs, seaside cliffs, several black sand beaches (there is one white sand beach on the northwest coast) and a large area affected by volcanic eruptions and lahars flows.

The island has a stated population of approximately 4,900 people. That population lives along the west coast and the northern sections of the island (2011 census, www.gov.ms), and over half of the island, the section south of the Belham River valley, is an exclusion zone which is not open for habitation. This zone varies in size with volcanic activity.

Modern Volcanic History

In July of 1995 the Soufriere Hills volcano became active again, and since that date there has been multiple dome growth and collapse events which caused both the evacuation and destruction of the capital city, Plymouth, and numerous smaller villages. The Soufriere Hills is an explosive stratovolcano, whose volcanic events are characterized by the extrusion of rock, gases and ash from vents on the dome, often in the form of pyroclastic flows. Eruptions produce billowing ash clouds reaching over 10,000 m into the air; avalanches of super-heated gases, rocks and boulders, which at times reach and travel across the sea; and dense deposits of fine ash particles on land and sea (www.mvo.ms). Repeated dome collapses and volcanic activity have buried extensive tracts of land and changed the contours of the coastline (Fig. 8.2). Dome collapses and lahars over the past 15 years have created and extended deltas of volcanic debris into the sea extensively on both the eastern and western southern sections of the island. Over 1 km^3 of material has been produced by the volcano with an estimated 75–90% of that material ending up in the sea (Trofimovs et al. 2006). Over 1 km^2 of new beaches and land have been formed, burying previously existing coastline and offshore habitat (Le Friant et al. 2009).

Montserrat's Marine Habitat

The island has approximately 40 km of coastline, mostly backed by towering sea cliffs with occasional beaches. The submarine shelf, at depths from 20 to 60 m, varies greatly in width around the island, extending approximately 5 km off the northern section but only 0.5 km off the southern coast (Le Friant et al. 2004). Habitat on the deeper section of the shelf is comprised of sand flats with scattered low profile rock platforms with coral coverage. Between the 20 m depth contour and the shoreline habitat density increases.

This typical inshore habitat (<30 m depth) is composed of erosional material from the sea cliffs (large boulders and varying sized rocks), ridges and low profile rock shelves and sand gullies and plains. The reef system of Montserrat is not created by coral growth exclusively but includes much coral growth on rock, boulders, and hard, low profile substrate Each of these zones support different marine life.

A. Myers (✉)
Montserrat Reef Project, P.O. Box 289, Brades, Montserrat, Eastern Caribbean or British West Indies
e-mail: montserrat.reef.project@gmail.com

C.R.C. Sheppard (ed.), *Coral Reefs of the United Kingdom Overseas Territories*, Coral Reefs of the World 4,
DOI 10.1007/978-94-007-5965-7_8, © Springer Science+Business Media Dordrecht 2013

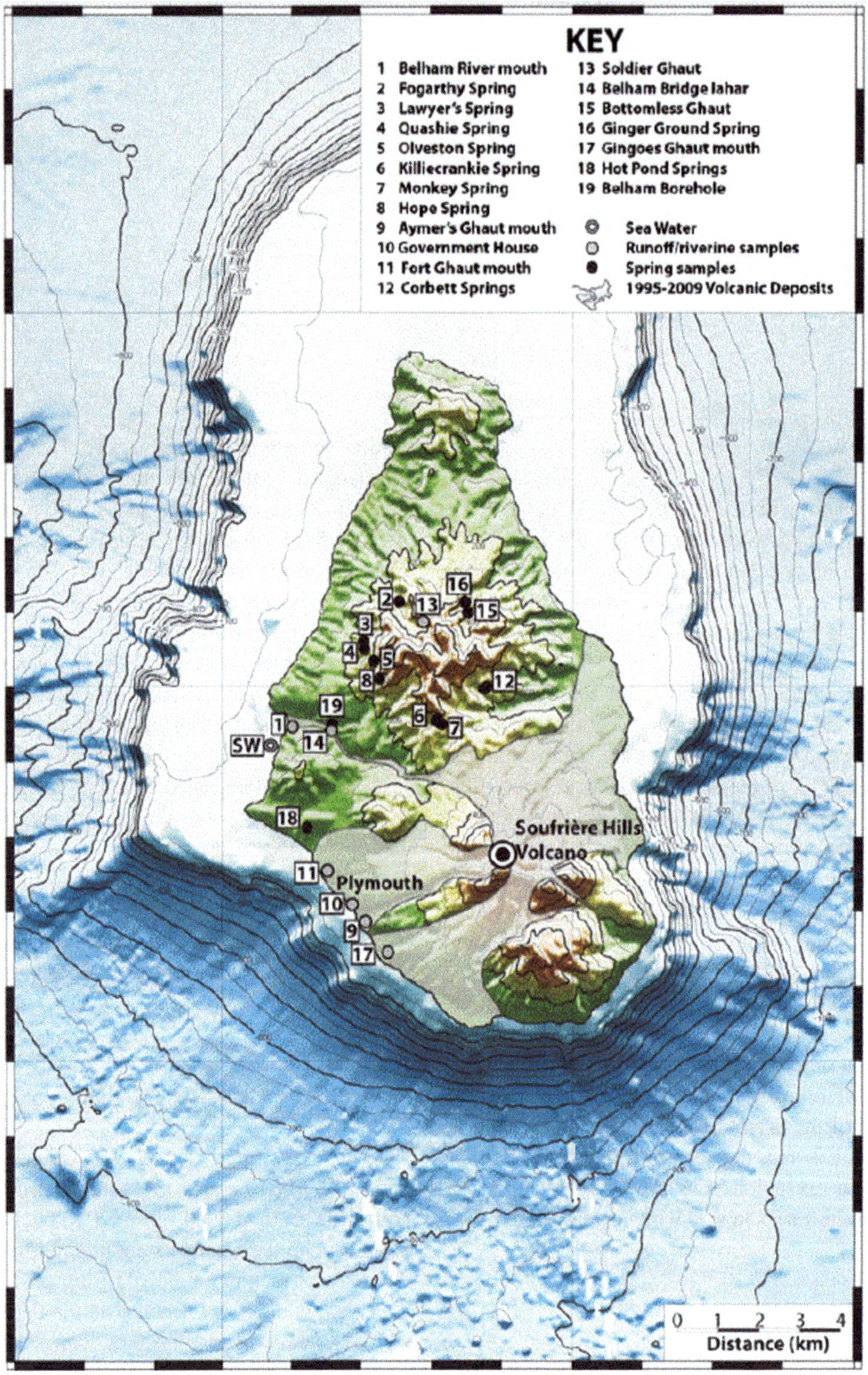

Fig. 8.1 Map view of Montserrat displaying bathymetry and the submarine shelf (From Jones et al. 2010, Springer book chapter)

Shallow Water Boulder and Rock Zone

Extending from shore to approximately the 10 m depth contour, much of the island is surrounded by the remnants of the erosional decay of the island, with varying sized rocks. Within this zone evidence of the island's volcanic roots are clearly apparent. As the softer material eroded/erodes from the sea cliffs an intricate marine topography and habitat is formed. Some of the boulders found here measure over 20 m across and 15 m tall (Fig. 8.3).

Fig. 8.2 Repeated dome collapses and volcanic activity have buried extensive tracts of land and changed the contours of the coastline. This photo was taken after the most recent partial dome collapse in February of 2010 which extended 600 m into the sea (© Henry Odbert)

Fig. 8.3 Example of the biodiversity found on the large boulder structures in shallow water. This one is at 5 m depth on the northwest coast (© Kim Guinn)

The resulting substrate hosts an extensive variety of corals, invertebrates, sponges, juvenile and adult fish species, marine creatures, and marine plants and algae which are typical of Caribbean coral reef areas. This zone appears to be significant in terms of providing juvenile fish nursery areas.

Ridges and Low Profile Rock Shelves

Further offshore, with increasing depths, the reef has a lower profile, with elevations and ridges typically less than 2–3 m high. Heavily sloping bottom contours create ridges in the

Fig. 8.4 Typical low profile reef biodiversity in 15 m of water; west coast of Montserrat (© Marcus Merrin)

15–20 m depth span, and within this zone "islands" of rock create scattered patch reef between 10 and 30 m depth (Fig. 8.4).

The reef substrate is often eroded into a honeycombed rock base. Though many of the same species of corals and sponges are found on these reefs as occur on the shallower ones, there is a greater abundance of gorgonians and far larger barrel sponges, again typical of reefs in the eastern Caribbean region. Pelagic species, such as jacks and mackerels, are most common in this depth zone.

Sandy Bottom Gullies and Plains

This zone supports species that live or feed off of the sand flats. Several reef inhabitants leave the shelter of the protective coral reef to forage for molluscs, crustaceans and marine plants within the sand flats near the reef. Within this zone, Montserrat's marine habitat supports a healthy population of southern stingrays, flying gunards, and spotted snake eels, as well as tobacco fish and conch.

Differences in Marine Habitat Around the Island

As noted earlier Montserrat's submarine shelf varies greatly in width from the northern and southern regions of the island. This factor, as well as volcanic activity and prevailing sea conditions, have created varying topography and health of the reef system (Fig. 8.5).

Northern Reef System

The hard substrate shelf extends to 5 km off shore in this region, and consistent wave action and open ocean currents have created bunkers within the reef that allow for protective areas for reef life. The corals of this area are exposed to constant water movement and, as a consequence, appear to be amongst the most healthy around the island. This region is affected only occasionally by sedimentation from the volcano.

Western Leeward Reef System

As the shelf narrows the extent of the reef from the shore reduces. Dense reef coverage is found within 300 m of shore, with patch reef extending to over 1 km offshore to depths of >30 m.

Coral health varies greatly along this coast with healthier reefs, in general, existing further from the volcanic runoff plains, but there are anomalies to this, where multiple, small reefs exist close to these runoff plains where they still support healthy reef habitat, possibly as a result of water currents redirecting sediment though the area.

Fig. 8.5 Satellite image depicting the distinct reef systems around Montserrat. Northern reef system in *red*; Western Leeward is *yellow*; Eastern Windward is *purple*. Volcanic zones shown in *black*; and southern reef system area is *green* (From Google Earth)

Eastern Windward Reef System

The reefs here are consistently subject to heavy wave action. Exploration of this region has been limited, although visits have found eroded rock substrate forming overhangs and 'swim throughs'. Corals are affected and stressed by frequent heavy sedimentation from volcanic runoff.

Southern Reef System

This zone is also only visited rarely because of sea conditions. The reefs are found close to shore and quickly drop to considerable depths. Visits have shown healthy corals and abundant fish populations. Though flanked by two volcanic plains, water currents direct sediment away from these reefs.

Volcanic Flanks of the Volcano: East and West Coast

Exploration of this area is often unsafe because of volcanic activity. Reefs experience heavy sedimentation and frequent burial.

Reef Species of Note

Pillar Coral. Montserrat has several large and healthy pillar coral colonies (*Dendrogyra cylindrus*) as well as numerous smaller colonies (Fig. 8.6). This species is listed as 'vulnerable' by IUCN. Elkhorn coral (*Acropora palmata*) also exists, and is listed as a critically endangered species. Throughout the Caribbean region huge tracts of Elkhorn have been lost to white band disease and tropical storm conditions, and what was once a very abundant coral in Montserrat is now rarely found healthy. Though white band disease has devastated most of the island's Elkhorn, Montserrat still has scattered healthy colonies within the northern and north-western region of the island (Fig. 8.7). As with many regions, Montserrat has abundant populations of barrel sponges, but whereas these are in some areas common mostly deeper than 15 m (Humann and Deloach 2002), in Montserrat they are common on reefs less than 8 m deep.

Invasive Species

The Lionfish is an Indo-Pacific predatory fish that was first reported within the region (Florida) in the early 1990s. Since that time sightings have been reported throughout the whole region. The lionfish is an aggressive predator of juvenile fish and a prolific breeder. Coupled with no natural predators the population quickly explodes once established (USGS/NAS website). The first sightings occurred in Montserrat the summer of 2011, and by 2012, multiple lionfish can be found on every dive.

The Orange Cup Coral is another Indo-Pacific introduction which has established itself within the region since the 1940s. It is believed to be the only stony coral introduced within the Caribbean/Western Atlantic. It is found abundantly on several shallow reefs, often in shaded areas, around Montserrat (Humann and Deloach 2002).

Coral Reef Stresses

Montserrat's reefs are experiencing many of the same stresses that found throughout the region, but in addition has the unique one (amongst the Overseas Territories) of volcanic activity, which, over the past 15 years has been destroying and damaging the island's marine ecosystem. The processes which created the island have buried extensive tracts of the inshore reefs around the lower half of the island. Heavy sedimentation, not only occurs during active volcanic dome growth and collapse but also occurs during non-active times through erosional runoff, and this sedimentation smothers

Fig. 8.6 A large healthy colony of pillar coral located off the northwest coast of Montserrat. Pillar coral is considered a vulnerable species (© Kim Guinn)

Fig. 8.7 A healthy specimen of critically endangered Elkhorn Coral in tin 3 m of water. This coral will be affected by planned development (© Andrew Myers)

Fig. 8.8 Green turtle killed by the February 17th 2010 partial dome collapse. Volcanic dome collapse can create pyroclastic flows across the surface for several kilometres offshore depositing super heated rocks and ash (©Henry Odbert)

beaches and its life including nesting turtles (Fig. 8.8) and offshore it buries corals.

Coral Diseases are also prevalent, as they are in much of the Caribbean, and extensive areas of dead Elkhorn coral can be found on the northwest coast. While healthy colonies exist, others exhibit White-band Disease (WBD) which, with other White Syndromes, has killed a majority of Elkhorn and Staghorn colonies within the region. Other diseases that most likely are affecting Montserrat's reefs are black-band, red-band, and yellow blotch diseases (reported sightings though not confirmed) (Humann and Deloach 2002).

Collateral damage from the volcano is occurring due to redevelopment in the northern end of the island as the nation rebuilds its lost infrastructure. With the loss of its capital Plymouth, both the port facility and most commercial and many residential buildings were lost. Currently there is a plan to build a new port facility (the second port development since the volcano) and a breakwater which will result in the loss of habitat.

Antiquated fishing practices, and un-regulated fishing and over fishing are prevalent. Montserrat fishermen use handmade fish traps or pots, gill nets and seine nets to catch most of the fish that are landed. Fish pots are often poorly placed on the reefs, left unchecked because of sea conditions, or become lost when the marker buoys get cut. This results in damage not only to the corals but also causes loss of fish that die due to these lost "ghost" pots that continue to kill. Because of volcanic activity and the change in the island's population base, accessible fishing areas have been reduced and overfishing in some areas is occurring (Fig. 8.9).

There are few fisheries laws and currently there is no enforcement of those that exist. According to the Caribbean Regional Fishing Mechanism (2006) the island does not export any fish and has 60 fishermen at this time (www.caricom-fisheries.com/members/montserrat.asp).

Marine Protective and Damage Mitigation Measures

Montserrat does not have marine protected areas, though discussions into the possibilities are being conducted. Currently a program known as the Montserrat Reef Project (MRP) is creating new reef habitat through the installation of designed artificial reefs. The project also identifies imperilled corals for propagation to the new reef system. The MRP is a grant funded project and is currently finishing the second phase of reef creation. The project has created over 240 reef structures known as Reef Balls since late 2010 whose intent is to generate new areas of hard substrate and bottom relief.

In summary, the coral reefs of Montserrat provide rich habitat for hundreds of reef species and typically support healthy and diverse corals and abundant juvenile fish, despite the elimination of substantial areas from volcanic action in the last 18 years. As with many of the other islands within the region, the reefs have suffered from various coral diseases and from damage caused by fishing and development issues. All will need to be addressed, and Montserrat with its small population has a great opportunity to make the necessary changes to optimise the condition of its reefs and the benefits they may accrue to the small island.

Fig. 8.9 The use of fish pots damage the reef system by poor placement, lack of recovery by the fishermen, and indiscriminately killing when lost. This fish pot is on the reef and filled with various reef fish; northwest Montserrat (© Andrew Myers)

References

Caribbean Regional Fisheries Mechanism (2006). www.caricom-fisheries.com/members/montserrat.asp.

Government of Montserrat. www.gov.ms

Humann P, Deloach N (2002) Reef set: Florida, Caribbean, Bahamas. New World Publications, Inc.IUCN (International Union for Conservation of Nature) Red List. www.iucnredlist.org

Jones M, Hembury D et al (2010) The weathering and element fluxes from active volcanoes to the oceans: a Montserrat case study. Bull Vulcanology 73:207–222

Le Friant A, Harford CL, Deplus C et al (2004) Geomorphological evolution of Montserrat (West Indies): importance of flank collapse and erosional processes. J Geol Soc 161:150–151

Le Friant A, Deplus C, Boudon G et al (2009) Submarine deposition of volcaniclastic material from the 1995–2005 eruptions of Soufriere Hills volcano, Montserrat. J Geol Soc, vol 166. Montserrat Volcano Observatory, London. www.mvo.ms

Trofimovs J, Amy L, Boudon G, et al (2006) Submarine pyroclastic deposits formed at the Soufriere Hills volcano, Montserrat (1995–2003): what happens when pyroclastic flows enter the ocean? Geology, p 550. United States Geological Survey, Non-indigenous Aquatic Species. www.nas.er.usgs.gov/default.aspx

The Reefs of the Turks and Caicos Islands

9

Alan Logan and Kathleen Sullivan Sealey

Introduction

The British Overseas Territory of the Turks and Caicos Islands lies at the southern end of the Bahamian archipelago in the tropical north-west Atlantic between 21° and 22°N and 71° and 72° 30″ W (Fig. 9.1) The territory itself is made up of three separate carbonate bank systems: the Caicos Bank, the Turks Island Bank and the Mouchoir Bank. The larger Bahamian archipelago includes territories of three countries: The Bahamas, the Turks and Caicos Islands and the Dominican Republic. All of the land areas in the archipelago are part of the Bahamas or the Turks and Caicos Islands, but the Dominican Republic claims the submerged coral reefs of the Silver and Navidad Banks at the extreme southeastern extent of the chain (Sealey 2006)

The Turks and Caicos Islands (TCI), along with the rest of the Bahamian archipelago, are characterized by clear, shallow waters overlying white carbonate sands. The banks cover about 90%, with the exposed landforms (islands) comprising only about 10%, of the area of the banks (Table 9.1). Each bank system is characterized by low-lying islands that are the result of high carbonate production, cycles of low and high sea level, and prevailing winds. The largest bank, Caicos Bank, has six major islands (West Caicos, Providenciales, North Caicos, Middle Caicos, East Caicos and South Caicos) arranged sequentially in an arc along the northern platform margin (Fig. 9.2). The smaller Turks Bank has two islands (Grand Turk and Salt Cay) situated on the western side of a much smaller platform. The two shallow carbonate banks are separated by a deep passage, the 35 km-wide Turks Island Passage, which reaches depths of 2,200 m, while the Caicos Bank is separated from Mayaguana Island and the southern Bahamas islands of Great and Little Inagua by the 70 km-wide Caicos Passage with maximum depths approaching 4,400 m. The Turks Islands are separated from the Dominican Republic and Haiti to the south-east by the Mouchoir and Silver Bank Passages. The margins of these banks are defined by sharp drop-offs into deep water on all sides and are fringed by coral reefs, particularly on the seaward side of the Caicos Islands. The coral habitat includes two major groups: true coral-dominated accreting reefs and hard-bottom non-reefal habitats that contain corals but are not accreting coral reefs.

In this chapter we give a general description of the carbonate bank geology, reef geomorphology and zonation, known biotic communities, coral reef fish and fisheries, marine parks, and assessments of reefs as related to anthropogenic and other threats. TCI is the least prosperous territory, and has the smallest population compared to other western Atlantic British Overseas Territories (Anguilla, Bermuda, Cayman Islands, Montserrat, and British Virgin Islands), but has the largest shallow water marine resources, including reefs. The chapter concludes with a short overview of the governance and regulatory structure that impact reef management within this island microstate.

Some of the first overviews of the coral reefs of the Turks and Caicos Islands were given by Wells (1988) and Koltes (1993), the latter based on one of the first coral reef research field stations established on Grand Turk. Since then there has been a paucity of synoptic quantitative studies on the reefs, as much of the area is inaccessible except by boat. Quantitative studies have been carried out in association with fisheries assessments, regional assessment of reefs or placement of marine parks (see Medley and Ninnes 1999; Gaudian and Medley 2000; Tupper and Rudd 2002; Dikou et al. 2009). Much of the field research has been carried out on the south-eastern side of South Caicos and nearby Long Cay at the School for Field Studies station on South Caicos (Sullivan et al. 1994, 1996; Chiappone et al. 1996; Steiner 1999; Dikou et al. 2009). Research has also been carried out on

A. Logan (✉)
Centre for Coastal Studies, University of New Brunswick, 5050, Saint John E2L 4L5, NB, Canada
e-mail: logan@unbsj.ca

K.S. Sealey
Department of Biology, University of Miami, Coral Gables, FL, USA
e-mail: ksealey56@gmail.com

C.R.C. Sheppard (ed.), *Coral Reefs of the United Kingdom Overseas Territories*, Coral Reefs of the World 4,
DOI 10.1007/978-94-007-5965-7_9, © Springer Science+Business Media Dordrecht 2013

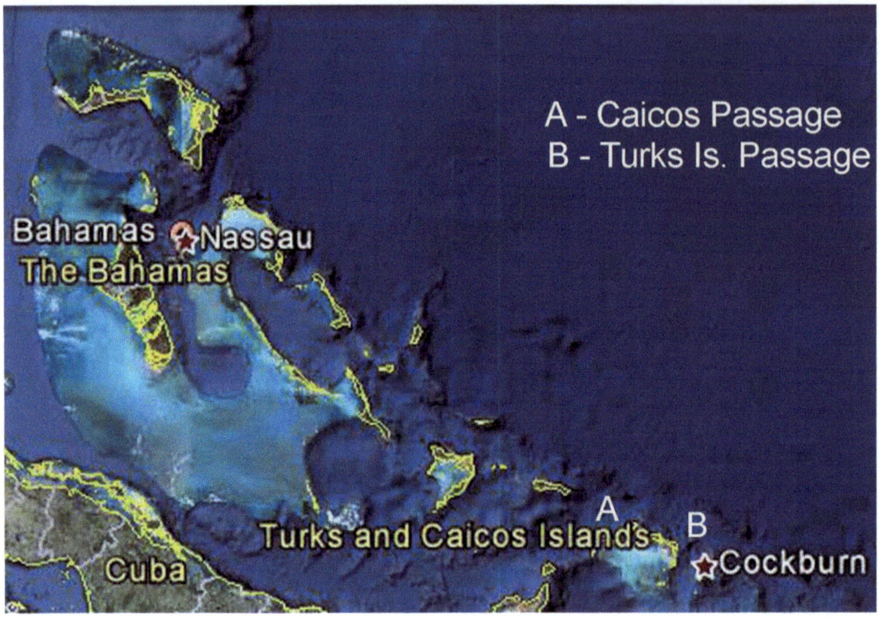

Fig. 9.1 Map of the Bahamian Archipelago showing islands, banks and deeps, with TCI and major passages shown at southern end of chain (Google Earth)

Table 9.1 Bank and Island Areas and Perimeters of the Turks and Caicos Islands: Area of the shallow banks and island areas are given in square kilometers. Bank perimeters and island shorelines are given in kilometers. The total shallow water bank area includes from the shoreline to the 200-m bathymetric contour. Land areas include the area of coastal mangroves, but not large creeks and bights

Bank type	Bank name	Bank area (sq. km)	Bank perimeter (km)	Island area (sq. km)	Shoreline (km)
Sheltered bank with Cays	***Caicos***	***6,856***	***375***	***489***	***568***
Total of all sheltered banks		***60,313***	***2,409***	***9,017***	***5,099***
Full-exposed banks	Silver Banks	2,833	226	0	0
	Mouchoir	***958***	***149***	***0***	***0***
	Turks Islands	***607***	***137***	***22.7***	***41.2***
	Navidad	434	83	0	0
Total of all exposed banks		***44,611***	***1,783***	***61.9***	***188.3***
Totals for the entire archipelago	***134,447***		***12,972***		

commercial SCUBA diving boats that operate around the banks, both as live-aboard and shore-based dive tours (Spotte et al. 1994; Schelten et al. 2006).

Geography and Geomorphology of Reefs

As previously mentioned the Turks and Caicos Islands form the southern extent of the Bahamas Archipelago which stretches from the Turks Islands north-west to Little Bahama Bank. The islands and banks of the islands are geographically, geologically and ecologically part of this single archipelago (Enos 2011), with Pleistocene carbonates, often in the form of oolites and eolianites, providing the foundation for Holocene reefs and sediments. Rankey et al. (2009) studied Holocene and Pleistocene shallow marine carbonates from Providenciales in the northwestern part of the Caicos Platform, using sedimentology, bottom observations, remote sensing and sub-bottom profiling. They conclude that such northeast-facing platform margins exhibit considerable facies variability and show the best developed reefs where wave-dominated conditions from the open Atlantic prevail. Furthermore, buried top-Pleistocene bedrock configuration appears to exert a strong control on the Holocene morphology of the reef complex in the study region. Differences between the carbonate bank systems are perhaps the most important aspect in describing the reef ecology of the archipelago. Classification of the carbonate bank environment is based on geomorphology, energy exposure and bank size, with a strong latitudinal gradient. The areas of banks and

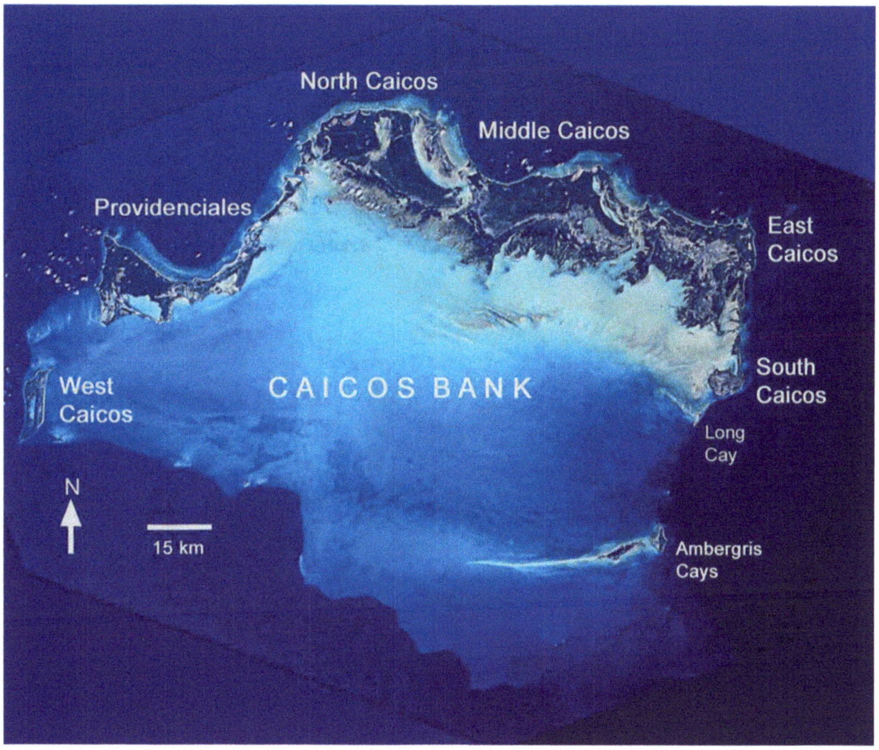

Fig. 9.2 NASA Space Shuttle image STS050-82-98, 1992 (modified) of Caicos Bank with major islands identified. Turks Bank and islands not shown

islands included in the territory are shown in Table 9.1. The Turks and Caicos platforms form a small part at the southern extent of a much larger geological and ecological system. The entire archipelago experiences a tropical dry climate, with the Turks and Caicos as the hotter and drier southern extreme. The archipelago stretches over 7° in latitude (1,270 km) and extends north-westwards from the tropical dry islands of the Turks and Caicos Islands to the subtropical Abacos, all influenced by frontal systems from North America. Bank systems, and their associated islands, are the fundamental components of biodiversity in the archipelago. There is roughly a 10:1 ratio of marine bank to terrestrial island areas in the archipelago (Table 9.1). The bank perimeter includes only platform margin length. The platform margin is characterized as the area of barrier and fringing reef growth, upwelling and sediment transport events critical in marine faunal distributions. The length of shoreline can be much larger than bank perimeter length due to convolutions and embayments on many of the islands. Near-shore reef features include patch reefs, non-reefal hard bottom and fringing reefs. The dry, tropical climate of the archipelago means there are no surface water resources, and no river run-off or sedimentation, resulting in extremely clear water, facilitating coral growth near the shore.

The sheltered banks, such as Caicos Bank, are dominated by long, often narrow, islands that stretch along the northern and eastern platform margins. The islands separate high energy, wind-blown environments along their eastern shore from protected coastal wetlands and beaches along the western shores. A typical island thus provides a barrier to wave energy, and creates extensive soft sediment habitats in its lee to the west. Caicos Bank is characterized by well-developed exposed reefs along the eastern platform margin (Fig. 9.3). The fully-exposed banks, like Turks and Mouchoir Banks, lack islands along the eastern platform margin, and are characterized by deep (mesophotic) reefs, patch reefs, and fringing reefs along very low-lying islands. The large tropical shallow-water marine environments of the Turks and Caicos banks support a wide variety of reefal systems, both true accreting coral reefs and non-reef coral habitats. The geomorphic classification of these reefal areas used here generally follows the published classification schemes of Mumby and Harborne (1999) and Andrefouet (2008).

The prevailing trade winds are from the east and north-east, resulting in rough seas on the east side of Caicos Bank but generally calm seas on the west side (Wells 1988). Reefs fringe the platform on the windward side of its eastern and north-eastern islands and on the leeward side of its northern and western islands (Steiner 1999). Hurricanes occasionally visit the islands from June to November (Gaudian and Medley 2000). Air temperatures average 25 °C in winter

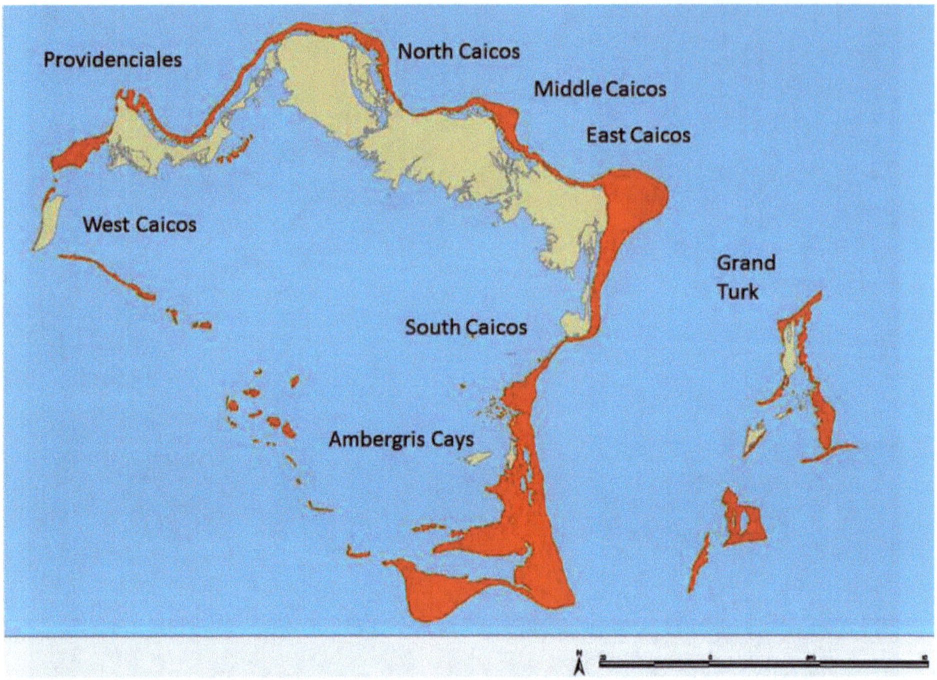

Fig. 9.3 Distribution of reefs on the Turks and Caicos banks (From REEFbase GIS 2012)

and 29 °C in summer, with concomitant surface water temperatures ranging from 22 °C in the winter to 28 °C in the summer. The islands of the Turks and Caicos are at the eastern edge of the wider Caribbean Sea, and experience unequal, semi-diurnal microtides. The tides are less than 1.5 m in amplitude, and there are two high tides and two low tides each day. The semi-diurnal tides create strong currents through channels and move water on and off the carbonate platforms. Salinity and temperature variability on the banks are thus impacted by radiative heating and evaporation (Buchan 2000).

The dry, tropical climate was historically (and continues to be today) attractive for solar evaporative production of salt. Perhaps the greatest ecosystem-wide changes to the islands occurred with the establishment of salt ponds on South Caicos, Grand Turk and Salt Cay in the eighteenth century with the salt trade (see the Turks and Caicos National Museum website for overview). In fact, historical uses of marine resources and land have likely led to significant changes on the coral reefs, starting with the occupation of Tiano Amerindians beginning some 700 years prior to the arrival of Columbus (Keegan et al. 2008; Sinelli 2010). Seals, turtles, seabirds and near-shore fauna were removed from areas of human occupation. Europeans followed with the deforestation and removal of mangrove wetlands on the small islands to facilitate solar salt production which became the basis of the colonial economy.

Coral Habitats and Reef Topography

Islands tend to occur at the platform margin, a narrow fore-reef shelf gradually sloping to about 20 m, with a steeper drop-off to 200 m depth. This is particularly well shown in the crescent of islands from West Caicos to East Caicos. The 20 m isobath approximates the transition from shallow to deep water, where slopes steepen to 45° or more at the edge of the platform margin. The shelf averages about 400 m in width but increases to over 9,000 m off the southeastern margin of the Caicos Bank. The narrow shelf, protected by islands, supports the true accreting reef communities in the Turks and Caicos banks. Coral reefs occur as patch reefs, fringing reefs, and bank-barrier reef systems most extensively on the larger Caicos Bank, but also on the Turks Bank.

The platform margin is characterized by two gently-sloping terraces. The shallow (or upper) terrace slopes gradually from either the shoreline or fringing reef to about 8–10 m depth, where there is an appreciable increase in slope to about 15 m depth coinciding with the seaward edge of spur-and-groove development, where present (Steiner 1999; Rankey and Reeder 2010). The deep (or lower) terrace has a depth of 15–20 m and consists of a narrow sand plain with isolated patch reefs or low-relief spur-and groove reefs, beyond which is a steep drop-off into deep water. These topographic zones, which will be described in more detail later, have some similarity to those seen in the Cayman Islands

(Logan 1994; Roberts 1994), Belize (Rützler and Macintyre 1982) and Roatan (Fenner 1993).

Throughout the Bahamian archipelago the transformation (diagenesis) of soft sediments into hard bottom or rock is the norm. The marine cementation of sedimentary material is usually the result of subsea cementation of aragonite around sands composed of ooliths or shell fragments. In areas where fresh groundwater joins the sea, a process of precipitation of calcium carbonate around these sand grains results in beach rock. This cementation process is rapid in this dry tropical climate (Rankey and Reeder 2010; Whitaker and Smart 1997). Sediments created by calcified algae found on the banks, especially *Halimeda*, *Penicillus*, *Udotea*, and *Rhipocephalus*, are cemented together through their root systems. This process is enhanced by the presence of sponges and corals to make for a hard bottom community. "Hard bar" or hard bottom communities are not true accreting coral reefs, but are important throughout the carbonate bank system.

Coral Reef Communities

There are six ecotypes comprising true (accreting) coral reefs: bank patch reefs; near-shore patch reefs; channel reefs; near-shore fringing reefs, platform margin bank-barrier reefs and platform margin deep reefs. Combinations of these types produce different reef zonation patterns from shallow to deep water. The wide range of hard and soft coral species in TCI reefs is shown in Table 9.2.

Bank Patch Reefs

Bank patch reefs are one of two types of patch reefs found in the Bahamian Archipelago and are common in leeward and lagoon environments (Alevizon et al. 1985; Sullivan et al. 1994). These patch reefs are distributed on the banks some distance from islands, and exposed to wave and wind energy of the open bank. Many bank patch reefs are found within 2 km of a margin, and are common along the exposed southeastern platform margin of Caicos Bank (Fig. 9.4). In addition to distance from any landmass, bank patch reefs differ from near-shore patch reefs in that patches tend to be clustered, rather than isolated circular patches. Massive head corals provide the framework of the bank patch reef and there can be great variability in the contributions of algae, sponges, gorgonians, and hard corals to the patch reef surface. Roughly circular in shape, patch reefs south of Ambergris Cay can be moderate (1,000 m^2) to large (+10,000 m^2) in size. Vertical relief is as high as 10 m and is due to large coral colonies, including acroporids (Fig. 9.5). Large patch reefs are important

Table 9.2 Species list of Cnidaria species recorded from reef communities in the Turks and Caicos Islands, compiled from Sullivan et al. (1994, 1996), Steiner (1999) and School for Field Studies, South Caicos, internal reports (http://www.fieldstudies.org/tci)

Order Actiniaria	
Actiniidae	*Condylactis gigantea*
Aiptasiidae	*Bartholomea annulata*
Aliciidae	*Lebrunia coralligens*
Aliciidae	*Lebrunia danae*
Aurelianidae	*Actinoporus elegans*
Corallimorphidae	*Ricordea florida*
Phymanthidae	*Phymanthus crucifer*
Stichodactylidae	*Stichodactyla helianthus*
Zoanthidae	*Palythoa caribaeorum*
Zoanthidae	*Zoanthus sociatus*
Order Gorgonacea	
Anthothelidae	*Erthropodium caribaeorum*
Briareidae	*Briareum asbestinum*
Gorgoniidae	*Gorgonia sp.*
Gorgoniidae	*Pseudopterogorgia americana*
Gorgoniidae	*Pseudopterogorgia spp.*
Gorgoniidae	*Pterogorgia anceps*
Gorgoniidae	*Pterogorgia citrina*
Gorgoniidae	*Pterogorgia guadalupensis*
Plexauridae	*Eunicea calyculata*
Plexauridae	*Eunicea mammosa*
Plexauridae	*Plexaura homomalla*
Plexauridae	*Plexaurella flexuosa*
Plexauridae	*Plexaurella spp*
Plexauridae	*Pseudoplexaura sp.*
Order Scleractinia	
Acroporidae	*Acropora cervicornis*
Acroporidae	*Acropora palmata*
Agariciidae	*Agaricia agaricites*
Agariciidae	*Agaricia tenuifolia*
Agariciidae	*Agaricia humilis*
Agariciidae	*Agaricia fragilis*
Agariciidae	*Agaricia grahamae/lamarcki*
Agariciidae	*Leptoseris cucullata*
Astrocoeniidae	*Stephanocoenia intersepta*
Caryophylliidae	*Eusmilia fastigiata*
Faviidae	*Colpophyllia natans*
Faviidae	*Diploria clivosa*
Faviidae	*Diploria labyrinthiformis*
Faviidae	*Diploria strigosa*
Faviidae	*Favia fragum*
Faviidae	*Manicina areolata*
Faviidae	*Montastraea annularis*
Faviidae	*Montastraea cavernosa*
Faviidae	*Montastraea faveolata*
Faviidae	*Montastraea franksi*
Meandriniidae	*Dendrogyra cylindrus*
Meandriniidae	*Dichocoenia stokesi*

(continued)

Table 9.2 (continued)

Meandriniidae	*Meandrina meandrites*
Milleporidae	*Millepora alcicornis*
Milleporidae	*Millepora complanata*
Mussidae	*Isophyllastrea rigida*
Mussidae	*Isophyllia sinuosa*
Mussidae	*Mussa angulosa*
Mussidae	*Mycetophyllia danaana*
Mussidae	*Mycetophyllia ferox*
Mussidae	*Mycetophyllia lamarckiana*
Mussidae	*Scolymia cubensis*
Pocilloporidae	*Madracis decactis*
Pocilloporidae	*Madracis formosa*
Pocilloporidae	*Madracis mirabilis* (now *M. auretenra*)
Poritidae	*Porites astreoides*
Poritidae	*Porites bournei*
Poritidae	*Porites colonensis*
Poritidae	*Porites divaricata*
Poritidae	*Porites furcata*
Poritidae	*Porites branneri*
Poritidae	*Porites porites*
Siderastreidae	*Siderastrea radians*
Siderastreidae	*Siderastrea siderea*
Stylasteridae	*Stylaster roseus*

fisheries habitats, especially for spiny lobster. It is possible that large deep patch reefs may also provide spawning aggregation sites for coral reef fishes.

Near-Shore Patch Reefs

Near-shore patch reefs are smaller, but more abundant along the bank or leeward side of islands of the Turks and Caicos Islands, occurring at 1–6 m depths (Chiappone et al. 1996). This patch reef type is distinguished from bank patch reefs primarily by its proximity to shore (<1 km). In addition, near-shore patch reefs tend to be adjacent to seagrass beds and are usually isolated and smaller than the clusters of bank patch reefs found further on the banks. Near-shore patch reefs are typically small, averaging 20–30 m in diameter, and roughly circular in shape, but may be quite variable in size (Chiappone et al. 1996). Near-shore patch reefs are constructed by massive frame-building corals but can exhibit substantial variability in the relative abundance patterns of algae, corals, sponges, and gorgonians (Sullivan and Chiappone 1992). Maximum vertical relief ranges from 1 to 1.5 m and is due to living or dead massive corals that comprise the structure of the patch reef (Fig. 9.6). Near-shore patch reefs are common in lagoonal areas near islands in protected environments with offshore fringing or bank barrier reefs in shallow waters, for example 2–5 km wide off Providenciales (Rankey et al. 2009). These lagoons have limestone floors with a thin veneer of mostly biogenic carbonate sediments, each area of which supports its own distinctive benthic biotic community. Sediment grain sizes vary from fine sands to coarse boulders, with sands the dominant sediment type. Coarse-grained deposits are often transported into the lagoon from the rubble zone of the fringing reef during storms. Sands are often colonized by marine grasses and algae, while patch reefs of various shapes and sizes occur in the lagoon, some being elongate reefal ridges with their long axis oriented normal to the shoreline. In between, relatively barren rippled or bioturbated sands cover large areas of the lagoons and are colonized by algae.

Channel Reefs

Channel reefs are prevalent in the Bahamian Archipelago, and, especially in the Turks and Caicos Islands, along the northern Caicos Bank between cays. Channels serve as major conduits between deep water and bank water. Essentially four bottom types can occur in channels (sand, seagrass, hard-bottom, or reef) and community composition is dependent upon the length, width, and depth of the channel (Sullivan et al. 1994; Sluka et al. 1996). Channels with coral reefs tend to be wider and deeper, and are dominated by massive coral species, often arranged in finger-like ridges parallel to the channel The sizes of channel reefs can vary substantially (<1 to >3 ha), but many are small. Major factors associated with the lack of substantial reef development in such systems are attributed to turbidity, sediment transport, and extreme fluctuations in water temperatures (Lang et al. 1988). Channel reefs can be important locations for spawning of fisheries target species such as the Queen conch *Strombus gigas* (Hesse 1979).

Near-Shore Fringing Reefs

Fringing reefs are the dominant platform margin reef type in the region. The presence of a fringing reef near sea level and spur-and-groove structure on the deeper reefs, both of which occur in TCI, comprises a reef complex that modifies the effects of waves, reducing their energy and providing relatively quiet-water conditions in the lee of the reefs, where lagoon and shore communities can develop. Fringing reefs are represented by three structural types: (1) those occurring immediately offshore on an island platform, (2) those that form ridges parallel to shore, and (3) those, both shallow (<5 m) and deeper (>10 m) with spur-and-groove development on the seaward side (Zankel and Schroeder 1972; Sullivan et al. 1994). Fringing reefs are often based on outcrops of the island platform, prevalent across bays offshore

Fig. 9.4 A typical deep bank patch reef on sloping platform margin of Caicos bank. Rounded heads of *Montastrea annularis* at *left*, large colonies of *Agaricia* at *right*, branching gorgonians and coiled sponges in between. Depth about 12 m (Photo by Cindy Lott)

Fig. 9.5 *Acropora palmata* (*large*) and *A. cervicornis* (*small*) growing on the top of a large bank patch reef south of Ambergris Cays, depth about 6 m (Photo by Sullivan Sealey)

off South Caicos and Providenciales. Spur-and-groove or buttress reefs consist of elongate coralline spurs or coral bars oriented perpendicular to shore (Shinn 2011). Few spur-and-groove reefs are reported from the TCI banks that are directly exposed to the Atlantic Ocean (Bunt et al. 1981). Spurs or coralline fingers are greater than 100 m in length in some reefs, with the spur surfaces typically found in 8 m to 16 m depth, or sometimes shallower. Spurs are separated by sand grooves from 13 to 18+ m depth. The deeper spur-and-groove sites extend to the fore reef escarpment, or drop-off zone, at

Fig. 9.6 Near-shore patch reef off South Caicos showing (*centre*) colonies of *Montastrea cavernosa* (*above*) and *Agaricia* sp. (*below*) with encrusting *Millepora* at *right*, yellow sponges, branching gorgonians in background, depth about 6 m (Photo by Cindy Lott)

20+ m depth. At several locations in the archipelago, spur-and-groove topography occurs on reef terraces, ranging from wide, gently sloping surroundings to narrow and steep (Zankel and Schroeder 1972; Bunt et al. 1981; Steiner 1999). The eastern Caicos Bank fringing reefs are characterized by spur-and-groove formations dominated by massive head corals (Sullivan et al. 1996; Chiappone et al. 1996). These spur-and-groove features extend towards the platform margin, and are characteristic of the southeastern margin that is not a precipitous drop-off (Fig. 9.7)

The active fringing reef, easily visible from the air, is the reef crest, the shallowest part of the reef complex. This linear wave-resistant reef, where present, separates the protected lagoonal area from the open ocean and is built along the apex of two surfaces sloping in opposite directions, one shorewards, the other seawards. It is well developed along the shoreline or across bays and lagoons such as the small bay east of Cockburn Harbour, South Caicos (Fig. 9.8). The reef crest lies less than 2 km distant from the shoreline off Northwest Point in Providenciales and comprises a mix of corals such as species of *Acropora, Montastrea, Diploria* and *Porites*, as well as the hydrozoan *Millepora* and a variety of red and green algae (Rankey et al. 2009). At low tide the top of the reef crest is at sea level or slightly emergent and waves break over it. It comprises mainly an *Acropora-Millepora* thicket, the reef crest zone in the islands spanning the *Palythoa-Millepora* and *Diploria strigosa-Acropora palmata* zones described for Caribbean reefs by Geister (2011) and typical of moderate wave energy conditions.

Platform Margin Bank-Barrier Reefs

Barrier reefs are found throughout the Bahamian Archipelago, and have historically presented the greatest challenge to exploration by sailing vessels in the past. Although not at far from shore or as extensive as the barrier reefs off Belize and Mexico, the bank barrier reefs occur as a line or "ribbon" along the northern platform margin of the Caicos Bank 3 km or more off shore (Fig. 9.9). Structurally, these reefs exhibit a similar biological community composition to fringing reefs such as in the reef crest or breaker zone. Barrier reefs at the platform margin may also have a spur-and-groove topography offshore (Fig. 9.10). While the distinction from fringing reefs is not always clear, barrier reefs differ from fringing reefs in their greater distance from shore and the presence of an extensive back reef environment separating the shoreline from the reef (e.g. Belize). Barrier reefs occur along the northern and eastern platform margin of the Caicos Bank, especially across embayments. Although the "fringing" reef acts as a protective barrier to inshore waters it is discontinuous and there are occasional gaps (cuts) that allow interchange of waters between the lagoon and the open sea. The dominant coral species on the reef crest, as elsewhere in the northern Caribbean, are *Acropora palmata* and *Millepora complanata,* the latter occurring in both bladed and encrusting growth forms. Gorgonians are common as subsidiary forms. *A. palmata* occurs as large arborescent colonies strong enough to withstand the vigorous surf, the upwardly inclined fronds preferentially oriented towards the open ocean. Like its congener *A. cervicornis*, this species has suffered a marked decline in recent years in many parts of the Caribbean (Lewis 1984) and the TCI is no exception. However, Gaudian and Medley (2000, p. 590) report that "large healthy stands can still be found at Salt Cay, and the exposed sides of Grand Turk , South Caicos and some parts of the Caicos Bank." Schelten et al. (2006) also noted that *A. palmata* populations are still in good condition off the coast of South Caicos. Where it is rare or absent its role is taken over by the hydrozoan *M. complanata* which forms a hedge of vertically-inclined, upwardly-flaring blades facing the direction of wave advance. Other common forms are species of *Diploria, Montastrea* and sponges, while pillar-like colonies of the

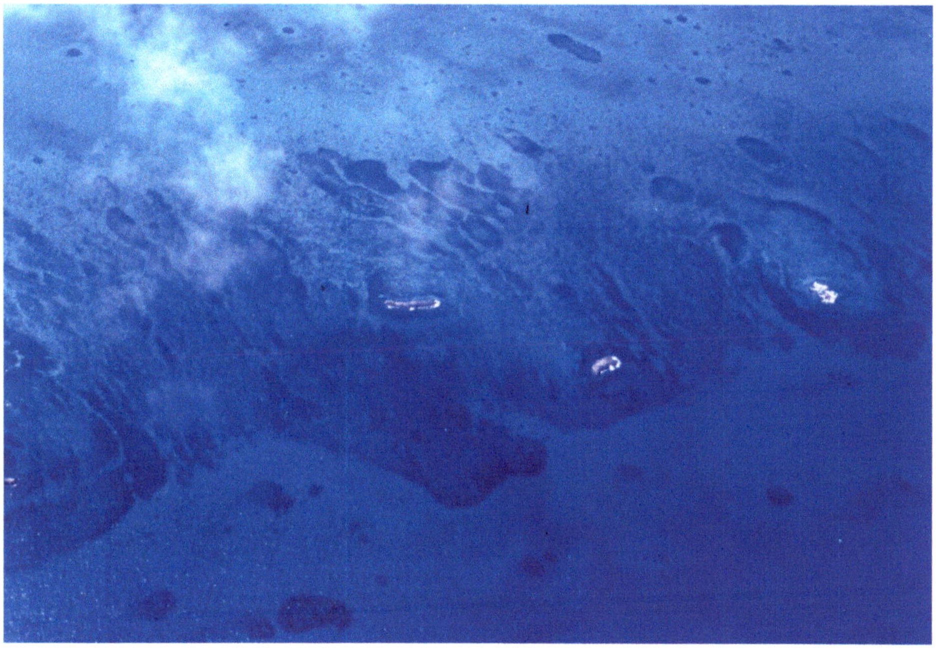

Fig. 9.7 Platform margin bank-barrier reefs along the southern Caicos Banks north of Ambergris Cays, showing spur-and-groove structure (Photo by Sullivan Sealey)

meandrenid coral *Dendrogyra cylindrus* are sometimes present (Fig. 9.11).

Platform Margin Deep Reefs

Wind and wave energies have predominantly influenced the production of the bank system and island formation. The interface of the archipelago and the ocean energies are, in many ways interrelated and represent a defining characteristic of the island system. The lack of surface water on the small islands contributes to very clear water, and deep light penetration energy, supporting coral growth to over 80 m. Such mesophotic coral habitats are characterized by the presence of light-dependent corals and associated communities at water depths greater than normal along the platform margins of TCI. The dominant communities providing a structural habitat in this depth zone are made up mainly of coral, sponge, and algal species. These deep reefs thus provide a transition area from the shallow bank systems to the open seas of the tropical Western Atlantic.

A sand plain at the base of the shallow spur-and-groove zone is very narrow (less than 500 m) along the windward (east) side of the Caicos Bank and terminates at the edge of the fore-reef escarpment between 15 and 30 m depth. (Chiappone et al. 1996). Typically there is a slight elevation before the edge of the escarpment is reached. This escarpment is a steep and, in some places, near-vertical wall (Fig. 9.12) extending to at least 85 m, the approximate limit of coral growth in the region. Narrow grooves form conduits for sediment transport downslope. The escarpment has fewer coral species than the shallower reefs. Remnants of spur-and-groove structure can still be recognized, the spurs forming enormous overhanging buttresses where many coral colonies, in response to reduced light, adopt a characteristic platy growth form. Examples include *Montastrea annularis* but its congener *M. cavernosa* generally retains the columnar growth form seen in shallower water. Large platy colonies of species of *Agaricia* spread laterally far beyond their narrow base of attachment, eventually becoming unstable and sliding downslope. Although Chiappone et al. (1996) recorded only 16 and 18 coral species, respectively, from 2 sites on eastern Caicos Bank, Steiner (1999) recorded 26 species of corals in this deep zone, the most important of which are species of *Montastrea, Agaricia, Siderastrea, Stephanocoenia* and *Porites*. Species of *Mycetophyllia, Colpophyllia* and *Eusmilia* are common subsidiary forms. Black corals belonging to *Antipathes* are present on vertical faces in deeper water (Fig. 9.12), while demosponges occur in a variety of growth forms, mostly whip-like and tubular and include *Agelas* sp. and *Aplysina archeri* (Steiner 1999). Enormous barrel sponges belonging to *Xestospongia muta* are present, as they are throughout the deep reefs of the Caribbean.

Non-Reefal Hard Bottom Communities

There are four types of non-reefal hard-bottom communities in the region: channels with algal-dominated benthos; channels with sponge/gorgonian-dominated benthos;

Fig. 9.8 (*Upper photo*) Oblique aerial photo of small bay looking southwest along southern shore of South Caicos towards Cockburn Harbour, showing waves breaking on reef crest, lagoon with nearshore sands (*light*) and grass beds (*dark*) behind reef; Long Cay in background (Photo by Sullivan Sealey). (*Lower photo*) Google Earth aerial image of the same bay, showing the following zones: *A* back-reef sands, grass beds and rubble; *B* reef crest; *C* platform margin hard bottom; *D* high relief spur-and-groove; *E* low relief spur-and-groove; *F* platform margin deep reefs

near-shore hard bottoms and platform margin with algal-dominated benthos. These communities on consolidated carbonate sediments are important habitats for corals, and can be rugose, with ledges, cracks and crevices, but are not actively-building or accreting with massive scleractinian corals.

Channels with Algal-Dominated Benthos

Tidal channels or cuts in the banks are created and maintained by unequal semi-diurnal tides. Channels moving water on and off the platform can be floored by sand, coral reefs or non-reefal hard bottoms. Channels dominated by hard

Fig. 9.9 Google Earth aerial view of platform margin bank-barrier reef off Whitby, North Caicos, with *arrow* showing direction of wave advance under influence of prevailing winds from the east. Waves breaking on reef crest, with back-reef linear ridges of rubble and sand from storm waves

Fig. 9.10 Google Earth aerial view of platform margin bank-barrier reef west off North Creek outlet, Grand Turk, showing back-reef sands and rubble, reef crest with waves breaking, and alternating parallel zones of spur-and-groove development seawards

bottoms or consolidated sediments are often algal-dominated. The substratum is typically scoured and very low profile, consisting of exposed and lithified oolite of Pleistocene or Holocene age. Maximum vertical relief is generally less than a meter. Algal-dominated channels are often characterized by rooted *Sargassum* or other Phaeophyceae such as *Turbinaria* or *Dictyopteris*.

Channels with Sponge-Gorgonian Benthos

Channels with hard bottoms can also be sponge-gorgonian dominated. The substrate is typically scoured, and very low profile, with rubble. The consolidated sediments are exposed Pleistocene oolite. Maximum vertical relief is generally less than 1 m.

Fig. 9.11 Platform margin bank barrier reef off western Grand Turk Island with pillar coral *Dendrogyra cylindrus*, *Montastrea cavernosa* and sponges, depth about 15 m (Photo by Cindy Lott)

Fig 9.12 Deep reef wall off Grand Turk Island, with the sponge *Agelas* and large branching colonies of the black coral *Antipathes* at about 30 m depth (Photo by Cindy Lott)

Near-Shore Hard Bottom Benthos

Several natural processes of cementation, lithification, and levels of crystallization result in a hard underwater surface. Along the rocky shores of the small islands and cays, near-shore hard bottom communities provide an important community, often dominated by tolerant, fast-growing coral species such as *Diploria strigosa, D. clivosa, Porites astreoides, Stephanocoenia intersepta* and *Dichocoenia stokesi*. Mixed facies of oolite with skeletal or coral sediments are the dominant base sediments in many areas, with low-relief sponge and coral colonies indicative of the high wave energy occurring in these areas. Near-shore hard bottom communities in the Bahamian archipelago are typically regarded as a recently-submerged extension of island platforms.

Platform Margin with Algal-Dominated Benthos

The platform margin or rims of bank systems in the Bahamian Archipelago consist of a matrix of bare sand, fringing or barrier reefs, and low-relief hard bottom substrates. Low relief hard bottom is the dominant, shallow-water (<20 m) community type found on the platform margin in this region, especially on the exposed sides of banks such as the eastern Caicos banks. The substratum consists of exposed, lithified sand-rock and is not of reefal origin, as in other locations such as the Florida Keys (Wanless and Dravis 1989). Platform margin hard bottoms are the least variable of the hard bottom community types, both reefal and non-reefal, throughout the Bahamian Archipelago. This community type is consistently dominated by algae with occasional patches of sand, and is also referred to as "hard-bar" or windward hard bottom (Sluka et al. 1996). The substrate is very low profile, although occasional ledges and fissures in the substratum surface may occur. Variations in relief are due principally to the presence of isolated coral heads (<0.5 m). This is the primary fisheries habitat on the Caicos Bank (Medley and Ninnes 1999; Tupper and Rudd 2002). Steiner (1999) recorded the algae *Dictyota* and *Lobophora*, plus species belonging to 3 genera of gorgonians, and a total of 17 species of corals from off eastern Long Cay at a 9 m-depth station in this zone, the most abundant corals being *Porites astreoides*, *Siderastrea siderea* and *Dichocoenia stokesi*. Similarly the hard bottom surveyed by Sullivan et al. (1994, 1996) and Chiappone et al. (1996) off the eastern Caicos Bank, and particularly South Caicos at 6–10 m depth, lies above the zones of high-and low-relief spur-and-groove and is characterized by the dominance of algae, gorgonians, sponges and 23 species of small encrusting stony corals, although percent coral coverage is low. This relatively high-energy zone may be equivalent to the sparsely-colonized "barren rock pavement" zone described seawards of the fringing reef crest in Cayman Brac and Little Cayman by Logan (1994) and the "shallow forereef " zone in Roatan by Fenner (1993).

Reef Community Ecology and Diversity

The dominant influence on shallow marine substrates and their communities in the area is the prolific growth of reef-building corals, which, together with calcareous algae and a host of invertebrate species, help establish coral reefs. Coral reefs, consisting of a consortium of corals and calcareous algae, occur at three depth levels. Each coral habitat type displays a zonation in response to light, wave energy and currents. The reef zonation scheme off the east coast of South Caicos, as described by Sullivan et al. (1994), Chiappone et al. (1996) and Steiner (1999) includes bank and lagoonal habitats, reef crest, fore reef and deep reef developments.

Due to very specific habitat requirements (e.g. water quality and temperature), corals are inherently sensitive to environmental variability. Community composition and population size structure are heavily influenced by environmental cues (Hughes and Connell 1987; Pandolfi et al. 1999), and environmental conditions also affect individual colony attributes, such as growth rates, fecundity, and survival (Coles and Fadlallah 1991; Dizon and Yap 2006). While offshore reef tracts are considered to supply "optimal" habitat requirements, corals are found in a wide variety of seemingly highly stressful habitats near to shore and land-based sources of pollution. Corals in the TCI are well established in shallow back-reef and hard-bottom areas, as well as on outcroppings and occasional substrates in seagrass beds, sand plains and artificial structures. Some corals can tolerate temperatures up to 36–40 °C, not unusual on the large shallow bank areas of TCI. A typical mosaic of coral habitat ecotypes is shown off Long Cay in the Admiral Cockburn National Park (Fig. 9.13). The large area of shallow banks with small islands along the margin allows for a diversity of depth, current, wave energy and substrate settings for coral recruitment and growth. Although these coral habitats include many of the same species of coral, the abundance and dominance changes between habitats.

Disturbance has long been recognized as an important factor contributing to community composition (Connell 1978). Hurricanes are known to play important roles in coral reef ecosystems, such as creating new space and restoring evenness to species abundances, as well as indirectly contributing to differential recruitment and population structure (Jackson and Hughes 1985). Hurricane *Irene* passed through TCI in August 2011, preceded by Hurricane *Ike* in October of 2008. Hurricanes present a regular disturbance regime, subsequently inducing population changes for some corals, including a higher incidence of colony fission (for example, larger *D. clivosa* colonies which can be reduced to smaller colonies) and fragmentation (for example *Porites porites* propagating from fragments). The richness in community diversity and structure of TCI reefs combined with the exceptionally clear water has made this area a text-book example in coral reef and benthic habitat mapping (Mumby and Harborne 1999).

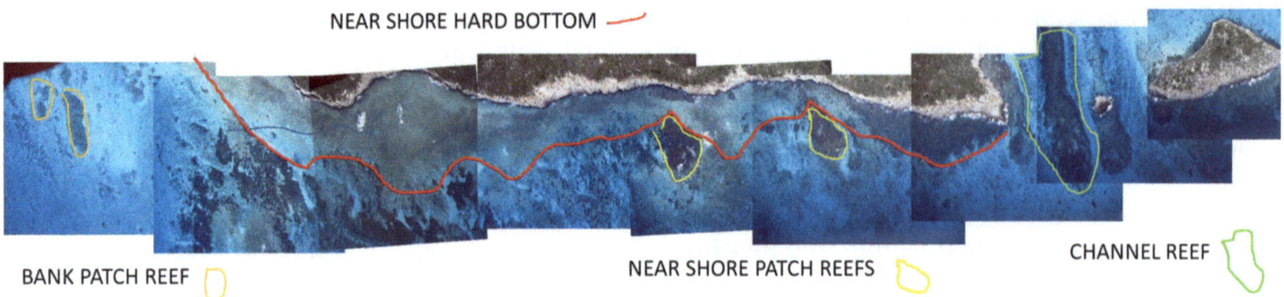

Fig. 9.13 Photo-mosaic illustrating the diversity of reef types in close proximity to Long Cay, Caicos Bank where there are near-shore hard bottoms, bank patch reefs, near-shore patch reefs, channel reefs and bank-barrier reefs to provide slightly different current and abiotic environments for corals and reef-associated fauna (Photos by Sullivan Sealey)

Coral Reef Fish and Fisheries

The Turks and Caicos Islands have historically depended and continue to depend on fisheries as high-value exports (spiny lobster and queen conch, for example) and ecotourism attractions (spear-fishing, flats fishing and deep-water sports fishing). The Division of Fisheries is part of the Ministry of Natural Resources, and is charged with the protection of marine resources to maintain not only biological diversity but also economic sustainability of fisheries through regulation and management of marine fisheries reserves (no-take zones). Fisheries regulations for the major target species: Nassau grouper (*Epinephalus striatus*), spiny lobster (*Panulirus argus*) and Queen conch (*Strombus gigas)* follow Caribbean Fisheries Council recommendations with region-wide size restrictions, gear limits and closed seasons (TCI Fisheries Limits Ordinance 1998). These species are wide-ranging; their management is coordinated with other countries and jurisdictions within the wider Caribbean to maintain sustainable stocks as part of the Caribbean Large Marine Ecosystem initiative.

South Caicos is the center of commercial fishing and fisheries exports. Spiny lobster, conch meat, finfish and even conch pearls are processed and exported from this settlement (Medley and Ninnes 1999). Fisheries officers are closely tied to the Protected Areas Division within the same ministry, and a series of marine fisheries reserves aim to protect key coral reef species from over-exploitation. These marine fisheries reserves serve as important regional research sites dedicated to maintaining healthy reefs and fish populations (Rudd 2001; Rudd and Tupper 2002; Tupper and Rudd 2002; Watson and Munro 2004).

Marine Parks and Protected Areas

Marine parks and protected areas have been implemented as part of the Turks and Caicos Islands resource strategic plan since the late 1980s (Mitchell and Barborak 1991; Zuidema et al. 2011). With the creation of the Turks and Caicos National Trust, along with the Ministry of Natural Resources' Division of Protected Areas, national parks and preserves are under co-management. The Turks and Caicos National Trust is a non-government organization that has legal authority to manage national parks, and is governed by council. The National Trust has been able to raise funds and bring public attention to coral reef conservation issues, particularly the installation of mooring buoys, vetting zoning in marine parks, and coral reef education.

The focus of marine parks has been two-fold: first, to provide protection for the natural resources of the Turks and Caicos Islands, and second, to provide management of marine areas to benefit tourism, fishing and boating. The newest marine national park, Columbus Landfall Marine National Park, was created in 2009, and covers over 428 ha (1,280 acres) from 1-m above the high water line to 92 m (300 ft) offshore along the platform margin escarpment. Other important parks include uninhabited small cays critical for seabird and turtle-nesting. The distribution of national parks and nature preserves is shown in Fig. 9.14.

Many of the National Parks, such as Princess Alexandra National Park on Providenciales, are located near resorts and tourism infrastructure. This national park, along with the new Columbus Landfall Marine Park, cover important coral resources and account for up to 90% of the recreational water sports activities for the country (with Grand Turk and Providenciales being the two main tourism islands).

Coral Reef Assessments and Threats

Regional reef assessments such as "Caribbean Reefs at Risk" (Burke et al. 2011) have long characterized the Turks and Caicos reefs as "low risk" to anthropogenic threats because of the low population density and isolation of the reefs. Assessment studies by researchers have sought to characterize reef geomorphology, obtain fisheries-independent information on target species, and assess countries for protected areas

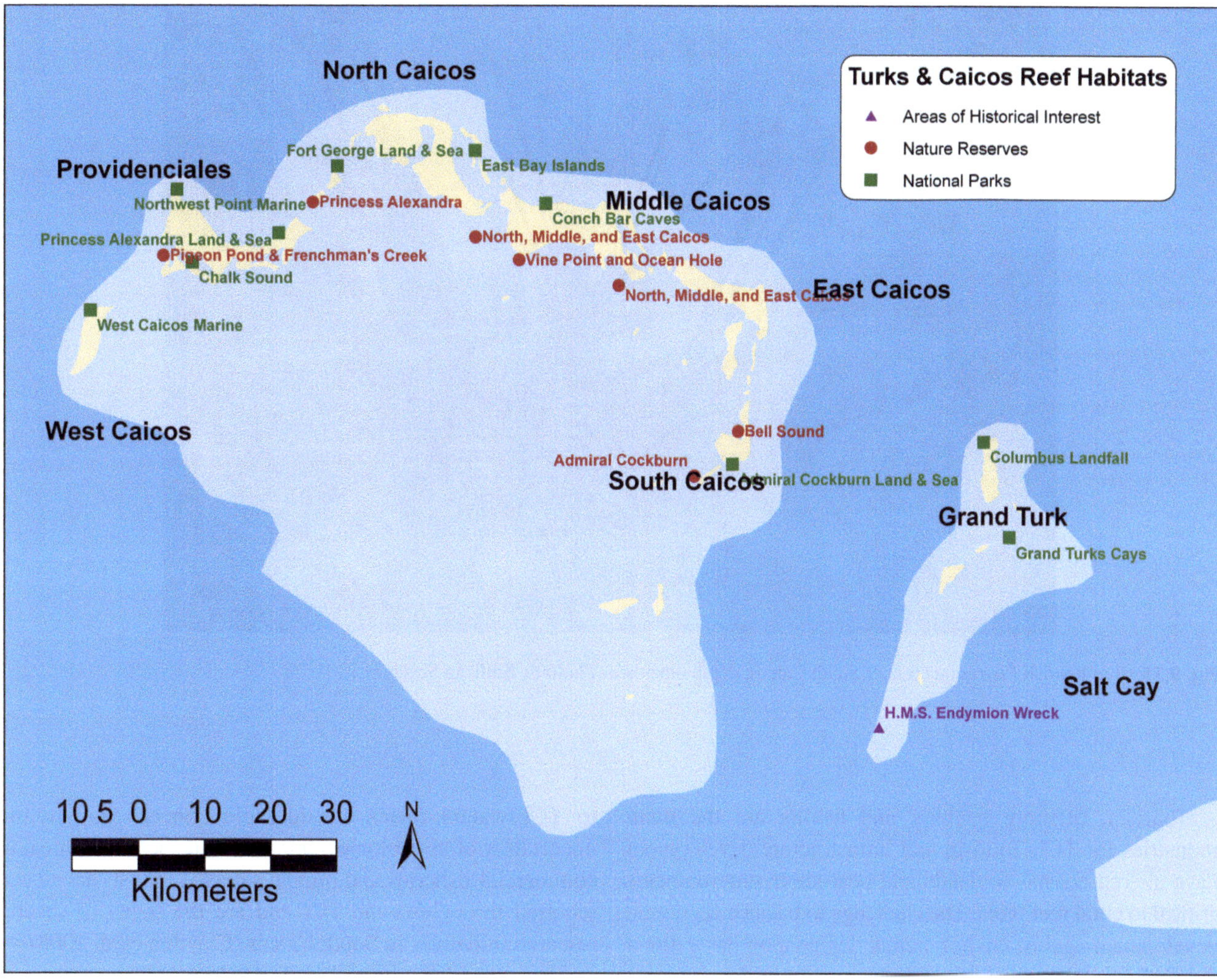

Fig. 9.14 Map of Turks and Caicos National Parks, Preserves and Historical Sites, that include coral habitats (Redrawn from Turks and Caicos National Museum)

planning. There have been several assessment studies on the reefs of the TCI (Riegl et al. 2003; Hoshino et al. 2003; Schelten et al. 2006; Goreau et al. 2008; Dikou et al. 2009) and the periodic overviews by the Global Coral Reef Monitoring Network on the status of coral reefs in the TCI. However, few studies have been able to address regional impacts from localized threats to coral reef resources. The islands of the wider Caribbean overall face regional threats of global climate change, African dust (Shinn et al. 2000) and eutrophication from land-based sources of nitrogen (Banks et al. 2012), as well as over-fishing (McClenachan 2009).

Regional threats are addressed by regional treaties and agreements, particularly for Small Island Developing States (SIDS). Although a territory of the UK, TCI faces the same challenges as other small island nations in the wider Caribbean. Three threats are particularly critical in the TCI: (1) Population growth and immigration, (2) Coastal development and (3) Capacity to manage reefs and other marine resources. The population of the TCI is one of the fastest growing in the region, with about 22,000 residents, but with ex-patriot workers and second-home owners, the population can exceed 44,000, with most people on Providenciales (67%), followed by Grand Turk (16%). Historically, people have immigrated to the TCI from both Haiti and the Dominican Republic; both countries are only 56 km to the south on the island of Hispaniola. Over the past two decades, the number of refugees and immigrants from this island to the south has contributed significantly to the high population growth rate. A staggering 18.4 million people live on the island of Hispaniola. TCI remains an important trans-shipment country for the movement of people and contraband from the Caribbean to the United States. The largest areas of reef development along the southern extents of the Caicos Bank, the Turks Bank and Mouchoir Bank are the more challenging areas to access and patrol to manage protected areas as well as fishing activity.

Fig. 9.15 Red lionfish *Pterois* on reefs of South Caicos, depth unknown (Photo by Sullivan Sealey)

Tourism, offshore banking and fishing are the main industries for TCI. Tourism and infrastructure development have altered coastal wetlands and near-shore environments critical to coral reef biota. The challenge to balance economic development against environmental degradation has a direct impact on coral reefs of the TCI. All aspects of development have coastal repercussions on these tropical, dry islands. For example, the construction of reverse osmosis plants for the production of freshwater from seawater, along with marina dredging, has negatively impacted ground water hydrology and near-shore reefs in Grace Bay (Rembert 1999).

The Turks and Caicos National Trust, along with the Division of Protected Areas, Department of Environment and Coastal Resources has embarked on an ambitious agenda to protect the important reefs and coral habitats of the country, relying on the large parks near tourist centers (Fig. 9.14) to generate funding for resource management and especially marine fisheries reserves in more remote areas. The regulatory structure for reef management is in place, but as is often the case, there are not enough resources to implement education, outreach, monitoring, and enforcement across large areas.

The accidental introduction of species of the venomous Indo-Pacific lionfish *Pterois spp.* (Fig. 9.15) to the Western Atlantic Ocean in 1992 brings into focus the substantial challenges faced by TCI to effectively manage high priority invaders and develop realistic prevention and early detection programmes for other exotics. The invasion of lionfish to TCI waters raises considerable concern due to the uncertainty of its ecological impact and its potential threat to commercial fisheries and human safety. Lionfish have been reported throughout the TCI and are the focus of several research initiatives in South Caicos (Claydon et al. 2008).

There have been several studies to look at the condition of reefs within the TCI in response to pressure from coastal development, diving tourists and fishers. Goreau et al. (2008), using spatially-extensive survey methods for reef surveys in 2006, reached some disturbing conclusions:

1. Live coral coverage is surprisingly low at all sites, usually between 10% and 20%, with a large fraction of the dead coral probably resulting from bleaching events in 1987 and 1990, along with mortality from coral diseases. Decreases in coral cover are consistent with reports of regional trends from macro-scale threats to reefs (Bryant 1998).
2. Elevated nutrient levels are suspected to have been derived from a combination of deep-water upwelling and land-based sources (exacerbated by increased use of water from reverse osmosis). Eutrophication, along with climate changes, are considered the top threats to coastal systems globally (Moss et al. 2011). For the TCI, chronic eutrophication could gradually change coral-dominated reefs into algae-dominated reefs.
3. High turbidity from dredging and cruise-ship propellers at certain localities such as the terminal on Grand Turk may cause increased local coral mortality downcurrent from the source (see also Zuidema et al. 2011). Increased

turbidity can also be a component of coastal eutrophication and loss of seagrass meadows.

4. Damage to shallow fringing reefs and the regional decline in *Acropora palmata* and *Acropora cervicornis* from disease, storm events and elevated sea-surface temperature is posing a threat to TCI lagoons and beaches.

Future management of reefs will require active strategies to protect coral reefs from some of the threats outlined above. In this respect reef restoration projects should be encouraged, notwithstanding the threat of damage from periodic hurricanes (Wells et al. 2010). The Mission Statement of the Department of Environment and Coastal Resources pledges "To ensure sustainable utilization of the natural resources of the Turks and Caicos Islands, protect and promote biodiversity and economic prosperity through a sustainable fishing industry and a protected areas system". This will remain a formidable challenge into the future to keep this unique reef environment intact.

Acknowledgements The authors appreciate the assistance of Dr. Hal Wanless, Mark Chiappone and Cindy Lott for reef description information and underwater pictures. Our thanks for years of work and policy development for the protection of reefs go to Ms Judith Garland Campbell, Permanent Secretary for the Ministry of Natural Resources Fisheries and the Environment, TCI.

References

Alevizon W, Richardson W, Pitts P, Serviss G (1985) Coral zonation and patterns of community structure in Bahamian reef fishes. Bull Mar Sci 36:304–317

Andrefouet S (2008) Coral reef habitat mapping using remote sensing: a user vs producer perspective implications for research, management and capacity building. J Spat Sci 53:113–129

Banks AC, Chimot J, Pina P, Donnadille J, Jeansou E et al (2012) A satellite ocean color observation operator system for eutrophication assessment in coastal waters. J Mar Syst 94:S2–S15

Bryant DG (1998) Reefs at risk: a map-based indicator of threats to the world's coral reefs. World Resources Institute, Washington, DC

Buchan KC (2000) The Bahamas. Mar Pollut Bull 41:94–111

Bunt JS, Williams WT, Chalker BE (1981) Coral associations at depths of 45 to 125 feet in the Bahamian region. In: Proceedings of the 4th international coral reef symposium, vol 1, Manila, pp 707–714

Burke L, Reytar K, Spalding MD, Perry A (2011) Reefs at risk revisited (Reefs at risk). World Resource Institute, Washington, DC, 114p

Chiappone M, Sullivan KM, Lott C (1996) Hermatypic scleractinian corals of the southeastern Bahamas: a comparison to western Atlantic reef systems. Carib J Sci 32:1–13

Claydon JAB, Calosso MC, Siri EJ (2008) The red lionfish invasion of South Caicos, Turks and Caicos Islands. In: Proceedings of the 61st Gulf and Caribbean fisheries institute, La Gosier, 10–14 Nov 2008

Coles SL, Fadlallah YH (1991) Reef coral survival and mortality at low temperatures in the Arabian Gulf—new species-specific lower temperature limits. Coral Reefs 9:231–237

Connell JH (1978) Diversity in tropical rain forests and coral reefs. Science 199(4335):1302–1310

Dikou A, Ackerman C, Banks C, Dempsey A, Fox M, Gins M, Hester P, Parnes A, Roach S, Rohde J, Spital C, Tapleshay M, Thomas L (2009) Ecological assessment to detect imminent change in coral reefs of Admiral Cockburn Land and Sea National Park, Turks and Caicos Islands. Mar Ecol 30:425–436

Dizon RM, Yap HT (2006) Effects of multiple perturbations on the survivorship of fragments of three coral species. Mar Pollut Bull 52:928–934

Enos P (2011) Bahamas. In: Hopley D (ed) Encyclopedia of modern coral reefs, structure, form and process. Springer Science, New York, 15p

Fenner DP (1993) Some reefs and corals of Roatan (Honduras), Cayman Brac and Little Cayman. Atoll Res Bull 388:30 p

Gaudian G, Medley P (2000) The Turks and Caicos Islands. In: Sheppard CHC (ed) Seas at the millennium—an environmental evaluation, vol 1. Elsevier Science, Amsterdam, pp 587–594

Geister J (2011) West Indian coral reef classification. In: Hopley D (ed) Encyclopedia of modern coral reefs, structure, form and process. Springer Science, New York, 6 p

Goreau TJ, Fisher T, Perez F, Lockhart K, Hibbert M, Lewin A (2008) Turks and Caicos Islands 2006 coral reef assessment: large-scale environmental and ecological interactions and their management implications. Int J Trop Biol 56:25–49

Hesse KO (1979) Movement and migration of the queen conch *Strombus gigas* in the Turks and Caicos Islands. Bull Mar Sci 29:300–303

Hoshino K, Brandt M, Manfrino C, Riegl B, Steiner SCC (2003) Assessment of the coral reefs of the Turks and Caicos Islands (part 2: fish communities). Atoll Res Bull 496:480–499

Hughes TP, Connell JH (1987) Population dynamics based on size or age? A reef coral analysis. Am Nat 129:818–829

Jackson JBC, Hughes TP (1985) Adaptive strategies of coral-reef invertebrates. Am Sci 73:265–270

Keegan WF, Fitzpatrick SM, Sealey KS, LeFebvre MJ, Sinelli PT (2008) The role of small islands in marine subsistence strategies: case studies from the Caribbean. Hum Ecol 36:635–654

Koltes KH (1993) Aspects of the reproductive biology and social structure of the Stoplight Parrotfish *Sparisoma viride* at Grand Turk, Turks and Caicos Islands, B.W.I. Bull Mar Sci 52:782–792

Lang JC, Wicklund RI, Dill RF (1988) Depth- and habitat-related bleaching of zooxanthellate reef organisms near Lee Stocking Island, Exuma Cays, Bahamas. In: Proceedings of the 6th international coral reef symposium, vol 3, Townsville, pp 269–274

Lewis JB (1984) The *Acropora* inheritance: a reinterpretation of the development of fringing reefs in Barbados, West Indies. Coral Reefs 3:117–122

Logan A (1994) Reefs and lagoons of Cayman Brac and Little Cayman. In: Brunt MA, Davies JE (eds) The Cayman Islands: natural history and biogeography. Kluwer Academic, Dordrecht, pp 105–124

McClenachan LE (2009) Historical change in marine animal populations and coastal ecosystems in the Caribbean and Florida Keys. Ph.D. dissertation, University of California at San Diego, La Jolla

Medley PAH, Ninnes CH (1999) A stock assessment for the Conch (*Strombus gigas* L.) fishery in the Turks and Caicos Islands. Bull Mar Sci 64:399–408

Mitchell BA, Barborak JR (1991) Developing coastal park systems in the tropics—planning in the Turks and Caicos Islands. Coast Manage 19:113–134

Moss B, Kostern M, Meerhoff M, Batterbee R, Jeppesen E, Mazzeo N, Havens K, Lacerot G, Liu Z, De Meester L, Paerl H, Scheffer M (2011) Allied attack: climate change and eutrophication. Inland Waters 1:101–105

Mumby PJ, Harborne AR (1999) Development of a systematic classification scheme of marine habitats to facilitate regional management and mapping of Caribbean coral reefs. Biol Conserv 88:155–163

Pandolfi JM, Llewellyn G, Jackson JBC (1999) Interpretation of ancient reef environments in paleoecological studies of community structure:

Curacao, Netherlands Antilles, Caribbean Sea. Coral Reefs 18:107–122

Rankey EC, Reeder SL (2010) Controls on platform-scale patterns of surface sediments, shallow Holocene platforms, Bahamas. Sedimentology 57:1545–1565

Rankey EC, Guidry SA, Reeder SL, Guarin H (2009) Geomorphic and sedimentologic heterogeneity along a Holocene shelf margin: Caicos Platform. J Sediment Petrol 79:440–456

REEFbase GIS (2012) REEFbase project. The WorldFish Center, Penang, Malaysia

Rembert TC (1999) Protecting paradise: the Turks and Caicos Islands gear up for responsible ecotourism. Environ Mag 10:46

Riegl B, Manfrino C, Hermoyian C, Brandt M, Hoshino K (2003) Assessment of the coral reefs of the Turks and Caicos Islands (part 1: stony corals and algae). Atoll Res Bull 49:459–479

Roberts HH (1994) Reefs and lagoons of Grand Cayman. In: Brunt MA, Davies JE (eds) The Cayman Islands: natural history and biogeography. Kluwer Academic, Dordrecht, pp 75–104

Rudd MA (2001) The non-extractive economic value of the spiny lobster *Panulirus argus* in the Turks and Caicos Islands. Environ Conserv 28:226–234

Rudd MA, Tupper MH (2002) The impact of Nassau Grouper size and abundance on scuba diver site selection and MPA economics. Coast Manage 30:133–151

Rützler K, Macintyre IG (1982) The habitat distribution and community structure of the barrier reef complex at Carrie Bow Cay, Belize. In: Rützler K, Macintyre IG (eds) The Atlantic barrier reef ecosystem at Carrie Bow Cay, Belize, Structure and communities. Smithsonian Institution Press, Washington, DC, pp 9–45

Schelten C, Brown S, Gurbisz CB, Kautz B, Lentz JA (2006) Status of *Acropora palmata* populations off the coast of South Caicos, Turks and Caicos Islands. Proc Gulf and Caribb Fish Inst 57:665–678

Sealey NE (2006) Bahamian landscapes: an introduction to the physical geography of the Bahamas. Media Publishing, Nassau, p 171

Shinn EA (2011) Spurs and grooves. In: Hopley D (ed) Encyclopedia of modern coral reefs, structure, form and process. Springer Science, New York, 4p

Shinn EA, Smith GW, Prospero JM, Betzer P, Hayes ML, Garrison V, Barber RT (2000) African dust and the demise of Caribbean coral reefs. Geophys Res Lett 27:3029–3032

Sinelli PT (2010) All islands great and small: the role of small cay environments in indigenous settlement strategies in the Turks and Caicos Islands. Ph. D. dissertation, University of Florida, 480p

Sluka R, Chiappone M, Sullivan KM, Wright R (1996) Habitat and life in the Exuma Cays, Bahamas: the status of groupers and coral reefs in the northern cays. Media Publishing, Nassau, 83p

Spotte S, Heard RW, Bubucis PM (1994) Pontoniine Shrimps (Decapoda: Caridea: Palaemonidae) of the northwest Atlantic. IV. *Periclimenes antipathophilus* new species, a Black Coral associate from the Turks and Caicos Islands and eastern Honduras. Bull Mar Sci 55:212–218

Steiner SCC (1999) Species presence and distribution of Scleractinia (Cnidaria: Anthozoa) from South Caicos, Turks and Caicos Islands. Bull Mar Sci 65:861–871

Sullivan KM, Chiappone M (1992) A comparison of belt quadrat and species presence and absence sampling for evaluating stony coral (Scleractinia and Milleporina) and sponge species patterning on patch reefs of the central Bahamas. Bull Mar Sci 50:464–488

Sullivan KM, Chiappone M, Lott C (1994) Abundance patterns of stony corals on platform margin reefs of the Caicos Bank. Bahamas J Sci 1:2–11

Sullivan KM, Chiappone M, Ninnes C (1996) Characteristics of hard-bottom assemblages for resource mapping of the Caicos Bank. In: Proceedings of the 44th Gulf and Caribbean fisheries institute, St. Croix, vol 44, pp 143–158

Tupper M, Rudd MA (2002) Species-specific impacts of a small marine reserve on reef fish production and fishing productivity in the Turks and Caicos Islands. Environ Conserv 29:484–492

Turks and Caicos Islands Fishery Limits Ordinance, revised May 15th, 1998, in: Laws of the Turks and Caicos Islands, Chapter 105

Wanless HR, Dravis JJ (1989) Carbonate environments and sequences of the Caicos platform: field trip guidebook T374. American Geophysical Union, Washington, DC, 75p

Watson M, Munro JL (2004) Settlement and recruitment of coral reef fishes in moderately exploited and overexploited Caribbean ecosystems: implications for marine protected areas. Fish Res 69:415–425

Wells SM (1988) Turks and Caicos. In: Wells SM (ed) Coral reefs of the world, vol 1. UNEP/IUCN, Cambridge, UK, pp 303–310

Wells L, Perez F, Hibbert M, Clerveaux L, Johnson J, Goreau TJ (2010) Effect of severe hurricanes on biorock coral reef restoration projects in Grand Turk, Turks and Caicos Islands. Int J Trop Biol 58:141–149

Whitaker FF, Smart PL (1997) Climatic control of hydraulic conductivity of Bahamian limestones. Ground Water 35:859–868

Zankel H, Schroeder JH (1972) Interaction of genetic processes in Holocene reefs of north Eleuthera Island, Bahamas. Geol Rundsch 61:520–541

Zuidema C, Plate R, Dikou A (2011) To preserve or to develop? East Bay dredging project, South Caicos, Turks and Caicos Islands. J Coast Conserv 15:555–563

10 Introduction to Bermuda: Geology, Oceanography and Climate

Kathryn A. Coates, James W. Fourqurean, W. Judson Kenworthy, Alan Logan, Sarah A. Manuel, and Struan R. Smith

Geographic Location and Setting

Bermuda is a subtropical island group in the western North Atlantic (Fig. 10.1a). A peripheral annular reef tract surrounds the islands forming a mostly submerged 26 by 52 km ellipse at the seaward rim of the eroded platform (the Bermuda Platform) of an extinct Meso-Cenozoic volcanic peak (Fig. 10.1b). The reef tract and the Bermuda islands enclose a relatively shallow central lagoon so that Bermuda is atoll-like. The islands lie to the southeast and are primarily derived from sand dune formations. The extinct volcano is drowned and covered by a thin (15–100 m), primarily carbonate, cap (Vogt and Jung 2007; Prognon et al. 2011). This cap is very complex, consisting of several sets of carbonate dunes (aeolianites) and paleosols laid down in the last million years (after Prognon et al. 2011, with reference to Vacher and Rowe 1997) (Fig. 10.2).

K.A. Coates (✉) • S.A. Manuel
Marine Ecology Section, Department of Conservation Services, 17 North Shore Road, Hamilton Parish, FL 04, Bermuda
e-mail: kathryn.coates@dilke.com; smanuel@gov.bm

J.W. Fourqurean
Department of Biological Sciences and SERC, Florida International University, 3000 NE 151 Street, North Miami, FL, 33181, USA
e-mail: fourqure@fiu.edu

W.J. Kenworthy
109 Holly Lane, Beaufort, NC, 28516, USA
e-mail: jud.kenworthy@gmail.com

A. Logan
Centre for Coastal Studies and Aquaculture, University of New Brunswick, Hazen Hall, 100 Tucker Park Road, Saint John, NB, E2L 4L5, Canada
e-mail: logan@unbsj.ca

S.R. Smith
Natural History Museum, Bermuda Aquarium, Museum and Zoo, Department of Conservation Services, 40 North Shore Road, Hamilton Parish, FL 04, Bermuda
e-mail: srsmith@gov.bm

Located at 32.4°N and 64.8°W, Bermuda lies in the northwest of the Sargasso Sea. It is isolated by distance, deep water and major ocean currents from North America, sitting 1,060 km ESE from Cape Hatteras, and 1,330 km NE from the Bahamas. Bermuda is one of nine ecoregions in the Tropical Northwestern Atlantic (TNA) province (Spalding et al. 2007).

Bermuda's national waters include pelagic environments and deep seamounts, in addition to the Bermuda Platform. The Bermuda Exclusive Economic Zone (EEZ) extends approx. 370 km (200 nautical miles) from the coastline of the islands. Within the EEZ, the Territorial Sea extends ~22 km (12 nautical miles) and the Contiguous Zone ~44.5 km (24 nautical miles) from the same baseline, both also extending well beyond the Bermuda Platform.

Geography, Settlement and Early Economies

Of more than 150 islands and islets only six are of any size, and these are connected by causeways and bridges to form a narrow fish hook-shaped island chain 34 km in length and 1.6 km in average width (3 km at the widest). The main islands have a land area of 53.6 km^2 and a shoreline of about 290 km (State of the Environment Report 2005). Approximately 66% of this land area is built upon.

Bermuda consists of a series of low rolling hills, generally with heights only 40–50 m above sea-level and a maximum of 79 m (State of the Environment Report 2005). Natural flat areas are absent except at a few sea-level marshlands in the middle. There are no rivers, streams, or freshwater lakes due to the very permeable limestone cap. Rain, collected from roof tops and stored in tanks, is the principal source of drinking water.

Bermuda was not populated prior to its inhabitation by the British. In 1609 the *Sea Venture* on its way to Jamestown, Virginia from Plymouth, England wrecked on Bermuda's eastern coral reefs, unintentionally delivering Bermuda's first colonists. Only two of the castaways remained on the

C.R.C. Sheppard (ed.), *Coral Reefs of the United Kingdom Overseas Territories*, Coral Reefs of the World 4,
DOI 10.1007/978-94-007-5965-7_10, © Springer Science+Business Media Dordrecht 2013

a

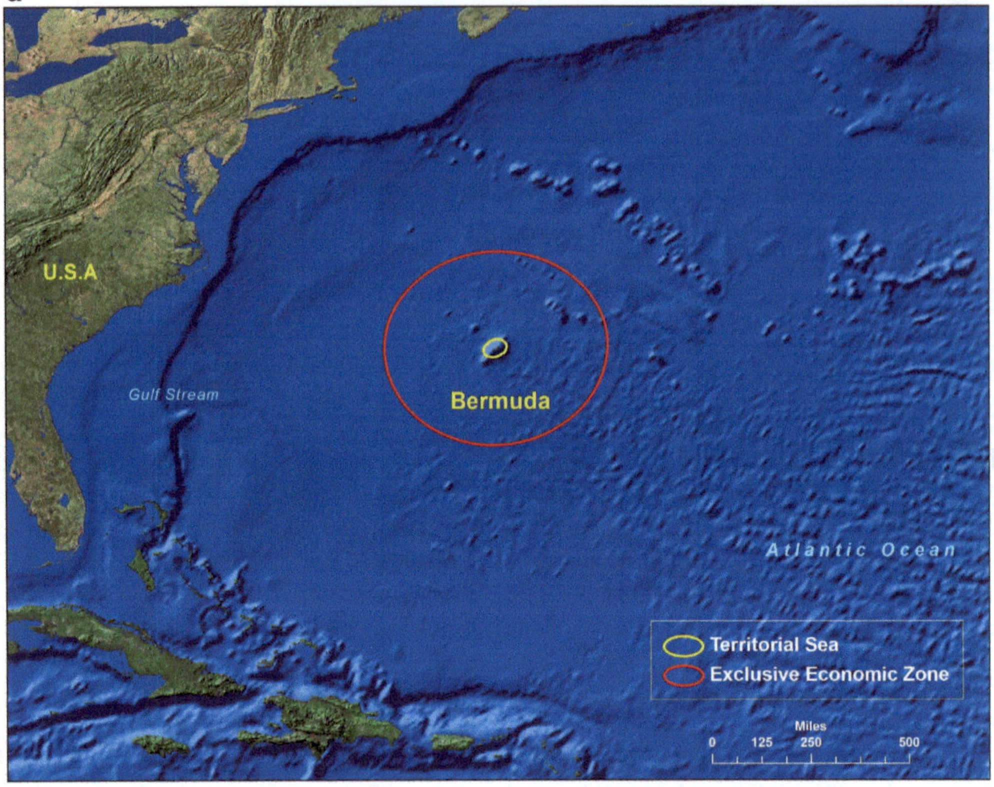

b

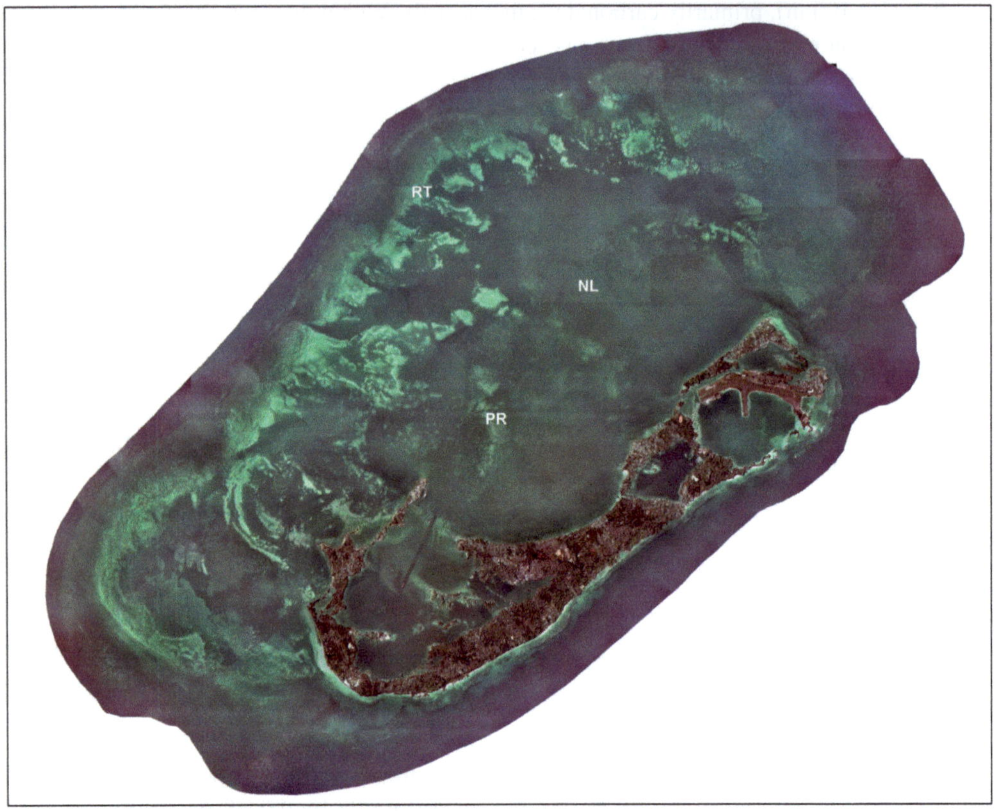

Fig. 10.1 (**a**) Location of Bermuda in the western North Atlantic; the Bermuda EEZ extends over the Bermuda Rise, and includes the deeply submerged Crescent Seamount to the NNW and Muir Seamount to the NNE. (**b**) The Bermuda Platform. Aerial photograph from 1997 (Copyright Bermuda Zoological Society) showing details to approx. 10 m depth; the islands lie parallel to the south-eastern edge of the Platform and a reef tract bounds the whole Platform (RT). Numerous coral patch reefs (PR) occur in the North Lagoon (NL); these are separated by sandy bottom at depths up to 23 m. Images are oriented with North to the top

Fig. 10.2 Aeolianite strata exposed along the South Shore of Bermuda

islands when the Bermuda built boats *Deliverance* and *Patience* set sail for Virginia in 1610. The first formal settlers, approximately 50, arrived from England in 1612. Today the population is about 64,000 (Bermuda Census 2010), with a population density of nearly 1,195 persons per km^2, one of the ten highest country-wide densities in the world, but still only a fraction of that of many similar-sized metropolitan areas. Early inhabitants mainly pursued maritime occupations but by the late 1800s agriculture had grown in importance (Hayward 1924) and later still tourism became the pillar of Bermuda's economy. The tourist industry peaked in 1980 (State of the Environment 2005) but now international business, including a wide variety of financial services, has become the mainstay of the economy. There are a few small manufacturing industries in Bermuda that cater primarily to locals.

There were 206 licensed commercial fishing vessels and 307 registered fisherman in 2011 (T. Trott, pers. com.). These fish primarily on the Bermuda Platform or along its edge although a very few venture further offshore. Baited lobster traps are deployed in-season but other nearshore fisheries primarily employ hook and line, with very limited long-lining. Nets are used primarily in shallow waters for the capture of bait fishes. Catches are discrete for the most part and there is little by-catch, although lobster traps seem to be effective for the capture of the invasive lion fishes. There is no fishery associated large-scale bottom destruction, such as would occur with bottom trawls. Anchor damage at highly frequented fishing sites has not been investigated.

Agriculture declined in importance during the first century of settlement, followed by a recovery based on different products and different markets in the mid 1800s, and there was an economically important production of exports into the early 1900s. However, since the 1940s, production has been primarily for local consumption, nonetheless about 90% of fresh produce is imported (State of the Environment Report 2005). Only a small portion of the island is under cultivation, with about 178 ha of arable fields or pastures designated, of which only 154 ha were being farmed commercially in 2001. Golf courses cover a greater area, 260 ha, and landscaped properties and playing fields total 669 ha. Fertilizers and pesticide use on all of these add to the nutrients and toxins that move into ground water and into the sea (State of the Environment Report 2005). Most foods are imported, as is all fuel and almost all other manufactured goods. Energy production is virtually all based on fossil fuels, but supplementation by solar and wind is beginning to be employed by a very few individuals and proposals are under development in 2012 for commercial-scale alternate energy supplies.

Geology

Bermuda has a complex history of volcanism. The Bermuda Pedestal is the subaerially eroded stump of a shield volcano formed during middle Eocene (less than 45 mya) and early Oligocene (33–34 mya) volcanic episodes (Vogt and Jung 2007).

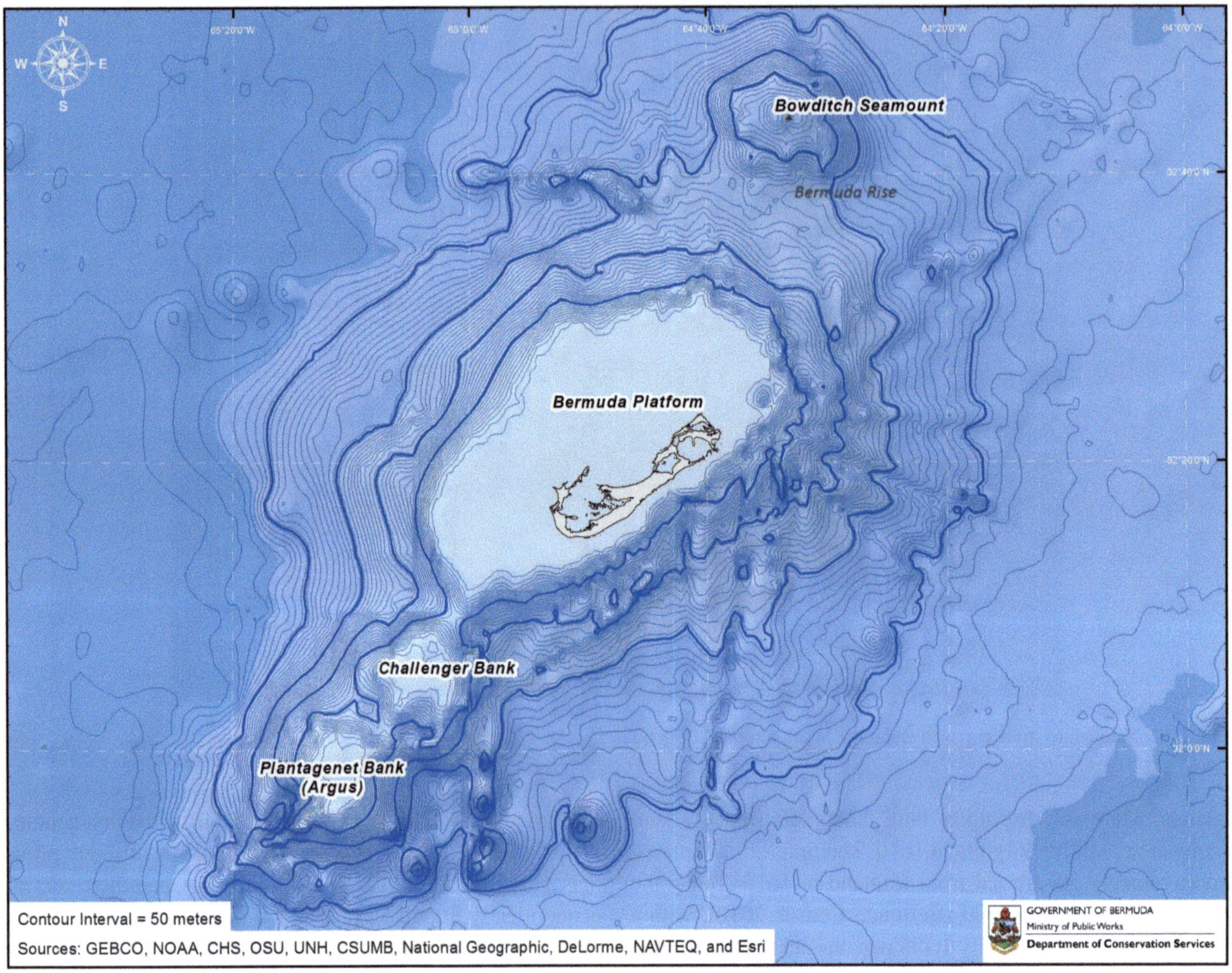

Fig. 10.3 Bathymetry of the Bermuda Pedestal and nearby seamounts; Plantagenet and Challenger to the southwest have relatively shallow platforms, about 50 m deep, which support photosynthetic organisms; Bowditch to the northeast is a deeply submerged volcanic peak

On the top of the eroded stump, or Pedestal, is the Bermuda Platform.

Bermuda is one of a line of four associated volcanos that runs NE over approximately 100 km. This line is located near the summit of the Bermuda Rise, a NE-trending oval swell of the ocean bottom, about 1,500 km long and 500–1,000 km wide, which at its summit rises about 800 m above the surrounding deep sea bottom (see Vogt and Jung 2007 and references therein). Although the volcanos were extinct by the early Oligocene, uplift of the Bermuda Rise continued into the Miocene, thus the overall uplift of the volcanos also continued into this period.

Bermuda is the largest and highest of the four volcanoes (Vacher and Rowe 1997) (Fig. 10.3). To the southwest of Bermuda are Plantagenet (also called Argus; Plantagenet has also been used for Irving Seamount) and Challenger Banks (or tablemounts). These were also eroded subaerially and have relatively flat and shallow (but fully submerged) platforms. They are located about 41 km and 27 km respectively from Bermuda and rise to within about 47 m and 42 m respectively of the sea surface. The fourth volcano, Bowditch Seamount, lies about 39 km NNE of the Bermuda Platform and ascends only to about 600–800 m below the sea surface (Seamount Biogeosciences Network 2012).

The platforms of Plantagenet and Challenger, lie at depths of about 50 m and have areas of 65.6 km^2 and 74.1 km^2, respectively. There is a sharp change in slope at about 60–70 m depth to the steepening slopes of the underlying seamounts. The platforms and upper slopes are shallow enough to support photosynthetic communities including zooxanthellate cnidarians (Cairns 1979; Calder 2000) and algae, which have been found as deep as about 137 m (SRS, pers. obs.; http://oceanexplorer.noaa.gov/explorations/11bermuda/).

Bermuda has the shallowest platform, at about 20 m, with the greatest platform area, about 623 km^2. Water depths in the central lagoon are 14–16 m at many sites, with a maximum depth of 23.9 m in one of the sounds. Around the rim, the 20–22 m isobath (Iliffe et al. 2011) separates the broad

Table 10.1 Classification of 90 tropical cyclones passing within 333.4 km (180 nautical miles) of Bermuda in the years 1899–1979. Intensity values are maximum sustained centre winds at the time of closest approach to Bermuda, here converted from knots to kilometers per hour (After Brand 2009)

Maximum intensity	May to July	Aug	Sept	Oct–Nov	Totals
Hurricane (>118 kph)	2	7	26	14	49
Intense tropical storm (89–117 kph)	0	1	5	8	14
Weak tropical storm (63–88 kph)	1	1	6	8	16
Tropical depression (<63 kph)	2	1	2	6	11
Totals	5	10	39	36	90

upper terrace of the Platform from a distinct seaward slope, rarely exceeding 10°. This shallow slope is followed by a markedly steeper slope commencing at about 55–65 m depth, which descends to almost vertical walls at about 100 m. A ridge, possibly a drowned reef, occurs at about 60 m (Stanley and Swift 1968; Iliffe et al. 2011).

Bermuda Climate and Marine Environment

Meteorology and Climate

The atmospheric pressure gradient over the western North Atlantic, the Gulf Stream that flows to the west of the island, and the prevailing conditions of the Sargasso Sea (Steinberg et al. 2001) strongly influence Bermuda's climate. Bermuda has a subtropical marine environment and climate despite its temperate latitude, experiencing relatively mild winters and moderately warm summers. Sea surface temperatures closely track air temperatures. Mean annual rainfall is about 1,410 mm, the driest months being April, May, and November and the wettest, on average, is October.

Tropical cyclones approach most years; indeed the initial colonizing of Bermuda relates to the survivors of one such storm in 1609 (Brand 2009). "Hurricane" season in Bermuda is officially June through November, with peak frequency being September and October (Neumann et al. 1985; Brand 2009). September storms are both the most numerous and the most intense (Brand 2009) (Table 10.1). Between 1871 and 1979, Neumann et al. (1985) report 127 storms passing within 333 km (180 nautical miles) of Bermuda, an average of more than one storm each year, of which, 49 were hurricane force, for an average of just more than one hurricane every second year.

Tropical cyclones typically approach Bermuda from the south or southwest (about 52% of storms for the period 1871–1979) and about 88% of all tropical storms approach from the west through the southeast (Brand 2009). Many of these have the strongest winds in their northwest quadrant, thus a passage to the east can have very different impacts than one to the west. Due to the common direction of approach and to the distribution pattern of reefs on the Bermuda platform, long period swells moving ahead of these storms have their highest impact on the southeastern coastline, where the breaking reefs are nearest shore (Smith Warner International 2004); these swells can exceed 10 m in height. To the west and north, breaking reefs lie far from shore, and lagoonal patch reefs further attenuate wave energy and reduce impacts on coastal areas. Maximum storm surge is usually less than 1.5–2 m, but wave energy and heights increase as waves move across shallow flats. Observations after recent storms suggest that the coral reefs, dominated by massive colonies of *Diploria* spp and *Montastraea* spp, do not show much direct wave damage (SRS, pers. obs.). Extreme weather events in summer can also bring heavy, prolonged, rainfall.

Incident solar radiation shows a large annual range. For the years 1983–2005 monthly means ranged from about 2.53 kW h m^{-2} day^{-1} in December to 7.03 kW h m^{-2} day^{-1} in July (NASA 2010). Day length is approximately 14.3 h at the summer solstice and only 10.0 h at the winter solstice and cloud cover is greatest from November to April, with overcast skies (7/8 obscured) on 35–45% of days, compared to only 15–20% of summer days (Morris et al. 1977).

Tidal and Oceanographic Features

Bermuda experiences semi-diurnal tides, with mean range of ~0.76 m and a mean diurnal range of ~0.86 m (NOAA Tides and Currents 2012 http://www.tidesandcurrents.noaa.gov/geo.shtml?location=2695540). Mean daily differences are 0.02 m and 0.08 m, respectively. Tide heights are infrequently affected by the slow passage of mesoscale-eddies that can elevate or attenuate tidal heights by 25 cm (Seigel et al. 1999; McGillicuddy et al. 2007). Harrington Sound, an enclosed body of water with a single, restricted, surface entrance has incomplete tidal loading and unloading resulting in a reduced tidal range of about 0.20–0.25 m (Morris et al. 1977) and delays in maximum tide levels of about 30 min.

Morris et al. (1977) calculated current speeds for a number of entry points into largely enclosed waters; these were all about 5–103 cm sec^{-1} (except at the entrance to Harrington Sound, with a maximum of 448 cm sec^{-1} and an average of about 215 cm sec^{-1}). Measured coastal and North

Lagoon current speeds were mostly 3–9 cm sec^{-1} and rarely greater than 15 cm sec^{-1} (Simmons and Johnson 1993). Current speeds on outer rim reef sites of the North Shore are typically 8–10 cm sec^{-1} but exceeded 25 cm sec^{-1} under winter storm conditions (Badgley et al. 2006). At a South Shore reef site measured current speeds were typically less than 15 cm sec^{-1}. There was no clear indication of predominant tidal forcing of surface currents at all the sites investigated by Simmons and Johnson (1993) and flow at some offshore sites was largely unidirectional. At the South Shore site currents also were not predominantly tidally influenced (Marine Environmental Program 2007).

Marine Physical Environment

The oceanography of the surrounding Sargasso Sea has been well characterized by a series of long term studies of physical, chemical and biological processes at National Science Foundation funded HydroStation S and the Bermuda Atlantic Time Series (BATS) site further offshore (Steinberg et al. 2001).

Since 2006 and the initiation of the Bermuda Benthic Mapping, Monitoring and Assessment Program (BMMAP), more than 17 sites on the Bermuda platform are monitored continuously for sea bottom temperature and other parameters, including oxygen concentration and saturation, *p*H, chlorophyll a, and various nutrients, were measured monthly until September 2012 (Boyer and Briceno 2010; KAC, JWF, WJK, SAM, unpub. obs.).

Sea Surface Temperature

The open water around Bermuda has an average annual sea surface temperature about 23°C (Locarnini et al. 2006) and average monthly minima and maxima of 19.2°C in March and 27.4°C in August, respectively; maximum surface temperatures range as high as 29°C (Steinberg et al. 2001). Mean annual surface water temperature on the Platform is similar but the extreme temperature range is greater, from 14°C to 15°C during winter to 30–31°C in summer, although the surface water temperature over the rim reefs rarely exceeds 29°C (Marine Environmental Program 2007; Boyer and Briceno 2010).

Temperature range variation across the Platform is biologically significant. From the rim to the shore, temperature extremes at the sea bottom increase and the annual range increases from about 12.5°C to about 17.5°C. Thus, inshore habitats experience a cold winter environment compared to offshore and mid-platform habitats, with low seawater temperatures extending over many weeks. Shallow inshore sites can also be warmer in summer, but the differences compared to rim sites are not as pronounced as those during winter months. Extreme summer weather events (hurricanes and tropical storms) cause rapid drops in sea water temperature through the entire water column of the whole platform but these changes are more extreme and long-lived at sites near the rim.

Salinity

Surface salinities in the offshore waters near Bermuda range annually from about 36.3 to 36.7 (Practical Salinity Scale), with an average annual value around 36.5 (http://ourocean.jpl.nasa.gov/AQUARIUS; Thacker and Sindlinger 2007). Gordon and Guilivi (2008) suggest that sea surface salinity in the Sargasso Sea near Bermuda has been slightly elevated since 1988 (at least up until 2005) and that a relatively salty period also occurred in 1963–1969. Steinberg et al. (2001) report anomalously high-salinity (36.8) water was present in the upper ocean from January to August of 1997 and attribute this to locally driven excess evaporation over precipitation.

On the Bermuda platform salinities range from about 36.4 to 36.8 (Boyer and Briceno 2010; KAC. JWF. WJK, SAM, unpub. obs.), with irregularly occurring highs and lows to above 37 or below 36. Despite the absence of streams and rivers, rain on the islands strongly influences salinity at specific locations due to surface runoff, and these may regularly have salinities of less than 36, with an extreme known minimum of 33.9 (Marine Environmental Program 2007). No clear seasonal pattern in salinity values has been recognized for platform sites nor is there a clear difference from rim to inshore sites, although the latter may be more variable through the year.

Dissolved Oxygen

Dissolved oxygen (DO) in the water column around Bermuda is high throughout the year with a mean % saturation of 108.4% (81.8–221.4%) (oxygen concentrations from about 5.29 mg l^{-1} to 14.65 mg l^{-1}) (KAC, JWF, JWK, SAM, unpub. obs., http://serc.fiu.edu/wqmnetwork/Bermuda/Data1.html) between March 2007 to April 2011. Saturation levels are generally lower in the summer than winter and lower offshore than inshore, although shallow inshore sites have greater ranges. (A number of monitored inshore sites are shallow seagrass beds.) Seasonal (June through August) anoxia at Shark Hole in Harrington Sound at depths greater than 15–20 m has been reported for the past 45 years (Morris et al. 1977), and more recently hypoxia was recognized in Little Sound (Marine Environmental Program 2007; KAC, SAM, unpub. obs.) at bottom depths greater than 20 m. Saturation levels well below the Platform average have also been observed at shallower locations in the Great Sound during summer months.

*p*H

Andersson and Mackenzie (2011) point out the importance of recognizing the variability of carbonate chemistry on shallow platforms. Surface seawater CO_2 chemistry in shallow water coastal areas is not predictable from air-equilibrium models and differs significantly from open ocean systems (Andersson and Mackenzie 2011). Many shallow water environments undergo large diurnal fluctuations in seawater chemistry associated with daily changes in benthic biological processes that produce and consume CO_2, such as photosynthesis/respiration and calcification/dissolution, as well as with water exchange owing to tidal cycles and changes in winds.

In 1976–1977, *p*H of the Sargasso Sea varied between 8.11 and 8.2 (Morris et al. 1977) and for the same period it was 7.94–8.23 at sites on the Platform, with a decrease in the summer related to increased water temperatures. Bates et al. (2010) indicate average *p*H is lower over coral reefs at the Platform rim (Hog Reef) than at nearby open water sites (BATS site; Bates et al. 2010). Some inshore sites have consistently lower *p*H than others, for example Harrington Sound compared to Great Sound; summer stratification in Harrington Sound may account for this difference (Morris et al. 1977; Andersson and Mackenzie 2011).

During daylight, the average *p*H of Platform waters is 8.06 (std. dev. 0.23; median 8.11), with a measured range from 6.67 to 8.74 (KAC, SAM unpub. obs., http://serc.fiu.edu/wqmnetwork/Bermuda/Data1.html). Surface *p*H at rim and lagoonal sites is generally less variable. There is no clear seasonal pattern, but 2010 was very different to either 2009 or 2011 at rim sites with *p*H below 8 measured on a number of dates.

Light in the Water Column

The light environment varies with daily, seasonal and probably other temporal cycles (Siegel et al. 1995). Deep water around Bermuda generally has high water clarity; diffuse attenuation coefficient, K_d, values of 0.045, 0.070 and 0.080 m^{-1} were reported by Morris et al. (1977). Siegel et al. (1995) report K_d for 488 nm wavelength light, generally the most penetrating wavelength, of between 0.02 m^{-1} and 0.05 m^{-1} over the upper 200 m of the water column. For wavelengths about 400–600 nm, those higher than 500 had increasingly higher K_d values (Siegel et al. 1995) and for shallower depths, up to 20 m, K_d was higher from fall to spring than in summer months.

Earlier studies of Platform waters reported K_d values ranging from 0.13 m^{-1} to 0.56 m^{-1} (Morris et al. 1977; McGlathery 1995). Recent studies (KAC, JWF, JWK, SAM, unpub. obs.) found summertime K_dPAR values from 0.025 m^{-1} to 0.55 m^{-1}, with a median of 0.12 m^{-1} ($n = 270$). The pattern from the rim to inshore waters is not a simple trend, and large areas of the North Lagoon, extending well into the lagoon, have on average lower K_dPAR than found over large areas of the north and east rim reefs.

Nutrients

Concern about anthropogenic effects in the protected inshore water stimulated study of several parameters including nutrient levels and dynamics (Morris et al. 1977; Barnes and von Bodungen 1978; von Bodungen et al. 1982). These showed nitrogen loading in the restricted inshore basins (Jickells et al. 1986; Lapointe and O'Connell 1989; Simmons and Lyons 1994; Boyer and Briceno 2010) and some of these studies implicated contaminated groundwater (Jones et al. 2010).

Geological History and Coral Reef Development

The maximum elevation of the Bermuda volcano, since its formation, was estimated as 2–4 km above sea-level (Pirsson 1914a, b; Vogt 1979), and it was the highest and the biggest volcano on the Bermuda Rise. Subaerial erosion to sea-level is estimated to have taken between 3 to 10 my (Vogt and Jung 2007); some of this may have occurred as early as the Middle Oligocene, about 30 mya, not long after the final periods of volcanic activity, for which period there is evidence of sea surface levels 50–75 m below present. Reef limestone began to accumulate on the eroded Platform somewhat more recently, probably in the Early Miocene (<25 mya). During the Miocene the global climate was moderately cool but by the Pliocene the earth was cooling and most of the accumulation of coral limestone on Bermuda is thought to have occurred during the Quaternary (Prognon et al. 2011) – a globally cold period. Coral limestone in Bermuda has been found as deep as 200 m below present day sea-level (Gees and Medioli 1970).

Pleistocene History, Sea-level and Coral Reef Development

"For more than a decade it has been recognized that the sedimentary and fossil record of Quaternary coral reefs has the potential to help decipher the role of history in the study of living reefs …" (Precht and Miller 2007). Most recently, it has been hoped that history will provide insights into what appear to be pressing issues for reef management arising from global changes and threats to modern reefs, for example, understanding the recent precipitious decline of acroporids throughout the Caribbean.

Because of its history, the Bermuda Pedestal can serve as a stable benchmark to determine Pleistocene sea-level changes (Hearty 2002; Hearty et al. 2007; Vogt and Jung 2007) both through submerged and emergent features (Hearty and Olson 2011). Sea-level fluctuations during the Pleistocene must have driven the location and rate of coral reef development and of species diversity on the Bermuda Platform and Pedestal. In the early Pleistocene, sea-level repeatedly rose to about present day sea-levels and then fell to levels up to 200 m below present.

The Bermuda islands are formed from several sets of carbonate dunes (aeolianites) and paleosols, which record the main glacial/interglacial cycles from about 430 kya to the present (Prognon et al. 2011). Bermuda is on the fringe of the marine "carbonate belt", with many coral species having slower growth rates than their more tropical conspecifics, so that limestone sediments accumulate relatively slowly. Several thousand years of submergence of the Platform may be required for development of the store of sediments sufficient to build dunes of the size of the Pleistocene record. Such a deposition has not happened yet during the Holocene submergence, and Bermuda remains in an erosional period, losing land mass during each major storm (Smith Warner International 2004).

Other emergent deposits include isolated sublittoral and beach deposits that record sea-level high stands and shorelines; however there are no emergent reef formations or individual in situ coral fossils. The beach deposits do provide evidence of species that occurred around Bermuda at particular times, but not the exact locations or elevations at which the species were living. Emergent fossil coastline formations in Bermuda have been correlated by aminostratigraphy and radiometric dating to at least three marine oxygen isotope stages (MIS), MIS 11, 9 and 5e (Olson et al. 2006). Odd numbered MIS all relate to interglacial periods when sea-level highstands occurred and which were possibly globally warmer. MIS 11, 9 and 5e are all recognized in Bermuda as warmer-than-present periods when sea-levels were higher than present (Hearty and Kaufman 2000; Muhs et al. 2002). Maximum sea level heights during two of these stages, have been estimated for Bermuda at 18–22 m above present for MIS 11 (about 427 kya) and 10 m above present for MIS 5e (130–133 kya). The islands of Bermuda are supposed not to have been fully submerged for close to 1 my and MIS 11 is the most recent highest known sea level highstand.

The total range from a highstand to a subsequent lowstand (corresponding to evenly numbered MIS) would be significant to growth and survival of any reefs that became established during sufficiently long and stable sea-level periods. The extreme low stand of MIS 12 (period start about 474 kya) of about 120–130 m below present sea level to the extreme high stand at MIS 11 (period start about 424 kya), of > +20 m represents an extreme range for the Pleistocene and perhaps a period of rapid sea level rise. MIS 11 has long been recognized as one of the longer and warmer Quaternary periods (Olson and Hearty 2009).

The modern sea-level curve for Bermuda reflects a postglacial rise of sea-level that slowed in the last 5 ky and reached "present" sea-level in the past 0.5–2 ky, with no intervening highstands. The rise was about 3.7 m ky^{-1} up to about 4,000 y BP, after which it rose at about 1 m ky^{-1} to its present position. Sea-level rise has likely been at 1.43–2.8 m ky^{-1} in the past 100 years (Ellison 1993; Pirazzoli 1987), greater than the rate of the past 4,000 years but less than in the early Modern Era.

Temperature has been considered the main control on reef distribution (Precht and Miller 2007) and temperatures during the last full interglacial period (MIS 5e) have been simulated using atmospheric general circulation models. Sea surface temperatures (Montoya et al. 1998) were inferred to have increased only ~1°C relative to present. Muhs et al. (2002) studies of molluscan and coral faunas dated to recent interglacials, MIS 3 and MIS 5, also suggest that temperatures were slightly warmer than in present-day Bermuda, based on the presence at those times of three extra-limital species, including the coral *Colpophyllia natans*.

During the last glacial maximum (LGM), which occurred about 18 kya, temperatures may have been between 1°C and 2.5°C (Crowley 2000; Trend-Staid and Prell 2002) or as much as 4°C or 5°C (Guilderson et al. 1994; Beck et al. 1997) lower than present. Nonetheless, Precht and Miller (2007; pp.263–264) indicate that temperatures … [in the subtropics were] … not low enough to terminate reef development during the [LGM].

The Under-Pinnings of Modern Reefs

The relatively stable period in the first half of highest most recent interglacial sea-level highstand, MIS 5e, could have corresponded to the establishment of the fringing reefs and coastline terraces (Hearty et al. 2007) that underlie present day reef tracts. Apparently, Holocene reef growth covers any surviving remnants of these older reefs in shallow water. Shallow patch reefs (2–10 m depth) of the North Lagoon and deeper soft sediment basins (~15–18 m) correspond to Pleistocene topographic features (Kuhn et al. 1981; Logan 1988).

Data from 240 km UNIBOOM seismic tracts (12 tracts) indicated Pleistocene foundations underneath reefs along the northwestern through southwestern margins of the Platform, and under patch reefs in Castle Harbour, but not under those in Harrington Sound (D. Meischner pers. comm. to AL; 17/04/2012). Cores from a few of the patch reefs in Castle Harbour, down to the Pleistocene substrates, showed the same hermatypic growth forms as seen in Holocene reefs of Bermuda. The seismic sections showed Pleistocene reefs of similar dimensions and extension to the overlying Holocene breakers and it seems reasonable that a reef line similar to the Holocene one extends under the whole of the recent rim-reef. A few Th/U dates from cores taken of the Pleistocene rim reefs rendered values about 125 kya.

Reefs of glacial periods would be deep, submerged, reefs. Submersible studies along the north slope (near North

Rock) revealed little of fossil reefs with the exception of a few vertically incised channels that offer profiles of stacked Pleistocene formations down to 160 m, but "Submersible observations around the platform edge by Meischner (pers. comm. to AL) indicate the presence of a supposed Wisconsin-age reef beginning at about 110 m and resting on even older reefs which can be seen in gully walls through the rhodolith cover down to 200 m depth" (Logan 1988; p. 5). The last glacial period, and other transitional periods, may be evidenced by sea bottom profiles between 60 m and 120 m depth that are now being explored (Iliffe et al. 2011).

Modern Reefs

Present broad reef zones are shown in Fig. 10.4. Bermuda has a fairly consistent diversity of coral species, representing a subset of those found in the greater Caribbean. However, species of *Colpophyllia* (Muhs et al. 2002) and *Cladocora* (Moore and Moore 1946), found as fossils, no longer occur in Bermuda. The predominant *Diploria-Montastraea-Porites* coral assemblage of the Caribbean also dominates Bermudian reefs. The genus *Acropora*, an important reef-builder in the Caribbean, is notably absent throughout the history of Bermuda.

The shallow system of rim reefs that borders the lagoon is formed primarily by either stony corals or by vermetid molluscs and coralline algae. The rim reef system is about 2–10 m deep and 1–1.5 km wide, and descends to the main terrace at about 20 m depth; below this terrace is the fore-reef slope with hermatypic corals extending to about 70 m. Scattered coral reef patches and coral communities occur within the North Lagoon and other inshore waters, intermixed with unconsolidated carbonate sediments. The coral patch reefs rise to within a few meters of the water surface. Coral communities that are spread across the Platform include meadows of seagrasses and extensive beds of calcareous green algae (KAC, JWF, WJK, SAM, unpub. obs.).

In surveys of the benthic habitats of the entire Bermuda platform, up to about 15–20 m outside the rim, soft corals and hard corals were found at 27.8% and 33.7% out of about 530 survey sites, respectively (Fig. 10.5a, b). Thus, roughly one third of the Platform is a coral community zone.

Modern Reef Types and Their Communities

There are two major reef-building communities in Bermuda: a coral-algal consortium responsible for most of the reefs on and around the Platform (Fig. 10.6a) and an algal-vermetid consortium, found mainly around the edge of the Platform and particularly along the South Shore. Sea rods and sea fans (soft corals or octocorals) are very prominent members of most reef habitat around Bermuda, which can cover up to 50% of the bottom on coral reefs. Cover values for soft corals are lower along more exposed southeast facing reefs than on reefs to the north, east or west.

Logan (1988) provided a detailed account of coral reef zones of Bermuda, which is summarized below.

Fore-Reef Slope

Fore-reef slope coral-algal reefs occur outside the margin of the platform from 20 m to 50 m depth (Logan and Murdoch 2011). Constructional coral growth along the southwest side extends only to about 30 m (Meischner and Meischner 1977). Total coral coverage ranges from about 50% (20 m) to 25% (30 m) in the shallow part of this reef zone (Logan 1992); coral cover and species diversity are reduced below 40–50 m (Focke and Gebelein 1978; Fricke and Meischner 1985). The presence of mobile rhodolith fields below 50–60 m may prevent the establishment of coral reefs (Fricke and Meischner 1985). The deep fore-reef has not been studied extensively, but submersible dives and recent mixed-gas diving studies have described a distinctive, depauperate, hermatypic coral fauna to 60 m (Fricke and Meischner 1985). Isolated remnants of Pleistocene reefs and patches of lithified rhodoliths (rhodolites) support an association of *Montastraea cavernosa*, *Agaricia fragilis*, *Scolymia cubensis*, antipatharians, sponges and deep water octocorals. The deepest hermatypic coral observed was a specimen of *M. cavernosa* found at 78 m (SRS and T. Iliffe, unpub. obs.). The dominant corals from 20 to 30 m are large overlapping shingle-like or platy colonies of *Montastraea franksi* (Fig. 10.6b) and domal heads of *Diploria strigosa* and *M. cavernosa*. These species account for over 85% of the total coral coverage. The bottom is highly irregular, with holes of 1–2 m relief between coral colonies. Understorey species include *Porites astreoides* and *Diploria labyrinthiformis,* but overall coral diversity compared to inshore reefs is low (Logan 1992). Octocorals are common, as is the encrusting growth form of the hydrozoan *Millepora alcicornis*. There is high, but seasonal, coverage by species of the fleshy phaeophytes *Lobophora, Dictyota* and *Stypopodium* (Logan 1998).

Main Terrace

The main terrace of reefs extends from 10 m to 20 m depth, seaward from the rim reefs. A series of reef ridges, separated by sand channels and forming an anastomosing pattern, similar to spur-and-groove structure (Upchurch 1970), extends from a narrow sediment apron at about 5 m depth to a relatively flat terrace at 15–20 m, which then merges into the fore-reef slope (Fig. 10.6a). These ridges are particularly well-developed along the western edge of the platform. Total coral cover is the highest in Bermuda, frequently reaching 50%, but coral diversity is low (Logan 1992; MEP 2007). The bottom has less relief than the fore-reef slope,

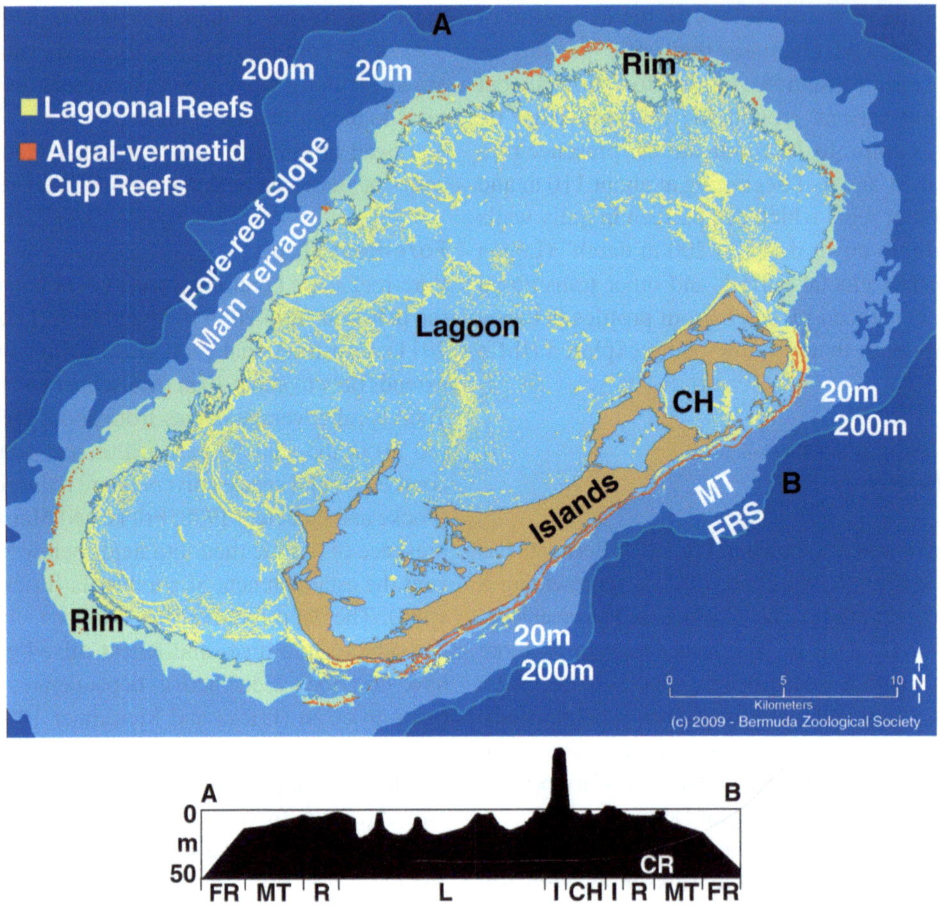

Fig. 10.4 Reef zones on the Bermuda Platform, interpreted from a geo-referenced aerial photomosaic (Copyright Bermuda Zoological Society), showing 20 m and 200 m isobaths. *Below*: NW to SE profile across the Bermuda Platform between points A-B indicated in the upper image; vertical exaggeration approx. 600 times. CH, Castle Harbour; CR, South Shore algal cup reef tract; FR and FRS, fore-reef slope; I, Bermuda Islands; L, lagoon; MT, main terrace; and R, rim. Figure after Logan and Murdoch (2011)

and is dominated by domal colonies of *Diploria* spp (64%), sheet-like or encrusting colonies of *Montastraea* spp (32%) and small hemispherical colonies of *Porites astreoides* (3%). Octocorals cover less than 6% of the bottom (MEP 2007).

Rim

The rim reefs, known locally as ledge flats, are developed on the elevated shallow shoals that encircle North Lagoon. The rim reefs project lagoonwards by lobate extensions (Fig. 10.6a). The reef tops lie between 2 m and 6 m depth and are dissected by ramifying sand channels of about 10–15 m depth. Reef tops show relief of about 1 m between coral heads. Coral coverage is about 20% (Dodge et al. 1982). Large octocorals are attached to the reef tops and channel sides, experiencing almost continuous surge from the open ocean. The *Diploria-Montastraea-Porites* assemblage is again predominant (Fig. 10.7), accounting for over 90% of the coral coverage, with *Diploria* spp alone accounting for over 65%. A wide variety of coral growth forms occur, from domal to encrusting to platy. Sponges, zoanthids, hydrozoans, anemones and corallimorphs are common, with smaller colonies of less common coral species (*Madracis* spp., *Stephanocoenia intersepta, Siderastraea radians, Agaricia fragilis*) present as understorey species. Diverse coelobite communities colonize shaded areas beneath coral heads or in caves and tunnels near the base of the reef (Logan et al. 1984).

Lagoonal Reefs

Reefs of the North Lagoon comprise patch reefs of many sizes and shapes (Logan 1988, 1992). Typical lagoonal patch reefs reach close to the sea surface, with steep flanks running down to 20–23 m. Coral coverage on the tops of these reefs is generally less than 20% (Dodge et al. 1982), although the flanks may have higher values (T. Murdoch, pers. comm.). Species of *Diploria* and *Porites* dominate the outer patch reefs, *Montastraea franksi* the central areas and *Madracis* spp the nearshore reefs (Murdoch 2007). The lagoonal reefs have more coral species than the outer platform reefs, and in

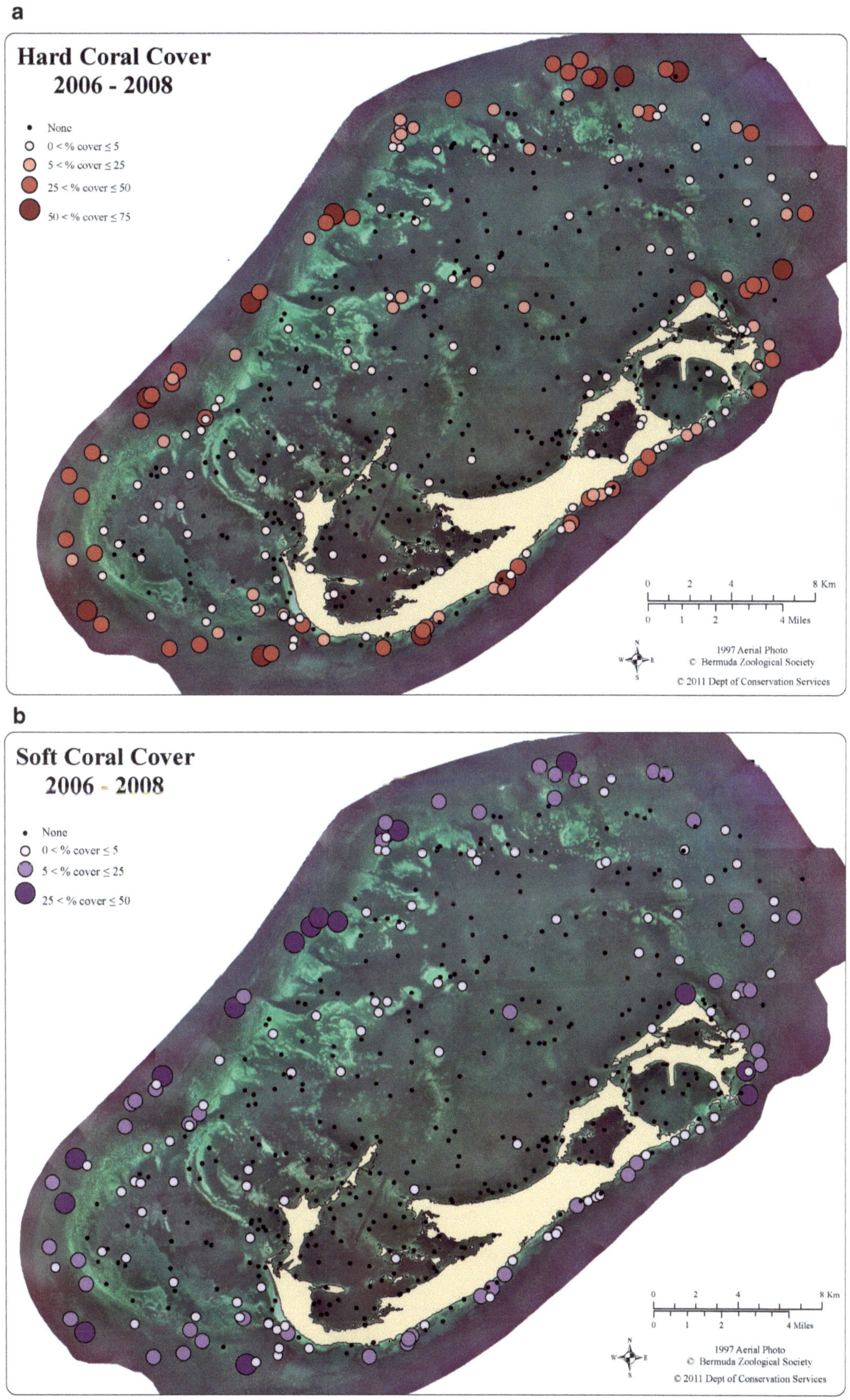

Fig. 10.5 Average percent bottom cover estimates based on 530 benthic transect sites of 50 m by 0.5 m for (**a**) Hard corals, including *Millepora* and (**b**) Soft corals. Based on KAC, JWF, WJK, SAM unpub. data

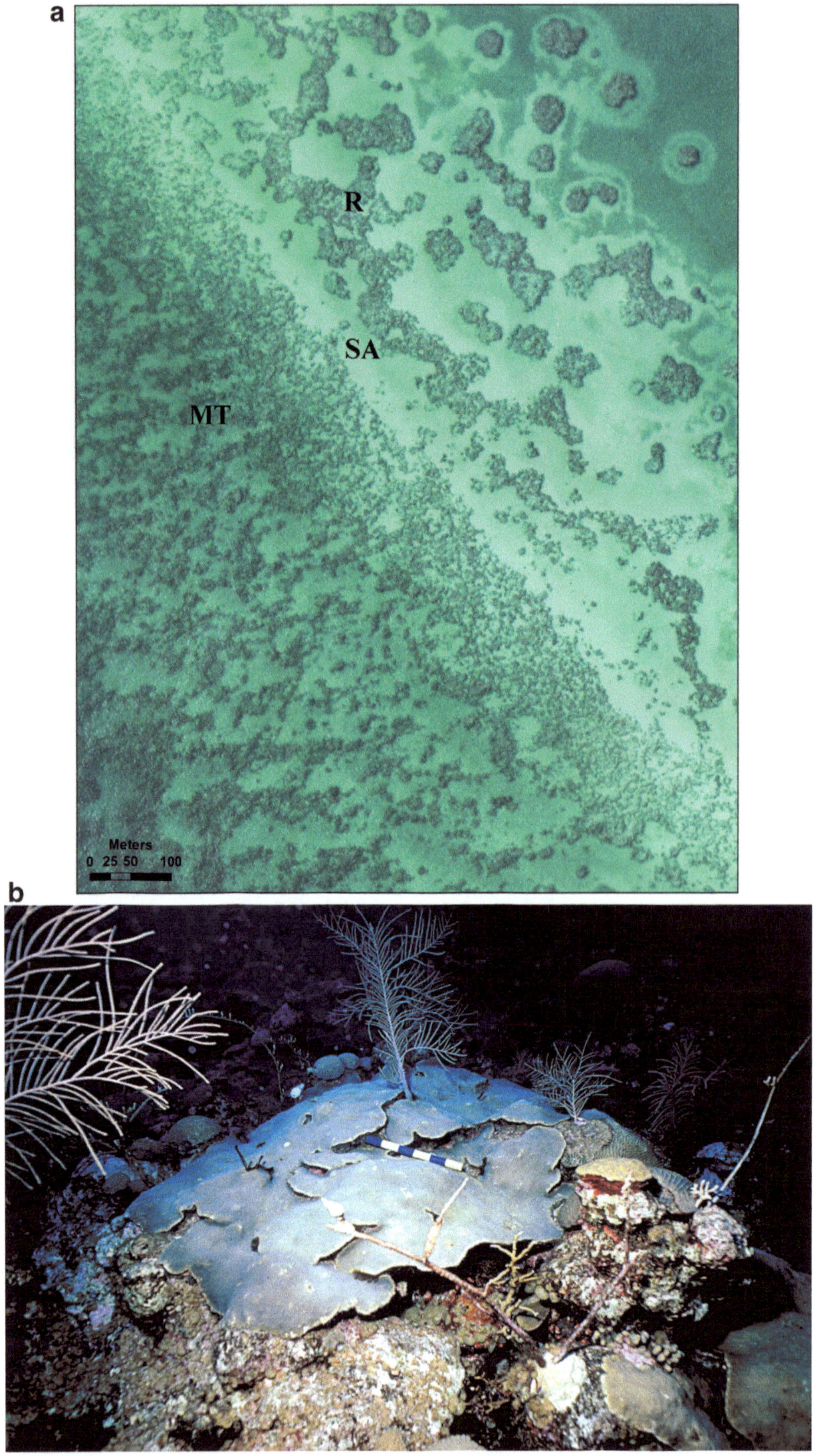

Fig. 10.6 (**a**) Aerial photograph (Copyright 1997, Bermuda Zoological Society) of lobate rim reefs (R) at the western end of Bermuda, in the area referred to as the Ledge Flats, a sediment apron (SA) and anastomosing reef spurs and sand channels of the descending main terrace reefs (MT). (**b**) A large colony of *Montastraea franksi* showing platy growth on the deep fore-reef slope north off North Rock, 32 m depth. Scale on the coral is 30 cm long

Fig. 10.7 Rim reefs with domal *Diploria* spp, *Porites astreoides* in the middle ground and octocorals, *Plexaura flexuosa* (purple), *Plexaura homomalla* (black), a single *Pseuopterogorgia* and probably *Antillogorgia* (foreground near centre, light purple), 5-8 m depth, on the western rim, July 2007

Fig. 10.8 Outer lagoon patch reef with large *Diploria strigosa*, branching *Millepora alcicornis* and octocorals, 5 m depth, Three-Hill Shoals, North Lagoon; queen parrotfish is about 50 cm in length

addition, support a rich sessile invertebrate biota of octocorals, zoanthids, sponges, anemones, tunicates and bivalves, as well as a variety of calcareous algae (Fig. 10.8).

Inshore

Hard bottom coral communities, with up to 5% coral cover, are common in Harrington, Great and Little Sounds. However, of Bermuda's inshore waters only Castle Harbour has significant coral reef development; linear reefs occur around the western and southern shorelines (Fig. 10.9) and steep-sided patch (pinnacle) reefs are present in the north-western and south-eastern areas.

Dredging for airport construction in 1941–1943 had deleterious effects on all reefs in Castle Harbour, reducing cover to only about 5% on linear reefs and 13% on pinnacle reefs (Dodge and Vaisnys 1977; Dryer and Logan 1978; Logan 1992). Prior to the dredging, the waters of Castle Harbour were pristine and supported healthy reefs (Dryer and Logan 1978). *Isophyllia sinuosa* and *D. labyrinthiformis* are the dominant corals on the tops of the pinnacle reefs, which have both low cover and diversity, whereas the steep flanks have relatively high cover by branching species of *Oculina* and *Madracis*. Recent surveys of Castle Harbour reefs by Cook et al. (1994) and Flood et al. (2005) indicate that *D. labyrinthiformis*, an efficient sediment-shedder, remains dominant on reef tops and there is active recruitment of *D. strigosa*, *Agaricia fragilis* and *Siderastraea* spp. Logan et al. (1994) showed that growth rates of post-dredging-age colonies of both *Diploria* spp are, surprisingly, higher in Castle Harbour than on lagoonal and platform-margin reefs. *Madracis auretenra* (misnamed in earlier studies as *M. mirabilis*) continues to rank high in coverage on the pinnacle reef flanks but *Oculina diffusa* appears to have declined since the 1978 survey (Flood et al. 2005).

Fig. 10.9 Castle Harbour, showing lobate fringing linear reefs along the northwestern shore of the Harbour; the airport runway, the Causeway with Longbird swingbridge are in the background. Dark blue areas of water are sites of dredging for the fill used to create the airport lands. The image is oriented with N at the top

Algal-Vermetid Cup Reefs

Algal-vermetid cup reefs (Fig. 10.10a, b) occur as a discontinuous tract on the outer edge of the platform rim, particularly along the south-eastern side where there are three distinct zones running more-or-less parallel to the shoreline. From the shore outwards, the first zone is bioconstructional lips attached to headlands, the second is the actively-growing tract at the edge of the near-shore platform and the third and oldest zone is drowned cup reefs lying at depths of 10–12 m (Meischner and Meischner 1977). These may have been at sea-level about 7,000 years ago and the bioconstructional lips will eventually become the actively-growing tract as headlands are eroded. Cup reefs are generally circular to oval in shape and less than 30 m in maximum dimension. In profile they have an elevated rim enclosing a shallow mini-lagoon with occasional small coral heads, and tapering to a narrow undercut base at 8–10 m depth (Logan 1992). Void space is high in these reefs at both micro- and macroscales (Logan et al. 1984). The main constructive agents are crustose coralline algae and the partially-embedded vermetid gastropod *Dendropoma corrodens*, with occasional encrusting *Millepora alcicornis*, all of which are adapted to turbulent conditions in high wave energy environments (Thomas and Stevens 1991). The algal-vermetid cup reefs represent an unusual reef type rarely found elsewhere in the world.

Other Major Members of the Coral Community Zone

As in many other tropical locations with coral reefs, seagrass beds and submerged macroalgae beds are closely associated spatially and ecologically with the corals. However, the spatial distribution of the reefs and contact potential of reef dwellers with seagrass and macroalgae habitats appears to be greater in Bermuda than elsewhere in the North Atlantic (JWK, pers. obs.). This coral community zone includes most of the Bermuda Platform and seagrasses and calcareous green macroalgae are widespread (Fig. 10.11a, b). Numerous studies document the ecological services provided by these species and communities as well as the ecological connectivities among them.

Fig. 10.10 (**a**) Near-shore algal-vermetid cup reef at low tide, about 6 m across, showing exposed rim and mini-lagoon, Elbow Beach, South Shore; note outer cup reef tract in background where waves are breaking. (**b**) Waves breaking over a line of algal-vermetid cup reefs in foreground, with rim reefs of nearshore platform behind. Dark smudge near centre of the image is Seabright sewage outfall

Seagrass Beds

Fewer seagrass species occur in Bermuda than in the Greater Caribbean; five genera and six species are reported for Bermuda: *Ruppia maritima, Thalassia testudinum, Syringodium filiforme*, *Halophila decipiens* and both *Halodule bermudensis* and *Halodule wrightii*. *Ruppia maritima* is restricted to land-locked brackish or marine ponds. In recent studies only four species are found in open waters, *S. filiforme*, *T. testudinum*, *H. decipiens* and *Halodule* sp. (not identified).

At least one seagrass species has been found at about 24% of about 530 sites that are distributed across the Platform. *Syringodium filiforme* was most commonly encountered, followed by *H. decipiens* and then *T. testudinum* and *Halodule* sp. (KAC, JWF. WFK, SAM, unpub. obs.). Prior to those studies the annual seagrass, *H. decipiens* was considered

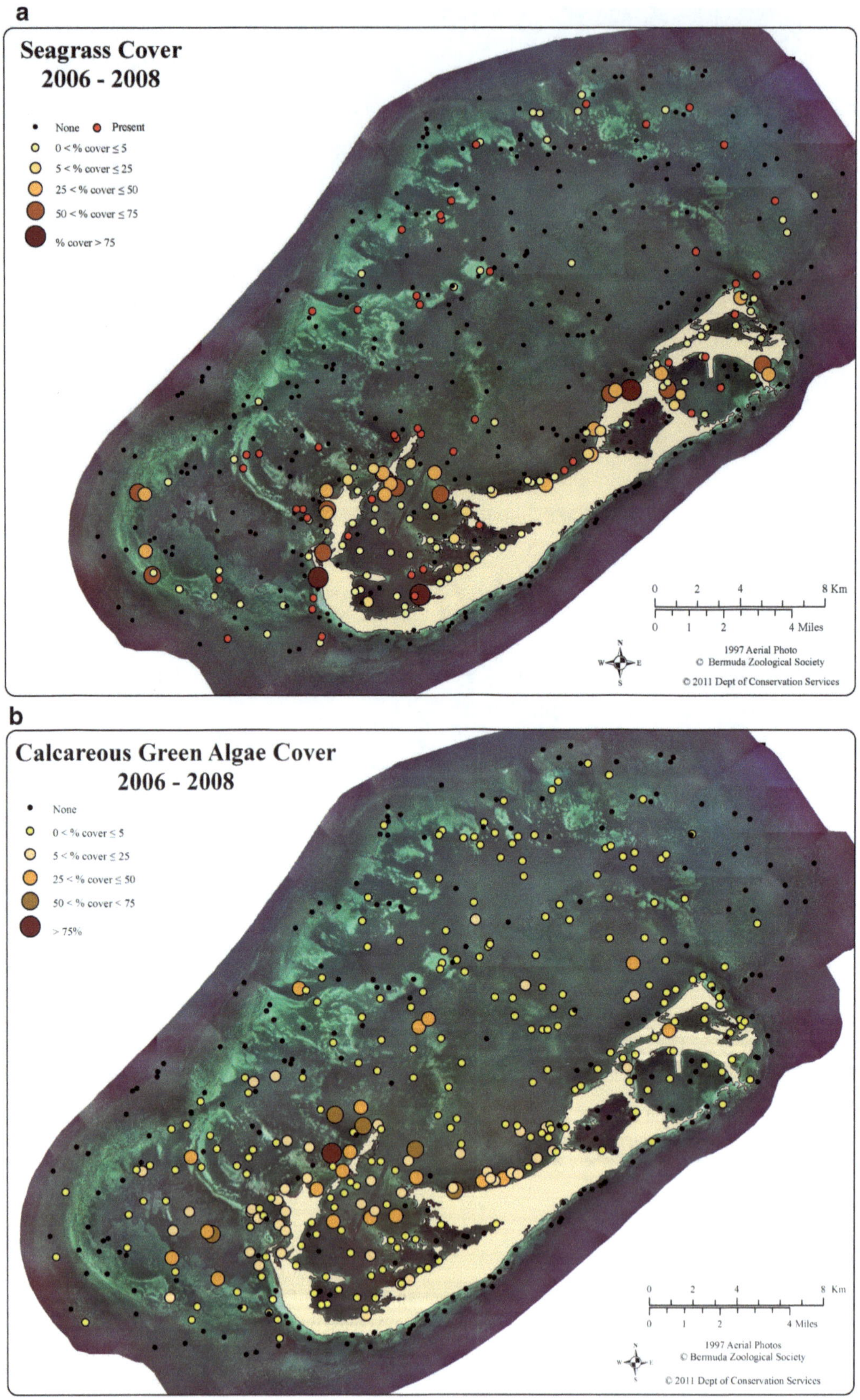

Fig. 10.11 Presence, absence and average percent bottom cover based on 530 benthic transect sites of 50 m by 0.5 m (**a**) Seagrass. Dense cover always includes either or both of *Thalassia testudinum* and *Syringodium filiforme*. (**b**) Calcareous green macroalgae. Based on KAC, JWF, JWK, SAM unpub. data

rare and *T. testudinum* was considered the dominant species. *Halophila decipiens* grows in deeper or murkier water than the other species.

Seagrasses form dense and spatially extensive beds in only a few locations (Fig. 10.11a). Murdoch et al. (2007) described the recent disappearance of reef-associated meadows in the North Lagoon.

Calcareous Green Algae Beds

Calcareous green algae form a second macrophyte dominated sea-bottom community. These were observed at 57% of about 530 survey sites on the Bermuda Platform (Fig. 10.11b) but were rarely encountered along the South Shore (KAC, JWF, JWK, SAM, unpub. obs.), a higher energy area with coarse mobile sediments. The most common genus was *Penicillus*, followed by *Udotea* and *Halimeda*. The last two sometimes formed dense beds with more than 75% cover.

Summary

Corals and coral reefs have played a central role in the complete history of Bermuda. They formed the platform and the islands, and then they protected those islands.

In some locations reefs have been considered nuisances and were removed, without much regard to the importance of the integrity of the reef system to its own viability and to the viability of Bermuda. Certainly, with more awareness of their past, present and future importance, destruction will decrease and protection will increase. Bermuda's corals occur at the very northern limits of coral reefs in the Atlantic Ocean so that they are surviving under conditions that may exemplify limiting environmental conditions for reef development. Understanding these reefs will become of critical importance in anticipating the effects of global climate change.

References

Andersson J, Mackenzie FT (2011) Ocean acidification: setting the record straight. Biogeosci Discuss 8:6161–6190

Badgley BD, Lipschultz F, Sebens KP (2006) Nitrate uptake by the reef coral *Diploria strigosa*: effects of concentration, waterflow and irradiance. Mar Biol 149:327–338

Barnes JA, von Bodungen B (1978) The Bermuda marine environment, vol II. Bermuda Biological Station for Research, special publication no. 17. Bermuda Biological Station, St Georges

Bates NR, Amat A, Andersson AJ (2010) Feedbacks and responses of coral calcification of the Bermuda reef system to seasonal changes in biological processes and ocean acidification. Biogeosciences 7:2509–2530

Beck JW, Recy J, Taylor F, Edwards RL, Cabioch G (1997) Abrupt changes in early Holocene tropical sea surface temperature derived from coral records. Nature 385:705–707

Bermuda Census (2010) Department of Statistics, Bermuda Government, 2010 Census of population and housing preliminary results. Hamilton, Pembroke Parish

Boyer JN, Briceno HO (2010) Water quality monitoring program for Bermuda's coastal resources. Tech Rept Southeast Environ Res Centre, Florida International University, Miami. http://serc.fiu.edu/wqmnetwork/Bermuda/home.html

Brand S (ed) (2009) Hurricane havens handbook. Naval Research Laboratory, Monterey. www.nrlmry.navy.mil?~cannon/tr8203nc/0start.htm

Cairns SD (1979) The deep-water Scleractinia of the Caribbean and adjacent waters. Stud Fauna Curaçao. 57:1–341

Calder DR (2000) Assemblages of hydroids Cnidaria) from the three seamounts near Bermuda in the western North Atlantic. Deep-Sea Res I 47:1125–1139

Cook CB, Dodge RE, Smith SR (1994) Fifty years of impacts on coral reefs in Bermuda. In: Ginsburg RN (ed) Proceedings on global aspects of coral reefs: health, hazards, and history, Rosenstiel School of Marine and Atmospheric Science, University of Miami, Miami, pp160–166

Crowley TJ (2000) CLIMAP SSTs re-visited. Clim Dynam 16:241–255

Dodge RS, Vaisnys JR (1977) Coral populations and growth patterns: responses to sedimentation and turbidity associated with dredging. J Mar Res 35:715–730

Dodge RE, Logan A, Antonius A (1982) Quantitative reef assessment studies in Bermuda: a comparison of methods and preliminary results. Bull Mar Sci 32:745–760

Dryer S, Logan A (1978) Holocene reefs and sediments of Castle Harbour, Bermuda. J Mar Res 36:399–425

Ellison JC (1993) Mangrove retreat with rising sea-level, Bermuda. Estuar Coast Shelf Sci 37:75–87

Flood VS, Pitt JM, Smith SR (2005) Historical and ecological analysis of coral communities in Castle Harbour (Bermuda) after more than a century of environmental perturbation. Mar Pollut Bull 51:545–557

Focke JW, Gebelein CD (1978) Marine lithification of reef rock and rhodolites at a fore-reef slope locality (–50m) off Bermuda. Geol Mijnbouw 52:163–171

Fricke H, Meischner D (1985) Depth limits of Bermudan scleractinian corals: a submersible survey. Mar Biol 88:175–187

Gees RA, Medioli F (1970) A continuous seismic survey of the Bermuda Platform, part I: Castle Harbour. Marit Sed 6:21–25

Gordon AL, Guilivi CF (2008) Sea surface salinity trends: over fifty years within the subtropical North Atlantic. Oceanography 21:21–29

Guilderson TP, Fairbanks RG, Rubenstone JL (1994) Tropical temperature variations since 20,000 years ago: modulating interhemispheric climate change. Science 263:663–665

Hayward WB (1924) Bermuda past and present. Dodd Mead & Company, New York

Hearty PJ (2002) Revision of the late Pleistocene stratigraphy of Bermuda. Sed Geol 153:1–21

Hearty PJ, Kaufman DS (2000) Whole-rock aminostratigraphy and Quaternary sea-level history of the Bahamas. Quat Res 54:163–173

Hearty PJ, Olson SL (2011) Preservation of trace fossils and molds of terrestrial biota by intense storms in mid–last interglacial (MIS 5c) dunes on Bermuda, with a model for development of hydrological conduits. Palaios 26:394–405

Hearty PJ, Hollin JT, Neumann AC, O'Leary MJ, McCulloch M (2007) Global sea-level fluctuations during the last interglaciation (MIS 5e). Quat Sci Rev 26:2090–2112

Iliffe TM, Kvitek R, Blasco S, Blasco K, Covill R (2011) Search for Bermuda's deep water caves. Hydrobiologia 677:157–168. doi:10.1007/s10750-011-0883-1

Jickells TD, Knap AH, Smith SR (1986) Trace metals and nutrient fluxes through the Bermuda inshore waters. Rapp P V Reun Cons Int Expl Mer 186:251–262

Jones RJ, Parsons R, Watkinson E, Kendall D (2010) Sewage contamination of a densely populated coral 'atoll' (Bermuda). Environ Monit Assess. doi:10.1007/s10661-010-1738-3

Kuhn G, Torunski H, Meischner D (1981) Reef growth and lagoon development on the Bermuda carbonate platform recorded by seismic reflection profiling and deep vibration coring. In: Proceedings of the fourth international coral reef symposium, Marine Sciences Centre, University of the Phillippines, Manila vol 1, p 597

Lapointe BE, O'Connell J (1989) Nutrient-enhanced growth of *Cladophora prolifera* in Harrington Sound, Bermuda: eutrophication of a confined, phosphorus-limited marine ecosystem. Estuar Coast Shelf Sci 28:347–360

Locarnini RA, Mishonov AV, Antonov JI, Boyer TP, Garcia HE (2006) In: Levitus S (ed) World ocean atlas 2005, volume 1: temperature, vol 61, NOAA Atlas NESDIS. US Government Printing Office, Washington, DC

Logan A (1988) Holocene reefs of Bermuda, vol XI, Sedimenta. Rosenstiel School of Marine and Atmospheric Science, University of Miami, Miami

Logan A (1992) Reefs. In: Thomas MLH, Logan A (eds) A guide to the ecology of shoreline and shallow-water marine communities of Bermuda, Bermuda Biological Station for Research special publication no. 30. Wm. C. Brown, Dubuque, pp 31–68

Logan A (1998) The high-latitude coral reefs of Bermuda: characteristics and comparisons. In: Viera Rodriguez MA, Haroun R (eds) Proceedings of the second symposium of fauna and flora of the Atlantic Islands, Las Palmas de Gran Canaria, 1996, Bol Mus Municipal do Funchal, Las Palmas, suppl. 5, pp 187–197

Logan A, Murdoch TJT (2011) Bermuda. In: Hopley D (ed) Encyclopedia of modern coral reefs. Springer, Dordrecht, pp 118–123

Logan A, Mathers SM, Thomas MLH (1984) Sessile invertebrate coelobite communities from reefs of Bermuda: species composition and distribution. Coral Reefs 2:205–213

Logan A, Yang L, Tomascik T (1994) Linear skeletal extension rates in two species of *Diploria* from high-latitude reefs of Bermuda. Coral Reefs 13:225–230

Marine Environmental Program (2007) Annual report: 2006 to 2007. Annual report submitted by the Bermuda Institute of Ocean Sciences to the Bermuda Department of Environmental Protection, Ministry of Environment and Sport. Bermuda Aquarium, Museum, and Zoo Pub #2230

McGillicuddy DJ, Anderson LA, Bates NR, Bibby T, Buesseler KO, Carlson CA, Davis CS, Ewart C, Falkowski PG, Goldthwait SA, Hansell DA, Jenkins WJ, Johnson R, Kosnyrev VK, Ledwell JR, Li QP, Siegel DA, Steinberg DK (2007) Eddy/wind interactions stimulate extraordinary mid-ocean plankton blooms. Science 316:1021–1026

McGlathery KJ (1995) Nutrient and grazing influences on a subtropical seagrass community. Mar Ecol Prog Ser 122:239–252

Meischner D, Meischner U (1977) Bermuda south shore reef morphology – a preliminary report In: Taylor DL (ed) Proceedings of the third international coral reef symposium, Rosenstiel School of Marine and Atmospheric Science, University of Miami, Miami, vol 2, pp 243–250

Montoya M, Crowley TJ, von Storch H (1998) Temperatures at the last interglacial simulated by a coupled ocean atmosphere model. Palaeo Oceanogr 13:170–177

Moore HB, Moore DM (1946) Preglacial history of Bermuda. Bull Geol Soc Am 57:207–222

Morris BF, Barnes J, Brown IF, Markham JC (1977) The Bermuda marine environment, vol I. Bermuda Biological Station for Research, special publication no. 15. Bermuda Biological Station for Research, St Georges

Muhs DR, Simmons KR, Steinke B (2002) Timing and warmth of the last interglacial period: new U-series evidence from Hawaii and Bermuda and a new fossil compilation for North America. Quat Sci Rev 21:1355–1383

Murdoch TJT (2007) A functional group approach for predicting the composition of hard coral assemblages in Florida and Bermuda. Ph.D. thesis, University of South Alabama, Mobile

Murdoch TJT, Glasspool AF, Outerbridge M, Ward J, Manuel S, Nash A, Coates KA, Pitt J, Fourqurean JW, Barnes PAG, Vierros M, Holzer K, Smith SR (2007) Large-scale decline in offshore seagrass meadows in Bermuda. Mar Ecol Prog Ser 339:123–130

NASA (2010) Surface meteorology and solar energy. A renewable energy resource web site (release 6.0), Stackhouse PW Jr, Whitlock CH, Kusterer JM (Administrators). http://eosweb.larc.nasa.gov/sse/

Neumann CJ, Cry GW, Caso EL, Jarvinen BR (1985) Tropical cyclones of the North Atlantic Ocean, 1871–1980. National Climatic Center, Asheville

NOAA (2012) Tides and currents. http://www.tidesandcurrents.noaa.gov/geo.shtml?location=2695540

Olson SL, Hearty PJ (2009) A sustained +21 m sea-level highstand during MIS 11 (400 ka): direct fossil and sedimentary evidence from Bermuda. Quat Sci Rev 28:271–285

Olson SL, Hearty PJ, Pregill GK (2006) Geological constraints on evolution and survival in endemic reptiles on Bermuda. J Herp 40:394–398

Pirazzoli PA (1987) Recent sea-level changes in the North Atlantic. In: Scott DB, Pirazzoli PA, Honig CA (eds) Later Quarternary sea-level correlation and applications. Kluwer, Boston, pp 153–167

Pirsson LV (1914a) Geology of Bermuda Island: the igneous platform. Am J Sci 38:189–206

Pirsson LV (1914b) Geology of Bermuda Island: petrology of the lavas. Am J Sci 38:331–344

Precht WF, Miller SL (2007) Ecological shifts along the Florida reef tract. In: Aronson RB (ed) Geological approaches to coral reef ecology. Springer, New York, pp 237–312

Prognon F, Cojan I, Kindler P, Thiry M, Demange M (2011) Mineralogical evidence for a local volcanic origin of the parent material of Bermuda Quaternary paleosols. Quat Res 75:256–266

Seamount Biogeosciences Network (2012) Seamount catalogue. Earthref.org/SC/

Seigel DA, McGillicuddy DJ Jr, Fields EA (1999) Mesoscale eddies, satellite altimetry, and new production in the Sargasso Sea. J Geophys Res 104:13,359–13,379

Siegel DA, Michaels AF, Sorensen JC, O'Brien MC, Hammer MA (1995) Seasonal variability of light availability and utilization in the Sargasso Sea. J Geophys Res 100:8695–8713

Simmons JAK, Johnson R (1993) Measurements of coastal currents at Tynes Bay and the North Lagoon, Bermuda. Bermuda Biological Station for Research, St Georges

Simmons JAK, Lyons WB (1994) The ground water flux of nitrogen and phosphorus to Bermuda's coastal waters. Water Res Bull 306:983–991

Smith Warner International (2004) Bermuda coastal erosion vulnerability assessment report. Submitted to the Government of Bermuda, Ministry of the Environment

Spalding M, Fos H, Allen G, Davidson N, Ferdaña Z, Finlayson M, Halpern BS, Jorge MA, Lomgana A, Lourie SA, Martin KD, McManus E, Molnar J, Recchia CA, Robertson J (2007) Marine ecoregions of the world: a bioregionalization of coastal and shelf areas. Bioscience 57:573–583

Stanley DJ, Swift DJP (1968) Bermuda's reef front platform – bathymetry and significance. Mar Geol 6:479–500

State of the Environment Report (2005) Ministry of the Environment Bermuda government. Hamilton, Pembroke Parish

Steinberg DK, Carlson CA, Bates NR, Johnson RJ, Michaels AF, Knap AH (2001) Overview of the US JGOFS Bermuda Atlantic Time-series Study (BATS): a decade-scale look at ocean biology and biogeochemistry. Deep-Sea Res II 48:1405–1447

Thacker WC, Sindlinger L (2007) Estimating salinity to complement observed temperature: 2 Northwestern Atlantic. J Mar Sci 65:249–267

Thomas MLH, Stevens J (1991) Communities of constructional lips and cup reef rims in Bermuda. Coral Reef 9:225–230

Trend-Staid M, Prell WL (2002) Sea surface temperature at the last glacial maximum: a reconstruction using the modern analog technique. Paleoceanography 17:1065–1083

Upchurch SB (1970) Sedimentation on the Bermuda Platform. U.S. Army Corps of Engineers, U.S. Lake Survey Research Report, RR 2-2, Detroit

Vacher HL, Rowe MP (1997) Geology and hydrogeology of Bermuda. In: Vacher HL, Quinn T (eds) Geology and hydrogeology of carbonate islands, 54 Developments in sedimentology. Elsevier, Amsterdam, pp 35–90

Vogt PR (1979) Volcano height and paleo-plate thickness. In: Tucholke BE, Vogt PR (eds) Initial reports of the deep sea drilling project, vol 43. US Government Printing Office, Washington, DC, pp 877–878

Vogt PR, Jung W-Y (2007) Origin of the Bermuda volcanoes and Bermuda Rise: history, observations, models, and puzzles. In: Foulger GR, Jurdy DM (eds) Plates, plumes, and planetary processes, Geological Society America special paper no. 430. Geological Society of America, Boulder

von Bodungen B, Jickells TD, Smith SR, Ward JAD, Hillier GB (1982) The Bermuda marine environment vol III: final report of the Bermuda inshore waters investigation. Bermuda Biological Station for Research special publication no. 19. St Georges, Bermuda

Biology and Ecology of Corals and Fishes on the Bermuda Platform

11

Struan R. Smith, Samantha de Putron, Thad J.T. Murdoch, Joanna M. Pitt, and Ivan Nagelkerken

Introduction

Bermuda's shallow coral reef ecosystem contains a variety of habitats inhabited by a diversity of algae, invertebrates and fish species derived from the Caribbean flora and fauna, as well as a number of endemic species. The following discusses reproduction, recruitment and growth of corals and fishes and their patterns of distribution and key ecological processes. The significance of a highly seasonal and spatially complex lagoon and rim reef is examined and comparisons are made with Caribbean reefs.

Coral Biology

Coral Reproduction

Regarding reproductive method, of the 20 described hermatypic coral species, nine species are known from elsewhere to be brooders and nine are broadcasters, with two species still unknown (Baird et al. 2009). Specific data from Bermuda is only available for a few species, although these have confirmed the same mode in each case (Wyers 1985; Wyers et al. 1991; Goodbody-Gringley and de Putron 2009; de Putron and Smith 2011). Thus, reproductive traits are similar to that seen in the Caribbean, which contrasts with the Indo-Pacific where broadcast spawning is massively dominant and where <20% of species are brooders (Baird et al. 2009). Soft corals in Bermuda are dominated by gorgonians (sea fans, sea rods, sea plumes) as in the Caribbean, but relatively little is known about gorgonian reproduction. Of 20 reported species from Bermuda, reproductive pattern is only known for the sea rods *Pseudoplexaura porosa* (de Putron and Ryland 2009) and *Plexaura flexuosa* (Pakes and Woollacott 2008), both of which are broadcast spawners. Of Bermudan species from other Caribbean locations eight are spawners, one species is a surface brooder (de Putron 2003), and reproductive mode is unknown for 11 species.

High-latitude reefs such as Bermuda, where coral species are at their distribution extreme, can provide interesting insights into the study of environmental factors controlling reproductive cycles since there is a wider range in parameters such as seasonal seawater temperature and photoperiod. Most scleractinians that broadcast do so over a discrete period of 1–2 months and the timing of spawning in Bermuda is similar to conspecifics in the Caribbean (van Woesik et al. 2006). Broadcast spawning occurs in Bermuda 7–9 day after the full moons of July to early September in *Diploria labyrinthiformis, D. strigosa, Montastraea cavernosa, and M. franksi* (Wyers 1985; Wyers et al. 1991). Gamete release for gorgonians overlaps with the scleractinians, occurring 5–8 day after the full moons of August-September (Pakes and Woollacott 2008; de Putron and Ryland 2009). In comparison, corals that brood planulae often do so over several months (Harrison and Wallace 1990; Baird et al. 2009), though planula release in Bermuda occurs over July-August/September for *Porites astreoides* (de Putron and Smith 2011), *Favia fragum* (Goodbody-Gringley and de Putron 2009), and

S.R. Smith (✉)
Natural History Museum, Bermuda Aquarium, Museum and Zoo, 40 North Shore Road, Hamilton Parish FL04, Bermuda
e-mail: srsmith@gov.bm

S. de Putron
Bermuda Institute of Ocean Sciences, 17 Biological Lane, St Georges GE01, Bermuda
e-mail: samantha.deputron@bios.edu

T.J.T. Murdoch
Bermuda Zoological Society, 40 North Shore Road, Hamilton Parish FL04, Bermuda
e-mail: tjmurdoch@gov.bm

J.M. Pitt
Department of Environmental Protection, Marine Resources Section, 3 Coney Island Road, St. Georges CR04, Bermuda
e-mail: jpitt@gov.bm

I. Nagelkerken
School of Earth and Environmental Sciences, University of Adelaide, Adelaide SA 5005, Australia
e-mail: ivan.nagelkerken@adelaide.edu.au

C.R.C. Sheppard (ed.), *Coral Reefs of the United Kingdom Overseas Territories*, Coral Reefs of the World 4,
DOI 10.1007/978-94-007-5965-7_11, © Springer Science+Business Media Dordrecht 2013

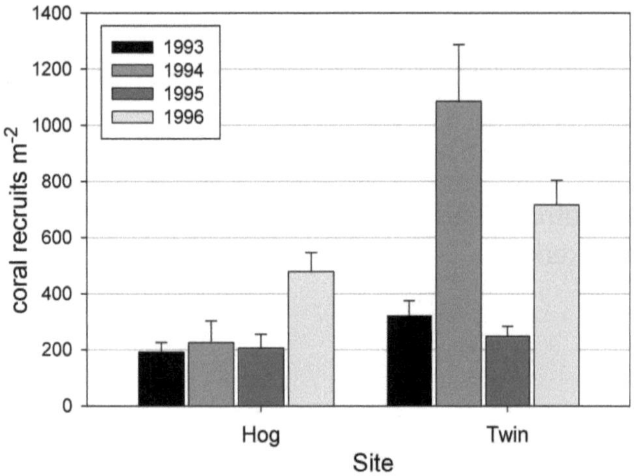

Fig. 11.1 Coral settlement on tile racks at two locations on the northern rim reef at 7–8 m depth, mean of five racks per site, +/– SE. Each rack consisted of eight grooved ceramic tiles (15 cm×15 cm; 4 vertical, 4 horizontal) with five racks per site. Racks were deployed for 12 months. (Smith unpublished data)

Siderastrea radians (de Putron unpublished data), and possibly earlier for *Agaricia fragilis* in May/June (Smith unpublished data). A similar restriction in reproductive season of gorgonian species which broadcast spawn is seen in the studied Bermuda species (de Putron and Ryland 2009). The shortened reproductive season in Bermuda means an overall lower annual reproductive effort, which may make these populations more sensitive to disruptions that occur during the reproductive months compared to species at lower latitudes.

Seasonally restricted reproduction in Bermuda can be explained by the wide annual temperature range and a narrow period of seawater temperature favorable for gamete development (Goodbody-Gringley and de Putron 2009; de Putron and Smith 2011). Even within the reproductive season, optimal seawater temperatures for several species is narrow (de Putron and Ryland 2009; de Putron and Smith 2011). The release of fewer planulae was correlated with higher temperatures and there was significant variation in the number of planulae released in colonies from sites across the Bermuda lagoonal seawater temperature gradient during a 2 year study (de Putron and Smith 2011). Reproductive effort of colonies of the gorgonian *P. porosa* also showed significant variation over summer months; however, in contrast to *P. astreoides*, there was a positive relationship between reproductive effort (gamete volume) of colonies and increasing seawater temperature (de Putron and Ryland 2009). The observed differences of these coral species across the Bermuda lagoonal seawater temperature gradient demonstrates the important role that temperature plays in controlling reproduction. However, turbidity also varies and is highest at the inner lagoon patch reef zone, and may cause a reduced reproductive effort there (de Putron and Smith 2011).

The lunar cycle is the primary environmental cue for the timing of gamete and planula release for many species. Broadcasting corals usually show tight synchrony, necessary for fertilization, whereas with brooding corals, planula release varies with species, varying from no pattern to tight synchrony over a few days (Harrison and Wallace 1990). Peak planulation of *F. fragum* in Bermuda occurred 6–12 days after the new moon, which coincides with peak release by conspecifics in the Caribbean (Goodbody-Gringley and de Putron 2009). Planula release of *P. astreoides* in Bermuda peaked several days before the new moon and diminished after the new moon, which is in slight contrast to colonies in Florida (McGuire 1998; de Putron and Smith 2011). However, this species showed spatial variation in this respect across different reef zones of the 18 km wide Bermuda platform, with colonies from the Inner Lagoon peaking a few days earlier than in the Outer Lagoon and Rim Reef zones (de Putron and Smith 2011). These differences may be influenced by environmental factors such as turbidity and sedimentation rates (de Putron and Smith 2011).

Coral Settlement and Recruitment

Coral recruitment is a key factor influencing coral community structure and is an index of reef health. Recruitment is the combined result of fecundity and the availability of planulae, settlement success, availability of suitable substrates, and of post-settlement mortality and survival. Various studies in Bermuda have examined each factor. Planula availability has been studied using modified sediment traps designed to capture settling planktonic organisms; these were deployed at the northern rim reef (Hog Breaker) over the summer reproductive season of July-September 1995 (de Putron unpublished data). Planulae flux was maximal at the end of July and decreased towards mid September. Mean planulae flux was estimated over the deployment period at 1168.4 planulae m^{-2}. The mean number of settled corals on tiles at the same site and year was 279.7 spat m^{-2}. This suggests that approximately 77% of the available planulae to this reef area do not successfully settle (de Putron unpublished data).

Settlement tile studies have been conducted in 1983 (Smith 1985), 1986 (Smith 1988), annually from 1993 to 1996 (Smith unpublished data, Fig. 11.1) and again in 2005 (Brylewska 2007). Racks with horizontal and vertical tiles (each 15 cm ×15 cm, approximately 0.045 m^2 total surface area per rack) were deployed for different time periods. In the early 1983 study, tiles were deployed for monthly periods at a ship grounding site on the northern rim reef of Bermuda. Settlement between June-September had a mean density of 162 spat m^{-2} (Smith 1985). Later, annual settlement studies done from 1993 to 1996 with tile racks deployed for 1 year showed significant inter-annual and between site differences, with higher rates than previously observed, ranging from 200 m^{-2} to 1,100 m^{-2}.

In 2005, a study was conducted on the influence of reef structure on settlement patterns across the various physiographic reef zones: terrace, rim, outer lagoon, inner lagoon and the enclosed basin of Castle Harbor (Brylewska 2007). There was inter-zone variability in settlement rates with a significantly greater mean number of recruits per tile on the rim, outer and inner lagoonal patch reefs of the northern platform (78–128 recruits m^{-2}). In comparison, inside the enclosed basin of Castle Harbor, and on the southern terrace reefs, settlement ranged from 2–32 recruits per m^2. Thus there is significant temporal and spatial variation in coral settlement in Bermuda.

Spatial variation in coral settlement does not appear to be related to coral abundance. The patch reefs of Castle Harbor have low coral cover (Dryer and Logan 1978; Flood 2004; Brylewska 2007) due to anthropogenic factors, and the terrace reef zone has the highest cover (MEP 2007; Murdoch et al. 2008a) but settlement rates did not differ significantly between them. Therefore, other factors also influence settlement rates, such as high sedimentation (e.g. in Castle Harbor), current patterns influencing availability of planulae, and post-settlement mortality. Overall settlement rates in Bermuda are comparable or higher to those recorded on other western Atlantic reefs: Barbados 79 m^{-2}, Bahamas 106 m^{-2} (Smith 1992).

Survival and growth of settled polyps to juvenile stage (defined here as those visible to the naked eye and up to 5 cm diameter) have been studied in Bermuda by visually recording the size of all recruits within permanent quadrats. Smith (1985, 1992) compared recruits at a ship grounding scar with adjacent disturbed and undisturbed reefs, noting progressive recruitment success over a 10 year period after the grounding. Webster and Smith (2000) identified decreased recruitment success adjacent to a sewage outfall, and recruitment also showed spatial variation across reef zones. Lowest recruitment was recorded in the enclosed basin of Castle Harbor (approximately 0.5–6.5 recruits m^2; Brylweska 2007), compared to approximately 15 recruits m^{-2} on the northern rim reefs (Smith 1992). Across all studies, the brooding species *P. astreoides, F. fragum, A. fragilis* along with *Siderastrea* spp. (which were not identifiable to species) were the dominant juvenile recruits seen in the quadrats. However, *Diploria* spp. and *Montastraea* spp. dominate the adult hard coral cover in Bermuda, indicating that these species are slow to recruit (Fig. 11.2) but suffer much lower rates of juvenile mortality (Smith 1992; Brylewska 2007; Murdoch et al. 2008a).

Coral Growth and Calcification Studies

The earliest assessments of coral growth in Bermuda involved determination of annual growth bands (Dodge and Thomson 1974) and recording of environmental variation in skeletal growth (Dodge and Vaisnys 1975). These authors used x-rays of corallum thin sections to reveal growth density bands and annual patterns of skeletal deposition. Bermuda's high latitude location with its seasonal variation of seawater temperature and solar insolation imposes constraints on coral growth rates (Logan and Tomascik 1991; Logan et al. 1994). Dodge (1978) showed that *Diploria strigosa* only grew 3.2–4.5 mm $year^{-1}$, less than the 3.5–10.0 mm $year^{-1}$ reported by Vaughan (1915) for Floridian and Bahamian corals. Logan and Tomascik (1991) and Logan et al. (1994) confirmed reduced growth rates for *Diploria strigosa*, *Diploria labyrinthiformis* and *Porites asteoides* compared to Caribbean conspecifics. *Scolymia cubensis* also showed reduced growth rates in Bermuda compared to Barbados (+38% vs. +52% change in polyp area per year, respectively, Tomascik and Logan 1990) which was related to reduced Bermuda winter temperatures.

Skeletal extension and density banding studies confirmed an annual growth rate of 3 ±0.5 mm $year^{-1}$ for *Diploria labyrinthiformis* from the Bermuda Terrace reef (Cohen et al. 2004). Rapid extension of the costae occurred in winter while septa and thecae grew fastest in the summer. Skeletal thickening (density) was also seasonal with greater thickening in summer. Over decadal time scales, skeletal density was shown to increase with increasing water temperature (Cohen et al. 2004). The slow annual growth and extension rates and distinct annual banding patterns have permitted the determination of remarkable coral ages in some species, with one *Montastraea cavernosa* exceeding 800 years (Patzold and Wefer 1992) and a *Diploria labyrinthiformis* colony exceeding 225 years (Goodkin et al. 2005).

The variations in temperature and sedimentation across the extensive reef lagoon and at the depth of the outer reef terrace allows for an assessment of growth rates across a range of environmental parameters (Logan and Tomascik 1991; Logan et al. 1994). These authors showed higher linear extension rates for *Diploria* spp. and *Porites astreoides* on shallow inshore reefs compared to corals growing on the rim and terrace reefs, attributed to higher light and more zooplankton in the shallower nearshore environment. In contrast, a retrospective assessment of growth banding patterns of *Diploria* spp. in Castle Harbour determined the negative impact of excessive sedimentation on growth and mortality, caused by dredging of Castle Harbour for the creation of an American airbase in the early 1940s (Dodge and Vaisnys 1977; Flood et al. 2005).

Other studies on coral growth in Bermuda have examined effects of anthropogenic impacts and climate change. Laboratory and field experiments have shown limited effects of oil and dispersed oil on *Diploria strigosa* (Dodge et al. 1984a, b). Growth of *Siderastrea radians* in simulated environment experiments showed no impact of elevated temperature and excess sedimentation on this species that naturally inhabits a wide variety of habitats in Bermuda including shallow inshore sites with large natural temperature variations and high sedimentation rates (Cody et al. 2010).

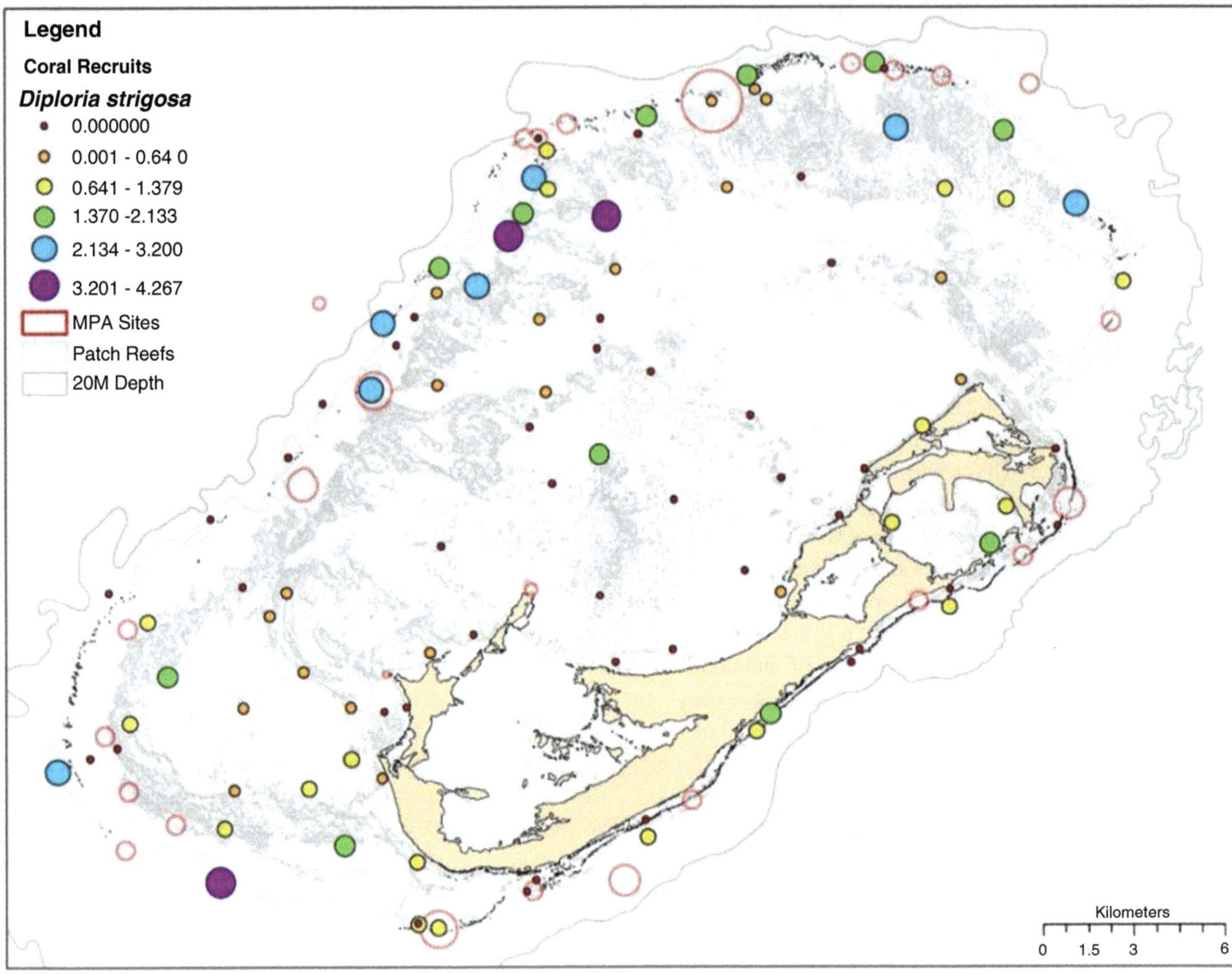

Fig. 11.2 Distribution of *Diploria strigosa* recruits per m^{-2} recorded in BREAM surveys from 2007 to 2008. Thirty 0.25 m^2 quadrats were randomly deployed at each site to census all juvenile corals (Murdoch et al. 2008a)

Other studies assessed the suitability of *Madracis* spp. growth rates as a monitoring tool (Smith et al. 1998).

Laboratory studies of *Favia fragum* and *Porites asteoides* have focused on skeletal formation characteristics in response to future predicted levels of ocean acidification, showing a clear reduction in calcification along with changes in crystal morphology in new recruits (Cohen et al. 2009; de Putron et al. 2011). Recent research within the 'Bermuda Ocean Acidification and Coral Reef Investigation (BEACON)' assessed growth and net calcification of individual adult colonies of *D. labyrinthiformis* and *P. astreoides* exposed to varying carbonate chemistry under both an *in situ* setting across the Bermuda reef platform as well as within mesocosm experiments (Andersson et al. 2011). *In situ* reefal scale studies of air/sea carbon dioxide fluxes over the Bermuda rim reef and fCO_2 have been done in the context of CO_2 flux studies in the Sargasso Sea to establish parameters of ecosystem metabolism and mass balance on the Bermuda reef (Bates et al. 2001; Bates 2002). Bates et al. (2010) describe a seasonal feedback system of productivity and calcification that may not be sufficient in maintaining net calcification under future ocean acidification scenarios (Kleypas et al. 1999). Continuing research on ocean acidification at Bermuda is important as it is at a critical end point with which to assess validity of models that predict coral reef responses to ocean acidification.

The growth of corals in Bermuda has also been examined in the context of biogeochemical signals of oceanographic and climate processes via the incorporation of elements in the skeletal matrix (Nozaki et al. 1978; Druffel 1989; Draschba et al. 2000; Cardinal et al. 2001; Kuhnert et al. 2002; Cohen et al. 2004; Creuger et al. 2006; Goodkin et al. 2008), the fidelity of proxy signals for seawater temperature (Fairbanks and Dodge 1979), and anthropogenic inputs (Dodge et al. 1984c; Shen and Boyle 1988; Kelly et al. 2009). One study on growth records in the axial skeleton of a soft coral, *Plexaurella dichotoma*, was performed to assess suitability for temperature proxy signals (Bond et al. 2005).

Coral Symbiosis, Nutrition and Physiology

Bermuda corals contain species-specific zooxanthellae clade distribution patterns similar to the Caribbean (Billinghurst et al. 1997; Savage et al. 2002a) and similar photosynthetic responses (Savage et al. 2002b). Photo-adaption responses were measured in *Montastraea franksi* (Lesser and Schick 1989) but zooxanthellae appear to have attenuated responses to factors that can induce coral bleaching in experimental conditions (Venn et al. 2008). Cook et al (1990) described the first coral bleaching event in Bermuda. Also, pollutant effects on coral photosynthesis have been assessed (Cook and Knap 1983; Owen et al. 2002, 2003; Yost et al. 2010). Bermuda's octocorals also share similar patterns of zooxanthellae clade distribution with their Caribbean conspecifics (Goulet and Coffroth 2004).

Nutrient fluxes in corals and partitioning with zooxanthellae has been evaluated (Piniak and Lipschultz 2004; Badgley et al. 2006), as well as amino acid uptake (Ferrier 1991) and nutrient sufficiency (Cook et al. 1994). Mills et al. (2004) assessed particulate feeding, and Johannes et al. (1970) studied zooplankton consumption rates.

Reef Fish Biology and Ecology

Fish Demographics

Bermuda's reef fish fauna is derived from the Caribbean (Smith-Vaniz et al. 1999) but for many species there are biological and ecological differences. Growth and life history patterns of reef fishes vary with environmental conditions at latitudinal, regional and even habitat scales (DeMartini and Anderson 1978; Choat and Robertson 2002) as a result of temperature effects on metabolism and reproduction, and variations in food availability (Warburton 1989; Ebeling and Hixon 1991). Cooler temperatures in combination with increased seasonality may reduce growth rates during early life history and thus delay maturation, yet the same factors often compress the reproductive season and allow greater allocation of energy to somatic growth later. Further, larger body sizes confer metabolic advantages in poikilotherms by reducing heat loss as a result of reduced surface area to volume ratio (Ebeling and Hixon 1991). The high latitude position of Bermuda's reef and the attendant strong seasonal seawater temperature variation has resulted in some distinctive traits in maximum size, growth patterns, growth rates and longevity across many taxa.

Many fish species attain significantly larger sizes in Bermuda (Smith-Vaniz et al. 1999). For example, the manytooth conger, *Conger triporiceps*, reaches 115 cm SL compared to 100 cm SL in the Caribbean, and the brown garden eel, *Heteroconger longissimus*, has reached 48.1 cm SL compared to 40.2 cm SL for Caribbean conspecifics. The ocean surgeonfish, *Acanthurus bahianus*, reaches 22.7 cm SL in Bermuda, while most Caribbean populations are less than 20 cm SL (Robertson et al. 2005b; Mutz 2006). Notably, a serranid, *Epinephelus guttatus*, attained a size of 72 cm FL, larger than all Caribbean conspecifics (Luckhurst et al. 1992). The larger sizes of Bermuda reefs fishes can be attributed partly to longevity (Luckhurst et al. 1992) and perhaps improved fisheries management that may increase survivorship (Luckhurst et al. 2000; Robertson et al. 2005b), but primarily reflect increased somatic growth associated with reduced reproduction in a colder climate.

Age and growth studies have been done on a number of fishes. The red hind, *E. guttatus* lagged in growth rates compared to Jamaican fish (Burnett-Herkes 1975). However, three acanthurids had comparable or greater growth rates compared to Western Atlantic conspecifics (Mutz 2006; Robertson et al. 2005a). Pitt et al. (2009) found that the blue-striped grunt, *Haemulon sciurus*, attained much greater ages (23 years vs. 12 years) than those in the Caribbean. However, several species show significantly older ages compared to Caribbean without concomitantly larger sizes, such as the coney, *Cephalopolis fulva* (Trott 2006) and the lane snapper, *Lutjanus synagris* (Luckhurst et al. 2000). Also, two deep water misty groupers, *Epinephelus mystacinus*, were aged at 135 years and 150 years, based on otolith annuli (Luckhurst and Dean 2009).

Reproduction

Bermuda's cool winter water appears to have displaced reproduction for many fishes into the warmer summer and fall months (Bardach et al. 1958; Burnett-Herkes 1975; Robertson 1991; Luckhurst et al. 2000; Trott 2006). The length of the spawning seasons may also be attenuated for some species (Luckhurst et al. 2000; Trott 2006) but extensive reproductive studies have been conducted on only a few commercially significant reef species.

Spawning aggregations were initially reported for groupers only on Bermuda's Plantagenet (Argus) and Challenger banks (Bardach et al. 1958). More recent studies on reproduction have focused on spawning aggregations of diverse fish taxa in deep and shallow waters around the main Bermuda platform (Luckhurst 2007). The earliest research on the red hind, *Epinephelis guttatus*, was conducted in seasonally closed aggregation areas (Burnett-Herkes 1975). Further studies were done to assess site fidelity and movement patterns via tag/recapture studies (Luckhurst 1998; Luckhurst 2010a). The closure of spawning sites for commercially vulnerable serranids has appeared to stabilize population declines over a 30 year period (Luckhurst and Trott 2008). Recently discovered spawning aggregations of the black grouper, *Mycteroperca bonaci,* have stimulated additional protection and research (Luckhurst 2010b), and implanted acoustic tags showed prolonged occurrence of the black grouper at spawning sites around the Bermuda platform from May to November (Trott et al. 2010). The blue-striped grunt, *Haemulon sciurus*, also forms spawning aggregations on a nearshore lagoonal environment, and this area is currently protected (Trott et al. 2009). Multi-species scarid spawning aggregations occur on shallow reefs (<20 m) along the South Shore (Luckhurst 2011). The limited information on spawning aggregations in Bermuda to date do show some distinctions from the Caribbean. For example, *M. bonaci* does not aggregate on deep promontories, as they do in Belize (Heyman and Kjerfve 2008).

Studies on reproduction of non-aggregating species are limited, with anecdotal reports of spawning activities, occurrence of gravid females and egg-laying by pomacentrids during the warm summer months. The coney, *Cephalopolis fulva*, breeds from April to July (Trott 2006). Colonies of the common western Atlantic garden eel (*Heteroconger longissimus*) were recently discovered in Bermuda (Tyler and Luckhurst 1994) and most members of the colonies occur as male–female pairs in closely adjacent burrows, whereas pairing is otherwise unknown in this species. These eels are gravid in September and October. Gonadal development in 14 serranids collected in spring and summer months did not present any temporal patterns (Smith 1958). Bardach et al. (1958) showed that of 18 "grouper" species, 9 spawned between June and August.

Fish Recruitment Patterns and Ecosystem Connectivity in Bermuda

There are few historical studies on the distribution of juvenile fishes in seagrass beds and mangroves in Bermuda (McRae 1997; Smith et al. 1998; Ward 1999) and limited anecdotal evidence on grouper recruitment in inshore areas (Bardach et al. 1958). Limited data have been published on larval fish settlement to any of Bermuda's reefs (Schultz and Cowen 1994). For larval fish across the North Lagoon and off the south shore, most reef-associated families showed a strong summer peak but others had larvae persistent throughout the year (Lutjanidae, Serranidae, Sygnathidae and Scomberesocidae) (Glasspool 1994).

However, patterns of juvenile fish recruitment have been studied at broad spatial and temporal scales across the Bermuda platform in mangrove areas, seagrass beds and rim reefs from 1999 to 2003 (Smith and Pitt 2002; Smith et al. 2003; MEP 2004). Sampling did not detect any fish recruitment between January and June, and subsequent sampling has shown strong seasonal peaks in recruitment across the lagoon and rim reef that tapers off by December (Smith et al. 2003). Juveniles of a range of families recruit more on lagoonal and near shore reefs than offshore rim and terrace reefs (Figs. 11.3 and 11.4). Murdoch et al. (2008a) discriminated juvenile fishes (<5 cm SL) in broad-scale reef fish surveys and detected distinctive patterns of disjunct distributions of juveniles and adults within species (and see below).

Nearshore seagrass beds and small mangrove areas are also significant areas for recruitment but the diversity of species utilizing these habitats is lower than that found on the nearshore and lagoonal reefs (Fig. 11.5). Grunts (*Haemulon* spp.) were conspicuously abundant in both these habitats and mojarras (*Eucinostomus* spp.) were very common at the mangrove sites.

Ontogenetic migration from shallow-water nursery habitats to deeper-water reef is a life history strategy used by a variety of species circumtropically. Critical nurseries are usually seagrass beds and mangroves in some reef systems (Nagelkerken et al. 2002, 2008; Verweij et al. 2008) but research on other Caribbean islands has shown that offshore seagrass beds hardly fulfill a nursery function and are only occupied by larger fish (Dorenbosch et al. 2007). In the western part of Bermuda, mangroves are restricted to the semi-enclosed lagoon of Ely's Harbour, while seagrass beds are found in this lagoon, along the shoreline, as well as offshore where the reef flat drops off. Here, the seagrass beds and mangroves at Ely's Harbour are likely the main nursery grounds in the western part of Bermuda for species that are known to be associated with these habitats during their juvenile stage (Huijbers et al. 2008). For some of these species, the patch reefs on the shallow shelf area directly bordering Ely's Harbour function as alternative nursery habitat for some species, probably related to the fact that this habitat resembles that of lagoonal habitat, in terms of water-depth, turbidity, closeness to other vegetated habitats, distance from reefs, etc.

Due to the strong association of "nursery fish species" with mangrove or seagrass juvenile habitats, their adults on shelf-break reefs showed a strong and significant decline in their densities with increasing distance from Ely's Harbour (Huijbers et al. 2008). The reefs investigated were located along a spatial gradient from Ely's Harbour across the lagoon to Cross Bay (i.e. a southward, followed by a eastward gradient). Nursery species that showed declining adult density along this gradient were: *Acanthurus chirurgus*, *Chaetodon capistratus*, *Haemulon aurolineatum*, *H. flavolineatum*, *H. sciurus*, *Lutjanus griseus*, *L. synagris*, *Ocyurus chrysurus*, *Scarus coeruleus*, *Scarus iserti*, and *Scarus guacamaia*, which are important species in terms of fishing (grunts and snappers) and ecosystem function (parrotfishes) (Nagelkerken unpublished data). For several of these species, the biomass of mature fish also declined along the distance gradient, showing that many fish do not disperse far away from their nurseries with increasing age. It appears that the nursery habitats studied along western Bermuda are of high importance to maintenance of adult reef populations and to reproductive

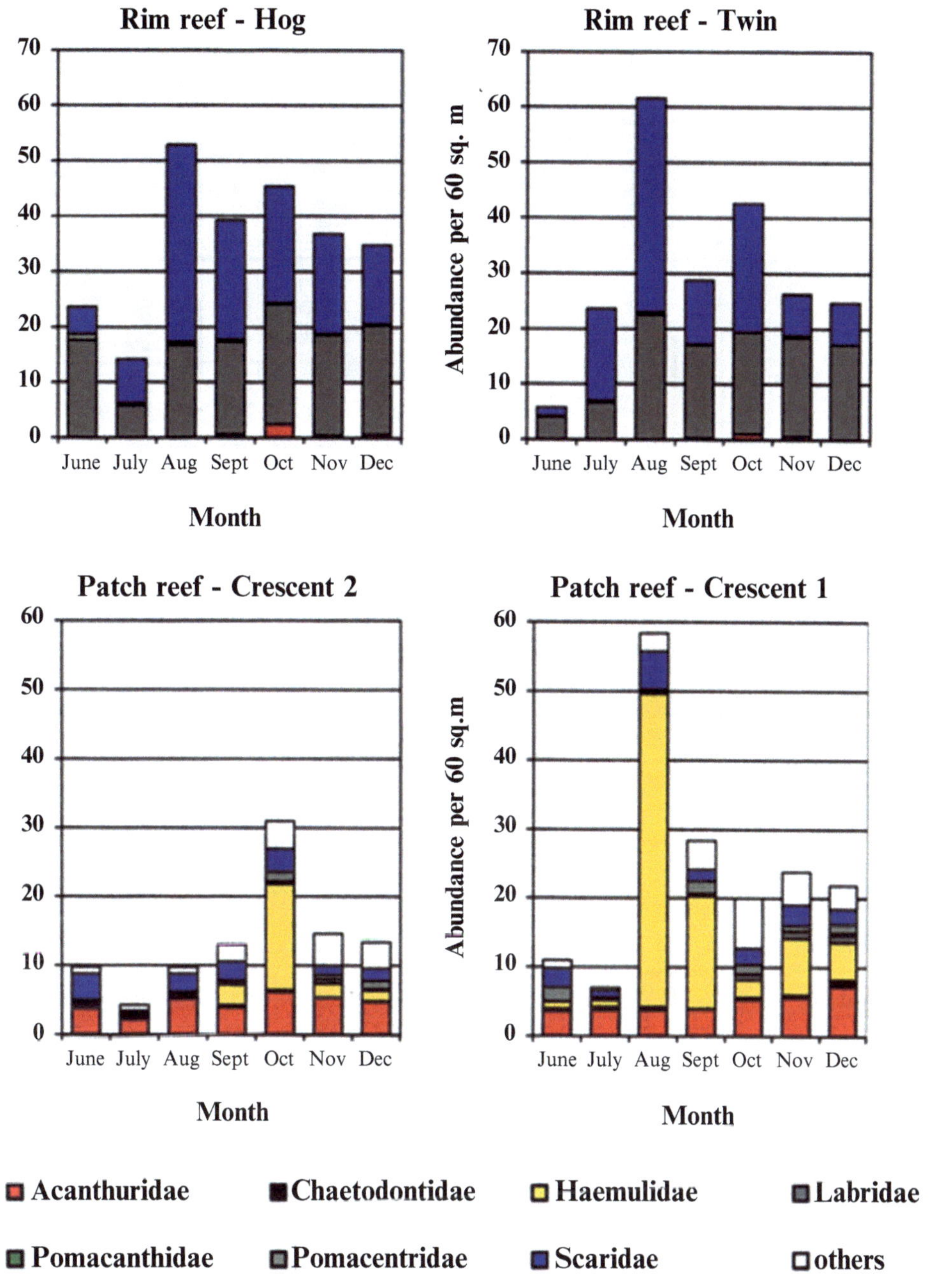

Fig. 11.3 Temporal patterns of juvenile fish recruitment by family at two rim reef and two patch reef sites in 2002. Five 30 m × 2 m transects were sampled monthly at each site (Smith et al. 2003)

fish stocks. Conservation of nursery habitat and connectivity-corridors between nurseries and coral reefs must be a high priority for further research.

Reef Mapping

Bermuda's significance as an important British naval outpost promoted extensive hydrographic surveys, especially the remarkably accurate survey of every single reef within the North Lagoon by Lt. Thomas Hurd from 1789 to 1793 (Hallet 2010). Satellite and aerial photos have been used to develop depth algorithms and to assess reef and seagrass distribution patterns in the North Lagoon (Vierros 1999). This work was superseded by a geo-referenced high-resolution photomosaic that has been extensively ground-truthed with synoptic surveys (Murdoch et al. 2008a). Limited LIDAR bottom profiling has been performed on the South shore reefs and several shipping channels. A complete geo-referenced multi-beam sonar map has been made of the

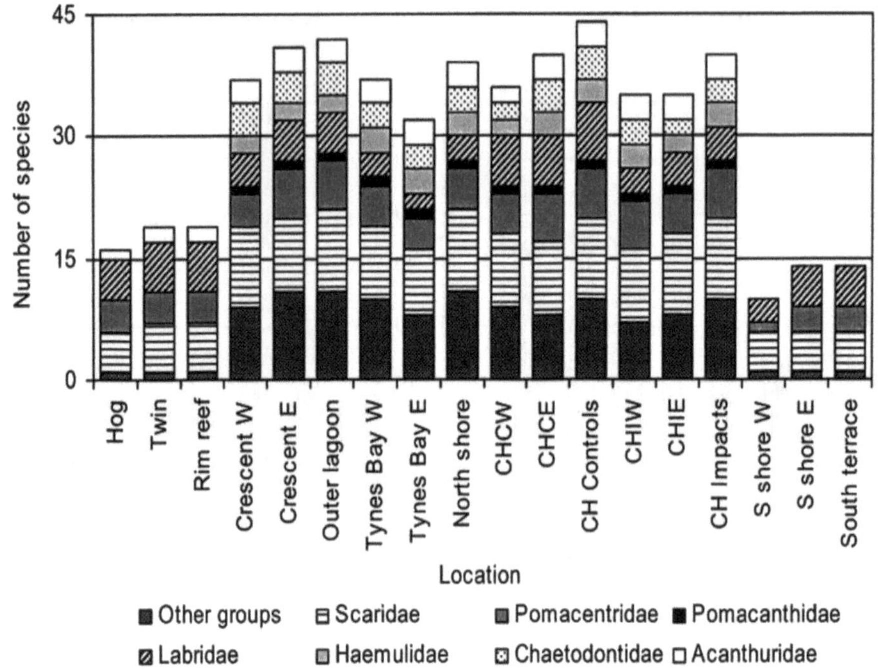

Fig. 11.4 Cumulative species richness of juvenile fishes (<6 cm SL) at 12 paired sites across the distinctive physiographic reef zone and total species richness for each zone. Rim Reef: Hog and Twin; Outer lagoon: Crescent W and Crescent E; North shore: Tyne's Bay W and Tyne's Bay E (near-shore reefs); CH Controls: CHW and CHE (protected lagoon); CH Impacts: CHIW and CHIE (protected lagoon); South terrace: S shore W and S Shore E. Monthly sampling from June–Dec in 2002 and 2003. Five 30 m × 2 m transects sampled monthly at each site (MEP 2004)

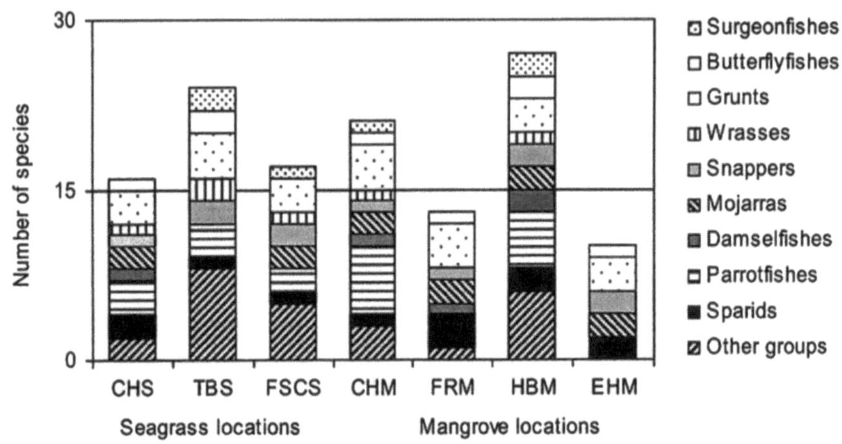

Fig. 11.5 Cumulative species richness of juvenile fishes (<6 cm SL) in 2002 and 2003 at seagrass locations in Castle Harbour (*CHS*), Tynes Bay (*TBS*) and Fort St. Catherine (*FSC*) and at mangrove locations in Ferry Reach (*FRM*), Hungry Bay (*HBM*) and Ely's Harbour (*EHM*). Monthly sampling from June–Dec in 2002 and 2003. Five 30 m × 2 m transects sampled monthly at each site, (MEP 2004)

deep reefs from 40 m to 150 m in 2010 (Iliffe et al. 2011). These various bathymetric surveys have been incorporated in a GIS database managed by the Bermuda Government's Department of Conservation Services. Species–specific survey information is now recorded in the GIS database.

The development of a GIS map of Bermuda's reefs by Murdoch et al. (2008a) shows several distinctive reef zones. This and other spatial and temporally explicit data sets have set the stage for integrated marine spatial planning where regulated activities can be high-lighted with respect to reef zones, critical habitats and ultimately species distributions (Figs. 11.6 and 11.7).

Reef Zonation and Reef Community Patterns

Coral Distribution Patterns

Zonation patterns within the Bermuda reef system have been summarized in Logan (1988). The high-energy outer rim and terrace reefs are dominated by massive and domal corals such

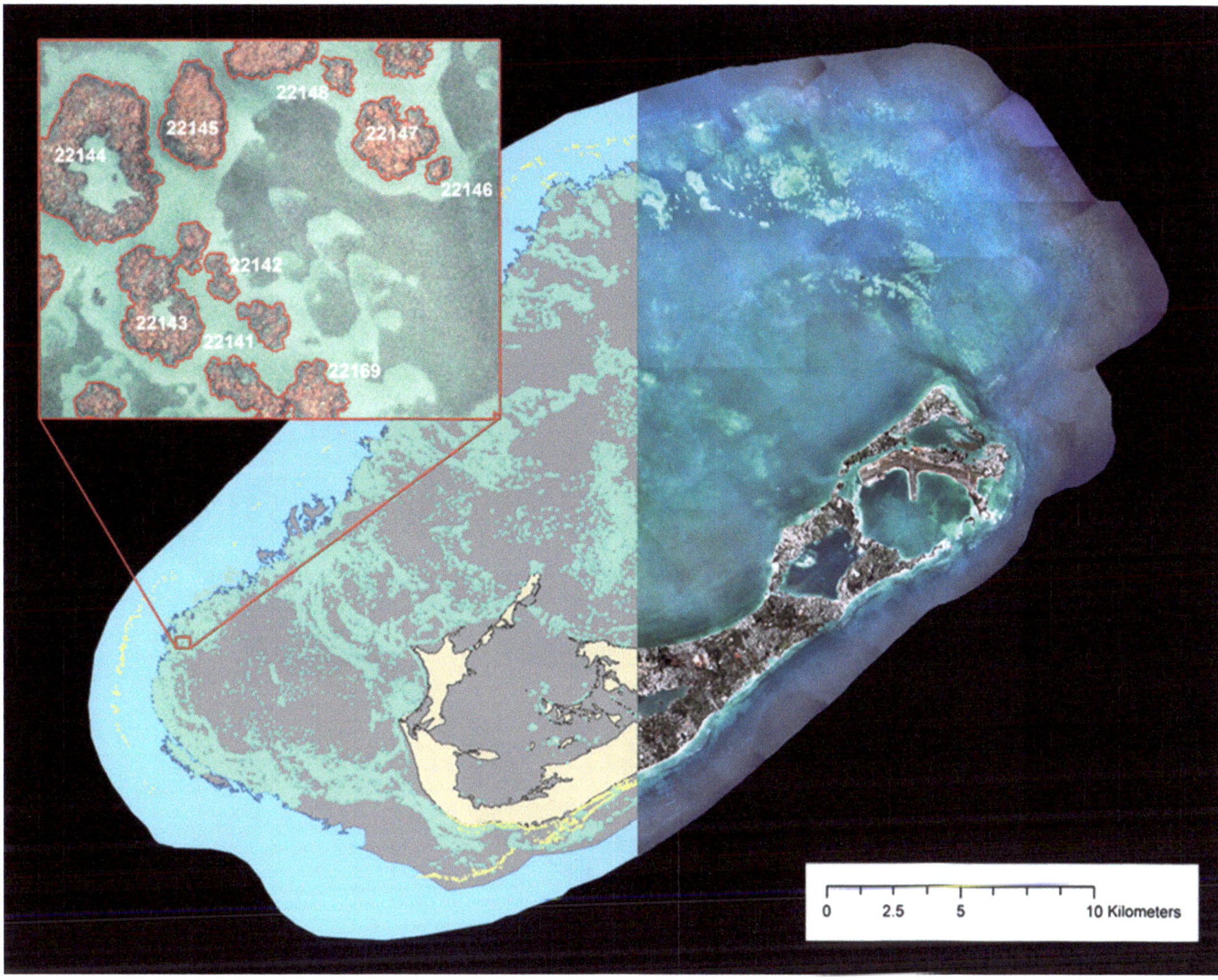

Fig. 11.6 The photo-mosaic of aerial images (shown on the *right side*) was used to generate a GIS map of all reef habitats across the Bermuda Platform, as shown on the *left side* of the image. The inset shows how GIS mapping allows for all ~35,000 patch reefs within the lagoon to be assigned their own unique identification number (Murdoch et al. 2008a)

as *Diploria* spp., *Montastraea* spp., and *Porites astreoides* (Fig 11.8, Dodge et al. 1982; Murdoch et al. 2008b). *Millepora alcicornis* is also prevalent on the rim reef but less so on the outer terrace. *Stephanocoenia intersepta*, *Agarica fragilis*, *Siderstrastrea radians*, *Dichocoenia stokesii* and *Meandrina meandrites* are uncommon or "understory" constitutents of these zones (Fig. 11.9). *Madracis* spp. are rarely found on the rim reef but are more common on the deeper terrace reefs. *Madracis auretenra* colonies transplanted onto the rim reef suffered from intensive injury by grazing parrotfishes (Smith and de Putron unpublished data).

Lagoon patch reefs have a higher coral diversity, supporting all the species found on the rim and outer terrace reefs but also include *Oculina diffusa*, *Porites porites* and very rare occurrences of *Siderastrea siderea*. More significantly, the branching corals *Madracis auretenra*, *M. decactis* and *Oculina diffusa* are very abundant on some reefs, particularly in the inner lagoon (Fig. 11.10). Also, *Millepora alcicornis* is a dominant species on the shallower parts of these lagoonal reefs. Murdoch (2007) described predicted responses of lagoonal patch reef coral assemblages in response to gradients of light, temperature, suspended particulate material and current flow. Three spatial patterns were discerned amongst lagoon patch reefs with the more nearshore reef characterized by a predominance of *Madracis* spp and the central offshore patch reefs dominated by *Montastraea franksi*, including many old very large colonies that have been fragmented into ramets over time (Murdoch et al. 2008a). Finally the eastern and western lagoon patch reefs are dominated by a *Diploria-Porites* assemblage.

Soft coral distribution is less well described for Bermuda reefs. Smith and Musik (1984) described patterns on the rim reef where *Gorgonia ventalina*, *Plexaura* spp., *Pseudoplexaura* spp. and *Pseudopterogorgia* spp. predominate. These same taxa occur on the outer terrace but at generally lower abundance. *Muricea* spp. may be more abundant

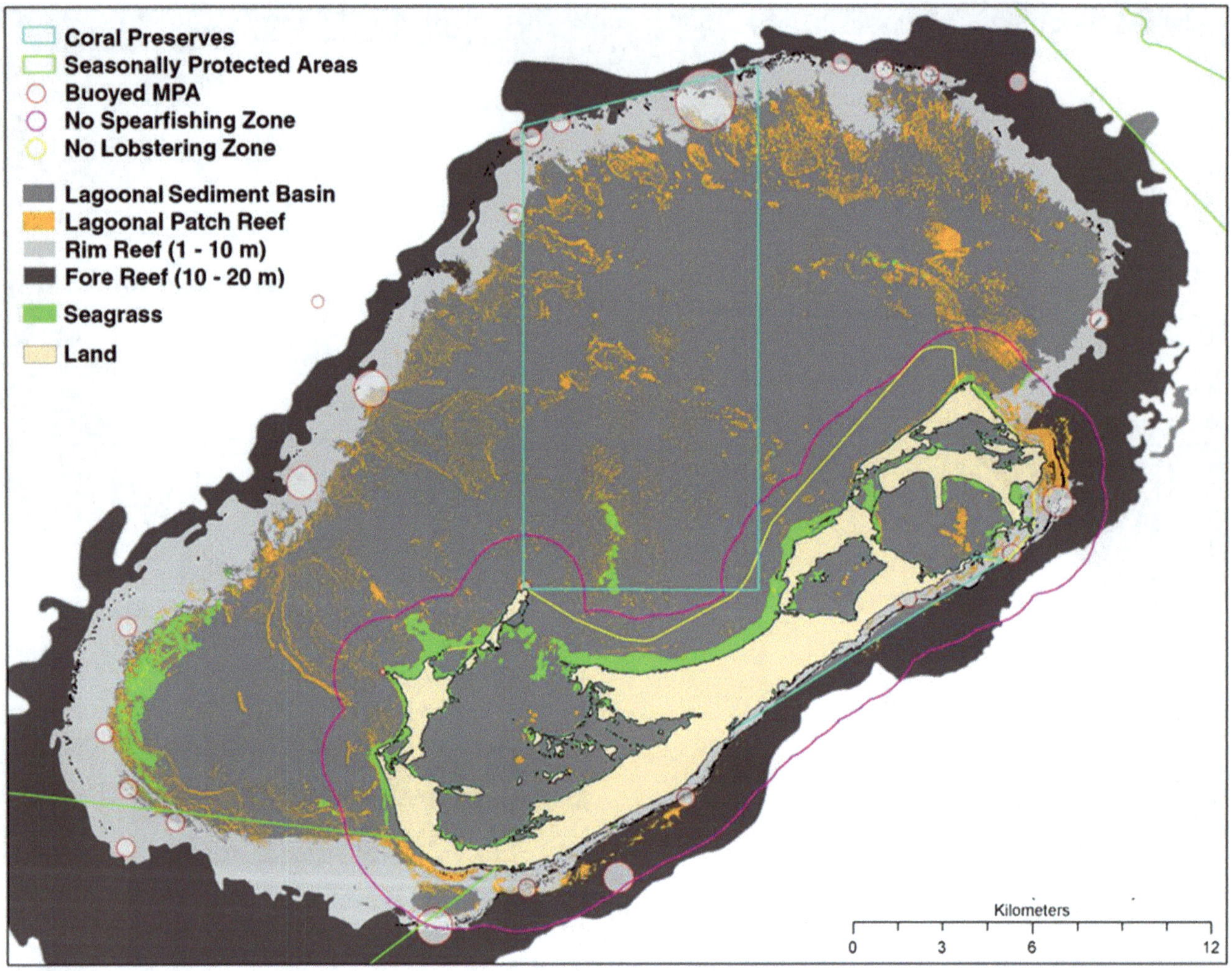

Fig. 11.7 Map of coral, seagrass and sediment habitats across the Bermuda reef platform, as well as the boundaries of the various kinds of marine protected areas (Murdoch et al. 2008a)

on the south shore reefs, *Eunicea* spp. are less common in all zones, *Plexaurella* spp. are more commonly encountered on lagoonal patch reefs, and a new gorgonian, *Leptogorgia setacea* was recently discovered in the sheltered inner lagoon (Locke et al. 2013). Murdoch (2006) described the distribution of the sea fan, *Gorgonia ventalina*, across the lagoon when investigating the impact of the gorgonian predatory nudibranch, *Tritonia hamnerorum*.

Reef Community Structure

Early reef geologists explosively dissected reefs and produced detailed descriptions of framework builders, algae, cavity dwellers, boring species and motile infauna (Scoffin and Garrett 1974). This is the most exhaustive study of species diversity in any of Bermuda's reef zones. The majority of subsequent reef studies have relied on various methods to quantify the most abundant species on reef surfaces (Dryer and Logan 1978; Dodge et al. 1982; Smith et al. 1998; Catell 2002; CARICOMP 2000; Flood 2004; MEP 2007; Murdoch et al. 2008a). Overall coral coverage and species diversity have remained quite stable on the rim reefs since the 1980s (Dodge et al. 1982; CARICOMP 2000; Linton and Fisher 2004; MEP 2007). However lagoon patch reefs have seen more dynamic changes in growth and mortality (Catell 2002; MEP 2007). Murdoch et al. (2008a, b) and Murdoch (2012) have the most recent and detailed spatially explicit assessments of major reef taxa, across the lagoon, at the rim reef (10 m) terrace reef (20 m) and fore-reef slope (30–40 m) (Fig. 11.11).

Synoptic studies of smaller sessile and motile reef invertebrates and algal diversity are largely lacking. Amongst conspicuous invertebrates, the herbivorous gastropod *Cerithium*

Fig. 11.8 Permanent study quadrat on the terrace reef at 20 m showing high percent cover by *Diploria* spp, *Montastraea* spp and *Porites astreoides*. Quadrat size is 95 cm × 65 cm (Photo: S.R. Smith)

Fig. 11.9 Rim reef dominated by brain corals *Diploria* spp and soft corals (Photo credit: S.R. Smith)

litteratum is ubiquitous in all reef zones reaching high densities (100s per m^2) on disturbed reefs (Smith 1990, Murdoch et al. 2008a; Murdoch 2013). The endemic hermit crab *Calcinus verrilli* uses cerithid shells as well as worm tubes and also attains high densities on reefs (Smith 1988; Rodrigues et al. 2000). Erect sponges, such as *Ircinia* spp. or *Pseudoceratina crassa* are not very common on outer rim or terrace reefs but species such as *Ircinia* spp, *Aplysinia* spp. and *Callyspongia vaginalis* are more common on lagoonal patch reefs.

Patterns in the Distribution and Abundance of Fishes

Bermuda's reef fishes have been studied intensively since the early nineteenth century (summarized in Smith-Vaniz et al. 1999) but quantitative assessments on distribution and abundance did not begin until scientific fish trapping (Bardach and Mowbray 1955; Bardach 1958) and visual surveys in the 1950s (Bardach 1959). There was a gap of effort in fishery-independent assessments of reef fish populations until Luckhurst (1994) initiated in-water fish censuses. Smith (1988, 1990) studied fish activity on a grounding site and on adjacent undisturbed reefs. Smith et al. (1998) examined fish populations on near-shore lagoonal patch reefs and inshore seagrass beds. The first

Fig. 11.10 Colonies of *Madracis auretenra* and *Millepora alcicornis* on a nearshore lagoonal patch reef, about 2 m deep, with a school of tomates, *Haemulon aurolineatum* (Photo: S.R. Smith)

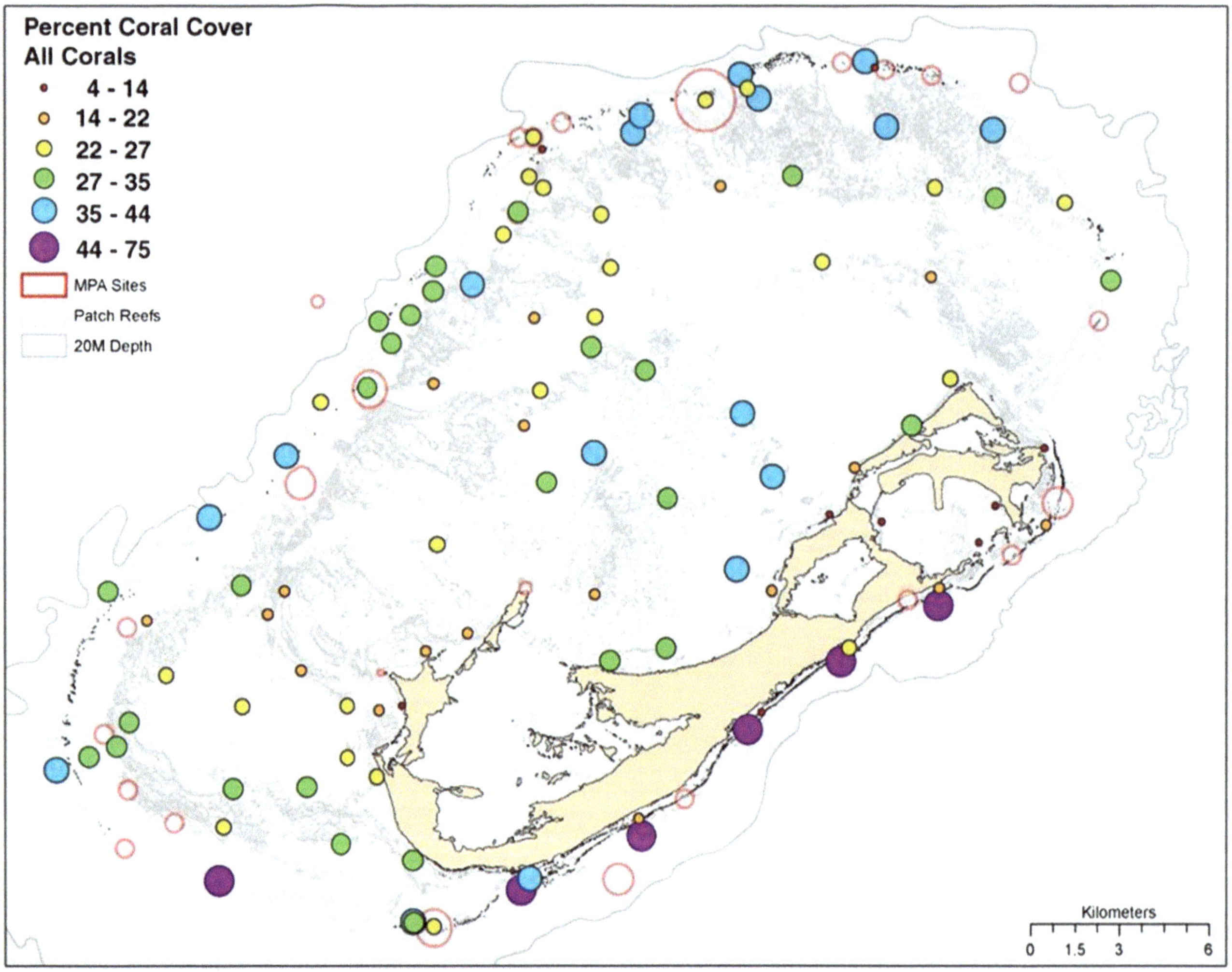

Fig. 11.11 A map of the average percent coral cover for all hard corals, derived from six 10 m long transects at each site using the AGGRA protocol (Murdoch et al. 2008a)

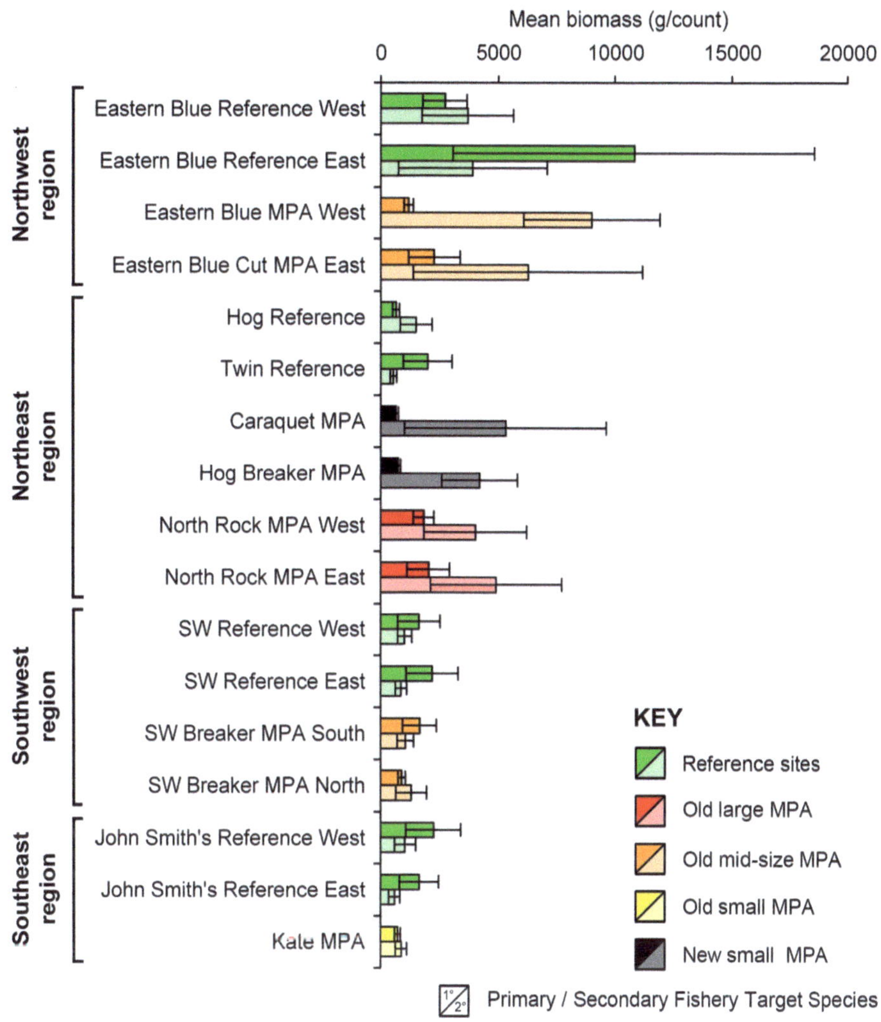

Fig. 11.12 Mean biomass of primary and secondary fishery target species at MPAs and reference sites around Bermuda. Reference sites are green, old MPAs of decreasing size are red, orange and yellow. New MPAs are black. The upper, darker-shaded bars are primary fishery species include groupers, snappers, grunts, porgies and hogfish. The lower, lighter-shaded bars are secondary species include triggerfishes, chub, barracuda and puddingwife. (MEP 2005)

broad-scale intensive fish surveys were conducted by J. Pitt and S. Patterson as part of a study of the effectiveness of Marine Protected Areas in comparison to unprotected areas, primarily within the rim reef zone (Fig. 11.12, MEP 2005). Although they could not detect an MPA effect they did discern spatial differences in fish distributions across the reef platform at family and species level. Likewise, Murdoch et al. (2008a) could not detect differences between MPAs and adjacent reference sites.

Murdoch et al. (2008a, 2012) investigated differences in fish populations across the entire lagoon, the rim, between MPA sites and adjacent comparable reefs, and extended the study to compare fish populations between 10 m fore-reef sites (10 m) to terrace reefs (20 m) around the entire reef platform. Many fish species showed habitat-specific patterns of distribution, particularly between the lagoonal patch reefs and offshore rim reefs. Adult and juvenile princess parrotfish, *Scarus taeniopterus*, were found primarily on the outer rim reef (Fig. 11.13). Adult yellowtail snappers, *Ocyurus chrysurus*, were found across the North Lagoon and on the rim reef but juveniles were primarily found on lagoonal patch reef. Other examples include the blue tang, *Acanthurus coerulus,* which was found only on the outer rim reef but the ocean surgeonfish, *Acanthurus bahianus* was found both on lagoonal patch reefs and the outer rim reef (Murdoch et al. 2008a).

More research is required to assess temporal changes in the reef fish communities and the connections that link shallow reefs (<20 m) to deep forereef and meso-photic reefs. The deeper reefs are targeted for commercial species (grouper, lobsters) and also appear to be a refugia for the invasive lionfish, *Pterios volitans*.

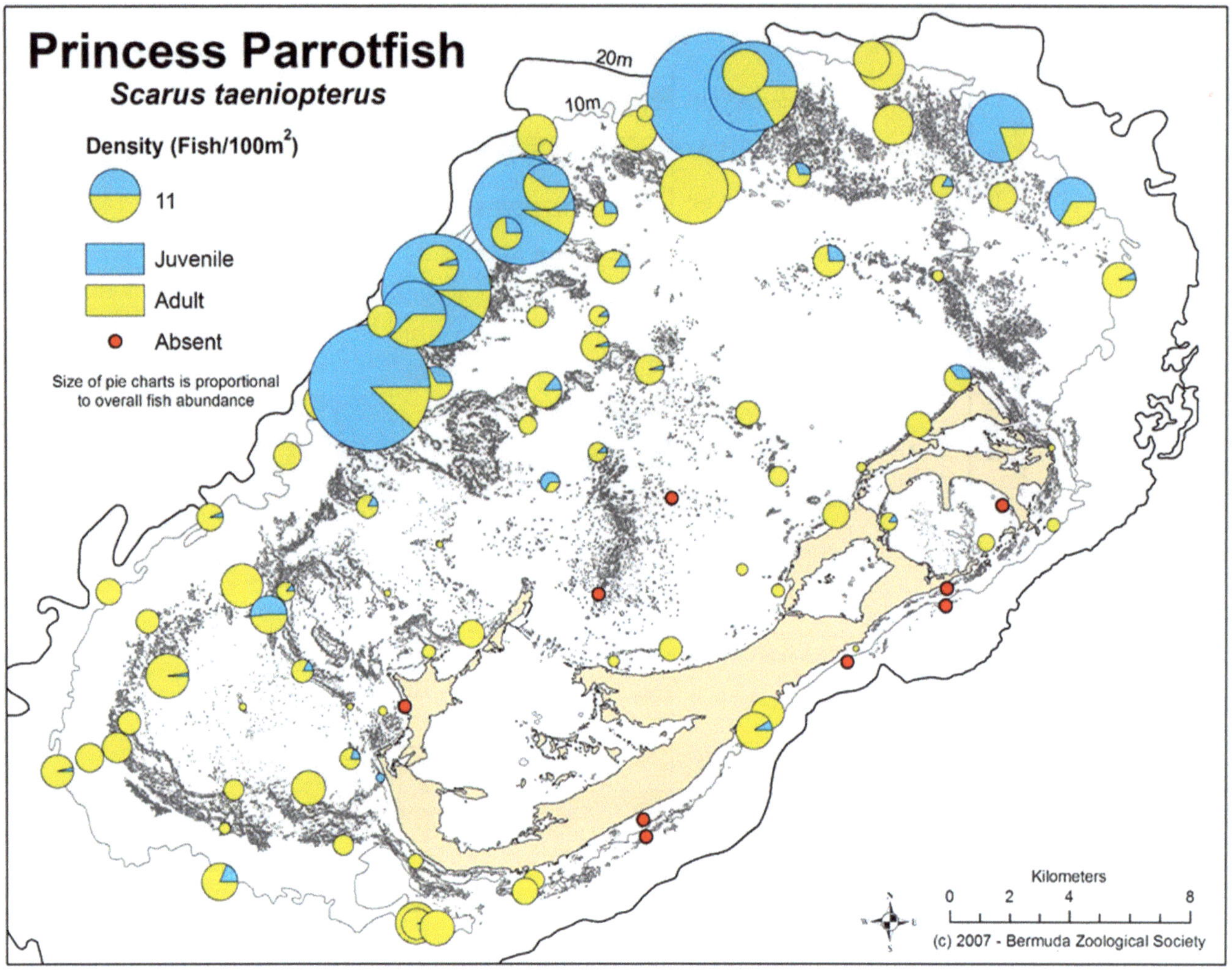

Fig. 11.13 Geospatial map illustrating juvenile and adult densities of *Scarus taeniopterus*, derived from AGGRA surveys

Acknowledgement This is Contribution #198 of the Bermuda Biodiversity Project (BBP), Bermuda Aquarium, Museum and Zoo, in the Department of Conservation Services.

References

Andersson AJ, de Putron SJ, Bates NR, Collins A, Garley R, Noyes T, Dexter E (2011) Bermuda ocean acidification and coral reef investigation: temporal and spatial changes in seawater carbonate chemistry and calcification. Abstract, IPCC meeting, Okinawa, Feb 2011

Badgley B, Lipschultz F, Sebens K (2006) Nitrate uptake by the reef coral *Diploria strigosa*: effects of concentration, water flow, and irradiance. Mar Biol 149:327–338

Baird AH, Guest JR, Willis BL (2009) Systematic and biogeographical patterns in the reproductive biology of scleractinian corals. Annu Rev Ecol Evol Syst 40:551–571

Bardach JE (1959) The summer standing crop of fish on a shallow Bermuda reef. Limnol Oceanogr 4:77–85

Bardach JE, Mowbray LS (1955) Bermuda fisheries research program progress report 2, 96pp

Bardach JE, Smith CL, Menzel DW (1958) Bermuda fisheries research program final report, Bermuda Trade Development Board, Hamilton, 59pp

Bates NR (2002) Seasonal variability of the effect of coral reefs on seawater CO_2 and air–sea CO_2 exchange. Limnol Oceanogr 47:43–52

Bates NR, Samuels L, Merlivat L (2001) Biogeochemical and physical factors influencing seawater fCO_2 and air-sea CO_2 exchange on the Bermuda coral reef. Limnol Oceanogr 46:833–846

Bates NR, Amat A, Andersson AJ (2010) Feedbacks and responses of coral calcification on the Bermuda reef system to seasonal changes in biological processes and ocean acidification. Biogeosciences 7:2590–2530

Billinghurst Z, Douglas AE, Trapido-Rosenthal HG (1997) On the genetic diversity of the symbiosis between the coral *Montastraea cavernosa* and zooxanthellae in Bermuda. In: Proceedings of the 8th international coral reef symposium, vol 2, Panama, pp 1291–1294

Bond ZA, Cohen AL, Smith SR, Jenkins WJ (2005) Growth and composition of high-Mg calcite in the skeleton of a Bermudian gorgonian (*Plexaurella dichotoma*): potential for paleothermometry. Geochem Geophys Geosyst 6:Q0810

Brylewska HRT (2007) An examination of a degraded coral reef system and the biological processes that may lead it to recovery. M. Phil. Thesis, Swansea University, Wales, p 232

Burnett-Herkes J (1975) Contribution to the biology of the red hind, *Epinephelus guttatus*, a commercially important serranid fish from the tropical Western Atlantic. Ph.D. dissertation, University of Miami, Miami, 154pp

Cardinal D, Hamelin B, Bard E, Patzold J (2001) Sr/Ca, U/Ca and $\delta^{18}O$ records in recent massive corals from Bermuda: relationships with sea surface temperature. Chem Geol 176:213–233

CARICOMP (2000) Status and temporal trends at CARICOMP cora lreef sites. In: Proceedings of the 9th international coral reef symposium, vol 1. Bali, pp 325–330

Catell N (2002) Dynamics of the coral and sponge populations, Tyne's Bay, Bermuda. Bermuda programme report, Bermuda Biological Station for Research, 46pp

Choat JH, Robertson DR (2002) Chapter 3. Age-based studies. In: Sale PF (ed) Coral reef fishes: dynamics and diversity in a complex ecosystem. Elsevier Science, Amsterdam, pp 57–80

Cody T, de Putron SJ, Murdoch TJT (2010) Distribution, population dynamics and growth of the scleractinian coral *Siderastrea radians* in Bermuda, BIOS REU report, 17pp.

Cohen AL, McCorkle DC, de Putron SJ, Gaetani GA, Rose KA (2009) Morphological and compositional changes in the skeletons of new coral recruits reared in acidified seawater: insights into the biomineralization response to ocean acidification. Geochem Geophys Geosyst 10:Q07005. doi:10.1029/2009GC002411

Cohen AL, Smith SR, McCartney MS, van Etten J (2004) How brain corals record climate: an integration of skeletal structure, growth and chemistry of *Diploria labyrinthiformis* from Bermuda. Mar Ecol Prog Ser 271:147–158

Cook CB, Knap AH (1983) Effects of crude oil and chemical dispersant on photosynthesis in the brain coral *Diploria strigosa*. Mar Biol 78:21–27

Cook CB, Logan A, Ward J, Luckhurst B, Berg CJ (1990) Elevated temperatures and bleaching on a high latitude reef: the 1988 bleaching event. Coral Reefs 9:45–49

Cook CB, Mueller-Parker G, Orlandini CD (1994) Ammonium enhancement of dark carbon fixation and nitrogen limitation in zooxanthellae symbiotic with the reef corals *Madracis mirabilis* and *Montastrea annularis*. Mar Biol 118:157–165

Creuger T, Kuhnert H, Patzold J, Zorita E (2006) Calibrations of Bermuda corals against large-scale sea surface temperature and sea level pressure pattern time series and implications for climate reconstructions. J Geophys Res 111:D23103

de Putron SJ (2003) The reproductive ecology of two corals and one gorgonian from sub-tropical Bermuda. U. of Wales, Swansea, 285pp

de Putron SJ, Ryland JS (2009) Effect of seawater temperature on reproductive seasonality and fecundity of *Pseudoplexaura porosa* (Cnidaria: Octocorallia): lattitudinal variation in Caribbean gorgonian reproduction. Invertebr Biol 128:213–222

de Putron SJ, Smith SR (2011) Planula release and reproductive seasonality of the scleractinian coral *Porites astreoides* in Bermuda, a high-latitude reef. Bull Mar Sci 87:75–90

de Putron SJ, McCorkle DC, Cohen AL (2011) The impact of seawater saturation state and bicarbonate ion concentration on calcification by new recruits of two Atlantic corals. Coral Reefs 30:321–328

DeMartini EE, Anderson ME (1978) Comparative survivorship and life-history of painted greenling (*Oxylebius pictus*) in Puget Sound, Washington and Monterey Bay California. Environ Biol Fishes 5:33–47

Dodge RE (1978) The natural growth records of reef building corals. Ph.D. dissertation, Yale University, New Haven, 237pp

Dodge RE, Thomson J (1974) The natural radiochemical and growth records in contemporary hermatypic corals from the Atlantic and Caribbean. Earth Planet Sci Lett 23:313–322

Dodge RE, Vaisnys JR (1975) Hermatypic coral growth banding as environmental recorder. Nature 258:706–708

Dodge RE, Vaisnys JR (1977) Coral populations and growth patterns: response to sedimentation and turbidity associated with dredging. J Mar Res 35:715–730

Dodge RE, Logan A, Antonius A (1982) Quantitative reef assessment studies in Bermuda: a comparison of methods and preliminary results. Bull Mar Sci 32:745–760

Dodge RE, Wyers SC, Frith HR, Knap AH, Smith SR, Cook CB, Sleeter TD (1984a) Coral calcification rates by the buoyant weight technique: effects of alizarin staining. J Exp Mar Biol Ecol 75:217–232

Dodge RE, Wyers SC, Frith HR, Knap AH, Smith SR, Sleeter TD (1984b) The effects of oil and oil dispersants on the skeletal growth of the hermatypic coral *Diploria strigosa*. Coral Reefs 3:191–198

Dodge RE, Jickells TD, Knap AH, Boyd S, Bak RPM (1984c) Reef-building coral skeletons as chemical pollution (phosphorus) indicators. Mar Poll Bull 15:178–187

Dorenbosch M, Verberk WCEP, Nagelkerken I, van der Velde G (2007) Influence of habitat configuration on connectivity between fish assemblages of Caribbean seagrass beds, mangroves and coral reefs. Mar Ecol Prog Ser 334:103–116

Draschba S, Patzold J, Wefer G (2000) North Atlantic climate variability since AD 1350 recorded in ^{18}O and skeletal density of Bermuda corals. Int J Earth Sci 88:733–741

Druffel ERM (1989) Decade time scale variability of ventilation in the North Atlantic: high-precision measurements of bomb radiocarbon in banded corals. J Geophys Res 94:3271–3285

Dryer S, Logan A (1978) Holocene reefs and sediments of Castle Harbour. Bermuda J Mar Res 36:399–425

Ebeling AW, Hixon MA (1991) Tropical and temperate reef fishes: comparison of community structure. In: Sale PF (ed) The ecology of fishes on coral reefs. Academic Press, San Diego, pp 509–563

Fairbanks RG, Dodge RE (1979) Annual periodicity of the $^{18}O/^{16}O$ and $^{13}C/^{12}C$ ratios in the coral *Montastrea annularis*. Geochim cosmochim Acta 43:1009–1020

Ferrier D (1991) Net uptake of dissolved free amino acidsby four scleractinian corals. Coral Reefs 10:183–187

Flood V (2004) Coral community structure and patterns of sedimentation in Castle Harbour, Bermuda. M.Sc. thesis, University of Georgia, XXpp

Flood V, Pitt JM, Smith SR (2005) Historical and ecological analysis of coral communities in Castle Harbour (Bermuda) after more than a century of environmental perturbation. Mar Poll Bull 51:545 557

Glasspool AF (1994) Larval distribution, population structure and gene flow in Bermuda's reef fish. Ph.D. dissertation, University College North Wales, Menai Bridge, 186pp and Appendices

Goodbody-Gringley G, de Putron SJ (2009) Planulation patterns of the brooding coral *Favia fragum* (Esper) in Bermuda. Coral Reefs 28:959–963

Goodkin NF, Hugen KA, Cohen AL, Smith SR (2005) Record of Little Ice Age sea surface temperatures at Bermuda using a growth-dependent calibration of coral Sr/Ca. Paleo Oceanogr 20:PA4016

Goodkin NF, Hugen KA, Curry WB, Doney SC, Ostermann DR (2008) Sea surface temperature and salinity variability at Bermuda during the end of the Little Ice Age. Paleo Oceanogr 23:PA3203

Goulet TL, Coffroth MA (2004) The genetic identity of dinoflagellate symbionts in Caribbean octocorals. Coral Reefs 23:465–472

Hallett J (2010) Changes to Bermuda's marine environment since Lt. Hurd's survey of 1789–1797: learning from the past to enhance conservation awareness. Bermuda Zoological Society report, 91 pp

Harrison PL, Wallace CC (1990) Reproduction, dispersal and recruitment of scleractinian corals. In: Dubinsky Z (ed) Coral reef ecosystems. Elsevier, Amsterdam, pp 133–207

Heyman WD, Kjerfve BJ (2008) Characterization of transient multi-species fish spawning aggregations at Gladden Spit. Belize Bull Mar Sci 83:531–551

Huijbers CM, Grol MGG, Nagelkerken IC (2008) Shallow patch reefs as alternative habitats for early juveniles of some mangrove/seagrass-associated fish species in Bermuda. Rev Biol Trop 56 (suppl 1):161–169

Iliffe TM, Kvitek R, Blasco S, Blasco K, Covill R (2011) Search for Bermuda's deep water caves. Hydrobiologia 677:157–168

Johannes RE, Coles SL, Kuenzel NT (1970) The role of zooplankton in the nutrition of some scleractinian corals. Limnol Oceanogr 15:579–586

Kelly AE, Reuer MK, Goodkin NF, Boyle EA (2009) Lead concentrations and isotopes in corals and water near Bermuda, 1780–2000. Earth Planet Sci Letts 283:93–100

Kleypas JA, Buddemeier RW, Archer D, Gattuso J-P, Langdon C, Opdyke BN (1999) Geochemical consequences of increased atmospheric carbon dioxide on coral reefs. Science 284:118–120

Kuhnert H, Patzold J, Schnetger B, Wefer G (2002) Sea-surface temperature variability in the 16th century at Bermuda inferred from coral records. Paleogeogr Paleoclimat Paleoecol 179:159–171

Lesser MP, Schick JM (1989) Photoadaptation and defenses against oxygen toxicity in zooxanthellae from natural populations of symbiotic cnidarians. J Exp Mar Biol Ecol 134:129–141

Linton D, Fisher T (eds) (2004) CARICOMP – Caribbean coastal marine productivity program: 1993–2003. Caribbean Coastal Data Centre, Centre for Marine Sciences, University of the West Indies, Mona

Locke JM, Bilewitch JP, Coates KA (2013) Scleractinia, Octocorallia and Antipatharia of Bermuda's reefs and deep-water coral communities: a taxonomic perspective including new records. In: Sheppard C (ed) Coral Reefs of the World Vol. 4. Coral Reefs of the UK Overseas Territories pp 189–200

Logan A (1988) Holcene reef of Bermuda, vol IX, Sedimenta. University of Miami, Miami

Logan A, Tomascik T (1991) Extension growth rates in two coral species from high-latitude coral reefs in Bermuda. Coral Reefs 10:155–160

Logan A, Yang L, Tomascik T (1994) Linear skeletal extension rates in two species of *Diploria* from high-latitude reefs in Bermuda. Coral Reefs 13:225–230

Luckhurst BE (1994) A fishery-independent assessment of Bermuda's coral reef fish stocks by diver census following the fish pot ban – aprogress report. Proc Gulf Caribb Fish Inst 46:309–323

Luckhurst BE (1998) Site fidelity and return migration of tagged red hinds (*Epinephelus gutttatus*) to a spawning aggregation site in Bermuda. Proc Gulf Caribb Fish Inst 50:750–763

Luckhurst BE (2010a) Long term site fidelity of tagged red hinds (*Epinephelus guttatus*) at two spawning aggregation sites in Bermuda. Proc Gulf Caribb Fish Inst 63:188–192

Luckhurst BE (2010b) Observations of a black grouper (*Mycteroperca bonaci*) spawning aggregation in Bermuda. Gulf Caribb Res 22:1–8

Luckhurst BE (2011) Observations at a multi-species parrotfish (Scaridae) spawning aggregation site at Bermuda with notes on the predation behavior of black grouper (*Mycteroperca bonaci*). Gulf Caribb Res 23:55–60

Luckhurst BE, Dean JM (2009) Age estimates of two large misty grouper, *Epinephelus mystacinus* (Serranidae) from Bermuda with a comparison of the age of tropical groupers in the western Atlantic. Gulf and Caribb Res 21:73–77

Luckhurst BE, Trott TM (2008) Seasonally-closed spawning aggregation sites for Red Hind (*Epinephelus guttatus*): Bermuda's experience over 30 years (1974–2003). Proc Gulf Caribb Fish Inst 61:331–336

Luckhurst BE, Barnes JA, Sadovy Y (1992) Record of an unusually large red hind, *Epinephelus guttatus* (Pisces: Serranidae) from Bermuda with comments on its age. Bull Mar Sci 51(2):267–270

Luckhurst BE, Dean JM, Reichert M (2000) Age, growth and reproduction of the lane snapper, *Lutjanus synagris* (Pisces: Lutjanidae) at Bermuda. Mar Ecol Prog Ser 203:255–261

McGuire MP (1998) Timing of larval release by *Porites astreoides* in the Northern Florida Keys. Coral Reefs 17:369–375

McRae PSD (1997) Structure of fish assemblages inhabiting Bermudian seagrass beds. B.Sc. thesis. St. Mary's University of Nova Scotia, 41pp

MEP (2004) Marine environmental program annual report, 2003–2004, Bermuda Institute of Ocean Sciences, 46pp

MEP (2005) Marine environmental program annual report, 2004–2005, Bermuda Institute of Ocean Sciences, 136pp

MEP (2007) Marine environmental program annual report, 2006–2007, Bermuda Institute of Ocean Sciences, 80pp

Mills MM, Lipschultz F, Sebens K (2004) Particulate matter ingestion and associated nitrogen uptake by four species of scleractinian corals. Coral Reefs 23:311–323

Murdoch TJT (2006) Population outbreak of the *Tritonia hamnerorum* nudibranch, an obligate grazer on the purple sea fan *Gorgonia ventalina*, across Bermuda's northern reefs in 2005. Bermuda Biodiversity Project special report 2006–003, 22pp

Murdoch TJT (2007) A functional group approach for predicting the composition of hard coral assemblages in Florida and Bermuda. Ph.D. dissertation, University of South Alabama, 298pp

Murdoch TJT (2013) Final report of the UK OTEP reef project (in prep)

Murdoch TJT, Hammond MP, Glasspool AF, Outer bridge M, Clee J, Shailer M, Colella M. (2008a) A multi-scale assessment of the benthic communities of Bermuda's shallow water platform focusing on existing MPA's & control sites to enhance local management practices. Bermuda reef ecosystem assessment and mapping (BREAM) report, 89pp

Murdoch TJT, Glasspool A, Colella, M (2008b) The results of the first comprehensive benthic assessment of coral reef habitats across the Bermuda Platform. In: Proceedings of 11th coral reef symposium Ft. Lauderdale, Poster

Mutz SJ (2006) Comparative growth dynamics of acanthurid fishes. M.Sc. thesis, James Cook University, Townsville

Nagelkerken I, Roberts CM, van der Velde G, Dorenbosch M, van Riel MC, de la Morinere EC, Nienhuis PH (2002) How important are mangroves and seagrass beds for coral-reef fish? The nursery hypothesis tested on an island scale. Mar Ecol Prog Ser 244:299–305

Nagelkerken I, Bothwell J, Nemeth RS, Pitt JM, van der Velde G (2008) Interlinkage between Caribbean coral reefs and seagrass beds through feeding migrations by grunts (Haemulidae) depends on habitat accessibility. Mar Ecol Prog Ser 368:155–164

Nozaki Y, Rye DM, Turekian KK, Dodge RE (1978) A 200 year record of carbon-13 and carbon-14 variations in a Bermuda coral. Geophys Res Lett 5:825–828

Owen R, Knap AH, Toaspern M, Carbery K (2002) Inhibition of coral photosynthesis by the antifouling herbicide Irgarol 1051. Mar Poll Bull 44:623–632

Owen R, Knap AH, Knap AH, Ostrander N, Carbery K (2003) Comparative acute toxicity of herbicides to photosynthesis of coral zooxanthellae. Bull Environ Contam Toxicol 70:0541–0548

Pakes MJ, Woollacott RM (2008) Reproduction of the gorgonian *Plexaura flexuosa* in Bermuda. J Exp Mar Biol Ecol 357:121–127

Patzold J, Wefer G (1992) Bermuda coral reef record of the last 1000 years. Fourth international conference on paleoceanography, Kiel, pp 224–225

Piniak GA, Lipschultz F (2004) Effects of nutritional history on nitrogen assimilation in congeneric temperate and tropical scleractinian corals. Mar Biol 145:1085–1096

Pitt JM, Trott TM, Luckhurst BE (2009) Bluestriped grunt in Bermuda: age, growth and reproduction studies. Proc Gulf Caribb Fish Inst 62:241–245

Robertson DR (1991) The role of adult biology in the timing of spawning of tropical reef fishes. In: Sale PF (ed) The ecology of fishes on coral reefs. Academic, San Diego

Robertson DR, Ackerman JL, Choat JH, Posada JM, Pitt JM (2005a) Ocean surgeonfish, *Acanthurus bahianus*. I. The geography of demography. Mar Ecol Prog Ser 295:229–24

Robertson DR, Choat JH, Posada JM, Pitt JM, Ackerman JL (2005b) Ocean surgeonfish, *Acanthurus bahianus*. II. Fishing effects on longevity, size and abundance? Mar Ecol Prog Ser 295:245–256

Rodrigues LJ, Dunham DW, Coates KA (2000) Shelter preferences in the endemic Bermudian hermit crab, *Calcinus verrilli* (Rathbun, 1901) (Decapoda, Anomura). Crustaceana 73:737–750

Savage AM, Goodson MS, Visram S, Trapido-Rosenthal H, Wiedenmann J, Douglas AE (2002a) Molecular diversity of symbiotic algae at the latitudinal margins of their distribution: dinoflag-ellates of the genus *Symbiodinium* in corals and sea anemones. Mar Ecol Prog Ser 244:17–26

Savage AM, Trapido-Rosenthal H, Douglas AE (2002b) On the functional significance of molecular variation in *Symbiodinium*, the symbiotic algae of Cnidaria: photosynthetic response to irradiance. Mar Ecol Prog Ser 244:27–37

Schultz ET, Cowen RK (1994) Recruitment of coral-reef fishes to Bermuda: local retention or long-distance transport? Mar Ecol Prog Ser 109:15–28

Scoffin TP, Garrett P (1974) Processes in the formation and preservation of internal structure in Bermuda's patch reefs. Proceedings of the 2nd international coral reef symposium, vol 2. Brisbane, pp 429–448

Shen GT, Boyle EA (1988) Determination of lead, cadmium and other trace metals in annually-banded corals. Chem Geol 67:47–62

Smith CL (1958) Hermaphroditism in some serranid fishes from Bermuda. Pap mich Acad Sci Arts Letts 44:111–119

Smith SR (1985) Reef damage and recovery after ship groundings on Bermuda. In: Proceedings of the 5th international coral reef symposium, vol 6. Tahiti, pp 497–502

Smith SR (1988) Recovery of a disturbed reef in Bermuda: influence of reef structure and herbivorous grazers on algal and sessile invertebrate recruitment. In: Proceedings of the 6th international coral reef symposium, vol 2. Townsville, pp 267–272

Smith SR (1990) The influence of herbivorous grazers on the recovery of a disturbed coral reef in Bermuda. Ph.D. dissertation, University of Georgia, 191pp

Smith SR (1992) Patterns of coral recruitment and post-settlement mortality on disturbed and undisturbed coral reefs in Bermuda: comparisons to the Caribbean and tropical Pacific. Am Zool 32: 663–673

Smith SR, Hellin DC, McKennna SA (1998) Environment Bermuda: marine environmental impact studies on the thermal effluent released from the Tynes Bay Incinerator and ash waste disposal in Castle Harbour. Bermuda Biological Station for Research Special Publication No. 35, 210pp

Smith SR, Musik KM (1984) Aspects of the distribution and ecology of shallow water scleractinian and octocorallian species on Bermudian reefs. Advances in Reef Science, Miami, pp 115–116

Smith SR, Pitt JP (2002) Summary report on the Benthic Monitoring Project to assess long term effects of waste disposal in Castle Harbour. Bermuda Biological Station for Research, p 34

Smith SR, Pitt JP, Webster G (2003) Summary report of the Benthic Monitoring Project, Bermuda Biological Station for Research, 41pp

Smith-Vaniz WF, Collette BB, Luckhurst BE (1999) Fishes of Bermuda: history, zoogeography, annotated checklist, and identification keys. Amer Soc Ichthy Herpet Spec Publ 4, p 424

Tomascik T, Logan A (1990) A comparison of peripheral growth rates in the recent solitary coral *Scolymia cubensis* (Milne-Edwards and Haime) from Barbados and Bermuda. Bull Mar Sci 46:799–806

Trott TM (2006) Preliminary analysis of age, growth and reproduction of Coney (*Cephalopholis fulva*) at Bermuda. Proc Gulf Caribb Fish Inst 57:385–399

Trott TM, Pitt JM, Luckhurst BE (2009) Occurrence and management of a spawning aggregation of bluestriped grunt (*Haemulon sciurus*) in Bermuda. Proc Gulf Caribb Fish Inst 62:260–267

Trott TM, Pitt JM, Luckhurst BE (2010) Investigations into the dynamics black grouper spawning aggregations in Bermuda. Proc Gulf Caribb Fish Inst 63:541 (poster abstract)

Tyler JC, Luckhurst BE (1994) Unusual features of the colonies of the common western Atlantic garden eel (Heterocongrinae), with a new record for Bermuda. NE Gulf Sci 13(2):89–100

van Woesik R, Lacharmoise F, Koksal S (2006) Annual cycles of solar insolation predict spawning times of Caribbean corals. Ecol Lett 9:390–398

Vaughan TW (1915) The geological significance of the growth rate of the Floridian and Bahamian shoal-water corals. J Wash Acad Sci 5:591–600

Venn AA, Wilson MA, Trapido-Rosenthal HG, Keely BJ, Douglas AE (2008) The impact of coral bleaching on the pigment profile of the symbiotic alga, *Symbiodinium*. Plant Cell Environ 29:2133–2142

Verweij MC, Nagelkerken I, Hans I, Ruesler SM, Mason PRD (2008) Seagrass nurseries contribute to coral reef fish populations. Limnol Oceanogr 53:1540–1547

Vierros, M (1999) Remote sensing and GIS applications in tropical marine ecosystems. Ph.D. dissertation, University College North Wales Bangor, 168pp

Ward JAD (1999) Bermuda's coastal seagrass beds as habitats for fish. M.Sc. thesis, University of Glasgow, 96pp

Webster G, Smith SR (2000) Reduced juvenile coral populations on reefs affected by sewage discharge in Bermuda. In: Proceedings of the 9th international coral reef symposium, vol 2. Bali, pp 1041–1046

Wyers SC (1985) Sexual reproduction of the coral *Diploria strigosa* (Scleractinia, Faviidae) in Bermuda: Research in progress. In: Proceedings of the 5th international coral reef symposium, vol 4. Tahiti, pp 301–305

Wyers SC, Barnes HS, Smith SR (1991) Spawning of hermatypic corals in Bermuda: a pilot study. Hydrobiologia 216(217):109–116

Yost DM, Jones RJ, Mitchelmore CL (2010) Alterations in dimethylsulfoniopropionate (DMSP) levels in the coral *Montastraea franksi* in response to copper exposure. Aquatic Tox 98:376–373

Biogeography, Biodiversity and Connectivity of Bermuda's Coral Reefs

12

Jan M. Locke, Kathryn A. Coates, Jaret P. Bilewitch, Lyndsey P. Holland, Joanna M. Pitt, Struan R. Smith, and Henry G. Trapido-Rosenthal

Biogeographical Region and Environmental Factors

Biogeographic provinces are "Large areas defined by the presence of distinct biotas that have at least some cohesion over evolutionary time frames. Provinces will hold some level of endemism, principally at the level of species…[and] … In ecological terms, provinces are cohesive units likely, for example, to encompass the broader life history of many constituent taxa, including mobile and dispersive species." Spalding et al. (2007). They are also "areas of relatively homogeneous species composition, clearly distinct… [but] endemism is not a key determinant in ecoregion identification." Herein, we explore the composition and genetics of the fauna of the coral reefs in Bermuda and discuss overall and taxon-specific diversity patterns, particularly as they relate to conservation and management of marine resources.

Bermuda is one of nine ecoregions in the Tropical Northwestern Atlantic (TNA) (Spalding et al. 2007), differentiated from adjacent regions by differences in sea surface temperature minima. Bermuda's marine biota is a reduced set of the biota of the TNA. Due to its northerly location, Bermuda has a distinct seasonality in temperature, incident light regimes, and ocean chemistry, which are all important factors in determining the distribution of corals and other marine species (Kleypas et al. 1999; S.A. Manuel and K.A. Coates, unpubl. data).

The Sargasso Sea is an important contributor to the biological diversity of Bermuda. *Sargassum* mats provide homes to many co-adapted species, and spawning grounds to many others that range more widely. Humpback whales (*Megaptera novaeaeangliae*) annually migrate through the Sea and congregate outside Bermuda's shallow rim reef (Stone et al. 1987). *Sargassum* and associated biota are frequently stranded on the shores of Bermuda (Butler et al. 1983) creating a diverse strandline community (Healy and Coates 2003), and other temporary residents of the Sea move into the shallow waters around Bermuda for some stages of their lives.

Ecoregion Forcing Agents for Bermuda

Of the dominant geographic and physical forcing agents on the biological diversity of Bermuda, those of most significance include isolation, temperature, currents, sediments, bathymetric or coastal complexity, and environmental seasonality, both now and in the past.

Isolation: Distance, Depth and Size

The nearest source populations for Bermuda's marine species are over 1,000 km away in the Caribbean and Florida. The Gulf Stream is recognized as the primary conduit in the recent era for the introduction of species and individuals

J.M. Locke (✉) • K.A. Coates
Marine Ecology Section, Department of Conservation Services,
17 North Shore Road, Hamilton Parish FL 04, Bermuda
e-mail: jml100@gmail.com; kathryn.coates@dilke.com

J.P. Bilewitch
School of Biological Sciences, University of Queensland,
Brisbane, Queensland, 4072, Australia
e-mail: jbilewitch@gmail.com

L.P. Holland
Biosciences, College of Life and Environmental Sciences,
University of Exeter, Stocker Road, Exeter EX4 4QD, UK
e-mail: lynzholland@yahoo.co.uk

J.M. Pitt
Marine Resources Section, Department of Environmental Protection,
3 Coney Island Road, St. Georges CR04, Bermuda
e-mail: jpitt@gov.bm

S.R. Smith
Natural History Museum, Bermuda Aquarium, Museum and Zoo,
Department of Conservation Services,
40 North Shore Road, Hamilton Parish FL04, Bermuda
e-mail: srsmith@gov.bm

H.G. Trapido-Rosenthal
Department of Biology, Division of Natural Sciences and
Mathematics, Chaminade University of Honolulu,
Honolulu, HI 96816, USA
email: henry.trapido-rosenthal@chaminade.edu

C.R.C. Sheppard (ed.), *Coral Reefs of the United Kingdom Overseas Territories*, Coral Reefs of the World 4,
DOI 10.1007/978-94-007-5965-7_12,

Table 12.1 Number of species of different groups with different coastline lengths in the Greater Caribbean region (Data adapted from Miloslavich et al. 2010)

Location	Coastline length (km)	Numbers of species Corals[a]	Sponges	Molluscs	Echinoderms
Guatemala	148	25	–	–	23
Cayman Islands	160	50	82	477	–
Costa Rica	212	37	64	638	23
Aruba, Bonaire, Curaçao	360	53	113	239	–
Trinidad, Tobago	362	29	–	–	55
Belize	386	40	193	580	134
Nicaragua	493	39	–	129	65
Puerto Rico	501	53	40	1,078	121
Honduras	644	55	–	580	95
Mexico (Yucatán)	911	47	118	733	182
Jamaica	1,022	52	169	824	86
Panama	1,295	55	146	587	155
Lesser Antilles	1,322	54	126	1,119	79
Colombia	1,880	52	142	1,168	180
Venezuela	2,722	58	144	664	124
Hispaniola	3,059	53	71	572	117
Cuba	3,735	54	255	1,300	145
Bermuda – low resolution	103	26	99	942	98
Bermuda – high resolution	296	26	99	942	98

[a]Has been changed based on review of original data

from these source populations to the marine communities of Bermuda. It is likely that species get carried into waters that encounter Bermuda's shallow platform via meso-scale eddies from the Gulf Stream. Genetic connectivity of various Bermuda populations with Caribbean, Bahamian and Floridian populations are evidence of ongoing and regular recruitment.

The Bermuda Platform is a very small area of shallow "habitable benthic environment" (see Kleypas 1997 and Pandolfi 1999), in a vast region of very deep water. The coastline is about 296 km (Meyer 2012) surrounded by a shallow marine area of about 620 km^2 (to 20 m depth). Similar length coastlines in the TNA can harbour very different numbers of species (Table 12.1) and only for corals does there seem to be a positive correlation between numbers of species and coastline length; for sponges, echinoderms and molluscs for example, there seems to be much less or no relationship (Miloslavich et al. 2010). Although Bermuda has a comparatively short coastline, that may not be a primary limitation to species diversity on the platform. Coastline length, area and configuration of shallow waters have changed many times during the development of coral reefs in Bermuda. Both extreme highstands (up to 22 m above present day sea level) and lowstands (to perhaps 140 m below present day sea level) occurring in the Pleistocene would have decreased the total area of shallow water on the platform (Hearty and Olson 2010). The most extreme recent sea level highstand, corresponding to an interglacial period, about 120 kya may have reached 9–10 m above present sea level (Hearty et al. 2007), probably drowning many areas of shallow reef. The last Pleistocene glaciation maximum and a corresponding extreme sea level lowstand about 120 m below present occurred about 18 kya. This would have limited shallow-water habitat to a narrow band on the sloping sides of the Bermuda Platform and nearby seamounts (Iliffe et al. 2011) (Fig. 12.1).

Climate Changes and Temperature

Environmental shifts during the Pleistocene glaciations could have resulted in other factors than simple loss of habitat constraining the diversity of reef communities; sea surface temperature shifts, for example, are correlated to periods of glacial minima and maxima. There are Bermudian fossils of two extant Caribbean coral taxa no longer found in Bermuda, *Cladocora* (Moore and Moore 1946) and *Colpophyllia* (Muhs et al. 2002). They are both now found in locations with higher average minimum sea surface temperatures than seen in Bermuda, and the known Bermudian fossils of *Colpophyllia* are dated to interglacial periods (Muhs et al. 2002) when sea surface temperatures may have been higher. However, species of *Acropora*, which are common in the Caribbean, never have occurred in Bermuda and this has also been interpreted as a temperature-related distribution even though acroporids occur in Florida where seawater temperature

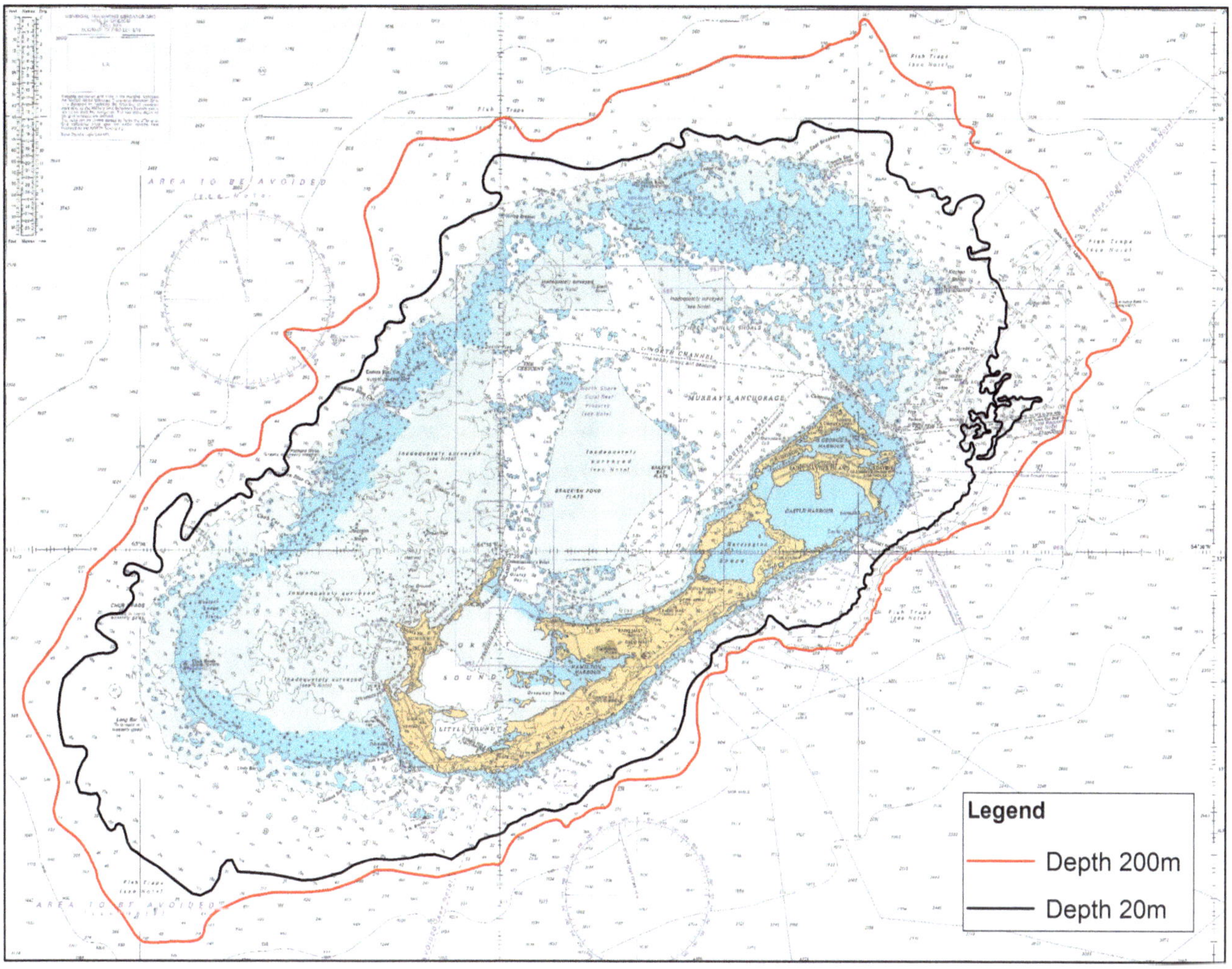

Fig. 12.1 Bathymetric contours at 20 m and 200 m that might represent Pleistocene shorelines at glacial maxima of about 80-90 kya and 18 or 424 kya, respectively. Based on Bermuda Admiralty Chart #334, reproduced by permission of the Controller of her Majesty's Stationary Office and the UK Hydrographic Office (www.ukho.gov.uk)

extremes are even greater (Precht and Miller 2007) than in Bermuda. Again, a number of factors may combine, that are related to Bermuda's northern latitude, to exclude some tropical species and not others, including duration of cooler temperatures, aragonite saturation levels (Kleypas et al. 1999; Bates et al. 2010), and low light during periods of cooler temperatures (S.A. Manuel and K.A. Coates, unpubl. data).

Habitat Complexity

There are no significant surface freshwater outflows in Bermuda, and no regular or high volume sources of terrigenous sediments, which is generally considered beneficial to coral growth. However, some highly productive and diverse tropical marine coastal communities, including mangroves and seagrasses, thrive in the finer organic sediments characteristic of terrigenous sources and estuaries. Significant and even sustaining connections exist between the communities of these habitats and of coral reefs (Nagelkerken et al. 2008) so that their reduction in relative size may decrease the success of some segments of the average coral reef community.

Seasonality in Incident Light

The low angle of the sun, combined with increased cloud cover, during the winter months limits total incident light at the sea surface, and there is a significant seasonal difference in light regime of southern versus northern exposed faces of the reefs (e.g., Murdoch 2007). This high seasonality may be a primary limiting factor to much marine life, including the depth distribution of seagrass (S.A. Manuel and K.A. Coates, unpubl. data). Neither Kleypas et al. (1999) nor Bates et al. (2010) found a strong correlation between coral growth and light for Bermuda but, they could only consider corals that were present and which must be tolerant of low light conditions. Corals that do not occur in Bermuda may well have different behaviours and physiologies, and light conditions may restrict them to lower latitudes.

Biological Diversity

A distinctive characteristic of the marine fauna and flora of Bermuda is the absence of species that are otherwise ubiquitous in the Tropical Northwestern Atlantic province. Bermuda does not show an increased diversity due to the mixing of species that survive at the boundaries of biogeographic realms, in this case the Tropical Atlantic and the Temperate Northern Atlantic. For fish, the Bermuda coastal marine fauna includes some eurythermic tropical species (Smith-Vaniz et al. 1999), which have geographic ranges crossing the tropical-temperate "boundary", but which are not restricted to the boundary region. Nonetheless, Bermuda's fish species diversity is reduced relative to both higher and lower latitude neighboring regions.

Bermuda's isolation suggests the possibility of a relatively high local level of endemism, and although Sterrer (1998) suggests Bermuda exhibits low marine endemism of about 3%, recent reviews of a number of taxa indicate much higher endemism among some major taxa that are common in coral communities, including sponges and polychaetes. Lists of species for these taxa can be accessed on the Bermuda Government, Department of Conservation Services, website www.conservation.bm.

Fig. 12.2 Bermuda's coral reef biodiversity as illustrated by the scleractinian and octocorallian species commonly found on the fore-reef terrace at Hog Breaker (Photo: JM Locke)

Scleractinia

Bermuda's reef building coral diversity fits a general idea that lower diversity characterizes areas located at the range limits of coral distribution. This is clearly visible on Bermuda's reefs, where the favids *Diploria* and *Montastraea* are the dominant scleractinian genera together with *Porites astreoides* (Fig. 12.2), and a total of 12 shallow-water genera commonly found in the Caribbean and USA are notably absent. However, a "fit" is not explanatory of this lower diversity and as noted earlier, Bermuda has similar numbers of coral species to equally small coastal areas that are much further south in the TNA – numerous factors control the species diversity of Bermuda's corals. To date, 20 species of zooxanthellate corals, in nine families and 13 genera, and six azooxanthellate corals, in four families and six genera, are known to inhabit shallow waters, including Bermuda's only endemic coral *Rhizopsammia bermudensis* (Locke 2009; Locke et al. 2013, and at www.conservation.bm/corals).

Given the similarity in diversity between Bermuda and its more northerly tropical con-provincial ecoregions in the Caribbean, it is notable that the Acroporidae, common on many Greater Caribbean reefs, is absent both from recent and fossil reefs. Ecological factors are presumed to have prevented the establishment of acroporids in Bermuda. This may also be true of other genera common to the Caribbean but not currently known from Bermuda. The species *Cladocora arbuscula* and *Colpophyllia natans* have, however, been documented from the fossil record (Moore and Moore 1946; Muhs et al. 2002). We can only speculate what led to their demise though, as noted earlier, fossil specimens of *C. natans* have been dated to the last interglacial maximum.

Limited study of the deep benthic habitats have provided brief insight into corals at and beyond the mesophotic zone. Investigations have documented 13 zooxanthellate species from 30 to 78 m with the genera *Agaricia, Scolymia* and *Montastraea* inhabiting depths greater than 60 m (the deepest is 78 m for *Montastraea cavernosa*) (Fricke and Meischner 1985; Venn et al. 2009).

Beyond the mesophotic zone, our knowledge of deep-water corals consists solely of records of species occurance. Collectively, expeditions have acquired azooxanthellate specimens at depths from 55 m to 1966 m. Bermuda's known deep-water azooxanthellate Scleractinia number 23 species from 10 families, many species being known only from single collections (Locke et al. 2013). Within the deep waters (>183 m) of the western Atlantic, 102 species of azooxanthellate corals are currently recorded (Cairns 2000). Considering this, it is difficult to interpret Bermuda as being a low diversity location; instead the number of species reported for Bermuda is likely a reflection largely of the limited sampling.

The 48 valid scleractinian species is a significant fraction, nearly 25%, of the 197 scleractinian species currently known from the western Atlantic (adapted from Cairns 2000); the majority reported from shallow-water.

Genetic Variation and Population Connectivity

Cairns and Chapman's (2001) analysis of deep-water azooxanthellate corals in Bermuda revealed components of both a warm temperate and a tropical fauna; with increasing depth, the western Atlantic cnidarian fauna becomes increasingly cosmopolitan (Cairns 1979), and deeper subhabitats of the Bermuda EEZ may be biogeographically more inclusive than those of shallow depths.

Genetic studies, utilizing various molecular markers, have produced plausible hypotheses concerning the genetic diversity and connectedness of a few scleractinian species found in Bermuda. These concern shallow-water zooxanthellate brooding species, *Favia fragum* and *Madracis auretenra*, and a single broadcast spawning species, *Montastraea cavernosa*. Gene flow was investigated at recent (Shearer 2004; Locke 2009; Nunes et al. 2009; Goodbody-Gringley et al. 2010, 2011) and historical scales (Locke 2009).

Studies reveal high levels of intraspecific genetic diversity of *M. cavernosa* and *M. auretenra*, and in some cases this diversity (i.e., numbers of haplotypes in a population from a single geographical site) exceeded that of conspecifics within other regions. Both species harbour unique (=private) and shared haplotypes for the molecular markers among the locations investigated (Locke 2009; Nunes et al. 2009; Goodbody-Gringley et al. 2010, 2011). In fact, more unique haplotypes were observed in Bermuda than in conspecific populations of any other geographic location (Locke 2009; Goodbody-Gringley et al. 2011).

In contrast, Bermuda's *F. fragum* population showed no intraspecific variation, with only one haplotype determined for each of two markers analyzed (*ITS* and *β-tubulin)*; although the single *ITS* haplotype was unique to Bermuda (Goodbody-Gringley et al. 2010). The extremely different levels of intraspecific variability reported for the two brooding species *F. fragum* and *M. auretenra* (*SRP54*) may simply be due to the different markers used in the two studies or other intrinsic, biological, factors. Studies documenting high levels of variation in *M. cavernosa* analyzed the same *β-tubulin* region as the *F. fragum* study (Nunes et al. 2009; Goodbody-Gringley et al. 2010, 2011); further evidence that these two species are experiencing gene flow regimes that may be attributed to their differing life histories and, perhaps, to their different histories in Bermuda. This does not however explain the high diversity seen in *M. auretenra.*

The greater level of genetic diversity in Bermuda coral populations contradicts predictions that oceanographically isolated reefs may be associated with lowered genetic diversity (Ayer and Hughes 2004; Miller and Ayre 2004, 2008; Nunes et al. 2009; Goodbody-Gringley et al. 2010). The unique haplotypes found in three studied Bermuda coral species may be relicts of early colonizers that were once present in US and Caribbean regions but have since been extirpated, or may be the result of post colonization diversification.

The genetic connectivity of *M. cavernosa* populations within Bermuda and to USA and Caribbean populations has been investigated on three accounts. Shearer (2004) determined, from four and five microsatellites, that Bermuda populations were differentiated from populations in the Flower Garden Banks, Florida Keys, and the Bahamas. Conversely, when compared across the Atlantic, using two nDNA *β-tubulin* regions and mtDNA *cox1-trnM*, Bermuda's *M. cavernosa* population was not significantly differentiated from populations in Puerto Rico, Belize and Panama (Nunes et al. 2009). The connection of Bermuda *M. cavernosa* populations with those in the USA and Caribbean was corroborated by Goodbody-Gringley et al. (2011) for populations in the Flower Garden Banks, Jamaica, Panama and Barbados using mtDNA *cox1-trnM*. However, based on *ITS* and a *β-tubulin* region, the same study revealed F_{st} values that indicated Bermuda populations were differentiated from Barbados and Jamaica respectively (Goodbody-Gringley et al. 2011).

Bermuda's shallow-water coral species are mostly brooding species. Studies of *F. fragum* and *M. auretenra*, showed that these species were genetically structured from USA and Caribbean conspecifics (Locke 2009; Goodbody-Gringley et al. 2010). More specifically, analyses of Bermudian *F. fragum* using nuclear regions *ITS* and *β-tubulin*, determined that gene flow was restricted between Bermuda and Jamaica, Panama and Barbados (Goodbody-Gringley et al. 2010). Likewise, analyses of a region of the nuclear gene encoding *SRP54* in *M. auretenra* populations determined that Bermuda corals were genetically structured from populations in the Florida Keys and Puerto Rico (Locke 2009).

Bermuda's connections at an evolutionary scale are investigated by Locke (2009). Shared ancestral haplotypes among the Bermuda, Puerto Rico and Florida Keys populations indicate that these regions might have shared an historic connection, but a distinctly Bermudian clade reveals that this connection may not be as prevalent as it was in the past. A recent division of Bermuda and Puerto Rico populations was proposed based on distinct haplotypes and phylogenetic clades for these regions. Shared phylogenetic clades for some Bermuda and Florida haplotypes demonstrate that the isolation of Bermuda's genetically differentiated populations may be broken in rare events.

The current theories of Bermuda coral connectivity, based on limited study only, indicate there is free exchange of genetic information among the reefs in Bermuda (Shearer 2004; Locke 2009; Goodbody-Gringley et al. 2010, 2011) (but see Shearer 2004 and Goodbody-Gringley et al. 2011 for exceptions). Brooding species appear to be self seeding with gene flow from other regions restricted to rare events, if at all, whereas, broadcast spawning coral species connectivity is at best contradictory with inferred connections between some but not all sampled Caribbean and USA populations. Opposing results for the same species and sampled locations provide an indication that these ambiguous outcomes may be related to marker resolution.

Octocorallia

Studies of deep-water diversity are deficient and existing reports rely heavily on anecdotal records and haphazard, accidentally-collected specimens (Cairns et al. 1986; Bilewitch 2008). A number of shallow-water studies exist (e.g. Chester 1913; Grode et al. 1983), but a wide-ranging field survey of octocoral diversity and abundance has only recently been attempted (L.P. Holland, unpubl. data), and thus any cumulative list of species may only represent the most obvious, non-cryptic representatives and a handful of circumstantial and anecdotal samples from the deep-seas.

We still know little of the vertical, bathymetric limits of any given species, and most zooxanthellate octocoral specimens have been obtained from less than 20 m depth, although anecdotal records indicate that certain species (e.g. *Plexaurella nutans* and *Antillogorgia acerosa*) may also be found at depths exceeding 50 m on the Challenger and Plantagenet seamounts. Study of the mesophotic zone will undoubtedly provide insights on environmental controls on octocoral distributions and phenotypic variation (e.g., Fricke and Meischner 1985; Venn et al. 2009) and overall species diversity may be high (Bridge et al. 2011). Most studies to date have been on either shallow-water or deep-water zones, and not this transition region.

With one exception, all shallow-water Bermudian octocorals are contained within the suborder Holaxonia Studer 1887, most being in the family Plexauridae Gray 1859 (Locke et al. 2013). Of the 24 shallow-water species, 18 are plexaurids, while five are Gorgoniidae. *Briareum asbestinum* represents the only known member of the Scleraxonia in Bermuda; the encrusting species *Erythropodium caribaeorum*, which is common in the Caribbean, is notably absent.

Most deep-sea species have been observed only once in Bermuda so that the lack of corroborating evidence for a species' presence and distribution from a number of records presents a serious impediment to the characterization of Bermuda's deep-sea octocoral community. Below 30 m depth we do not know whether sole specimens or observations represent one of a few or one of many such colonies in the deep-seas surrounding Bermuda, but new records of primnoids, isidids and paramuriceids, based largely on haphazard and accidental collections, suggest that many more octocorals await discovery (Locke et al. 2013).

Genetic Variation and Population Connectivity

The intraspecific genetic diversity of octocorals in Bermuda has been studied only twice (Bilewitch 2006; Bilewitch et al. 2010) and only one of these studies examined population-level differences in some depth. Bilewitch (2006) examined species-level (and above) octocoral molecular systematics but also provided indirect evidence of intraspecific genetic variation in the four shallow-water species *Plexaura flexuosa*, *Pseudoplexaura porosa*, *Gorgonia ventalina* and *Briareum asbestinum*. Of these, all except *P. flexuosa* contained some genotypes that were unique to Bermuda (Bilewitch 2006). Bilewitch et al. (2010) further examined the population variation of *B. asbestinum* and found both unique and shared genotypes in the Bermudian population. The star-like genetic diversification of *B. asbestinum*, where the wide-ranging genotypes are central to the region-specific genotypes is consistent with a pattern of Caribbean ancestry with subsequent limited allopatric diversification in Bermuda. The Bermudian population thus appears to be semi-isolated to the extent that allows some endemic genotypes to develop, but either the age of the population is too young or the influx of genotypes from the Caribbean is too frequent to have resulted in enough genetic drift to lead to speciation (Bilewitch et al. 2010). Although *B. asbestinum* may have been introduced into Bermuda through anthropogenic means, the population likely exceeds a century in age and remains genetically and morphologically similar to conspecifics in the Caribbean.

Zooxanthellae

Many reef-associated marine invertebrates including hard and soft corals, hydrozoans, sponges, and molluscs, have symbiotic associations with *Symbiodinium*; a diverse dinoflagellate assemblage comprising nine distinct lineages, or 'clades' (Clade A to Clade I). As these clades vary in physiological response to fluctuating environmental parameters, assessing diversity of *Symbiodinium* can be used as a proxy for predicting coral resilience to detrimental events such as bleaching and climate change.

In Bermuda, early research suggested that local symbiont diversity was low. Boschma (1925) concluded that symbionts within the corals *Isophyllia dipsacea* (=*Isophyllia sinuosa*) and *Siderastraea radians*, the zoanthid *Zoanthus sociatus* and the anemone *Condylactis passiflora* (=*Condylactis gigantea*) were probably the same species, an assertion based upon morphological uniformity. However, following the development of DNA-based methods such as PCR-RFLP in the 1990s, it was demonstrated that *Symbiodinium* was, in fact, extremely diverse (Rowan and Powers 1991), and one host could harbour mixed molecular variants. Generally, in Metazoa, clades A, B and C predominate on Caribbean reefs, whereas C and D are more prevalent in Pacific reefs (Baker 2003), and some clades are associated predominantly with certain host taxa (e.g., F, H and I in soritid foraminifera, Pochon and Gates 2010).

Using PCR-RFLP, the diversity of zooxanthellae has been examined in anthozoan hosts in Bermuda over a depth gradient (Billinghurst et al. 1997), at latitudinal limits (Savage et al. 2002), in Bermudian and Caribbean conspecifics (e.g., Holland 2006), and at varying temporal and spatial scales (Loram et al. 2007a, b; Venn et al. 2008). Additional data from Bermudian hosts stems from a few samples included in larger studies of *Symbiodinium* diversity in the Greater Caribbean. At cladal level, the diversity of symbionts in Bermuda is similar to the Caribbean, with most zooxanthellae belonging to clades A, B and C. Clade D has not yet been found in Bermuda. Most symbiont assemblages examined in Bermudian Anthozoa are characterised by the occurrence of a single clade, although mixed infections are also characteristic of symbioses in certain host taxa; presence of mixed infections may vary spatially or temporally (e.g., Loram et al. 2007b; Venn et al. 2008), and may differ between Bermuda and the Caribbean (e.g., Savage et al. 2002). For example, between conspecifics, a reduced diversity is observed in Bermuda corals; *Diploria labyrithiformis* contains B in Bermuda, not B and C as in the Caribbean, *Montastraea franksi* harbours A and B, as opposed to A, B, C and E, and *Porites astreoides* contains A, and not A and C (Savage et al. 2002). Several octocoral species exhibit a similar pattern; *Plexaurella dichotoma* and *Briareum asbestinum* contain clade B or C in the Caribbean (Goulet and Coffroth 2004), yet only clade B in Bermuda (Holland 2006, unpublished data).

In all octocorals sampled in Bermuda, only clade B has been found (Holland 2006; Goulet et al. 2008), with the exception of one report of clade A in the sea fan *Gorgonia ventalina* ($n = 1$, LaJeunesse 2001).

The anemone *Condylactis gigantea* harbours Clades A and B *Symbiodinium*. In offshore sites, where water is cooler and less variable, it may harbour Clade B alone, while at more thermally variable inshore sites it contains either solely Clade A, or mixed Clades A and B. As water temperatures rise and fall with seasonal changes, *C. gigantea* with mixed infections can alter the proportions of Clade A and B symbionts, with populations of Clade A increasing with increasing water temperature, and Clade B increasing with declining temperature (Loram et al. 2007b; Venn et al. 2008). These variations have physiological correlates in that more photosynthate is incorporated into the host's lipid and amino acid pools when their symbiont population is dominated by the thermally-tolerant Clade A algae, than when dominated by thermally-sensitive Clade B algae (Loram et al. 2007b).

ITS-Level Symbiodinium Diversity in Bermuda

At present, symbiont diversity is not adequately represented by the denotation as clades A, B, C, etc. in phylogenies based on small or large subunits of rDNA genes. There is substantial intra-cladal variation revealed by sequence variation of spacer (ITS) regions that evolve significantly faster than rDNA sub-units. Therefore, the past decade has seen an increase in the use of rDNA spacer regions (i.e., ITS1 or ITS2) to delineate symbiont genotypes and to assess the correlation of resultant haplotypes (typically described as 'types', e.g., type B1 *sensu* LaJeunesse 2001) to ecological characteristics.

Although ITS markers have revealed finer divergence within each clade, caution must be exercised when using them. Ribosomal DNA (including both spacer regions) is multicopy, therefore several variants may co-exist within an individual *Symbiodinium* cell (see Stat and Gates 2011). Selective use of one ITS spacer may present a flawed strategy resulting in underrepresentation of isolated haplotypes (see Holland 2006). Nevertheless, the ITS region (ITS1-5.8s-ITS2) has been sequenced to identify symbionts in various anthozoan hosts in Bermuda (LaJeunesse 2001; Savage et al. 2002; Holland 2006). A proliferation of subsequent research focusing solely on ITS2 haplotypes has made it difficult to directly compare known Bermuda symbiont types to those found within the Caribbean, primarily due to incomparable datasets (see Holland 2006), although where possible the corresponding symbiont 'types' identified are listed in Table 12.2.

There is little evidence for endemism in Bermuda. There are several generalist symbiont types, exemplified by the prevalence of redundant haplotypes obtained from taxonomically and geographically disparate hosts and a lack of Bermuda-specific symbiont clusters in phylogenetic analyses (Holland 2006). Based solely upon ITS2 data, it appears that type B1 (*sensu* LaJeunesse 2001) may be the predominant symbiont in octocorals in Bermuda (Goulet et al. 2008); this warrants further investigation.

Microsatellite markers have recently been employed to assess symbiont diversity within the sea fan *Gorgonia ventalina* across the Greater Caribbean (Andras et al. 2011). Haplotypic diversity within the *Symbiodinium* B1/B184

Table 12.2 List of all host species with characterized *Symbiodinium* in Bermuda. *Type/Strain*: Letters refer to cladal designation, whereas alphanumeric identities refer to ITS genotype designation (where available e.g., B2.1). '+'=mixed infections, e.g., A+B. N: number of host colonies sampled. *Methods:* LSU=large sub unit rDNA sequence data, ssuRFLP=small sub unit restriction fragment length polymorphism, ITS1-ITS2=entire ITS array rDNA sequence data including 5.8S region. Numbers in superscript refer to relevant reference and those in parentheses refer to number of individual colonies with the preceding *Symbiodinium* genotype (where available)

Host species	Type/Strain	N	Method	Reference
Order Alcyonacea	–	–	–	–
Family Plexauridae	–	–	–	–
Plexaura homomalla	B	5	LSU, ITS1-ITS2	1
Plexaura flexuosa	B	6	LSU,ITS1-ITS2	1
Pseudoplexaura porosa	B	17[1], 1[2]	ssRFLP[2],LSU, ITS1-ITS2[1]	1,2
Pseudoplexaura flagellosa	B	1	ssRFLP	2
Eunicea sp.	B	8	LSU, ITS1-ITS2	1
Eunicea tourneforti	B	1	ssRFLP	3
Plexaurella dichotoma	B, B1	1	LSU,ITS1-ITS2	1
Muricea laxa	B	1	ssRFLP	3
Family Gorgoniidae	–	–	–	–
Antillogorgia americana	B	3	LSU,ITS1-ITS2	1
Gorgonia ventalina	B & B1[1], B1/B184 & A[a,4],A2[5]	11[1],116[4],1[5]	LSU, ITS-ITS2[1,5], Microsatellites[4]	1,4,5
Family Briareidae	–	–	–	–
Briareum asbestinum	B	<10	LSU,ITS1-ITS2	1
Order Scleractinia	–	–	–	–
Family Astrocoeniidae	–	–	–	–
Stephanocoenia intersepta	B (3), C (2)	5	LSU	6
Family Oculinidae	–	–	–	–
Oculina diffusa	B2.1[5], B[6]	1[5], 11[6]	LSU, ITS1-ITS2	5,6
Family Faviidae	–	–	–	–
Favia fragum	A+B (1), B (7)[6], B (1)[1]	9	LSU,ITS1-ITS2	6,1
Diploria labyrinthiformis	B	10	LSU	6
Diploria strigosa	B	4	LSU	6
Montastraea cavernosa	C	16[6],62[7]	LSU[6], ssRFLP, allozymes[7]	6,7
Montastraea franksi	B (44), A (2)	46	LSU, ITS1-ITS2	6
Family Mussidae	–	–	–	–
Isophyllia sinuosa	B	2	LSU	6
Scolymia sp.	C	2	LSU	6
Family Siderastreidae	–	–	–	–
Siderastrea radians	B	12	LSU	6
Family Poritidae	–	–	–	–
Porites astreoides	A	11	LSU	6
Porites porites	A	11	LSU	6
Family Pocilloporidae	–	–	–	–
Madracis auretenra[a]	B[6], B[1], B7[1]	4	LSU, ITS1-ITS2	6
Madracis decactis	B[1], B[6], B+C[8]	1[1], 3[6],n/a[8]	LSU	1,6,8
Family Agariciidae	–	–	–	–
Agaricia sp.	C	18	LSU	6
Family Meandrinidae	–	–	–	–
Dichocoenia stokesi	B[6], B+C[8]	1 [6],n/a[8]	SSU	6,8
Meandrina meandrites	B[6], B+C[8]	1 [6],n/a[8]	SSU	6,8
Order Actinaria	–	–	–	–
Family Aiptasiidae	–	–	–	–
Aiptasia pallida	B, B2	32	LSU, ITS1-ITS2	6
Aiptasia tagetes	B1	1	ITS1-ITS2	5
Bartholomea annulata	A	25	LSU	6

(continued)

Table 12.2 (continued)

Host species	Type/Strain	N	Method	Reference
Family Actiniidae	–	–	–	–
Condylactis gigantea	A (8), B (7), A+B (2)[6], A, B, A+B[9]	17[6], 179[9]	LSU[6], qPCR based on LSU & AFLP[9]	6,9
Family Aliciidae	–	–	–	–
Lebrunia danae	C	8	LSU, ITS1-ITS2	6
Order Rhizostomeae	–	–	–	–
Family Cassiopeidae	–	–	–	–
Cassiopeia xamachana	A, A1	12	LSU, ITS1-ITS2	6
Order Coronatae	–	–	–	–
Family Linuchidae	–	–	–	–
Linuche unguiculata	A4	1	ITS1-ITS2	5

[a]This species was referred to as *Madracis mirabilis* in Holland (2006) and Savage et al. (2002) (See Locke et al. 2007)
References: [1]Holland (2006), [2]Goulet et al. (2008), [3]Goulet and Coffroth (2004), [4]Andras et al. (2011), [5]LaJeunesse (2001), [6]Savage et al. (2002), [7]Billinghurst et al. (1997), [8]Venn et al. (2009), [9]Venn et al. (2008)

lineage is extremely high in symbionts in *G. ventalina* which again highlights the concept that *Symbiodinium* diversity and functionality are best explained by use of faster evolving and more informative markers.

Overall, most clades found in the Caribbean are also found in Bermuda, but mixed assemblages occur less frequently in Bermuda. At present, there is no evidence of a single lineage of symbionts evolving in Bermuda. Apparently holobionts that colonized Bermuda are taxonomically indistinguishable from their (likely) source pool to the south. This is in contrast to the data for the hosts themselves, which have unique Bermudian haplotypes for more than one genetic marker.

Fish

The ichthyofauna of Bermuda is less diverse, at both the species and genus level, than at other locations in the Tropical Northwestern Atlantic (Smith-Vaniz et al. 1999). Most fishes have a western Atlantic distribution, including 62% of the nearshore fishes, but amphi-Atlantic and more widely distributed species are also common (Smith-Vaniz et al. 1999, reprinted here as Table 12.3).

Regional Faunal Similarities

A coefficient of faunal similarity (CFS) following the Sorensen index (Magurran 1988) was calculated by Smith-Vaniz et al. (1999) to compare Bermuda's ichthyofauna to other localities with closely related faunas. Included in the comparison were the Carolinian Bight, which approximates the area of the US east coast included in the Warm Temperate Northwest Atlantic (WTNA) province which represents the northerly extent of many tropical fish species in the western Atlantic; the Florida Keys and the Bahamas. Smith et al. (2002), in their analyses of the fish fauna of the Tropical Western Central Atlantic, determined that the southern coastal area of Florida harboured the most species (442–552 fish species) per cell of 0.5° lat by 0.5° long, making it a potential source for most Bermudian fish species.

The 58 families of fishes used by Smith-Vaniz et al. (1999) for estimating CFS were represented in Bermuda by 283 species, in the Carolinian Bight by 598, in the Florida Keys by 566 and in the Bahamas by 523 species. The CFS was 0.54 between Bermuda and the Carolinian Bight (238 shared species), 0.58 between Bermuda and the Florida Keys (245 shared species), and 0.63 between Bermuda and the Bahamas (253 shared species) (Smith-Vaniz et al. 1999). Bermuda is about equally similar to all these sites, although appearing slightly more similar to the Bahamas but only because the Bahamas had the fewest total species. The ichthyofauna of Bermuda is dominated by eurythermic tropical species that are distributed in the Caribbean and along the southeastern coast of the United States and shared among all the locations considered.

Many Caribbean species are strongly associated with continental shelves and are generally absent from island platforms isolated by deep water (Smith et al. 2002). Smith-Vaniz et al. (1999) remark on a few notable occurrences of some such species in Bermuda, including *Orthopristis chyrsoptera* (pigfish), *Pareques umbrosus* (cubbyu), *Holacanthus bermudensis* (blue angelfish), *Gymnothorax saxicola* (honeycomb moray) and *Mycteroperca microlepis* (gag).

Reef Fish Diversity

Smith-Vaniz et al. (1999) report 430 described plus two presumptive species of fishes native to Bermuda, with 362 species from 79 families being significantly associated with coral reefs and adjacent platform habitats, to depths of 200 m. Of

Table 12.3 Biogeographic analysis of the native ichthyofauna of Bermuda based on adult occurrences. (Updated from Table 7 in Smith-Vaniz et al. 1999, reproduced with the kind permission of the American Society of Ichthyologists and Herpetologists)

Distribution	No. of species	of total species	No. of established nearshore species	of total established nearshore species
Western Atlantic only	232	53.6	210	61.6
Amphi-Atlantic	77	17.8	56	16.4
Cosmopolitan	67	15.5	26	7.6
Atlantic and Indo-Pacific	24	5.5	19	5.6
Western and Central Atlantic	18	4.2	16	4.7
Atlantic and Eastern Pacific	4	0.9	3	0.9
Bermuda endemic	9+2	2.5	9+2	3.2
Total	433	–	341	–
Oceanic species	67	15.5	–	–
Nearshore species	366	84.5	–	–
Established nearshore species	341	78.7	–	–

these, 25 species are considered waifs or vagrants (Smith-Vaniz et al. 1999), leaving 337 species from 72 families as established components of the local nearshore ichthyofauna (Smith-Vaniz et al. 1999). Important amendments to this list include the recently described endemic halfbeak, *Hyporhamphus collettei* Banford 2010, formerly considered *Hyporhamphus unifasciatus*, and a new record of the spotted spoon-nose eel, *Echiophis intertinctus* (Richardson, 1848) (Smith 2010).

An additional 13 species of fishes are known locally only from juveniles (Smith-Vaniz et al. 1999). These are considered waifs and they do not appear to survive to maturity, either because adult habitat or dietary requirements are not met or because the numbers of larvae or juveniles never reach the critical mass required for a population to develop. During the twentieth century, there were a number of attempts to introduce fishery species from the Caribbean (Smith-Vaniz et al. 1999), however it appears that none of these species became established.

The Pacific lionfish, *Pterois volitans* has been present in Bermuda since 2000 (Whitfield et al. 2002). This brings the number of fish species known from shallow marine habitats to 339, and the total number described from Bermuda to 432. This introduced invasive is excluded from any calculations of endemism rates and biogeographic indices, as is the introduced freshwater species *Gambusia holbrooki*.

Families of fishes with no or poor representation in Bermuda, but which are found in potential source locations along the US east coast and in the Caribbean, are often small-bodied (Blennioidei, Engraulidae, Opisthognathidae) or odd-shaped, poor swimmers (Achiridae, Pleuronectidae, Batrachoididae, Uranoscopidae, Ogcocephalidae, Tetraodontiformes) (Fig. 12.3). However, filefishes (Monacanthidae) and triggerfishes (Balistidae), which fall into the latter category, are well represented.

Hermaphroditism could be considered an advantage for establishing and maintaining a population derived from small and rare initial recruitments. Families such as the protogynous Scaridae, Labridae and, to some degree, the Serranidae are well represented in Bermuda.

Endemics and Endemism

There are nine fish species endemic to Bermuda (Table 12.4) (Smith-Vaniz et al. 1999; Smith 2010; Banford 2010). A number are small to medium-sized and a study of larval distributions (Glasspool 1994) found larvae of these families only within the lagoon, indicating strong retention of larvae in these taxa, and also no evidence of regular off platform/long distance recruits. Larval retention may be significant to local speciation within these families.

An endemism rate of 2.7% for Bermuda is at the low end of the 3–25% endemism seen for shallow-water fish faunas of other isolated oceanic islands in the Atlantic and Eastern Pacific (Smith-Vaniz et al. 1999; Robertson 2001). When only reef fishes are included, rates of endemism for these types of locations range from 1 to 13% (Floeter et al. 2008), suggesting that reef-dwelling fishes have high connectivity among sites even though their habitat is discontinuous. These low levels of endemism are consistent with suggestions that shallow water marine communities were largely extirpated during Pleistocene glacial maxima (Floeter et al. 2008). Subsequent allopatric speciation would require both a change in the recruitment patterns that allowed the initial recolonization and very rapid speciation. Although time is clearly an important factor, it remains likely that dispersal is significant to the lack of endemism seen in the reef ichthyofauna of Bermuda.

Other Notable Fish Taxa in Bermuda

A number of diversification events exist. Bermuda is one of three locations where intergeneric hybrids of the coney,

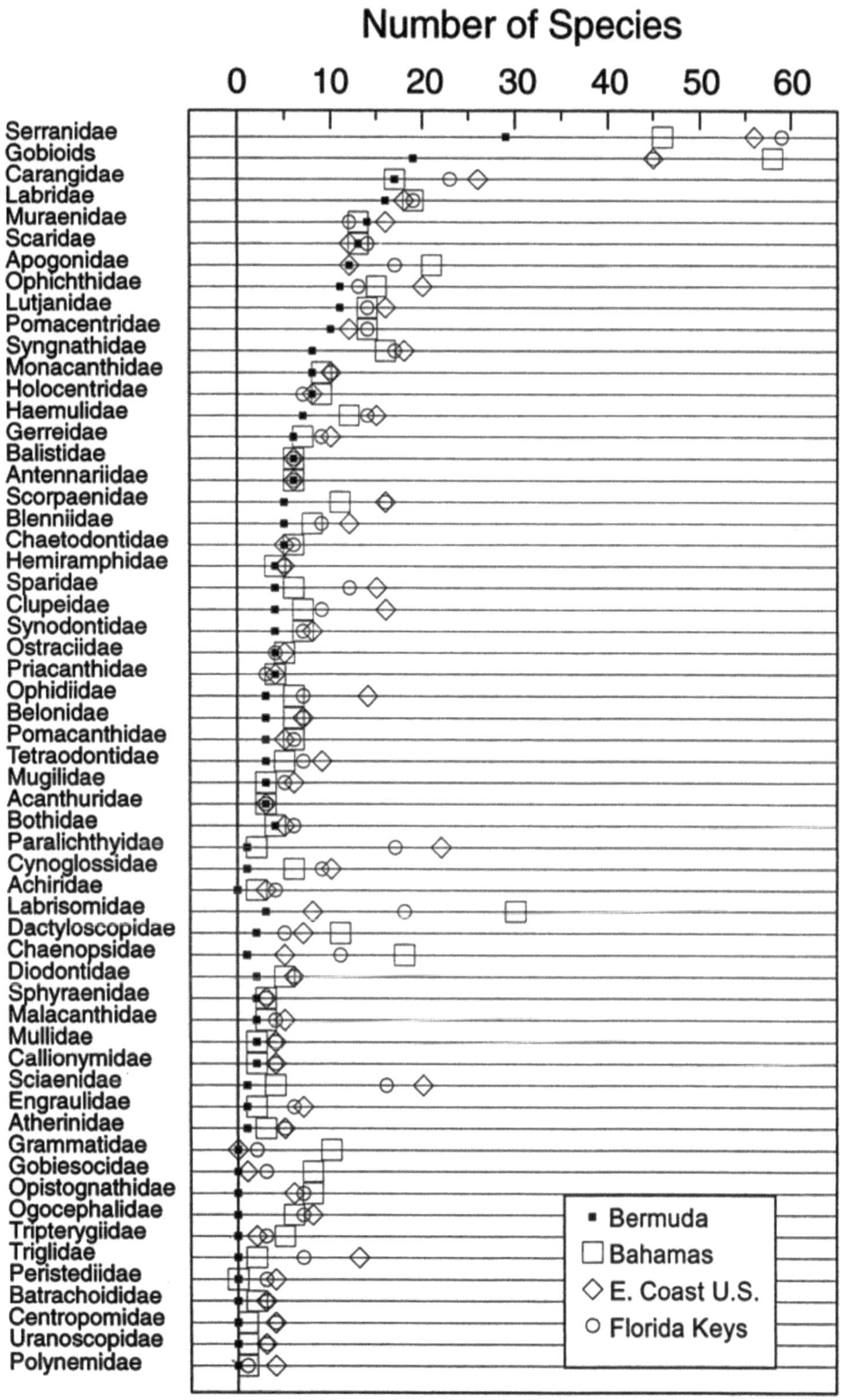

Fig. 12.3 Comparison of total number of species in selected families present in Bermuda, Bahamas, Carolinian Bight and the Florida Keys. (Figure 86 from Smith-Vaniz et al. 1999, reproduced with the kind permission of the American Society of Ichthyologists and Herpetologists)

Table 12.4 Described fish species endemic to Bermuda including key characteristics and museum catalogue numbers for collections held within the Bermuda Natural History Museum (BAMZ) and the Smithsonian National Museum of Natural History (USNM)

Species	Key characteristics	Catalogue number
Family Fundulidae	–	–
Fundulus bermudae Günther, 1874	Small; occurs in saltwater ponds, formerly inshore	BAMZ 1998.170.005
Fundulus relictus Able and Felley, 1988	Small; occurs in saltwater ponds, formerly inshore	BAMZ 1998.171.009
Family Hemiramphidae	–	–
Hemiramphus bermudensis Collette, 1962	Medium size; occurs offshore and inshore	BAMZ 1998.170.014
Hyporhamphus collettei Banford, 2010	Medium size; occurs inshore	USNM 391198 Holotype
Family Engraulidae	–	–
Anchoa choerostoma (Goode, 1874)	Small; occurs inshore	BAMZ 1990.083.012
Family Sparidae		
Diplodus bermudensis Caldwell, 1965	Medium size; occurs inshore	BAMZ 1995.124.013
Family Gobiidae		
Lythrypnus mowbrayi Bean, 1906	Small; occurs inshore, lagoon and banks	BAMZ 1998.171.004
Family Malacanthidae		
Caulolatilus bermudensis Dooley, 1981	Moderate size; occurs rim reef	BAMZ 1989.047.005
Family Serranidae		
Parasphyraenops atrimanus Bean, 1912	Moderate size, occurs on deep reefs	USNM 74085

Cephalopholis fulva, and the barber or creole-fish, *Paranthias furcifer*, have been reported (Bostrom et al. 2002). Hybrid specimens from Bermuda were first described by Poey (1860, 1875) as two species of a new serranid genus but later meristic and morphometric analyses suggested the individuals were intergeneric hybrids (Smith 1966, in Bostrom et al. 2002). The most recent analysis of 15 new specimens collected from Bermuda reported morphological characters intermediate between the two parent species and genetic patterns indicating hybridization between a female *C. fulva* and a male *P. furcifer* (Bostrom et al. 2002).

Another hybrid, the Townsend angelfish – a hybrid of *Holacanthus bermudensis* and *H. ciliaris* – is very common (Smith-Vaniz et al. 1999), and possibly the entire Bermuda population may be hybrids (Pyle et al. 2010). Small populations, concurrent restricted spawning seasons, and spatially limited spawning habitat could play a role in the frequency of hybridization; however, if the most recent supposition about the angelfish proves correct, it may be that for this species only the hybrid has been able to colonize Bermuda, perhaps an example of superior performance by a hybrid.

Mitochondrial DNA studies of the slippery dick, *Halichoeres bivittatus*, have revealed a 'subtropical' genotype occurring in Bermuda, Florida and the Gulf of Mexico. It is found in close proximity to a 'tropical' genotype, but in habitats exposed to cooler water temperatures (Rocha et al. 2005a). In Bermuda, all *H. bivittatus* collected inshore where greater temperature extremes occur, were of the 'subtropical' type, whereas half of the specimens collected on offshore reefs were of the 'tropical' type. No significant morphological differences have been recognized between the 'tropical' and 'subtropical' types. Thus the full, documented, diversity of this species, from across its geographical range, is present in Bermuda.

Finally, and uniquely, Bermuda is home to a strikingly different colour morph of the yellowhead wrasse, *Halichoeres garnoti* (Smith-Vaniz et al. 1999; Rocha 2004) (Fig. 12.4) which dominates the local population. An analysis of mtDNA cytochrome b showed variation correlated with colour differences in other *Halichoeres* species, but no similar differences in the Bermuda colour morph of *H. garnoti* (Rocha 2004).

Factors Influencing Fish Diversity in Bermuda

The absence of certain groups of reef fishes has been ascribed to their physiological, biological and ecological characteristics – narrower temperature tolerances, limited dispersal capabilities (e.g., brooders and sedentary, demersal spawners with short larval durations, poor swimming capabilities, absence of rafting behaviours), the absence of symbiotic or commensal partners (e.g., massive sponges are not available for sponge-dwelling gobies) (Smith-Vaniz et al. 1999) – and to lower overall habitat diversity in Bermuda. The absence of many common Caribbean coral species such as branching acroporids may limit the diversity of niches (see Smith-Vaniz et al. 1999). Similarly, the Sparidae, a family of medium-sized fish that are strong swimmers with moderately long larval durations and known to exhibit a range of hermaphroditic strategies – all of which are characteristics that would support their success in Bermuda – are not well represented, presumably because of the reduced availability of the inshore, reef-associated, habitat types.

Fig. 12.4 Colour morphs of *Halichoeres garnoti*. (**a**) Bermuda colour morph and (**b**) a typical Caribbean colour morph, photographed in Bermuda (Photo: TJT Murdoch)

Conversely, the strong representation in Bermuda of some unexpected taxa, such as poor swimmers, is attributed to interactions of behavior and the environment. Such taxa include triggerfishes and filefishes, whose larvae and juveniles are well known for their rafting behaviours.

Genetic Variation and Population Connectivity

Among biological factors that control connectivity, pelagic larval duration (PLD) has an obvious significance, and species with longer PLDs are generally widely distributed and well mixed; isolated locations, such as Bermuda, are more commonly inhabited by fish species with PLDs longer than 45 days (Brothers and Thresher 1985) while species with shorter PLDs (<15 days) often have genetically structured sub-populations (Thresher et al. 1989). Indeed, PLD is often used as a proxy for dispersal and colonizing capacity.

For families with PLD data (in Victor 1986; Wellington and Victor 1989; Sponaugle and Cowen 1994; Bergenius et al. 2002), Bermudian species have similar larval duration to their Caribbean congenerics, however, species known to have long larval periods all occur in Bermuda, and Bermuda lacks a number of shorter PLD species. A classic example is the Goldspot goby, *Gnatholepis thompsoni*, which has a very long larval duration (~89 days) (Sponaugle and Cowen 1994). The Bermuda population of the species, along with St. Croix and Brazil, falls into a genetically uniform, very widespread, western Atlantic grouping, established ~115,000–139,000 years ago (Rocha et al. 2005b), indicating consistent gene flow between western Atlantic populations over geological time and implying regular larval dispersal to Bermuda.

Earlier studies, in which electromorph variation at several loci for a variety of fish species was examined among Bermudian and Caribbean populations (Glasspool 1994), showed that genetic differentiation generally covaried with distance between populations but that the relationship to PLD was more variable. The Caribbean population of the French grunt, *Haemulon flavolineatum*, with a PLD of 2 weeks did not appear to be panmictic. In individuals from Bermuda, a rare allele was present at one electromorph loci and a common allele was absent (Glasspool 1994). In contrast, the pomacentrid *Abudefduf saxatilis*, with a slightly longer PLD of 3 weeks (as well as demersal eggs), was panmictic. Species of *Abudefduf* generally have broad distributions but short PLDs, and there is evidence for pelagic rafting by pre-juveniles. These remain planktonic by associating with floating objects such as *Sargassum* (Jokiel 1990; Wellington and Victor 1989; Glasspool 1994; Casazza and Ross 2008). The bluehead wrasse, *Thalassoma bifasciatum*, which had the longest PLD also appeared to be panmictic in Bermuda, Miami and the Bahamas, but with somewhat more overall genetic separation than seen in *A. saxatilis*. Thus behavioral differences are plausible explanations for the different connectivities of these species.

Levels of genetic differentiation among populations allowed the calculation of probable migration rates among populations (Glasspool 1994). These were 4–10 migrants per generation for *H. flavolineatum*, 28–250 (mean of 41) for *A. saxatilis* and between these two for *T. bifasciatum*. Various studies suggest that even a very low migration rate, on the order of only one individual per generation, is enough to prevent speciation (Wright 1951; Slatkin 1985), and rare individuals dispersing long distances are thought to reduce the genetic differences between geographically separated populations (Palumbi 1995; Shanks 2009); thus we see genetic differences in Bermudian populations, but not distinct species.

Based on microsatellites, black grouper, *Mycteroperca bonaci*, from Bermuda are significantly different from those in southeastern United States, Gulf of Mexico and the western Caribbean, while there are no differences among those locations (Chapman et al. 2002). For this species, synchronized, aggregative spawning, large-size, and a long life span may be significant traits promoting self-recruitment, and genetic discontinuities (Swearer et al. 2002).

Even within species, there is little evidence of the direct significance of PLD to the establishment of successful populations. Schultz and Cowen (1994) found no indication that Bermuda populations were recruited from individuals with a propensity for a longer than average pelagic phase. Rather, the Bermuda individuals had a significantly shorter pelagic larval duration (PLD) in two cases, a significantly longer PLD in two cases, and a typical PLD in the remaining two cases, when compared to populations in South Florida and the Caribbean.

Genetic Variation and Population Connectivity of Other Reef Associated Fauna

Mollusca

The genetic variation and connectivity of Bermuda's molluscan populations has been investigated only briefly. Allozyme studies have shown that populations of the queen conch, *Lobatus gigas*, were reproductively isolated from a comparatively panmictic Caribbean gene pool (Mitton et al. 1989). Likewise, Bermuda's populations of the direct developing bivalve *Lasaea adansoni* sensu latu was determined to be genetically similar to, but still distinct from, Florida populations based on sequences of mitochondrial DNA. For both these taxa, overall genetic diversity was higher in the Bermudian populations than in those to which Bermuda was compared (Mitton et al. 1989; Foighil and Jozefowicz 1999; Park and Foighil 2000). However, studies based on sequences of *COI*, 28S and ITS1 of *Brachidontes exustus* (scorched mussel), which has been recognized as a species complex, indicate no significant genetic differentiation among populations of a Bahamas clade, found in Florida Keys, Bahamas and Bermuda (Lee and Foighil 2005).

Arthropoda

These studies are limited to the spiny lobster, *Panulirus argus*. Studies of allozymes and mitochondrial DNA (mtDNA) indicate high levels of gene flow between Bermudian populations and those of Florida and the Caribbean, as far south as Venezuela (Hatley and Sleeter 1993; Silberman et al. 1994). Laboratory rearing studies indicate larval durations of 5–7 months (Goldstein et al. 2008) and estimates from earlier studies fit with the higher end of this range (Silberman et al. 1994; references in Naro-Maciel et al. 2011). Studies of speciation patterns of Pacific species of *Panulirus* (Pollock 1992) suggested that individuals might remain planktonic for up to 4 years, settling and metamorphosing only when appropriate environmental cues are detected. Thus, for this taxon, with its unusually long planktonic period, the lack of genetic structure between Bermudian and Caribbean populations is easy to comprehend in the context of general assumptions about pelagic phase duration and dispersal. Recent studies (Naro-Maciel et al. 2011) of mitochondrial DNA sequence variation among spiny lobsters of Florida, the Bahamas, the Caribbean and Brazil, but excluding Bermuda, which also did not detect significant levels of genetic structure, did identify two divergent COI lineages, which overlapped geographically. The possibility that these lineages might represent cryptic species was acknowledged (Naro-Maciel et al. 2011).

Echinodermata

Bermuda is home to two genetically (mtDNA [*COI*]) distinct lineages of the very widely distributed ophiuroid *Ophiactis savignyi*, one of which is very widely distributed and seems most likely to be present in the Atlantic due to anthropogenically influenced introduction (Roy and Sponer 2002). The other lineage has a western Atlantic distribution– it is reported only from Bermuda, Florida, the Caribbean and Brazil. The Bermuda individuals of the western Atlantic lineage display one numerically common, shared, haplotype and one unique Bermudian haplotype. Interestingly, one Bermudian individual also displayed a unique haplotype in the widespread (Pacific and Atlantic) lineage. Molecular studies of the echinoid *Lytechinus variegatus* (Zigler and Lessios 2004, mtDNA [*COI*], protein [bindin], isozyme) indicated there was no significant genetic structure among populations from Brazil, Florida, other locations in the Caribbean and Bermuda.

Conservation and Management of Biodiversity in Bermuda

The success of future efforts in marine resource management and conservation in Bermuda depends on valid information on the biogeography and population connectivity of the island's marine species. At this time, we have data from a few molecular studies on population structures of even fewer species that can be applied to conservation and management. Bermuda's marine species are shared with the other ecoregions within and bordering its biogeographical province, allowing theories of population connectivity to be applied to management efforts. However, uncertainty about life histories, dispersal capabilities and behaviours, basic population statistics and ecology, of many marine species preclude formulation of an effective and responsive coral reef management plan.

Gene Flow

Studies agree that within Bermuda marine populations are panmictic and self-seeding. Due to the major currents, gene flow is one-way – into Bermuda only. On an ecological time scale, Bermudian populations of certain octocorallian and scleractinian species, molluscs and reef fish appear to be somewhat isolated from neighbouring USA and Caribbean conspecifics. Other studies of species of scleractinian, crustacean, bivalves, and echinoderms and of reef fishes concluded that Bermuda populations were panmictic with populations in the USA and Caribbean. Even rare dispersal into Bermuda for some of these species is likely sufficient to maintain genetic similarities. For certain species, the strength of connections with neighboring reef ecosystems appear to be related to reproductive mode, pelagic larval duration or specific reproductive behaviours (i.e., brooding with no pelagic larval stage but with rafting). However, reproductive pattern or history in itself is not a reliable predictor of genetic connectivity (Weersing and Toonen 2009) and any correlations may be purely coincidental.

Given the limitations of current species-level information, a best practice from the conservation point of view would be to assume characteristics for each species that would dictate the most conservative management measures and to adjust conservation plans on a species by species basis, as information is acquired. However, an argument has recently been presented that overly protected habitats lose their abilities to adjust to normal and natural events and become increasingly susceptible to these natural perturbations (Côté and Darling 2010).

Endemics, by their nature, have to be managed on a local scale, but even if carefully protected locally they may still be subject to events that are widespread and not species specific, for example diseases.

Population Maintenance

Studies of the scleractinian, *M. auretenra*, and the octocoral, *B. asbestinum*, suggest genetic divergence in the largely self-seeding Bermudian populations could be constrained by the periodic breakdown of allopatric boundaries (Locke 2009; Bilewitch et al. 2010). These species brood larvae until they are competent to settle, then release planulae that in *B. asbestinum* are known to have low-dispersal potential and typically settle in a philopatric pattern (Brazeau and Lasker 1990). It would seem nearly impossible for these species even to initially colonize Bermuda from the nearest populations in the Bahamas – over 1,300 km away. However, recent evidence of rafting in a brooding coral (Hoeksema et al. 2012) indicates there is true potential for regular, although probably infrequent, genetic exchange even among what would generally be considered poor dispersers. The transport of *Sargassum* spp. from the Gulf of Mexico to the Sargasso Sea (Gower and King 2011) provides a mobile habitat that could bridge the larval and demersal stages of a wide range of fish species (Jokiel 1990; Coston-Clements et al. 1991; Casazza and Ross 2008).

An ecological connection between the Caribbean and Bermuda, as inferred from certain population genetic studies, is directly apparent in the presence in Bermuda of recent episodic diseases shortly after they appeared in the Caribbean (e.g., diseases affecting sponges, *Diadema*, and corals). The recent colonization of Bermuda by the invasive Pacific lionfish (Whitfield et al. 2002) also seems incontrovertible evidence of immigration/recruitment on an ecological time scale. Establishment of the lionfish in Bermuda was only a few years after their invasion of the east coast of the United States (Whitfield et al. 2002; Ahrenholz and Morris 2010).

In some regards, being well-connected may not be beneficial. Thus, it is necessary to have strategies that protect local individuals and to collaborate to manage distant populations whose health can impact Bermudian populations.

Resilience

The high levels of genetic variability are suggestive of genetic resilience, and Bermuda's populations may possess an increased probability of survival during environmental or biological changes. For example the widespread and severe bleaching event of 2005, which drastically impacted the Caribbean reefs of Puerto Rico and Florida, had minimal impact in Bermuda (Wilkinson and Souter 2008, J.M. Locke, pers. obs.). Similarly, Bermudian individuals of *Diploria labyrinthiformis* are resistant to black band disease (Jones et al. 2012). Haplotypes unique to Bermudian coral and molluscan populations may be evidence of local adaptation or they could be relicts that have been lost in other regions.

The unique biological diversity of Bermuda is important to conservation strategies for the entire TNA. However, until more information and Caribbean-wide integrated management is in place, conservative, local management measures are an imperative. Currently protection is afforded to some reef species through the dedication of marine reserves and marine protected areas/no take zones, and seasonally protected areas. All corals are individually protected from the general public but habitat destruction is not as easily governed. Current legislation for these protections include the Fisheries Act 1972, (Protected Species) Order, 1978, the Fisheries (Protected Areas) Order, 1990 and 2005, the Coral Reef Preserves Act, 1966, and the Protected Species Act 2003 (www.bermudalaws.bm).

Special Concerns for Local Management

Fishery Species

Even for Bermudian marine species with high genetic connectivities, investigation into the relative importance of locally produced larvae versus immigrant recruits is warranted, particularly for species that are exploited by commercial and recreational fisheries.

Briareum asbestinum – an invader or a Native

If the Bermudian population represents a natural, rare, recruitment event from Caribbean source populations, then the endemic genetic diversity of the population, as documented by Bilewitch et al. (2010), warrants protection. Although currently not under threat, its restricted distribution and small population size in the northeastern islands of Bermuda would make it highly susceptible to loss of significant genetic diversity. On the other hand, the species may have been introduced sometime in the nineteenth century. The centre of *B. asbestinum* distribution in Bermuda lies along St. George's Channel, between Smith's and Paget Islands. This channel was the primary northeastern shipping route into St. George's Harbour throughout the nineteenth century until the opening of the 'town cut' channel to the north in 1917 (Stranack 1990). If the *B. asbestinum* population does represent a centuries-old introduction, its conservation as a naturally occurring species would obviously be unwarranted. However, the population's apparent inability to expand beyond its current distribution over at least the past decade (J.P. Bilewitch, pers. obs.), and probably more than three decades (see Cairns et al. 1986) indicate it is not highly invasive and, combined with its historical interest, it merits maintenance.

Reef Fish

It has been suggested that the black grouper, *Mycteroperca bonaci*, is predominantly self-recruiting, although the population is not genetically distinct from Caribbean populations. Thus, current management measures are conservative, based on size, distribution and behavior of the Bermuda populations (Pitt and Shailer 2010). Likewise, management measures for the two endemic species of Hemiramphidae, which are used as bait, may be warranted.

Symbiodinium

In Bermuda, symbiont clades are identical to those found in the Caribbean and not to those of the eastern Atlantic (Savage et al. 2002), with little divergence of haplotypes or endemic *Symbiodinium* types (Holland 2006). Furthermore, population-level markers suggest that shared alleles and clustering between Bermuda and Florida *Symbiodinium* in the octocoral *Gorgonia ventalina* are the result of connectivity between the two (Andras et al. 2011). It remains impossible to determine the extent and frequency of dispersal of *Symbiodinium* to Bermuda, and if this occurs via host larvae or during free-living phases. Given that up to 85% of cnidarian species acquire symbionts exogenously from the surrounding environment (Schwarz et al. 2002), it is likely that a

(continued)

(continued)

local reservoir, coupled with host specificity, maintains the diversity of available *Symbiodinium* in Bermuda.

Diversity of cnidarian-algal symbioses is probably determined by local availability of the alga in addition to host-specificity. Physiological attributes of each clade are an important consideration in connection with changing temperature and light regimes, although how the *Symbiodinium* complement in Bermuda will protect or endanger corals in climate change scenarios remains unclear. The lack of clade D and the predominance of clade B could render Bermudian reefs vulnerable in a high temperature bleaching event (Savage 2001). Clade B symbionts have high regional abundance in the Caribbean (Santos et al. 2003) and are prevalent at higher latitudes, suggesting that they are generalists adapted to low irradiance and cooler seas (Holland 2006). In contrast, as they are restricted to shallower depths of the range of some hosts, clade B zooxanthellae have also been described as 'sun-specialists' (e.g., *Montastrea* spp., Rowan 1998) and 'narrowly-adapted specialists' (Toller et al. 2001). There is some suggestion that the colder temperature range in Bermuda may have resulted in coral communities with lower upper thermal tolerances than their tropical counterparts (Cook et al. 1990). Furthermore, offshore scleractinians may be more susceptible than those inshore, as more bleaching was recorded in them following the 1988 event (Cook et al. 1990). This was attributed to adaptation by inshore corals to increased temperatures, and/or the mitigating effects of turbidity in the lagoon.

As pre- and post-bleaching symbiont identities were not measured during past bleaching events, short to long-term changes in symbiont composition of anthozoans in Bermuda remains untested. Nevertheless, clade B should by no means predispose the hosts to increased susceptibility to environmental fluctuations; clade B itself is a diverse assemblage and each type is likely to vary in its response to bleaching (e.g., Santos et al. 2003; Holland 2006). Furthermore, hosts may acquire novel zooxanthellae from the environment as well as regulating the frequencies of their *in hospite* community, such as the octocoral *Briareum asbestinum* which has switched symbionts following experimentally induced bleaching (Lewis and Coffroth 2004).

Acknowledgements We wish to acknowledge the assistance of Lisa Greene, Bermuda Natural History Museum; Alison Copeland and Mandy Shailer, Bermuda Department of Conservation Services; Wolfgang Sterrer and Katie Dilke. This is a contribution of the Bermuda Biodiversity Project BBP#199.

References

Ahrenholz DW, Morris JA (2010) Larval duration of the lionfish, *Pterois volitans*, along the Bahamian Archipelago. Environ Biol Fish 88:305–309. doi:10.1007/s10641-010-9647-4

Andras JP, Kirk NL, Harvell CD (2011) Range-wide population genetic structure of *Symbiodinium* associated with the Caribbean Sea fan coral, *Gorgonia ventalina*. Mol Ecol 20:2525–2542

Ayre DJ, Hughes TP (2004) Climate change, genotypic diversity and gene flow in reef-building corals. Ecol Lett 7:273–278

Baker A (2003) Flexibility and specificity in coral-algal symbiosis: diversity, ecology and biogeography of *Symbiodinium*. Ann Rev Ecol Syst 34:661–689

Banford HM (2010) *Hyporhamphus collettei*, a new species of inshore halfbeak (Hemiramphidae) endemic to Bermuda, with comments on the biogeography of the *Hyporhamphus unifasciatus* species group. Proc Biol Soc Wash 123:345–358

Bates NR, Amat A, Andersson AJ et al (2010) Feedbacks and responses of coral calcification on the Bermuda reef system to seasonal changes in biological processes and ocean acidification. Biogeosciences 7:2509–2530

Bergenius MAJ, Meekan MG, Robertson DR, McCormick MI (2002) Larval growth predicts the recruitment success of a coral reef fish. Oecologia 131:521–525

Bilewitch JP (2006) The phylogenetic systematics of octocorals with a taxonomic review of five Bermudian species. M.Sc. thesis, University of Toronto

Bilewitch JP (2008) Octocoral biodiversity in Bermuda. Report for the Bermuda zoological society, Bermuda, BAMZ #1915, 9pp

Bilewitch JP, Coates KA, Currie DC, Trapido-Rosenthal HG (2010) Molecular and morphological variation supports monotypy of the octocoral Briareum Blainville, 1830 (Octocorallia: Alcyonacea) in the Western Atlantic. Proc Biol Soc Wash 123:93–112

Billinghurst Z, Douglas AE, Trapido-Rosenthal HG (1997) On the genetic diversity of the symbiosis between the coral *Montastraea cavernosa* and zooxanthellae in Bermuda. In: Proceedings of the 8th international coral reef symposium, Panama vol 2. pp 1291–1294

Boschma H (1925) On the symbiosis of certain Bermuda coelenterates and zooxanthellae. Proc Am Acad Arts Sci 60:451–460

Bostrom MA, Collette BB, Luckhurst BE, Reece KS, Graves JE (2002) Hybridization between two serranids the coney *Cephalopholis fulva* and the creole-fish *Paranthias furcifer* at Bermuda. Fish Bull 100:651–661

Brazeau DA, Lasker HR (1990) Sexual reproduction and external brooding by the Caribbean gorgonian *Briareum asbestinum*. Mar Biol 104:465–474

Bridge TCL, Fabricius KE, Bongaerts P, Wallace CC, Muir PR, Done TJ, Webster JM (2011) Diversity of Scleractinia and Octocorallia in the mesophotic zone of the Great Barrier Reef, Australia. Coral Reefs 32:179–189

Brothers EB, Thresher RE (1985) Pelagic duration, dispersal, and the distribution of Indo-Pacific coral-reef fishes. In: Reaka ML (ed) The ecology of coral reefs, vol 3, Symposia series for undersea research, NOAA undersea research program. NOAA, Rockville

Butler JN, Morris BF, Cadwallader J, Stoner AW (1983) Studies of *Sargassum* and the *Sargassum* community, vol 22, Special publication. Bermuda Biological Station for Research, St Georges, Bermuda

Cairns SD (1979) The deep-water Scleractinia of the Caribbean and adjacent waters. Stud Fauna Curaçao 57:1–341

Cairns SD (2000) A revision of the shallow-water azooxanthellate Scleractinia of the Western Atlantic. Nat Hist Caribbean Reg 75:1–240

Cairns SD, Chapman RE (2001) Biogeographic affinities of the North Atlantic deep-water Scleractinia. In: Proceedings of the first international symposium on deep-sea corals, ecology action centre and Nova Scotia Museum, Halifax, Nova Scotia, p 30–57

Cairns SD, den Hartog JC, Arneson C, Reutzler K (1986) Class anthozoa (Corals, Anemones). In: Sterrer W (ed) Marine fauna and flora of Bermuda. A systematic guide to the identification of Marine organisms. Wiley, New York

Casazza TL, Ross SW (2008) Fishes associated with pelagic *Sargassum* and open water lacking *Sargassum* in the Gulf Stream off North Carolina. Fish Bull 106:348–363

Chapman RW, Sedberry GR, McGovern JC, Ball AO, Zatcoff MS, Luckhurst BE (2002) Genetic identification of scamp, *Mycteroperca phenax*, black grouper, *Mycteroperca bonaci*, and red grouper, *Epinephelus morio*, in the western Atlantic. MARFIN Tech Rep NA87FF0423, SC Department of Natural Resources, Charleston

Chester WM (1913) The structure of the gorgonian coral *Pseudoplexaura crassa* Wright and Studer. Proc Am Acad Arts Sci 48:737–774

Cook CB, Logan A, Ward J, Luckhurst B, Berg CJ Jr (1990) Elevated temperatures and bleaching on high latitude coral reef: the 1988 Bermuda event. Coral Reefs 9:45–49

Coston-Clements L, Settle LR, Hoss DE, Cross FA (1991) Utilization of the *Sargassum* habitat by marine inverterates and vertebrates – a review. NOAA Technical Memorandum. NMFS-SEFSC-296, 30pp

Côté IM, Darling ES (2010) Rethinking ecosystem resilience in the face of climate change. PLoS Biol 8(7):e1000438. doi:10.1371/journal.pbio.1000438

Floeter SR, Rocha LA, Robertson DR et al (2008) Atlantic reef fish biogeography and evolution. J Biogeogr 35:22–47

Foighil DÓ, Jozefowicz CJ (1999) Amphi-Atlantic phylogeography of direct-developing lineages of *Lasaea*, a genus of brooding bivalves. Mar Biol 135:115–122

Fricke H, Meischner D (1985) Depth limits of Bermudan scleractinian corals: a submersible survey. Mar Biol 88:175–187

Glasspool AF (1994) Larval distribution, population structure and gene flow in Bermuda's reef fish. Ph.D. dissertation, University of Wales

Goldstein J, Matsuda H, Takenouchi T, Butler MJ IV (2008) The complete development of larval Caribbean spiny lobster *Panulirus argus* (Latreille, 1804) in culture. J Crustacean Biol 28:306–327

Goodbody-Gringley G, Vollmer SV, Woollacott RM, Giribet G (2010) Limited gene flow in the brooding coral *Favia fragum* (Esper, 1797). Mar Biol 157:2591–2602

Goodbody-Gringley G, Woollacott RM, Giribet G (2011) Population structure and connectivity in the Atlantic scleractinian coral *Montastraea cavernosa* (Linnaeus, 1767). Mar Ecol. doi:10.1111/j.1439-0485.2011.00452.x

Goulet TL, Coffroth M-A (2004) The genetic identity of dinoflagellate symbiontsin Caribbean octocorals. Coral Reefs 23:465–472

Goulet TL, Simmons C, Goulet D (2008) Worldwide biogeography of *Symbiodinium* in tropical octocorals. Mar Ecol Prog Ser 355:45–58

Gower JFR, King SA (2011) Distribution of floating *Sargassum* in the Gulf of Mexico and the Atlantic Ocean mapped using MERIS. Int J Remote Sens 32:1917–1929

Grode SH, James TR, Cardellina JH II (1983) Brianthein Z, a new polyfunctional diterpene from the gorgonian *Briareum polyanthes*. Tetrahedron Lett 24:691–694

Hatley JG, Sleeter TD (1993) A biochemical genetic investigation of spiny lobster (*Panulirus argus*) stock replenishment in Bermuda. Bull Mar Sci 53:993–1006

Healy B, Coates KA (2003) A preliminary investigation of the decomposition of *Sargassum* stranded on Bermuda beaches, and the particular role of Enchytraeidae in its decomposition. Newsl Enchytraeidae 8:13–24

Hearty PJ, Olson SL (2010) Geochronology, biostratigraphy, and changing shell morphology in the land snail subgenus *Poecilozonites* during the quaternary of Bermuda. Palaeogeogr Palaeoclim Palaeoecol 293:9–29

Hearty PJ, Hollin JT, Neumann AC, O'Leary MJ, McCulloch M (2007) Global sea-level fluctuations during the last interglaciation (MIS 5e). Quat Sci Rev 26:2090–2112

Hoeksema BW, Roos PJ, Cadée GC (2012) Trans-Atlantic rafting by the brooding reef coral *Favia fragum* on man-made flotsam. Mar Ecol Prog Ser 445:209–218

Holland LP (2006) The molecular diversity of *Symbiodinium* (Seussiales: Dinoflagellata) within Alcyonacea of Bermuda and the Caribbean. M.Sc. thesis, University of Toronto

Iliffe TM, Kvitek R, Blasco S, Blasco K, Covill R (2011) Search for Bermuda's deep water caves. Hydrobiologia. doi:10.1007/s10750-011-0883-1

Jokiel PL (1990) Long-distance dispersal by rafting: re-emergence of an old hypothesis. Endeavour New Ser 14(2):66–73

Jones RJ, Johnson R, Noyes T, Parsons R (2012) Spatial and temporal patterns of coral black band disease in relation to a major sewage outfall. Mar Ecol Prog Ser 462:79–92

Kleypas JA (1997) Modeled estimates of global reef habitat and carbonate production since the last glacial maximum. Paleoceanog 12:533–545

Kleypas JA, McManus JW, Menez LAB (1999) Environmental limits to coral reef development: where do we draw the line? Am Zool 39:146–159

LaJeunesse TC (2001) Investigating the biodiversity, ecology, and phylogeny ofendosymbiotic dinoflagellates in the genus *Symbiodinium* using the ITS region: in search of a "species" level marker. J Phycol 37:866–880

Lee T, Foighil DO (2005) Placing the Floridian marine genetic disjunction into a regional evolutionary context using the scorched mussel, *Brachidontes exustus*, species complex. Evolution 59:2139–2158

Lewis CL, Coffroth MA (2004) The acquisition of exogenous algal symbionts by anoctocoral after bleaching. Science 304:1490–1492

Locke JM (2009) *Madracis auretenra* (Scleractinia: Pocilloporidae) – testing our knowledge of systematics, biology and connectivity in the western North Atlantic. Ph.D. dissertation, University of Puerto Rico

Locke JM, Bilewitch JP, Coates KA (2013) Scleractinia, Octocorallia and Antipatharia of Bermuda's reefs and deep-water coral communities: a taxonomic perspective including new records. In: Sheppard C (ed) Coral reefs of the world, vol 4. Coral Reefs of the UK Overseas Territories. Springer, Dordrecht, pp 189–200

Locke JM, Weil E, Coates KA (2007) A newly documented species of *Madracis* (Scleractinia: Pocilloporidae) from the Caribbean. Proc Biol Soc Wash 120:214–226

Loram JE, Boonham N, O'Toole P, Trapido-Rosenthal HG, Douglas AE (2007a) Molecular quantification of symbiotic dinoflagellate algae of the genus *Symbiodinium*. Biol Bull 212:259–268

Loram JE, Trapido-Rosenthal HG, Douglas AE (2007b) Functional significance of genetically different algae *Symbiodinium* in a coral reef symbiosis. Mol Ecol 16:4849–4857

Magurran AE (1988) Ecological diversity and its measurement. Princeton University Press, Princeton

Meyer EL (2012) Integrating ecology, natural history, and regional management for conservation of tropical intertidal gastropod fisheries. Ph.D. dissertation, University of California, Berkeley

Miller KJ, Ayer DJ (2008) Protection of genetic diversity and maintenance of connectivity among reef corals within marine protected areas. Conserv Biol 22:1245–1254

Miller KJ, Ayre DJ (2004) The role of sexual and asexual reproduction in structuring high latitude populations of the reef coral *Pocillopora damicornis*. Heredity 92:557–56

Miloslavich P, Diaz JM, Klein E, Alvarado JJ, Diaz C, Gobin J, Escobar-Briones E, Cruz-Motta JJ, Weil E, Cortés J, Bastidas AC, Robertson R, Zapata F, Martin A, Castillo J, Kazandjian A, Ortiz M (2010) Marine biodiversity in the Caribbean: regional estimates and distributional patterns. PLoS One 5:1–25 and PLoS One 5(8):e11916. doi:10.1371/ journal.pone.0011916

Mitton JB, Berg CJ, Orr KS (1989) Population structure, larval dispersal, and gene flow in the queen conch, *Strombus gigas* of the Caribbean. Biol Bull 177:356–362

Moore HB, Moore DM (1946) Preglacial history of Bermuda. Bull Geol Soc Am 57:207–222

Muhs DR, Simmons KR, Steinke B (2002) Timing and warmth of the last interglacial period: new U-series evidence from Hawaii and Bermuda and a new fossil compilation for North America. Quaternary Sci Rev 21:1355–1383

Murdoch TJT (2007) A functional group approach for predicting the composition of hard coral assemblages in Florida and Bermuda. Ph.D. dissertation, University of South Alabama

Nagelkerken I, Blaber SJM, Bouillon S, Green P, Haywood M, Kirton LG, Meynecke J-O, Pawlik J, Penrose HM, Sasekumar A, Somerfield PJ (2008) The habitat function of mangroves for terrestrial and marine fauna: a review. Aquat Bot 89:155–185

Naro-Maciel E, Reid B, Holmes KE, Brumbaugh DR, Martin M, DeSalle R (2011) Mitochondrial DNA sequence variation in spiny lobsters: population expansion, panmixia, and divergence. Mar Biol 158:2027–2041

Nunes F, Norris RD, Knowlton N (2009) Implications of isolation and low genetic diversity in peripheral populations of an Amphi-Atlantic coral. Mol Ecol 18:4283–4297

Palumbi SR (1995) Using genetics as an indirect estimator of larval dispersal. In: McEdward L (ed) Ecology of Marine invertebrates. CRC Press, Boca Raton

Pandolfi JM (1999) Response of pleistocene coral reefs to environmental change overlong temporal scales. Am Zool 39:113–130

Park J, Foighil DÓ (2000) Genetic diversity of oceanic island *Lasaea* (Mollusca: Bivalvia) lineages exceeds that of continental populations in the Northwestern Atlantic. Biol Bull 198:396–403

Pitt JM, Shailer ML (2010) GIS applications in marine resources management: examples of spatial management measures from Bermuda. In: Calado H, Gil A (eds) Geographic technologies applied to marine spatial planning and integrated coastal zone management. Universidade dos Açores, Ponta Delgada

Pochon X, Gates RD (2010) A new *Symbiodinium* clade (Dinophyceae) from soritid foraminifera in Hawai'i. Mol Phylogenet Evol 56:492–497

Poey F (1860) Memorias sobra la historia natural de la isla de Cuba, acompañadas de sumarios latino y extractos in francés. La Habana 2:97–336

Poey F (1875) Enumeratio piscium Cubensium. An Soc Española Hist Nat 4:75–112

Pollock DE (1992) Palaeoceanography and speciation in the spiny lobster genus *Panulirus* in the Indo-Pacific. Bull Mar Sci 51:135–146

Precht WF, Miller SL (2007) Chapter 9: ecological shifts along the Florida reef tract: the past is a key to the future. In: Aronson RB (ed) Geological approaches to coral reef ecology. Springer, New York, pp 237–212

Pyle R, Myers R, Rocha LA, Craig MT (2010) *Holacanthus ciliaris*. In: IUCN 2011. IUCN red list of threatened species. Version 2011.2. www.iucnredlist.org. Accessed 8 June 2012

Robertson DR (2001) Population maintenance among tropical reef fishes: inferences from small-island endemics. Proc Nat Acad Sci 98(10):5667–5670

Rocha LA (2004) Mitochondrial DNA and color pattern variation in three Western Atlantic *Halichoeres* (Labridae), with the revalidation of two species. Copeia 2004:770–782

Rocha LA, Robertson DR, Roman J, Bowen BW (2005a) Ecological speciation in tropical reef fishes. Proc R Soc B 272:573–579. doi:10.1098/2004.3005

Rocha LA, Robertson DR, Rocha CR, Van Tassell JL, Craig MT, Bowen BW (2005b) Recent invasion of the tropical Atlantic by an Indo-Pacific coral reef fish. Mol Ecol 14:3921–3928. doi:10.1111/ j.1365-294X.2005.02698.x

Rowan R (1998) Review – diversity and ecology of zooxanthellae on coral reefs. J Phycol 34:407–417

Rowan R, Powers DA (1991) A molecular genetic classification of zooxanthellae and the evolution of animal-algal symbioses. Science 251:1348–1351

Roy MS, Sponer R (2002) Evidence of a human-mediated invasion of the tropical western Atlantic by the 'world's most common brittlestar'. Proc R Soc Lond B 269:1017–1023

Santos SR, Gutierrez-Rodriguez C, Lasker HR, Coffroth MA (2003) *Symbiodinium* sp.associations in the gorgonian *Pseudopterogorgia* elisabethae in the Bahamas: high levels of genetic variability and population structure in symbiotic dinoflagellates. Mar Biol 143:111–120

Savage AM (2001) Genetic diversity and photosynthetic characteristics of zooxanthellae (*Symbiodinium*). Ph.D. dissertation, University of York

Savage AM, Goodson MS, Visram S, Trapido-Rosenthal HG, Wiedenmann J, Douglas AE (2002) Molecular diversity of symbiotic algae at the latitudinal margins of their distribution: dinoflagellates of the genus *Symbiodinium* in corals and sea anemones. Mar Ecol Prog Ser 244:17–26

Schultz ET, Cowen RK (1994) Recruitment of coral-reef fishes to Bermuda: local retention or long-distance transport? Mar Ecol Prog Ser 109:15–28

Schwarz JA, Weis VM, Potts DC (2002) Feeding behaviour and acquisition of zooxanthellae by planula larvae of the sea anemone *Anthopleura elegantissima*. Mar Biol 140:471–478

Shanks AL (2009) Pelagic larval duration and dispersal distance revisited. Biol Bull 216:373–385

Shearer TL (2004) Reef connectivity: genetic analysis of recruitment and gene flow among Caribbean scleractinian corals. Ph.D. dissertation, University of New York, Buffalo

Silberman JD, Sarver SK, Walsh PJ (1994) Mitochondrial DNA variation and population structure in the spiny lobster *Panulirus argus*. Mar Biol 120:601–608

Slatkin M (1985) Gene flow in natural populations. Ann Rev Ecol Syst 16:393–430

Smith CL (1966) *Menephorus* Poey, a serranid genus based on two hybrids of *Cephalopholi fulva* and *Paranthias furcifer*, with comments on the systematic placement of *Paranthias*. Am Mus Nov 2276:1–1966

Smith SR (2010) The search for knowledge and understanding of Bermuda's biological diversity. Envirotalk 78:2–6

Smith ML, Carpenter KE, Walker RW (2002) An introduction to the oceanography, geology, biogeography, and fisheries of the tropical and subtropical Western Central Atlantic. In: Carpenter KE (ed) The living marine resources of the Western Central Atlantic vol. 1, vol 5, FAO species identification guide for fisheries purposes and American Society of Ichthyologists Herpetologists special publication. Food and Agriculture Organization of the United Nations, Rome, pp 1–23

Smith-Vaniz WF, Collette BB, Luckhurst BE (1999) Fishes of Bermuda: history, zoogeography, annotated checklist, and identification keys. Amer Soc Ichthy Herpet Spec Publ. 4, 424 pp. Allen Press, Lawrence

Spalding M, Fos H, Allen G, Davidson N, Ferdaña Z, Finlayson M, Halpern BS, Jorge MA, Lomgana A, Lourie SA, Martin KD, McManus E, Molnar J, Recchia CA, Robertson J (2007) Marine ecoregions of the world: a bioregionalization of coastal and shelf areas. Bioscience 57:573–583

Sponaugle S, Cowen RK (1994) Larval durations and recruitment patterns of two Caribbean gobies (Gobiidae): contrasting early life histories in demersal spawners. Mar Biol 120:133–143

Stat M, Gates RD (2011) Clade D *Symbiodinium* in scleractinian corals: a "nugget" of hope, a selfish opportunist, an ominous sign, or all of the above? J Mar Biol 2011:9, Article ID 730715

Sterrer WE (1998) How many species are there in Bermuda? Bull Mar Sci 62:809–840

Stone GS, Katona SK, Tucker EB (1987) History, migration and present status of humpback whales, *Megaptera novaeangliae*, at Bermuda. Biol Conserv 42:133–145

Stranack I (1990) The Andrew and the onions. The story of the royal navy in Bermuda 1795–1975, 2nd edn. University of Toronto Press, Toronto

Swearer SE, Shima JS, Hellberg MA, Thorrold SR, Jones GP, Robertson DR, Morgan SG, Selkoe KA, Ruiz GM, Warner RR (2002) Evidence of self-recruitment in demersal marine populations. Bull Mar Sci 70:251–271

Thresher RE, Colin PL, Bell LJ (1989) Planktonic duration, distribution and population structure of western and central Pacific damselfishes (Pomacentridae). Copeia 1989:420–434

Toller WW, Rowan R, Knowlton N (2001) Zooxanthellae of the *Montastraea annularis* species complex: patterns of distribution of four taxa of *Symbiodinium* on different reefs and across depths. Bio Bull 201:348–359

Venn AA, Loram JE, Trapido-Rosenthal HG, Joyce DA, Douglas AE (2008) Importance of time and place: patterns in abundance of *Symbiodinium* clades A and B in the tropical sea anemone *Condylactis gigantea*. Biol Bull 215:243–252

Venn AA, Weber FK, Loram JE, Jones RJ (2009) Deep zooxanthellate corals at the high latitude Bermuda Seamount. Coral Reefs 28:135

Victor BC (1986) Larval settlement and juvenile mortality in a recruitment – limited coral reef fish population. Ecol Mono 56:145–160

Wellington GM, Victor BC (1989) Planktonic larval duration of one hundred species of Pacific and Atlantic damselfishes (Pomacentridae). Mar Biol 101:557–567

Weersing K, Toonen RJ (2009) Population genetics, larval dispersal, and connectivity in marine systems. Mar Ecol Prog Ser 393:1–12

Whitfield PE, Gardner T, Vives SP, Gilligan MR, Courtenay WR Jr, Carleton Ray G, Hare JA (2002) Biological invasion of the Indo-Pacific lionfish *Pterois volitans* along the Atlantic coast of North America. Mar Ecol Prog Ser 235:289–297

Wilkinson CR, Souter D (2008) Status of Caribbean coral reefs after bleaching and hurricanes in 2005. Global Coral Reef Monitoring Network and Reef and Rainforest Research Centre, Townsville

Wright S (1951) The genetical structure of populations. Ann Eugen 15:323–354

Zigler KS, Lessios HA (2004) Speciation on the coasts of the new world: phylogeography and the evolution of bindin in the sea urchin genus *Lytechinus*. Evolution 58:1225–1241

13 Threats to Coral Reefs of Bermuda

Struan R. Smith, Samia Sarkis, Thad J.T. Murdoch, Ernesto Weil, Aldo Croquer, Nicholas R. Bates, Rodney J. Johnson, Samantha de Putron, and Andreas J. Andersson

Introduction

Bermuda's coral reefs and associated inshore and lagoonal habitats have endured 400 years of deleterious human interference. Although early conservation laws were passed to protect specific species (sea turtles and fry), it is important to recognize the extent of human impacts at both system-wide and habitat scales. In this chapter we will review some of the historic impacts on Bermuda's reef and indicate the level of success in reducing or mitigating the effects of these problems and the extent to which they remain as current or future threats to the health of Bermuda's reefs. We divide the threats into three categories: (1) direct or indirect anthropogenic impacts; (2) effects of invasive species and chronic biological problems (such as diseases); and (3) global climate-related problems.

Recent analyses have characterized Bermuda's challenges as "very significant" at a coarse level (Pandolfi et al. 2003; Burke et al. 2011) primarily because of the high reef to island ratio and not because of direct threats that have caused measureable declines in reef health. Assessments of the health and character of Bermuda's reefs and associated habitats began in the 1970s (Garrett et al. 1971; Dodge and Vaisnys 1977; Dryer and Logan 1978; Dodge et al. 1982; Flood et al. 2005; MEP 2007), and some inferences of change can be discerned from accounts of early naturalists (Verrill 1907; Agassiz 1894), but we lack knowledge on how significantly the baseline has shifted since human occupation of Bermuda in the early 1600s. Recent studies indicate stability in coral cover since the 1990s in some reef zones (CARICOMP 2000; Linton and Fisher 2004; MEP 2007).

Bermuda's reefs have certainly experienced the spectrum of contemporary challenges to the health of the reefs (e.g. sedimentation, diseases, bleaching, over-fishing; Cook et al. 1994). They appear to have persisted with relatively limited change, over the past three decades, and have benefited from pro-active management to reduce deleterious anthropogenic impacts, especially fishing pressure. The near-shore and lagoonal reefs appear to be affected by maritime activities and chemical contaminants and remain vulnerable to future threats. Here we outline briefly the scope of the challenges that remain.

Anthropogenic Influences

Tourism and Maritime Commerce

The increasing needs and developments associated with a strong international business sector, a growing cruise ship

S.R. Smith (✉)
Natural History Museum, Bermuda Aquarium, Museum and Zoo, 40 North Shore Road, Hamilton Parish FL 04, Bermuda
e-mail: srsmith@gov.bm

S. Sarkis
Department of Conservation Services, 17 North Shore Road, Hamilton Parish FL 03, Bermuda
e-mail: scsarkis@gov.bm

T.J.T. Murdoch
Bermuda Zoological Society, 40 North Shore Road, Hamilton Parish FL 04, Bermuda
e-mail: tjmurdoch@gov.bm

E. Weil
Department of Marine Sciences, University of Puerto Rico, 9000, Mayagues, PR 00680, USA
e-mail: ernesto.weil@upr.edu

A. Croquer
Departamento de Estudios Ambientales, Universidad Simon Bolivar, Caracas 1080 A, Venezuela
e-mail: acroquer@usb.ve

N.R. Bates • R.J. Johnson • S. de Putron
Bermuda Institute of Ocean Sciences, 17 Biological Lane, St Georges GE01, Bermuda
e-mail: nick.bates@bios.edu; rod.johnson@bios.edu; samantha.deputron@bios.edu

A.J. Andersson
Scripps Institution of Oceanography, 9500 Gilman Dr, San Diego, CA 92093, USA
e-mail: aandersson@ucsd.edu

C.R.C. Sheppard (ed.), *Coral Reefs of the United Kingdom Overseas Territories*, Coral Reefs of the World 4, DOI 10.1007/978-94-007-5965-7_13, © Springer Science+Business Media Dordrecht 2013 industry and a high level of consumerism, pose potential threats to the environment. Specifically, the reliance of Bermuda on imported goods via maritime transport and a changing cruise ship industry require a re-development and/or modifications of channels and docks. These factors will necessitate dredging and coastal developments, which have a direct impact on the reef platform ecosystem. Two reports examined the scope of impacts of the development of infrastructure and cruise ship operation on the environment (Sarkis 1999; Price 2006).

There are currently four channels for ship passage into Bermuda's ports (North Channel, South Channel, Town Cut and Two Rock Passage). The South Channel has been the most frequently used route and accommodates container ship traffic into the Hamilton port. The use of the North Channel through the centre of the lagoon has recently increased for larger cruise ships that can only berth in Dockyard. Town Cut is the access to the Town of St. George and Two Rock Passage provides access to the City of Hamilton.

Bermuda has been a popular destination for cruise ships since the 1930s and the development of navigation channels through the reef platform has brought vessels to the City of Hamilton during the cruise ship season from April to October. A significant level of naval traffic has been associated with the British Naval Dockyard since its development in the early 1800s and in particular the development and operation of the US Naval Operations Base in the Little Sound in 1941–42 resulted in intensive naval traffic during World War II into the 1950s. The frequent passage of ships in both the North and South channels has resulted in re-suspension of sediments that potentially affect adjacent lagoonal patch reefs (Waltham 1998; Jones 2011). Although no demonstrable effects on reef communities have been measured to date (Smith et al. 1998; MEP 2007), we have no estimates of the character of the lagoonal reefs prior to the 1930s and how they may have been impacted by intensive ship activity in these periods.

Bermuda was served primarily by dedicated small cruise ships since the 1950s, berthing in Hamilton and St. George's, and in the 1980s vessels began to visit more frequently. The re-development of the former British Naval Dockyard as a dedicated port for Panamax and Post-Panamax ships was completed in 2005. Should the existing pressure placed by the cruise ship industry and its use of very large vessels continue, maritime infrastructure development will be necessary to accommodate their requirements. The size of cruise ships continues to increase (Post-Panamax 1, 2 and now 3) and they require more space to maneuver safely, especially under adverse weather conditions. The potential threats associated with channel modification and port re-development are:

(a) Destruction of reefs for easier passage through shipping channels
(b) Increased pollution and sedimentation on reefs adjacent to shipping channels and within ports
(c) Increased potential of vessel groundings
(d) Impact on recreational and commercial reef fisheries due to reef degradation
(e) Tourism repercussion- quality of visitor experience declines with increasing reef impacts.

The Bermuda Government commissioned a recent study to address the requirements of Post-Panamax vessels and access to the three ports (Ministry of Transport 2011). The most immediate threat to Bermuda's coral reefs is the perceived need for shipping channel modifications to ensure safe passage of these larger boats. The North Channel was used infrequently since World War II and this passage is considered to be surrounded by relatively healthy coral reefs. Although these ships have been using this channel since 2005 without any recorded incidents, there is concern for the safety of the vessels as the passage in the North Channel is narrow and extremely difficult in windy conditions. In addition, ship agents are pushing for a more direct access to the docking berths that would require cutting through a section of the reefs. The advent of the larger ships is therefore triggering a perceived need for modifications to the North Channel, imposing potentially severe negative impact or direct destruction of some of the surrounding reefs and within the North Lagoon.

Bermuda has a long history of vessel groundings on its reefs (Cook et al. 1994) and only after a series of near-catastrophic accidents with grounded oil tankers in the early 1980s were charts, navigational aids and vessel warning systems improved significantly. Nonetheless a modern cruise ship grounded in a main shipping channel in 2006 due to pilot error (MEP 2007) and visiting yachts and fishing vessels ground nearly every year. It is anticipated that increased maritime traffic and increased size of ships will result in the increased frequency of groundings. Such ship groundings often result in severe localized biological and physical damage, including the dislodgement of corals, pulverization of coral skeletons, displacement of sediment deposits, and loss of three dimensional complexity, as well as chemical contamination by the ship's antifouling paint (Jaap 2000; MEP 2007). Given the low recruitment rates and slow growth of corals in Bermuda, the reefs scars are anticipated to take decades to centuries to recover (Smith 1985).

Scuba diving and snorkeling on Bermuda's reefs are very popular activities for both residents and visitors to Bermuda but their impacts on reef health have not been directly assessed. The reefs have immense economic value to Bermuda and this has motivated many conservation measures. The economic value of reef-associated tourism activities to the local economy averages US$405 million per year with an additional recreational and cultural value of $36 million per year (Sarkis et al. 2010 and Chap. 15). Comprehensive legislation protects all corals and many important reef invertebrates and fish species. Many popular dive sites are within Marine Protected Areas, some of which are also historic wreck sites, and mooring balls are available, which reduces impacts from anchoring. It was the early appreciation in 1988 that diving and snorkeling had greater direct economic value compared to commercial fishing (calculated at the time at $9 million and US$3 million, respectively) that drove major changes in fisheries legislation to reduce overfishing and restore reef fish populations (Butler et al. 1993).

Water Quality

The quality of the inshore and nearshore waters surrounding Bermuda is generally very good because of the high rate of water exchange with the surrounding Sargasso Sea and the very limited amount of run-off from the karstic limestone islands (Morris et al. 1977). However, restricted bays and harbours, as well as offshore shipping channels as noted above, have experienced both acute and chronic impacts due to human activities (Jickells et al. 1986; Jones 2010; Jones et al. 2010).

Bermuda lacks a centralized sewage system and domestic wastes are held in soak-away pits built adjacent to houses. Larger commercial sites use deep-sealed bore holes (~40 m deep) to accommodate large discharge volumes. Bermuda's harbours and bays are subject to nutrient pollution due to the movement of nutrient-enriched groundwater and run-off from the surrounding land (Simmons and Lyons 1994; Jones et al. 2010). Phosphorous – in the form of phosphate – binds to the carbonate rock but groundwaters are heavily enriched in nitrate (>200 μM) that eventually enter inshore waters through tidal pumping and karst caves. Many old houses were built around these enclosed bays and direct sewage contamination from their cesspits probably occurs following heavy rains or during storm surges. The increased nitrogen levels in the enclosed bays and harbours support elevated plankton levels, but harmful algal blooms have not been detected (Bodungen et al. 1982; MEP 2007; Boyer and Briceno 2010). Bermuda also fails to control direct discharge of sewage wastes from recreational vessels. The end result is that faecal bacteria contamination occurs in most enclosed bays (Jones et al. 2010), although most locations still meet EU bathing water standards (European Union 2006).

Elevated nutrients are not usually seen on the offshore reefs (MEP 2007; Boyer and Briceno 2010) except at municipal sewage outfalls on the south shore. These outfalls appear to have limited impacts, due to high flushing rates, and do not appear to have grossly contaminated adjacent reefs. However, elevated phosphorus levels in coral skeletons on reefs adjacent to the outfalls have been detected, and coral recruitment appears reduced (Dodge et al. 1984; Webster and Smith 2002). Faecal contamination on near-shore reefs adjacent to a municipal outfall on the south shore was detectable (Jones et al. 2010; Jones et al. 2012) but did not result in an increase in coral diseases. Stable isotope ratios in sea fan skeletons determined a gradient of enriched nitrogen (D. Baker, unpubl. data) from the inshore waters to the lagoon. However, Bermuda has seen a significant decline in the number of direct offshore sewage outfalls over the past 40 years as the result of the adoption of deep injection sewage wells, which ostensibly dilute and disperse the effluent more effectively (Jones et al. 2010).

Water quality is also reduced by trace metal inputs in enclosed bays because of maritime activity and the presence of high densities of recreational vessels and boatyards (Jickells and Knap 1984; Jickells et al. 1986). Connelly et al. (2001) detected tributyltin and Irgarol in Hamilton Harbour, contaminants associated with cruise ships and large vessels, and measureable biological effects on bivalves were detected in a GEEP workshop (Burns et al. 1990; Widdows et al. 1990). Irgarol also inhibited coral photosynthesis at environmentally relevant concentrations in Bermuda (Owen et al. 2002). The Bermuda Government subsequently banned the importation of antifouling paints that contain Irgarol or Diuron. Bermuda began a nearshore solid waste dumping programme, primarily bulk metal and vehicles, in Castle Harbour in the 1970s, and measureable contamination and biological effects have been detected (Burns et al. 1990; Leavitt et al. 1990; Widdows et al. 1990; Knap et al. 1991). The dump site expanded dramatically in 1996 after Bermuda initiated mass incineration of domestic garbage and began depositing cement-stabilized ash material on the face of the existing bulk metals dump in Castle Harbour (Smith et al. 1998; Flood et al. 2005). Mobilization of trace metals and trace organics into marine species has been detected in the vicinity of the dumping area (Gunther 1999; Quinn et al. 2005; Morgan et al. 2005). A more recent study has fully documented the clear gradients of trace metal and trace organic compounds that exist around the waste deposit site (Jones 2010).

Coastal Developments

Bermuda's small size and high population density has resulted in intensive coastal development over the past 400 years of settlement. The largest and most significant changes were imposed by the development of naval bases and the airfield (Hayward et al. 1981). The most significant threats today are a new cycle of port and channel re-development, noted above, and the expanding scale of marina development in the protected bays and harbours. Both threats will potentially exacerbate the water quality problems described above. However, the re-location of recreational vessels from free-swinging moorings to marinas or land-based storage systems may allow for recovery of inshore seagrass beds (Murdoch et al. 2007).

As Bermuda is highly dependent on rainwater harvesting to meet potable water demands the adoption of reverse-osmosis systems as a source of fresh water has increased significantly since 1995. Three new outfalls have been installed that can periodically discharge several million litres of hypersaline water into the shallow North Lagoon each day. No long-term environmental impact studies have been conducted to date at any of the outfalls.

Fisheries Management

Bermuda has a long and detailed legislative history of the control of fisheries resources, dating back to 1620 (Butler et al. 1993; Luckhurst and Ward 1996). However, the post-World War II expansion in tourism, business and resident population resulted in very significant population declines of many commercial species of groupers and snappers by the mid-1980s (Luckhurst 1996a). The primary tool of commercial fishermen in this era was the Antillean fish trap. The Bermuda Government took the unprecedented step in 1989 to ban the use of fish pots entirely and compensate commercial fishermen with a one-time ex-gratia payment, based on previous pot use and reported catches (Butler et al. 1993). By 1990, commercial fishermen were left with several options:

(a) To use hand lines for regulated reef species (bag and size limits, seasonal closures)
(b) Enter a very tightly regulated lobster trap fishery
(c) Pursue pelagic fishes on the offshore banks
(d) Enter the sport fishing industry
(e) Retire from fishing

Since 1990 all of these options have been taken up and the net result has been a reduction in fishing pressure on the reefs and the recovery of some reef fish stocks (Luckhurst 1996b; MEP 2005). Populations of a wide range of reef species (scarids and acanthurids) that had been heavily harvested in the fish pots, appear to have stabilized. The recovery of high value species, such as the black grouper (*Mycteropera bonaci*) appears to be motivating illegal fishing practices that cannot be controlled with current enforcement capabilities so that the gains made since 1990 are at risk. The detection of the black grouper spawning grounds in 2008 (Luckhurst 2010) and subsequent seasonal closure of these areas will provide some protection during the reproductive season.

It is important to note that many species of top predators, in particular groupers, have not recovered and are placed on a protected species list (www.conservation.bm/protected-species). Annual shark catch rates were between to 4–9 mt per year from 1997 to 2006 and then rose to 18 and 20 mt in 2007 and 2008 respectively, but very little species-specific data exists (Environment Statistics Compendium 2009). Reef sharks (primarily *Carcharhinus galapagensis*) are caught intensively and these populations remain low. It is clear that Bermuda is a birthing location for this seamount-associated species (C. Eddy, unpubl. data).

Biological Perturbations

Invasive Species

The invasive Pacific lionfish, *Pterois volitians* became established in Bermuda in 2000 (Whitfield et al. 2002) and has been steadily increasing in abundance (Ocean Support Foundation (www.oceansupport.org)). Their occurrence on shallow rim reefs and lagoonal patch reefs is very sparse. However, they appear to be more abundant on the outer terrace reefs and the deep fore-reef slope, down to 80 m (G. Maddocks, pers. comm.). The lionfish appear to be voracious predators of juvenile fishes. Culling activities have been conducted, using trained and licensed spear-divers, and that effort has increased on the deep reefs in 2011, due to an increase in the number of trained technical and re-breather divers (Ocean Support Foundation 2012). At present there are no other threatening invasive marine species in Bermuda.

Coral Bleaching

The first reported occurrence of coral bleaching in Bermuda was reported in 1988, a year after the mass bleaching in the Caribbean (Cook et al. 1990). The distinction in timing events underlines the fundamental differences in the prevailing climatological conditions in Bermuda at its higher latitude position with respect to the Caribbean. Episodes of bleaching have recurred predictably since 1988, following elevated or rapid increases in seawater temperature in several years in the last decade (Cook et al. 1994; McKinney 1998; MEP 2007). Bermuda did not suffer an intensive bleaching event in 1998, compared to the Caribbean, and no extensive mortality was reported. Nevertheless, many species of corals in 1998 had up to 20% of bleached (pale or blotched areas) colonies and the hydrocoral *Millepora alcicornis* was the most affected, with over 80% bleaching prevalence (McKinney 1998). A less intense bleaching event occurred in 2000 (see Fig. 13.4), with prevalence levels of 13%, with many coral (*Montastraea* spp., *Diploria* spp., *Porites* spp., *Siderastrea* spp., *Favia fragum*, etc.) and octocoral (*Pseudoterogorgia americana*, *Plexaurella nutans and Pseudoplexaura porosa*) species affected. Bleaching occurred again in 2003 with prevalence jumping to 20% and over 90% bleaching of *M. alcicornis* (Brylewska unpublished data, 2003). Bleaching prevalence was significantly higher from 1999 to 2003 compared to later years. It is interesting that some level of bleaching has been occurring on these reefs every summer-fall since 1999 (Table 13.1).

Table 13.1 List of the common diseases affecting corals and other reef organisms in Bermuda. Acronym (ACR), year reported/observed (Year), pathogen/agent (P/A) identified=Y, No=N, number of taxa with disease signs in corals (CO), octocorals (OC), hydrocorals (HY), sponges (SP), zoanthids (ZO) and crustose coralline algae (CCA), depth distribution (DE), minimum and maximum average community-level prevalence (PR) and average tissue mortality rate (TM) (Updated from Weil and Rogers (2011) and data from surveys in Bermuda's reefs since 1999 by E. Weil, G. Smith, A. Croquer and D. Anderson, and other sources)

Disease	ACR	Year	P/A	CO	OC	HY	ZO	SP	CCA	DE (m)	PR (%)	TM (mm/day)
Black band disease	BBD	1971	Y	8	1					0–25	0.1–0.8	3–10
Bleaching	BLE	1998	N	16	9	1	1	1		0–20	0.5–12	–
Growth anomalies	GAN	1999	N	4	4					0–12	–	–
White plague disease-II	WPD	1999	Y	11						3–25	0.1–1.3	20–35
Caribbean yellow band	CYBD	2000	Y	4						3–20	0.1–1.0	0.1–0.4
Caribbean white syndromes[a]	CWS	2000	N	9		1	1	3		2–25	–	–
Dark spots disease	DSD	2001	N	2						1–25	–	–
Other coral health conditions[b]	OCHC	2001	N	8						1–25	0.4–3.1	–
Caribbean ciliate infection	CCI	2007	Y	3						2–25	–	–
Aspergillosis	ASP	1999	Y		10					1–25	0.4–9.3	–
Red band disease	RBD	2006	Y	1	2					2–20	–	–
Other octocoral health condi.[b]	OOHC	2006	N		10					3–20	0.6–7.0	–
Sea fan purple spots[c]	SFPS	2009	Y		1					5–18	–	–
Crustose-Coralline white b	CCWB	2006	N						2	1–20	–	–

[a]White syndromes include a group of health problems that produce patterns of tissue loss exposing bands, stripes, blotches, or irregular shapes of clean skeleton (different from the other "white" diseases)
[b]Other coral and octocorals health conditions include unhealthy-looking tissues with some degree of mortality
[c]SFPS produced by an unknown protozoan (Labyrinthulomycote)

Each bleaching episode has consistently been attenuated temporally (from July through September) and they have not appeared to impose significant mortality (MEP 2007). *Millepora alcicornis* has shown the earliest bleaching response, usually in late July or early August and *Diploria* spp and *Montastraea* spp displaying bleaching weeks later and over a wider depth range. Anecdotal observations indicate that partially bleached corals may persist through the fall and are assumed to have recovered zooxanthelle densities by the winter/spring of the following year (Smith, pers. obs.). No intensive follow-up studies on bleaching mortality have been done in Bermuda but long term studies have not detected any declines in coral coverage or reduction in species-specific abundance across the reef platform since the early 1990s (Linton and Fisher 2004; MEP 2007).

Coral Diseases

Black band disease was first described from Bermuda in 1975 (Garret and Ducklow 1975) followed by experimental studies to assess the infection processes and characterize the constituent microbial communities (Antonius 1982; Ruetzler et al. 1983). Assessment programmes were established in the 1990s and have continued ever since (McKinney 1998; Weil et al. 2002; MEP 2004–2009; Weil and Croquer 2009; Croquer and Weil 2009). The first extensive surveys to document the status and prevalence of diseases in corals and other reef organisms in Bermuda were done on five reefs in 1998 (McKinney 1998), followed by three years of surveys (1999–2001) in three reef localities to assess their spatial and temporal variability and their impact on the community (Weil et al. 2000). The Marine Environmental Program (MEP) of the Bermuda Institute of Ocean Sciences (BIOS) started a long-term video and coral condition monitoring program in 2003 to assess annually the conditions and changes in coral reefs influenced by anthropogenic and natural stressors.

A four year survey program supported by GEF and the Bermuda Government Department of the Environment, started in 2005 to: (1) characterize the number and distribution of coral diseases, species affected, and; (2) assess the spatial and temporal variability in prevalence and incidence of the main diseases affecting reef organisms, and (3) their impact on coral populations and communities. Early quantitative surveys of 1999–2001 were done in three shallow reefs (maximum depth 8 m), located in the north rim (Hog Breaker), in the exposed southeast coast (Cathedral) and the third in the protected sound of Castle Harbor (Pinnacles), but qualitative observations were conducted in several other localities also. From 2005 to the present, quantitative surveys have continued at Hog Breaker and at another three widespread reef sites, with a more extended depth profile (down to 20 m): Chub Cut on the west rim, Snake Pit on the north rim and Rita Zoretta wreck site on the south-east coast (Fig. 13.1). Quantitative surveys in other studies (MEP 2004, 2005, 2006, 2007) covered a greater number of reef localities and have reported consistently low patterns of disease prevalence in hard and soft corals

Overall, at least 16 species of scleractinian corals were susceptible to at least one biotic disease and 16 were susceptible to bleaching (Table 13.1 and Fig. 13.2). Twelve species of octocorals were affected by at least one disease and none were susceptible to bleaching (Fig. 13.3). Biotic diseases and bleaching affected a hydrocoral and three sponges, and one disease affected at least two crustose coralline algae (CCA) species (Table 13.1, Figs. 13.2 and 13.3). Most coral species were susceptible to more than one disease. During the early surveys, total coral disease prevalence at the community level (including all species surveyed) increased significantly from 1.3% to 4.3% between 1999 and 2001 (Fig. 13.4) but these disease prevalence values were lower than localities surveyed in the northern and southern Caribbean during the same times (Weil et al. 2002). Only black band disease (BBD) showed decreasing prevalence (0.59–0.46%) between 1999 and 2001, with other diseases showing high temporal variability. Jones et al. (2012) also found a decrease in BBD over time (2004–2009) and distinguished patterns across reef zones: it was highest on the outer rim reef (0.3–1.9%), followed by the outer lagoonal patch reefs (0.05–0.8%), and the lowest levels occurred on the deeper terrace reefs (0.1–0.2%).

White plague (WPD) and Caribbean yellow band (CYBD) prevalence were highly variable (0.23–1.31% and 0.06–0.58% respectively) with no particular trend (Figs. 13.4 and 13.5). However, CYDB ranged from 3–16% on *Montastraea franksi* from 2004–2007 at 26 sites, where over 1,000 colonies were surveyed each year (MEP 2007). Total coral disease prevalence continued to increase from 2005 to 2009 (4.97%). Prevalence of BBD was low and variable, and WPD and CYBD showed increasing trends in prevalence from 2005 to 2009 (0.06–1.1% and 0.54–1.0% respectively, Fig. 13.4). Other coral health problems (OCHP) increased significantly from 1999 to 2001, but then, decreased in 2005, and remained low thereafter. Growth anomalies (GA) have shown an increasing trend in prevalence in the last 10 years (0.03–0.23%, data not presented), with at least five species (*D. strigosa,, D. labyrinthiformis, M. franksi* and *M. cavernosa*) showing signs of this condition (Fig. 13.3 g, h). White syndrome and other coral health conditions include conditions poorly understood for which no pathogens have been identified. These conditions affected several of the main coral species but their prevalence was usually low (<1%).

Within the octocorals, the prevalence of total diseases varied between 5.1% and 10.5% over the years, with a declining trend from 2001 to 2009 (Fig. 13.5). The most common disease affecting octocorals is the fungal

Fig. 13.1 Ports and channels in Bermuda and site locations of six reefs surveyed between 1999 and 2009. Image courtesy of the Bermuda Zoological Society

condition aspergillosis (ASP), which has been observed infecting at least 10 species (Smith and Weil 2004). Aspergillosis prevalence at the community level was high between 1999 and 2001 (7.6–8.3%), but declined significantly from 2001 to 2009, varying between 0.7% and 3.0%. In contrast, other unidentified health conditions (OUHC) producing tissue mortality in a number of species showed a declining trend from 1999 to 2001, but increased significantly thereafter (Fig. 13.5). Octocorals were also affected by BBD and red band disease (RBD) at very low prevalence levels and only occasionally. The most recent new disease observed affecting octocorals was sea fan purple spots (SFPS) in *G. ventalina*, a ciliate (Labyrinthulomicota) infection producing dark, purple spots over the colony but with no signs of tissue mortality (Weil and Hooten 2009).

All three species of sea fans, *Gorgonia ventalina, G. flabellum* and *G. mariae* were highly susceptible to ASP and showed high prevalence levels (data not show). The most common and abundant sea fan *G. ventalina*, showed the highest prevalence levels between 1999 and 2001 (14–30%), significantly declining in later years (3.2 –13.0%) (Fig. 13.5). Many colonies lost extensive areas of live tissues over the years. The branching octocorals *Pseudoterogorgia americana, Plexaurella nutans* and *Pseudoplexaura porosa* were also highly susceptible to ASP and OUHC (Weil et al. 2002; Smith and Weil 2004; Croquer and Weil 2009), and during the bleaching event of 2000, many colonies showed signs bleaching.

The northern-most location and partial geographic isolation of Bermuda has not prevented the development of bleaching events and/or the onslaught of infectious diseases affecting the most important reef-building species and other important reef organisms. Even though sea water temperatures continue to rise and disease seasonality has disappeared in many localities in the Caribbean (Harvell et al. 2009; Weil et al. 2002), Bermuda still has a marked seasonal change in water temperatures, and some of the most damaging diseases (BBD, WPD and CYBD) still respond by lowering rates of advance (virulence) or disappearing all together from the colonies. CYBD and BBD probably are the only conditions that could be considered at epizootic levels in the populations of the corals *M. franksi* and *D. strigosa*, and ASP in the sea fan *G. ventalina*. High species-level disease prevalence, and significant tissue and colony mortality have occurred and continue occurring in populations of these species around the island.

Fig. 13.2 Photographs of the most common disease problems affecting the major reef-building species in Bermuda. Starting infection of BBD in *M. franksi* (**a**). Advanced stages of BBD in *D. strigosa* (**b**) showing a colony with two infected focal areas (**c**). Close-up of BBD in *M. franksi* (**d**) and a wide band of the bacterial mat in a small colony of *D. strigosa* (**e**). Caribbean yellow band disease infecting two large colonies of *M. franksi* (**f**, **g**) and a colony of *D. labyrinthiformis* (**h**). Tagged colony of *M. franksi* with CYBD (**i**). White band disease on *D. strigosa* (**j**, **n**), *D. labyrinthiformis*, which also has some bleaching signs (**k**) and *M. cavernosa* (**l**, **m**) (Photos E. Weil)

Fig. 13.3 Photographs of the other disease problems affecting corals and octocorals in Bermuda. Bleached colonies of *D. stokesi* (**a**), *M. cavernosa* (**b**) and *M. franksi* (**c**). White syndrome on *D. strigosa* (**d**) and other unidentified health problems producing fast tissue mortality in *M. cavernosa* (**e**) and *S. intersepta* (**f**). Growth anomalies in *D. strigosa* (**g**) and *D. labyrinthiformis* (**h**). Caribbean ciliate infection, the most recent disease identified on Bermudian corals, infecting *D.labyrinthiformis* (**j**) and *D. strigosa* (**k**), and dark spots disease on *M. franksi* (**l**) and *S. siderea* (**m**). Aspergillosis in the sea fan *G. ventalina* (**n**) and a gall produced by *P. americana* in response to ASP infection (**o**) (Photos E. Weil)

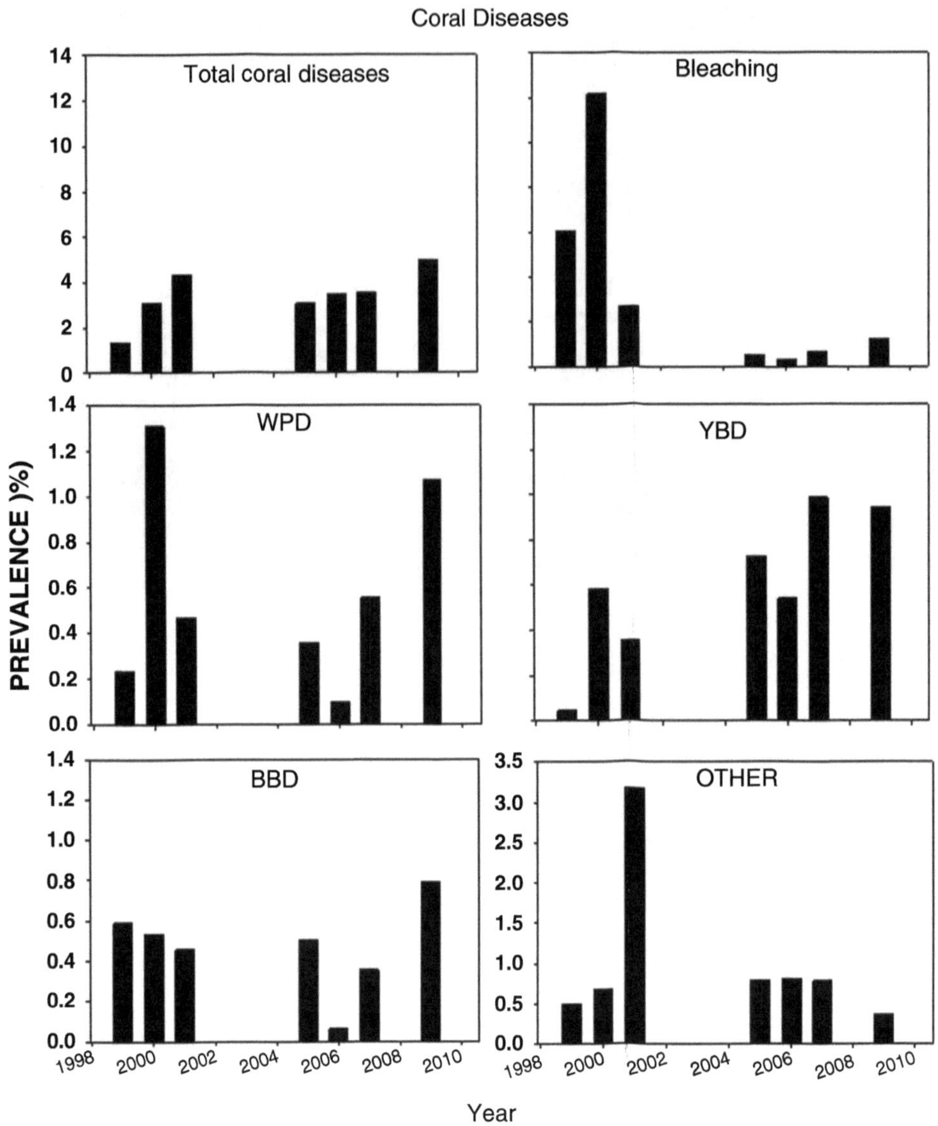

Fig. 13.4 Temporal dynamics of coral diseases in Bermudian reefs. Community level prevalence of total diseases (all diseases affecting corals pooled), bleaching, white plague disease (WPD), Caribbean yellow band disease (CYBD), black band (BBD), and other coral health conditions (OCHC=OTHER) from 1999 to 2009

Environmental Conditions

Potential Climate Change Impacts on the Bermuda Coral Reef Ecosystem

Bermuda's shallow marine waters are entirely surrounded by open-ocean waters of the North Atlantic subtropical gyre. As such, they are highly influenced by the physical state of the western North Atlantic Ocean, its circulation patterns and other factors such as climate variability (e.g. Hurrell and Deser 2009). Potential threats to Bermuda's marine waters and ecosystems, such as climate change and ocean acidification, are thus primarily influenced by basin-wide and global environmental change. Climate change is likely to influence the Bermuda marine environment primarily through warming, increase in salinity, ocean acidification (Bates et al. 2012), sea level rise, and changes in stratification and circulation of the upper ocean of the surrounding North Atlantic Ocean.

Wind and tidal processes allow continuous exchange between onshore lagoonal and offshore waters with the surrounding North Atlantic Ocean (e.g., Morris et al. 1977; Bates et al. 2001; Bates 2002). Freshwater from terrestrial runoff and atmospheric precipitation are relatively inconsequential contributors to the marine water budget of Bermuda. There is a wide range of marine water residence times relative to replenishment from the North Atlantic Ocean. The

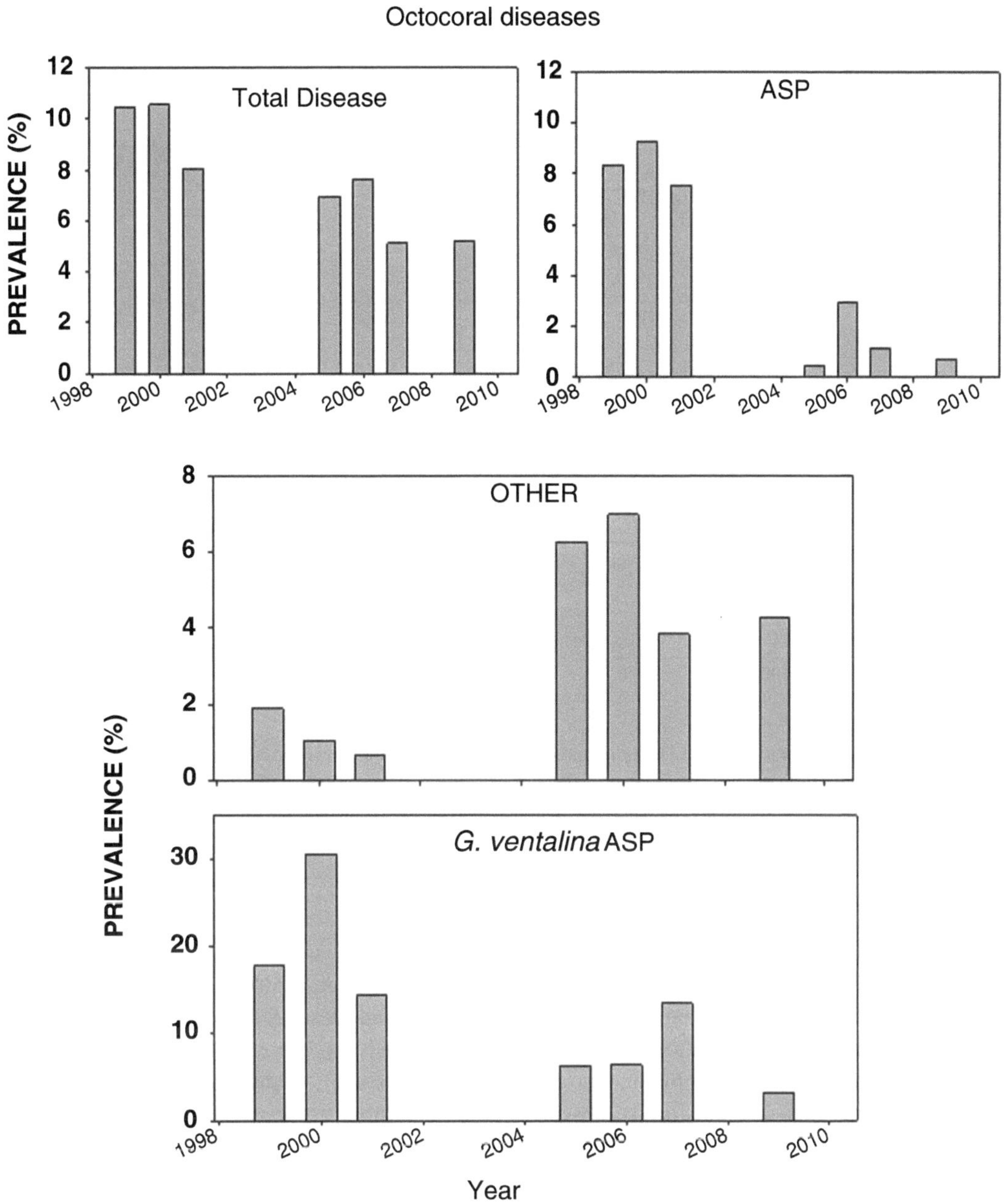

Fig. 13.5 Temporal dynamics of octocoral diseases in Bermudian reefs. Community level prevalence of total diseases (all diseases affecting octocorals pooled), aspergillosis and other octocoral health conditions (OOHC=OTHER), ASP on G. ventalina from 1999 to 2009

typical residence time of water overlying the rim and terrace reefs is approximately 1–4 days, while water residence times are longer in the North Lagoon (~5–12 days) (Venti et al. *in press*) and in those marine lagoons with restricted seawater exchanges (e.g., 100–150 days in Harrington Sound; Morris et al. 1977). Thus the influence of the surrounding North Atlantic Ocean lessens in those marine waters with longer residence times.

The evidence for climate change and it's attribution to anthropogenic greenhouse gas emissions has grown (e.g., IPCC 2007). Detecting and understanding emerging patterns of long-term ocean climate change, as evidenced through changes in the heat and salinity budgets of the upper ocean, has become increasingly important. Analysis of data from the Bermuda ocean time-series sites reveal significant long-term trends (Johnson et al. 2008; Bates et al. 2012). Within the upper 400 m of waters surrounding Bermuda, temperature and salinity are increasing at rates of ~0.01°C year^{-1} and 0.002 year^{-1}, respectively (Fig. 13.6). Similar long-term changes in surface temperatures have been observed basin-wide in the North Atlantic Ocean (e.g., Joyce et al. 1999; Grist et al. 2010). Furthermore, a rise in salinity near Bermuda is also evident

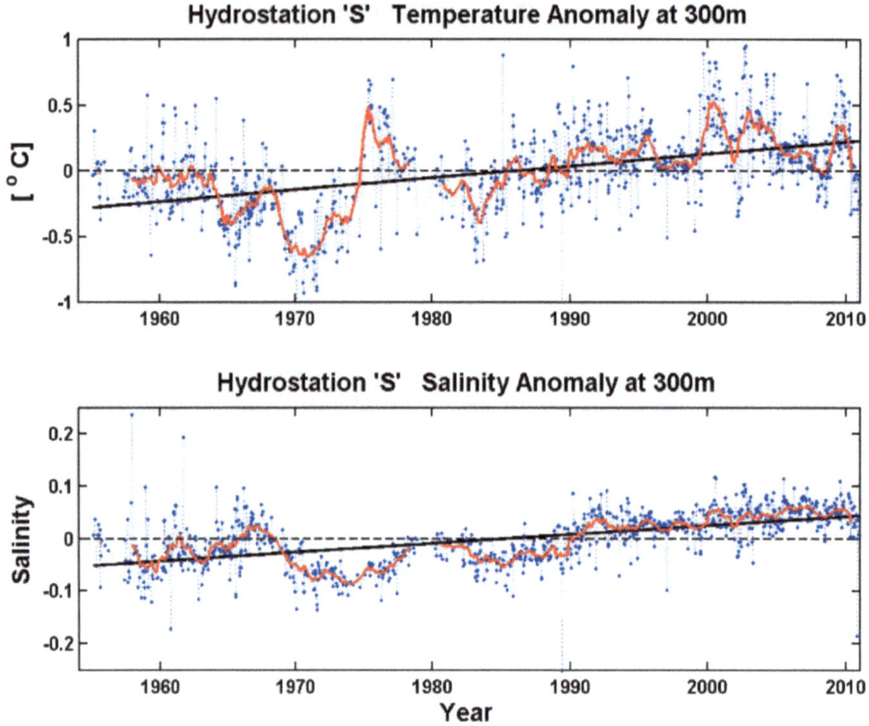

Fig. 13.6 Time-series plots of temperature (**a**) and salinity anomaly (**b**) at 300 m (STMW) at Hydrostation 'S' near Bermuda for the period 1955–2011. Temperature and salinity anomalies are computed by subtracting long term mean at this depth. The *red line* shows a 1-year central running mean and the observed data is shown as *blue dots*. Long-term trends for temperature and salinity are determined as 0.009°C $year^{-1}$ ($p < 0.01$) and 0.002 $year^{-1}$ ($p < 0.01$), respectively

across the North Atlantic subtropical gyre (Zhang et al. 2011), and is a response to the reorganization of the global water cycle and geographic patterns of evaporation and precipitation as a result of climate change (IPCC 2007). Given the strong control of Bermuda's marine waters by the surrounding subtropical gyre it confirms that the locally observed trends in seawater temperature and salinity are responses to climate change. As a potential threat, warming is likely to increase the incidence of bleaching events in Bermuda coral reef species (Cook et al. 1994). Furthermore, changes in nutrient fluxes and pelagic phytoplankton and zooplankton biomass in the surrounding subtropical gyre of the North Atlantic (e.g., Tilstone et al. 2009; Lomas et al. 2010) are also likely to influence the supply of offshore nutrients and pelagically sourced organic matter to Bermuda marine waters and its coral reef ecosystem.

In addition to warming and salinity rise, anthropogenically driven changes in climate variability, storm events mesoscale eddies and sea-level rise are linked threats to Bermuda's marine waters. The dominant long-term mode of climate variability in the North Atlantic Ocean is associated with the North Atlantic Oscillation (NAO; Hurrell 1995; Hurrell and Deser 2009) and El Nino Southern Oscillation (ENSO). The NAO index is typically defined as the difference in the normalized sea level pressures between Lisbon, Portugal and Stykkisholmur, Iceland (Hurrell and Deser 2009). During a positive NAO phase, anomalously high pressures are observed across the subtropical North Atlantic Ocean, while during negative phases, more intense winter storms pass over the subtropical gyre. These atmospheric changes induce strong changes in the upper ocean (Rogers 1990) and the intensity of the North Atlantic Ocean gyre circulation (Curry and McCartney 2001). The NAO was mostly negative from 1950 to 1970 and has since shifted to a mostly positive state (e.g., Hurrell and Deser 2009). During negative NAO years, winter mixing tends to be deeper with colder sea surface temperatures whereas relatively shallow warm winter mixed layers are typically observed during positive phases (Bates 2001), while La Niña phases of ENSO tend to induce local increase in salinity of the upper ocean (Bates 2001). Any long-term changes in NAO and ENSO conditions are likely to enhance warming and salinity increase of both North Atlantic subtropical gyre and Bermuda marine waters. Bermuda corals preserve good proxy records of the NAO and centennial scale changes in oceanographic conditions in their skeletons (Cohen et al. 2004; Kuhnert et al. 2005; Goodkin et al. 2008) and this

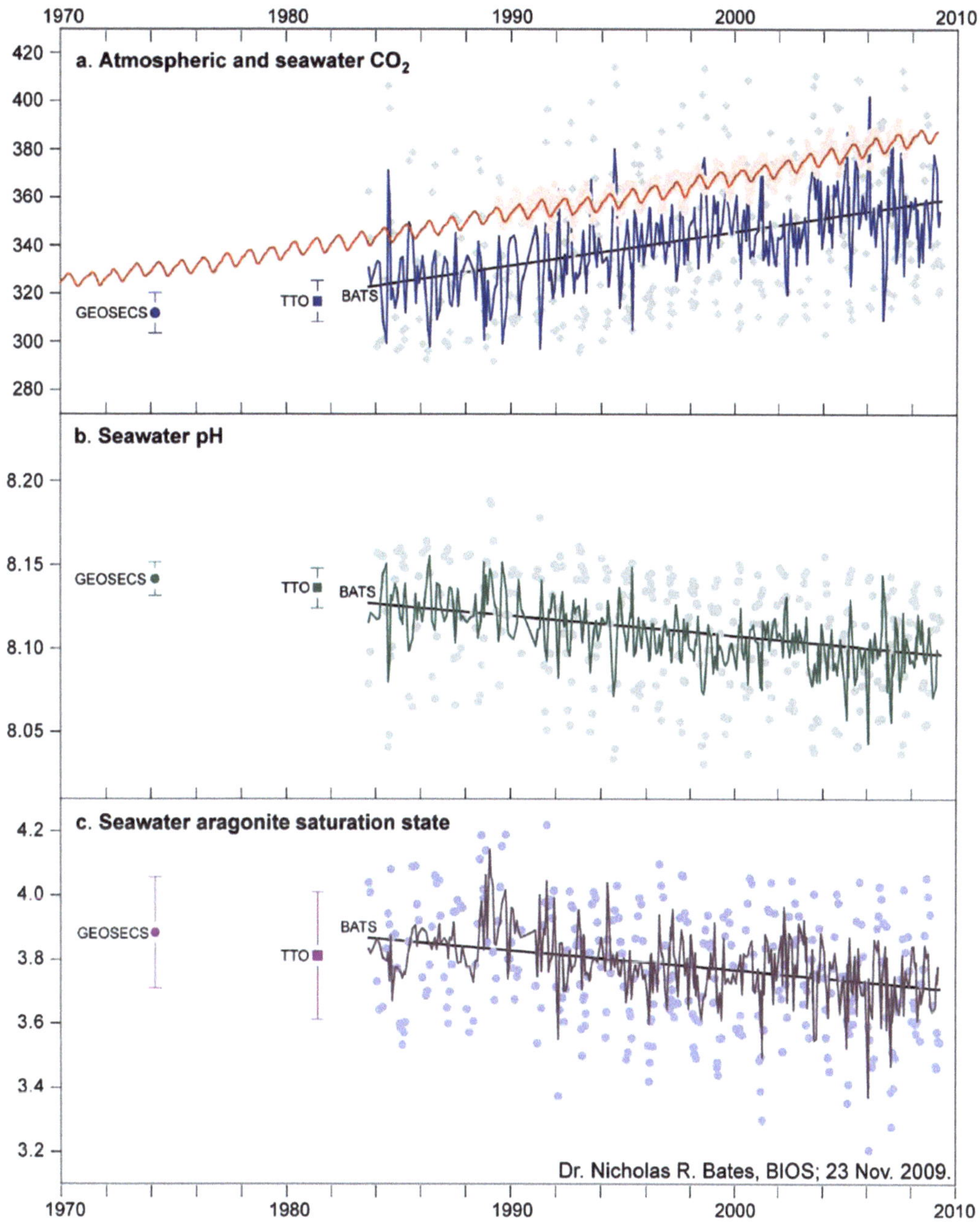

Fig. 13.7 Changes in atmospheric and seawater CO_2, pH and aragonite saturation state in Bermuda waters 1970–2010

may prove very useful in understanding how corals respond to changing oceanographic and climactic conditions.

Observations of seawater dissolved inorganic carbon (DIC), and computed pCO_2, pH and saturation states for aragonite ($\Omega_{aragonite}$) at the Bermuda Atlantic Time-series Study (BATS) site ~80 km south-east of Bermuda constitute the longest time-series record of ocean acidification anywhere in the global ocean (Fig. 13.7, Bates 2001, 2002, 2012; Bates et al. 2012). Ocean acidification is a significant challenge to Bermuda's corals that already show lower growth rates because of their high latitude position (Kleypas et al. 2001). Direct measurements of the dynamics of carbonate saturation state of Bermuda's reef water (Bates et al. 2001, 2011) show that the seasonal patterns of calcification on the reefs have shown positive correlation between available dissolved carbonate ions and the seawater aragonite saturation state, which partly controls the growth rate of the dominant coral *Diploria labyrinthiformis*. The observed relationships between seawater aragonite saturation state and calcification rates (as well as calcification rates from coral cores) suggest a ~37% decrease in calcification rates of *D. labyrinthiformis* for the 1959–1999 period, declining from a high of 4.5 g cm^{-3} $year^{-1}$ in 1959 to a low of 3 g cm^{-3} $year^{-1}$ in 1999 (Cohen and Jacowski, unpub. data; Cohen et al. 2004; Bates et al. 2011). Thus it appears that ocean acidification has already impacted calcification rates for a few coral species over the recent past, even within the apparently healthy coral reef ecosystem of Bermuda (Bates et al. 2001, 2011; Bates 2002). However, the recent BEACON project (BEACON: Bermuda ocean acidification and coral reef investigation; Andersson et al. 2010) indicates that *D. labyrinthiformis* and other Bermuda coral species have highly variable responses to natural variations in seawater carbonate chemistry and ocean acidification (Andersson et al. 2010, 2011). Juvenile coral recruits in Bermuda also experience reduced corallite formation rate under future aragonite saturation states (de Putron et al. 2011). The prospective impacts of ocean acidification on other calcifying reef species, such as the crustose coralline algae, have not been studied directly in Bermuda.

Further potential threats include sea level rise (IPCC 2007) that will increase vulnerability to shoreline erosion and coral reef damage during storms and cyclones, due to wave action and sea-level updoming during these events. In addition, naturally occurring mesoscale eddies in the subtropical gyre may amplify the effects of storm events. Mesoscale eddy phenomena in the North Atlantic subtropical gyre include cold core rings (Richardson et al. 1978) and mid-ocean mesoscale eddies (McGillicuddy and Robinson 1997; McGillicuddy et al. 1999, 2007) that have diameters of >100 km, lifetimes of months to years and propagation speeds of 3–5 km day^{-1}. The anticyclonic mid-ocean mesoscale eddies can updome sea level by 20–30 cm relative to mean sea-level, which if located over Bermuda during tropical cyclone events, amplifies the impact of anthropogenic sea-level rise, thereby increasing the storm surge vulnerability of Bermuda's coastal marine environment.

Summary

Although Bermuda has addressed significant challenges to the health of its reefs through management and research over the past several decades the list of threats is still significant.

Fishing pressure continues to threaten commercial species and the scope of illegal harvesting of protected species is not known. Recreational fishing is still largely unregulated, apart from species-specific daily catch and size limits and the seasonal closures for some species. In fact the magnitude of recreational fishing pressure and catch has never been quantified as no licenses are required and no catch statistics are reported.

Chemical contamination, nutrient pollution and faecal pollution will remain long-term threats and no immediate steps are planned for the reduction or mitigation of these issues. The great concern of course is that ongoing local anthropogenic stressors will have unpredictable interactions with natural and climate change factors, and the consequences are very uncertain for corals specifically. Great uncertainty exists with respect to the capacity of Bermuda's slow growing corals to cope with rising sea level and ocean acidification and that the reef system will deliver the valuable ecosystem services and essential economic benefits into the next century.

Acknowledgements This is Contribution #200 of the Bermuda Biodiversity Project (BBP), Bermuda Aquarium, Museum and Zoo, in the Department of Conservation Services.

References

Agassiz A (1894) Visit to the Bermudas. Bull Mus Comp Zool 26:209–281

Andersson AJ, Bates NR, and de Putron SJ (2010) BEACON: Bermuda ocean acidification and coral reef investigation – overview and early results. ASLO/AGU Meeting, Portland, Feb 2010

Andersson AJ, de Putron SJ, Bates NR, Collins A, Garley R, Noyes T, Dexter E (2011) Bermuda ocean acidification and coral reef investigation: temporal and spatial changes in seawater carbonate chemistry and calcification. Abstract, IPCC meeting, Okinawa, Feb 2011

Antonius A (1982) The 'band' diseases in coral reefs. In: Proceedings of the 4th international coral reef symposium, vol 2. Manila, pp 7–14

Bates NR (2001) Interannual variability of oceanic CO_2 and biogeochemical properties in the Western North Atlantic subtropical gyre. Deep-Sea Res II 48:1507–1528

Bates NR (2002) Seasonal variability of the impact of coral reefs on ocean CO_2 and air-sea CO_2 exchange. Limnol Oceanogr 47(1):43–52

Bates NR (2012) Multi-decadal uptake of carbon dioxide into subtropical mode waters of the North Atlantic Ocean. Biogeosciences 9:2649–2659. doi:10.5194/bg-9-2649-2012

Bates NR, Samuels L, Merlivat L (2001) Biogeochemical and physical factors influencing seawater fCO_2 and air-sea CO_2 exchange on the Bermuda coral reef. Limnol Oceanogr 46(4):833–846

Bates NR, Amat A, Andersson AJ (2011) Feedbacks and responses of coral calcification on the Bermuda reef system to seasonal changes in biological processes and ocean acidification. Biogeosciences 7:1–22

Bates NR, Best MH, Neely K, Garley R, Dickson AG, Johnson RJ (2012) Indicators of anthropogenic carbon dioxide uptake and ocean acidification in the North Atlantic Ocean. Biogeosciences 9:2509–2522. doi:10.5194/bg-9-2509-2012

Bodungen BV, Jickells TD, Smith SR, Ward JAD, Hillier GB (1982) The Bermuda Marine Environment – Vol. III. The final report of the Bermuda Inshore Waters Investigation. B.B.S.R. Spec. Publ. No. 18., p. 123

Boyer JN, Briceno HO (2010) Water quality monitoring program for Bermuda's coastal resources. Technical report #T-473 of the Southeast Environmental Research Center (FIU), prepared for Bermuda Government Department of Conservation Services, 57pp

Burke L, Reytar K, Spaulding M, Perry A (2011) Reefs at risk revisited. World Resources Institute, Washington, DC, p 130

Burns KA, Ehrhardt MG, MacPherson J, Tierney JA, Kananen G, Connelly D (1990) Organic and trace metal contaminants in sediments,

seawater and organisms from two Bermudan harbours (GEEP Workshop, September 1988). JEMBE 138:9–34
Butler JN, Burnett-Herkes J, Barnes JA, Ward J (1993) Bermuda fisheries, a tradgedy of the commons averted? Environment 35:7–33
CARICOMP (2000) Status and temporal trends at the CARICOMP reef site. In: Proceedings of 9th international coral reef symposium, vol 1, Bali. pp 325–330
Cohen AL, Smith SR, McCartney MS, van Etten J (2004) How brain corals record climate: an integration of skeletal structure, growth and chemistry of *Diploria labyrinthiformis* from Bermuda. MEPS 271:147–158
Cook CB, Logan A, Ward J, Luckhurst BE, Berg CJ Jr (1990) Elevated temperatures and bleaching on a high-latitude reef: the 1988 Bermuda event. Coral Reefs 9:45–49
Cook CB, Dodge RE, Smith SR (1994) Fifty years of impacts on coral reefs in Bermuda. In: Proceedings of the colloquium on coral reefs, 1993, R.S.M.A.S, Miami. pp 160–166
Connelly DP, Readman JW, Knap AH, Davies J (2001) Contamination of the coastal waters of Bermuda by organotins and the triazine herbicide Irgarol 1051. Mar Pollut Bull 42:409–441
Croquer A, Weil E (2009) Spatial variability in distribution and prevalence of Caribbean scleractinian coral and octocoral diseases. II. Genera-level analysis. Diseas Aquat Orgs 83:209–222
Curry RG, McCartney MS (2001) Ocean gyre circulation changes associated with the North Atlantic Oscillation. J Phys Oceanogr 31:3374–3400
de Putron SJ, McCorkle DC, Cohen AL, Dillon AB (2011) The impact of seawater saturation state and bicarbonate ion concentration on calcification by new recruits of two Atlantic corals. Coral Reefs 30:321–328
Dodge RE, Vaisnys JR (1977) Coral populations and growth patterns: responses to sedimentation and turbidity associated with dredging. J Mar Res 35:715–730
Dodge RE, Logan A, Antonius A (1982) Quantitative reef assessment studies in Bermuda: a comparison of methods and preliminary results. Bull Mar Sci 32:745–760
Dodge RE, Jickells TD, Knap AH, Boyd S, Bak RPM (1984) Reef-building coral skeletons as chemical pollution (phosphorus) indicators. Mar Poll Bull 15:178–187
Dryer S, Logan A (1978) Holocene reefs and sediments in Castle Harbour, Bermuda. J Mar Res 36:399
Environment Statistics Compendium (2009) Bermuda Government Department of Statisitics. Bermuda Government 90pp
European Union (2006) Directive 2006/7/EC of the European Parliament and of the Council of 15th February 2006 concerning the management of bathing water quality and repealing Directive 76/160/EEC
Flood V, Pitt JM, Smith SR (2005) Historical and ecological analysis of coral communities in Castle Harbour (Bermuda) after more than a century of perturbation. Mar Poll Bull 51:545–557
Garret P, Ducklow H (1975) Coral diseases in Bermuda. Nature 253:349–350
Garrett P, Smith DL, Wilson AO, Patriquin D (1971) Physiography, ecology and sediments of two Bermuda patch reefs. J Geol 79:647–668
Goodkin NF, Hugen KA, Curry WB, Doney SC, Ostermann DR (2008) Sea surface temperature and salinity variability at Bermuda during the end of the little ice age. Paleooceanography 23(PA3202):1–13
Grist JP, Josey SA, Marsh R, Good SA, Coward AC, de Cuevas BA, Alderson SG, New AL, Madec G (2010) The roles of surface heat flux and ocean heat transport convergence in determining Atlantic Ocean temperature variability. Ocean Dyn 60(4):771–790. doi:10.1007/s10236-010-0292-4
Gunther JA (1999) Bioaccumulation of metals in the turkey-wing mussel, *Arca zebra*, off the coast of Bermuda. M.Sc. thesis, Villanova University, 59pp
Harvell CD, Altizer S, Cattadori IM, Harrington L, Weil E (2009) Climate change and wildlife diseases: when does the host matter the most? Ecology 90(4):912–920
Hayward SJ, Gomez VH, Sterrer WE (eds) (1981) Bermuda's delicate balance. Bermuda National Trust, Bermuda, p 402
Hurrell JW (1995) Decadal trends in the North Atlantic Oscillation: regional temperatures and precipitation. Science 269:676–679
Hurrell JW, Deser C (2009) North Atlantic climate variability: the role of the North Atlantic Oscillation. J Mar Syst 79(3–4):231–244. doi:10.1016/j.marsys.2009.11.002
IPCC (2007) Climate change 2007: the physical science basis. In: Solomon S, Qin D, Manning M, Chen Z, Marquis M, Averyt KB, Tignor M, Miller HL (eds) Contribution of working group I to the fourth assessment report of the intergovernmental panel on climate change. Cambridge University Press, Cambridge/New York, 996pp
Jaap WC (2000) Coral reef restoration. Ecol Eng 15:345–364
Jickells TD, Knap AH (1984) The distribution and geochemistry of some trace metals in the Bermuda coastal environment. Estuar Coast Shelf Sci 18:245–262
Jickells TD, Knap AH, Smith SR (1986) Trace metal and nutrient fluxes through the Bermuda inshore waters. Rapp P-v Reun Cons Int Explor Mer 186:251–262
Johnson RJ, Knap AH, Bates NR, Whitefield JD, Kadko D, Lomas MW (2008) Coordinated change in the heat, salinity and CO_2 budgets of the mesopelagic zone at the Bermuda time-series sites. Abstract ASLO/AGU Ocean Sciences meeting, Orlando, March 2008
Jones RJ (2010) Environmental contamination associated with a marine landfill ('seafill') beside a coral reef. Mar Pol Bull 60:1993–2006
Jones RJ (2011) Environmental effects of the cruise tourism boom: sediment resuspension from cruise ships and the possible effects of increased turbidity and sediment deposition on corals (Bermuda). Bull Mar Sci 87:659–679
Jones R, Parsons R, Watkinson E, Kendall D (2010) Sewage contamination of a densely populated coral 'atoll' (Bermuda). Environ Monit Assess. doi:10.1007/s10661-010-1738-3
Jones RJ, Johnson R, Noyes T, Parsons R (2012) Spatial and temporal patterns of coral black band disease in relation to a major sewage outfall. MEPS 462:70–92
Joyce TM, Pickart TS, Millard RC (1999) Long-term hydrographic changes at 52 and 66 degrees W in the North Atlantic subtropical gyre and Caribbean. Deep-Sea Res II 46(1–2):245–278. doi:10.1016/S0967-0645(98)00102-7
Kleypas JA, Buddemeier RW, Gattuso J-P (2001) The future of corals reefs in an age of global change. Int J Earth Sci 90:426–437
Knap AH, Cook CB, Cook SB, Simmons JAK, Jones RJ, Murray AE (1991) Marine environmental studies to determine the impact of the mass burn incinerator proposed for Tynes Bay, Bermuda. BBSR report to Bermuda Government Ministry of Works & Engineering, Bermuda. 256pp
Kuhnert H, Cruger T, Patzold J (2005) NAO signature in a Bermuda coral Sr/Ca record. Geochem Geophys Geosys 6(Q04004):11
Leavitt DF, Lancaster BA, Lancaster AS, Capuzzo JM (1990) Changes in the biochemical composition of a subtropical bivalve, *Arca zebra*, in response to contaminant gradients in Bermuda (GEEP Workshop, September 1988). J Exp Mar Biol Ecol 138:85–98
Linton D, Fisher T (eds) (2004) CARICOMP – Caribbean coastal marine productivity program: 1993–2003. Caribbean coastal data centre, centre for marine sciences, University of the West Indies, Mona
Lomas MW, Steinberg DK, Dickey TD, Carlson CA, Nelson NB, Condon RH, Bates NR (2010) Increased ocean carbon export in the Sargasso Sea is countered by its enhanced mesopelagic attenuation. Biogeosciences 7(1):57–70
Luckhurst BE (1996a) Trends in commercial fishery landings of groupers and snappers in Bermuda from 1975 to 1992 and associated

fishery management issues. In: Arreguin-Sanchez F, Munro JL, Balgos MC, Pauly D (eds) Biology, fisheries and culture of tropical groupers and snappers. ICLARM conference proceedings, vol 48, Manila. pp 277–288
Luckhurst BE (1996b) A fishery-independent assessment of Bermuda's coral reef fish stocks by diver census following the fish pot ban – a progress report. Proc Gulf Caribb Fish Inst 46:309–323
Luckhurst BE (2010) Observations of a black grouper (*Mycteroperca Bonaci*) spawning aggregation in Bermuda. Gulf Caribb Res 22:43–49
Luckhurst BE, Ward JAD (1996) Analysis of trends in Bermuda's fishery statistical database from 1975 to 1990 with reference to fishery management measures implemented during this period. Proc Gulf Caribb Fish Inst 44:306–324
McGillicuddy DJ, Robinson AR (1997) Eddy induced nutrient supply and new production in the Sargasso Sea. Deep Sea Res 44: 1427–1450
McGillicuddy DJ, Johnson RJ, Siegel DA, Michaels AF, Bates NR, Knap AH (1999) Mesoscale variations of biogeochemical properties in the Sargasso Sea. J Geophys Res 104:13381–13394
McGillicuddy DJ, Anderson LA, Bates NR, Bibby T, Buesseler KO, Carlson CA, Davis CS, Ewart C, Falkowski PG, Goldthwait SA, Hansell DA, Jenkins WJ, Johnson R, Kosnyrev VK, Ledwell JR, Li QP, Siegel DA, Steinberg DK (2007) Eddy/wind interactions stimulate extraordinary mid-ocean plankton blooms. Science 316(5827):1016–1021. doi:10.1126/science.1136256
McKinney C (1998) Characterization of coral diseases and coral bleaching in Bermuda (Sept–Oct 1998). Research Experience for Undergraduates (REU) Report. Bermuda Biological Station for Research Inc. 50pp
MEP (2004) Marine environmental program annual report, 2003–2004. Bermuda Institute of Ocean Sciences, 46pp
MEP (2005) Marine environmental program annual report, 2004–2005. Bermuda Institute of Ocean Sciences, 136pp
MEP (2006) Marine environmental program annual report, 2005–2006. Bermuda Institute of Ocean Sciences, 128pp
MEP (2007) Marine environmental program annual report, 2006–2007. Bermuda Institute of Ocean Sciences, 80pp
Ministry of Transport (2011) Study of Bermuda's shipping channels to accommodate larger cruise ships. Public presentation, Bermuda Government, (www.gov.bm) 49pp
Morgan MB, Edge SE, Snell TW (2005) Profiling differential gene expression of corals along a transect of waters adjacent to the Bermuda municipal dump. Mar Pollut Bull 51:524–533
Morris B, Barnes J, Brown F, Markham J (1977) The Bermuda marine environment, vol 15. Bermuda Biological Station Special Publication, St. George's West
Murdoch TJT, Glasspool AF, Outerbridge M, Manuel S (2007) Bermuda seagrass status report 2004. BREAM project. Bermuda Biodiversity Project Special Publication 2007–002. 84pp
Owen R, Knap A, Toaspern M, Carberry K (2002) Inhibition of coral photosynthesis by the antifouling herbicide Irgarol 1051. Mar Poll Bull 44:623–632
Pandolfi JM, Bradbury RH, Sala E, Hughes TP, Bjorndal KA, Cooke RG, McArdel D, McClenachen L, Newman MJH, Paredes G, Warner RR, Jackson JBC (2003) Global trajectories of the long-term decline of coral reef ecosystems. Science 301:955–958
Price ARG (2006) Cruise ships and sustainability in Bermuda: a preliminary evaluation. Report prepared for the Bermuda National Trust. Industrial Economics, Inc., p. 44
Quinn B, Gagne F, Weber J-P, Blaise C (2005) Ecotoxicological effects of a semi-submerged municipal dump (Castle Harbour, Bermuda) on the calico scallop *Argopecten gibbus*. Mar Poll Bull 51:534–544
Richardson PL, Cheney RE, Worthington LV (1978) A census of Gulf Stream rings, spring 1975. J Geophys Res 83:6136–6144
Rogers JC (1990) Patterns of low-frequency monthly sea level pressure variability (1899–1986) and associated wave cyclone frequencies. J Climate 3:1364–1379
Ruetzler K, Santavy DL, Antonius A (1983) The black band disease of Atlantic reef corals. The black band disease of Atlantic reef corals. III. Distribution, ecology and development. PSZNI Mar Ecol 4:329–358
Sarkis S (1999) Report on the potential impacts of cruise ships on Bermuda's environment. Bermuda National Trust, p. 46
Sarkis S, van Beukering PJH, McKenzie E (2010) (eds) Total economic value of Bermuda's coral reefs: valuation of ecosystem services. Technical report. Department of Conservation Services, Bermuda Government, (www.conservation.bm), 199pp
Simmons JAK, Lyons WB (1994) The groundwater flux of nitrogen and phosphorous to Bermuda's coastal waters. Water Res Bull 30:984–991
Smith SR (1985) Reef damage and recovery after ship groundings on Bermuda. In: Proc 5th Int Coral Reef Symp vol 6. Tahiti, pp 497–502
Smith GW, Weil A (2004) Aspergillosis in gorgonians. In: Rosenberg E, Loya Y (eds) Coral reef health and diseases. Springer, Berlin, pp 279–286
Smith SR, Hellin DC, McKenna SA (1998) Marine environmental impact studies on the thermal effluent released from the Tynes Bay Incinerator and ash waste disposal in Castle Harbour. BBSR Spec Publ vol 35. Bermuda Biological Station for Research, Ferry Reach, 210pp
Tilstone G, Smyth T, Poulton A, Hutson R (2009) Measured and remotely sensed estimates of primary production in the Atlantic Ocean from 1998 to 2005. Deep-Sea Res 56(15):918–930. doi:10.1016/j.dsr2.2008.10.034
Venti A, Kadko D, Andersson A, Langdon C, Bates NR (*in press*) A multi-tracer approach to estimate reef residence times. Global Biogeochem Cycles
Verrill AE (1907) The Bermuda Islands. Part IV. Geology and paleontology, and part V. An account of the coral reefs. Trans Connect Acad Arts Sci 12:45–348
Waltham A (1998) The effects of shipping traffic on adjacent coral reefs along the North Shore, Bermuda. Undergraduate thesis, Southampton University, 52pp
Webster G, Smith SR (2002) Impact of a sewage outfall on juvenile coral populations in Bermuda. In: Proc 9th Int Coral Reef Symp vol 2. Bali, pp 1041–1045
Weil E, Croquer A (2009) Spatial variability in distribution and prevalence of Caribbean scleractinian coral and octocoral diseases. I. Community-level analysis. Diseas Aquat Orgs 83:195–208
Weil E, Hooten A (2008) Underwater cards for assessing coral health on Caribbean coral reefs. GEF-CRTR-Currie Comm, Australia, 30pp
Weil E, Urreiztieta I, Garzon-Ferreira J (2002) Geographic variability in the incidence of coral and octocoral diseases in the wider Caribbean. In: Proc 9th Int Coral Reef Symp vol 2. Bali, pp 1231–1237
Whitfield PE, Gardner T, Vives SP, Gilligan MR, Courtenay WR Jr, Ray GC, Hare JA (2002) Biological invasion of the Indo-Pacific lionfish *Pterois volitans* along the Atlantic coast of North America. Mar Ecol Prog Ser 235:289–297
Widdows J, Burns KA, Menon NR, Page DS, Soria S (1990) Measurement of physiological energetics (scope for growth) and chemical contaminants in mussels (*Arca zebra*) transplanted along a contaminant gradient in Bermuda (GEEP Workshop, September 1988). JEMBE 138:99–117
Zhang LP, Wu LX, Zhang JX (2011) Freshwater loss/salinification: simulated response to recent freshwater flux change over the Gulf Stream and it's extension: coupled ocean–atmosphere adjustment and Atlantic-Pacific teleconnection. J Climate 24(15):3971–3988. doi:10.1175/2011JCL14020.1

14

Scleractinia, Octocorallia and Antipatharia of Bermuda's Reefs and Deep-Water Coral Communities: A Taxonomic Perspective Including New Records

Jan M. Locke, Jaret P. Bilewitch, and Kathryn A. Coates

Introduction

Ongoing research continues to reveal more diversity and endemism in Bermuda's coral reefs and coral communities than previously recognized (Sterrer 1986, 1998). Deeper water biota in both Bermuda and the Greater Caribbean are poorly explored relative to shallow water, but new Bermudian records are presented here for both deep and shallow coral communities. The few deep water collections made within the Bermuda Exclusive Economic Zone (EEZ) indicate the presence of many species, including tropicals and species with ranges extending into temperate regions.

The Bermuda Natural History Museum (BAMZ) holds incomplete collections of Bermudian species and important collections, e.g., type specimens, are held in several major museums, including the Natural History Museum, United Kingdom (NHMUK), Smithsonian National Museum of Natural History (USNM), Harvard University Museum of Natural History (Museum of Comparative Zoology) and the Yale Peabody Museum (YPM).

Short histories of the exploration of diversity of primary reef taxa, scleractinians and octocorallians, are presented. Where the details of historic studies are reported, the species names used in the original documents are presented along with the current, accepted, name for the species; where possible. The current name will normally follow the original Bermuda name, bracketed. Current usages have been checked against the World Register of Marine Species (WoRMS) database (Appeltans et al. 2012) and the Integrated Taxonomic Information System (ITIS Retrieved August 10, 2011, http://www.itis.gov), and the primary literature.

J.M. Locke (✉) • K.A. Coates
Marine Ecology Section, Department of Conservation Services,
17 North Shore Road, Hamilton Parish FL 04, Bermuda
e-mail: jml100@gmail.com; kathryn.coates@dilke.com

J.P. Bilewitch
School of Biological Sciences, University of Queensland,
Brisbane, QLD 4072, Australia
e-mail: jbilewitch@gmail.com

Shallow-Water Zooxanthellate Scleractinia

History of Species Documentations

Most of Bermuda's shallow-water scleractinian diversity was recognized by the end of the nineteenth century. Contributions of AE Verrill are apparent in many of the early studies of the islands' species diversity, including his comprehensive works on the Bermuda islands (1900–1907). Prior to these, Verrill (1864) published the initial record of Bermudian Scleractinia naming *Diploria cerebriformis* Milne Edwards and Haime, 1849 (=*Diploria labyrinthiformis* Linnaeus, 1758), *Isophyllia dipsacea* Agassiz, MS (=*Isophyllia sinuosa* [Ellis and Solander, 1786]) and *Isophyllia rigida* Verrill (later corrected to *Mussa rosula* Verrill, 1907 [=*I. sinuosa*]) (Verrill 1907:230). A more comprehensive list for Bermuda published by JM Jones (1868) [mis-referenced as 1869 by Verrill 1900] added *Oculina diffusa* Lamarck, 1816, *Oculina varicosa* Lesueur, 1821, *Oculina valenciennesi* Milne Edwards and Haime, 1850, *Siderastrea radians* (Pallas, 1766), *Mycedium fragile* Dana (=*Agaricia fragilis* [Dana, 1846]), and *Porites clavaria* Lamarck, 1816 (=*Porites porites* [Pallas, 1766]). *Maeandrina cerebriformis* Lamarck (=*D. labyrinthiformis* Linnaeus, 1758) was also included but had been previously documented. Verrill identified most of the specimens included by Jones (1868) and, in Verrill's words, the collection was "incomplete" (1900:551). Dana's (1872:114) species list, also compiled with Verrill's assistance, noted the presence of *Oculina pallens* Ehrenburg, 1834 (=*O. diffusa* Lamarck, 1816), which agreed with previous accounts considering current synonymies. In 1871, Pourtalès noted that specimens of both *Lithophyllia cubensis* Milne Edwards and Haime (=*Scolymia cubensis* Milne Edwards and Haime, 1849) and *Isophyllia multilamella* Pourtalès (=*I. sinuosa*) from Bermuda were in the collections of the Museum of Comparative Zoology at Harvard College. In 1877, Brüggemann's work on specimens housed at the British Museum, now the Natural History Museum, United

C.R.C. Sheppard (ed.), *Coral Reefs of the United Kingdom Overseas Territories*, Coral Reefs of the World 4,
DOI 10.1007/978-94-007-5965-7_14,

Kingdom, noted *Scolymia lacera* as occurring in Bermuda based on Pourtalès' collections. The occurrence of this species in Bermuda has not been documented elsewhere.

In 1886, Quelch reported on the shallow-water corals collected during the HMS Challenger Expedition (1872–1876). He documented the first specimens of *Maeandrina strigosa* Dana (=*Diploria strigosa* [Dana, 1846]), *Astraea ananas* Ellis and Solander and *Astraea coarcta* Duchassaing and Michelotti (both=*Favia fragum*) and *Madracis decactis* (Lyman, 1859). Other species Quelch listed that have since been synonymized are *Isophyllia strigosa* Duchassaing and Michelotti (=*I. sinuosa*), *Isophyllia knoxi* Duchassaing and Michelotti (=*I. sinuosa*), *Isophyllia cylindrica* Duchassaing and Michelotti (=*I. sinuosa*), *Isophyllia australis* Edwards and Haime (=*I. sinuosa*), *Isophyllia fragilis* Dana (=*I. sinuosa*), *Oculina speciosa* Milne Edwards and Haime (=*O. diffusa*), *Oculina bermudensis* Duchassaing and Michelotti (=*O. valenciennesi*), *Siderastrea galaxea* Ellis and Solander (=*S. radians*), *Maeandrina labyrinthica* Ellis and Solander (=*D. labyrinthiformis*), *Diploria cerebriformis* Lamarck (=*D. labyrinthiformis*), and *Maeandrina sinuossima* Milne Edwards and Haime (=*D. labyrinthiformis*). The following of Quelch's records are of dubious validity and uncertain synonymy, *Oculina coronalis* n.sp. (suggested to be *O. diffusa* [J.W. Wells, pers. comm. to S. Cairns, 1977]; observation of type material would confirm this) and *Isophyllia marginata* Quelch (=?*I. sinuosa*) (see Verrill 1907:229). Besides the species obtained by the Challenger, Quelch (1886) noted in his report that both *Lithophyllia cubensis* Milne Edwards and Haime, 1857 (=*Scolymia cubensis*) and *Lithophyllia lacera* Pallas (=*Scolymia lacera* [Pallas, 1766]) as well as *Isophyllia multilamella* Pourtalès (=*I. sinuosa*) and *Isophyllia spinosa* Edwards and Haime (=*I. sinuosa*) were known to occur in Bermuda. These claims were undoubtedly based on the publications of Pourtalès (1871) and Brüggemann (1877). Quelch's (1886) collection is housed at the NHMUK and his report doubled previous species accounts; however Verrill (1900) synonymized certain species, including six *Isophyllia* species, reducing this number considerably.

The following year Rathburn (1887) recorded *Porites astraeoides* Lamarck, 1816 (sic, *Porites astreoides*) as occurring on Bermuda's reefs. In 1888, Heilprin noted the occurrence of *Diploria stokesii* Milne Edwards and Haime, 1849 (=*D. labyrinthiformis*), *Isophyllia guadeloupensis* Pourtalès (=?*I. sinuosa*; see Verrill 1900:223) and *Oculina recta* Quelch (of dubious validity and uncertain synonymy). Of the 19 species now recognized in Bermuda, eight were considered spurious or mere varieties (Verrill 1900) reinforcing previous accounts but not expanding what was known. In 1900, Verrill added *Orbicella annularis* Dana (=*Montastraea annularis* [Ellis and Solander, 1786]), *Orbicella cavernosa* (=*Montastraea cavernosa* [Linnaeus, 1767]), *Plesiastraea goodei* Verrill, 1900 (=*Stephanocoenia intersepta* [Esper, 1795]) and *Siderastrea siderea* (Ellis and Solander, 1786); at this point, 17 valid coral species had been recorded for Bermuda. Verrill (1901a, 1907) reported *Mussa (Symphyllia) annectens* sp. nov. and *Mussa rosula* sp. nov. (both=*I. sinuosa*) from Bermuda. The latter had been misidentified in 1864 as *I. rigida* (see Verrill 1907:230).

Over 50 years passed before the full complement of Bermuda's known diversity was documented again. In 1966, Laborel conducted the most recent comprehensive field survey of species, reporting both *Meandrina meandrites* (Linnaeus, 1758) and *Dichocoenia stokesi* Milne Edwards and Haime, 1848 for the first time. He noted that he could not find *Siderastrea siderea* and the occurrence of *Isophylliastrea rigida* (sic, *Isophyllastrea rigida*) (=*Isophyllia rigida* [Dana, 1848]) was questionable. In 1978, Dryer and Logan, list *Madracis mirabilis* (=*Madracis auretenra* Locke, Weil and Coates, 2007) as occurring in Bermuda. Although the solitary coral species *S. cubensis* and *S. lacera* were mentioned by Quelch (1886) (and mistakenly thought by Verrill [1907] to be young *Mussa* [=*Isophyllia*]), *S. cubensis* was not noted again in the literature until 1985 by Fricke and Meischner. *Scolymia lacera* has not been documented since Quelch (1886), and his report of the species may be the result of misinterpretation of prior literature; Brüggemann's specimen is the key to solving this question.

Current Studies of Shallow-Water Zooxanthellate Scleractinia

The most current previously published inventory of Bermuda's scleractinian species is Cairns et al. (1986). Changes to that list as presented here are due to the synonymy of *Stephanocoenia michelinii* with *Stephanocoenia intersepta*, revision of the genus *Montastraea* and revival of *Montastraea franksi* (Weil and Knowlton 1994), description of *Madracis auretenra* and re-evaluation of all shallow-water specimens identified as *Madracis mirabilis* (Locke et al. 2007), and a new record of *Phyllangia americana americana* (Logan 1988).

Based upon Cairns et al. (1986), the reference collection at BAMZ, literature reviews, and personal observations (J.M. Locke), a total of 20 species of valid zooxanthellate scleractinian corals are currently well documented for Bermuda's shallow waters (Locke 2009) (Table 14.1). Records have been verified from specimens deposited in Bermuda and at other museums.

Other, recent, novel records of extant shallow-water zooxanthellate coral species in Bermuda are dubious. Veron (2000) reported 25 species of zooxanthellate corals as occurring in Bermuda based on historical records; the following of which are unsubstantiated: *Diploria clivosa*, *Eusmilia fastigiata*, *Isophyllia rigida*, *Manicina areolata*, *Mycetophyllia lamarckiana*, and *Astrangia poculata*. A few new distribution records for Bermuda (Venn et al. 2009; Frade et al. 2010) are

Table 14.1 Current list of valid shallow-water zooxanthellate scleractinian species in Bermuda

Species
Family Agariciidae
Agaricia fragilis Dana, 1846
Family Astrocoeniidae
Stephanocoenia intersepta Lamarck, 1816
Family Faviidae
Diploria labyrinthiformis (Linnaeus, 1758)
Diploria strigosa (Dana, 1846)
Favia fragum (Esper, 1795)
Montastraea cavernosa Linnaeus, 1767
Montastraea franksi (Gregory, 1895)
Family Meandrinidae
Dichocoenia stokesi Milne Edwards and Haime, 1848
Meandrina meandrites (Linnaeus, 1758)
Family Mussidae
Isophyllia sinuosa (Ellis and Solander, 1786)
Scolymia cubensis (Milne Edwards and Haime, 1849)
Family Oculinidae
Oculina diffusa (Lamarck, 1816)
Oculina varicosa (Lesueur, 1821)
Oculina valenciennesi Milne Edwards and Haime, 1850
Family Pocilloporidae
Madracis auretenra Locke, Weil and Coates, 2007
Madracis decactis (Lyman, 1859)
Family Poritidae
Porites porites (Pallas, 1766)
Porites astreoides Lamarck, 1816
Family Siderastreidae
Siderastrea radians (Pallas, 1766)
Siderastrea siderea (Ellis and Solander, 1786)

also untenable due to a lack of archived and supporting materials (including specimens and taxonomically useful images, A. Venn, pers. comm., 2009); these are: *Madracis carmabi*, *Madracis formosa* (Bermuda extracts indistinguishable genetically from extracts from specimens from other countries that were identified, when collected, as *M. carmabi*, *M. formosa*, *M. senaria* and *M. pharensis*, in Frade et al. 2010), and *Madracis senaria* (no sequence obtained from Bermuda extracts, in Frade et al. 2010).

Table 14.2 Current list of valid azooxanthellate scleractinian coral species in Bermuda

Species
Family Caryophylliidae
Caryophyllia ambrosia ambrosia Alcock, 1898
Caryophyllia sarsiae Zibrowius, 1974
Coenocyathus goreaui Wells, 1972[a]
Deltocyathus calcar Pourtalès, 1874
Deltocyathus eccentricus Cairns, 1979
Deltocyathus italicus (Michelotti, 1838)
Deltocyathus moseleyi Cairns, 1979
Desmophyllum dianthus (Esper, 1794)
Lophelia pertusa Linnaeus, 1758
Oxysmilia rotundifolia (Milne Edwards and Haime, 1848)
Paracyathus pulchellus (Philippi, 1842)
Phyllangia americana americana Milne-Edwards and Haime 1849[a]
Soleosmilia variabilis Duncan, 1873
Tethocyathus cylindraceus (Pourtalès, 1868)
Family Dendrophylliidae
Enallopsammia rostrata (Pourtalès, 1878)
Rhizopsammia bermudensis Wells, 1972[a]
Family Flabellidae
Javania cailleti (Duchassaing and Michelotti, 1864)
Polymyces wellsi Cairns, 1991
Family Fungiacyathidae
Fungiacyathus symmetricus (Pourtalès, 1871)
Family Guyniidae
Guynia annulata Duncan, 1872[a]
Family Oculinidae
Madrepora carolina (Pourtalès, 1871)
Madrepora oculata Linnaeus, 1758
Family Pocilloporidae
Madracis asperula Milne-Edwards and Haime, 1849
Madracis myriaster (Milne Edwards and Haime, 1849)
Family Rhizangiidae
Astrangia solitaria (Lesueur, 1817)[a]
Colangia immersa Pourtalès, 1871[b]
Family Stenocyathidae
Stenocyathus vermiformis (Pourtalès, 1868)
Family Turbinoliidae
Deltocyathoides stimpsonii (Pourtalès, 1871)

[a]Known in Bermuda from shallow depths (0–3 m)
[b]Known from both shallow and deep depths

Shallow-Water Azooxanthellate Scleractinia

Six species of azooxanthellate corals are currently known from Bermuda's shallow waters. The earliest published record is of *Astrangia solitaria* (Lesueur, 1817) listed in Smith (1948) although it might have been recognized in 1877 by GB Goode (see Verrill 1901a:183). Otherwise the first notable records of azooxanthellate corals in Bermuda were made by Wells (1972) who reported *A. solitaria* as well as *Coenocyathus goreaui* Wells, 1972, *Guynia annulata* Duncan, 1872 (range 3–653 m) and *Rhizopsammia bermudensis* Wells, 1972 from reef cavities (0–6 m). Thus far, *Rhizopsammia bermudensis* is known only from Bermuda (Wells 1972; Cairns 2000) (Fig. 14.1). *Colangia immersa* Pourtalès, 1871 also inhabits reef cavities and although only collected from 6 to 8 m depths in Bermuda, this species has been reported from depths up to 347 m (Cairns et al. 1986; Cairns 2000). *Phyllangia americana americana* Milne-Edwards and Haime, 1849, first recorded from Bermuda by Logan (1988), commonly colonizes submerged man-made objects at depths up to 53 m (Cairns 2000) (Table 14.2).

Fig. 14.1 Bermuda's only endemic coral *Rhizopsammia bermudensis* (**a**) insitu within a reef cavity at Eastern Blue Cut, with polyps extended and (**b**) with polyps withdrawn. (**c**) Bermuda Natural History Museum specimen BAMZ 2009 266 013 collected from Eastern Blue Cut, 6 m. All scales equal 5 mm (Photo's **a** and **b**: I Murdoch, Photo **c**: L Greene)

The cryptic habitats occupied by these solitary and smaller species are poorly studied and further investigation may increase the number of known species. Records have been verified from BAMZ collections.

Deep-Water Scleractinia

Forays into the deep marine environments surrounding the Bermuda Pedestal and the seamounts within Bermuda's EEZ began with the deep-water dredging conducted during the HMS Challenger Expedition, and resumed over a century later with the submersible exploration of Fricke and Meischner (1985); the Royal Ontario Museum, Bermuda Underwater Institute-SDL Cruise (1997); the Ocean Projects Ltd-Innovator Cruise (2006); the Ocean Genome Legacy-Pacific Guardian Cruise (2007); and the Bermuda Institute for Ocean Science (BIOS)-Octopus Cruise (2007).

Collectively, these expeditions acquired azooxanthellate scleractinian specimens at depths from 55 m to 1,966 m. Many species are known only from single collections. The Bermuda specimens are housed at BAMZ, NHMUK, USNM, and Hessisches Landesmuseum (HLM), Darmstadt.

History of Species Documentations

The earliest records of deep-sea Scleractinia from Bermuda are those of Moseley (1881) from the HMS Challenger Expedition. He reported *Madracis asperula* Milne Edwards and Haime, 1849 (55 m), *Axohelia dumetosa* Pourtalès, 1874 (=*Madracis myriaster* [Milne Edwards and Haime, 1849]) (796 m), *Deltocyathus italicus* Milne-Edwards and Haime (1,966 m) (see Cairns 1979), *Deltocyathus italicus* var. *calcar* (=*Deltocyathus calcar* Pourtalès, 1874) (366 m), *Caryophyllia communis* Moseley, 1881 (=*Caryophyllia ambrosia ambrosia* Alcock, 1898) (1,262 m), *Caryophyllia cylindracea* Reuss? (=species indeterminate, S. Cairns, pers. comm., 2012) (796 m), and *Bathyactis symmetrica* Moseley, 1881 (=*Fungiacyathus symmetricus* [Pourtalès, 1871]) (1,966 m). The whereabouts of Moseley's single specimen of *D. italicus* var. *calcar,* "off Bermuda 200 fms", is unknown (S. Cairns and A. Cabrinovic [NHMUK], pers. obs.) and the lot of specimens included in Moseley's *D. italicus* has been determined by Cairns (1979) to consist of *D. italicus*, *Deltocyathus moseleyi* Cairns, 1979 and *Deltocyathus eccentricus* Cairns, 1979 (S. Cairns, pers. comm., 2012). To date, the deepest collection record for a living specimen for the Bermuda EEZ is 1,966 m from the report of Moseley (1881). Of the species listed by Moseley, only *M. myriaster* has been observed in and collected from Bermuda waters since the HMS Challenger Expedition.

A century later Fricke and Meischner (1985) explored Bermuda's mesophotic zone and provided the first photographs of corals living in Bermuda's deeper reef habitats. As well as revealing 13 zooxanthellate coral species living at depths between 30 m and 78 m their efforts documented new azooxanthellate records of *Madrepora carolina* (Pourtalès, 1871), *Oxysmilia rotundifolia* (Milne Edwards and Haime, 1848) and an unidentified species of *Polycyathus*. The latter two have not been observed again in Bermuda.

More recent studies of North Atlantic Scleractinia with bathymetric ranges extending below 200 m, report, cumulatively, 19 species occurring in Bermuda (Cairns 2000; Cairns and Chapman 2001; Dawson 2002). Of these, *Premocyathus cornuformis* (Pourtalés, 1868) and *Caryophyllia ambrosia caribbeana* Cairns 1979 were incorrectly listed for Bermuda (S. Cairns, pers. comm., 2012); in contrast, *Caryophyllia ambrosia ambrosia* Alcock, 1898 (=*Caryophyllia communis* in Moseley [1881]) was omitted from the list. The location data for *C. ambrosia ambrosia* and *C. ambrosia caribbeana* in Cairns (1979:58) do not support the occurrence of either subspecies in Bermuda. However, Moseley's (1881) record of *Caryophyllia communis*, considered by S. Cairns (pers. obs., pers. comm., 2012) to be a synonym of *C. a. ambrosia*, is evidence of the presence of this species. As of 2002 there were 17 known azooxanthellate coral species in Bermuda with bathymetric ranges exceeding 200 m. Although listed in Cairns and Chapman (2001), there are no confirmable records of collections of *Javania cailleti* (Duchassaing and Michelotti, 1864) prior to 2007 (BIOS-Octopus Cruise).

New Records of Species Diversity

Collections at BAMZ and USNM include species not documented in the literature as occurring in Bermuda. The following species were identified by Stephen Cairns (Smithsonian Institution) and represent new records for Bermuda: *Madrepora oculata* (221 m), *Tethocyathus cylindraceus* (Pourtalès, 1868) (354 m), *Polymyces wellsi* Cairns, 1991 (850 m), *Javania cailleti* (340 m), *Stenocyathus vermiformis* (Pourtalès, 1868) (700–900 m) and *Deltocyathoides stimpsonii* (Pourtalès, 1871) (850 m). These new records are the first documentation of the families Flabellidae, Stenocyathidae, and Turbinoliidae (Table 14.2) in Bermuda. The identifications of the following specimens have not been resolved: *Carophyllia* sp. (USNM 1114321, USNM 1129938) (354 m, 700–920 m), *Caryophyllia* n. sp. (BAMZ 2011.276.005) (850 m), and *Tethocyathus* n. sp. (USNM 1158159) (850 m) from Family Caryophylliidae. This family is represented in Bermuda by several other species.

These new records increase the number of known Bermuda deep-water azooxanthellate Scleractinia from 17 to 23 species and from 7 to 10 families (Table 14.2). The majority of these species occur in deep water (>200 m). Two of these species are known to inhabit depth ranges (~300–654 m) greater than 200 m in the western Atlantic, but thus far have only been found in Bermuda in shallow waters (<3 m). This, coupled with the 102 deep-water species of azooxanthellates currently known in the western Atlantic (Cairns 2000), indicates that a good understanding of Bermuda's deep-water Scleractinia diversity still eludes us.

Shallow-Water Octocorallia

History of Species Documentation

The documentation of shallow-water octocorals in Bermuda has a long and rich history, beginning in the nineteenth century. The earliest overview was produced by Verrill (1864), who included *Pterogorgia acerosa* Ehrenberg, 1834 (=*Antillogorgia* acerosa [Pallas, 1766]), *Rhipidogorgia flabellum* Valenciennes, 1855 (probably *Gorgonia ventalina* Linnaeus, 1758), *Plexaura homomalla* (Esper, 1792) and *Plexaura crassa* Lamouroux, 1816 (=*Pseudoplexaura porosa* [Houttuyn, 1772]). Shortly thereafter, Jones (1868) added *Plexaura flexuosa* Lamouroux, 1812 and *Pterogorgia americana* Ehrenberg, 1834 (=*Antillogorgia americana* [Gmelin, 1791]) to Verrill's list. Heilprin (1889) listed *Plexaurella dichotoma* (Esper, 1791), *Plexaura purpurea* (Pallas, 1766) (possibly a morphotype of *P. flexuosa*?) and *Eunicea pseudo-antipathes* (Lamarck, 1816), but the latter two taxa are of dubious validity and uncertain synonymy (see Verrill 1900:568, 1907:317).

The extensive effort of the Challenger Expedition in the nineteenth century surprisingly added nothing to the known shallow-water octocoral diversity, although it did yield new deep-sea records from the area (see section below). Wright and Studer (1889) mention only three shallow water species from Bermuda in their expedition report: *Plexaura valenciennesi* (=*P. flexuosa*), *P. crassa* (=*P. porosa*), and *G. flabellum* (=*G. ventalina*); all of which were already documented by Jones (1868).

The twentieth century brought the greatest contributions to our knowledge of Bermudian octocoral diversity. Addison E. Verrill was, in his later years, involved in significant expansions of octocoral records in Bermuda. He added *Gorgonia turgida* (Ehrenberg, 1834) (=?*Eunicea clavigera* Bayer, 1961) in Verrill (1869), *Gorgonia citrina* Esper, 1794 (=*Pterogorgia citrina* [Esper 1792]), *Eunicea grandis* Verrill, 1900 (=*E. calyculata* [Ellis and Solander, 1786]), *Muricea muricata* (Pallas, 1766) and *Eunicea rousseaui* Milne-Edwards and Haime, 1857 (=*E. tourneforti* forma *typica* Milne Edwards

Fig. 14.2 The encrusting form of *Briareum asbestinum* insitu at Paget Island. In Bermuda, the species is known only from certain islands located in the northeastern region. (Photo: LP Holland, JP Bilewitch)

and Haime, 1857) in Verrill (1900) and *Eunicea atra* Verrill, 1901 (=*E. tourneforti* forma *atra*) in Verrill (1901c) and finally *Plexaura esperi* Verrill, 1907 (=*Pseudoplexaura flagellosa*) in Verrill (1907). It is noteworthy that his collections (currently housed at YPM) also include the earliest known Bermudian specimen of the encrusting form of the scleraxonian *Briareum asbestinum* (Pallas, 1766), although it was not recognized in these collections and identified until the 1990s (E. Lazo-Wasem, pers. comm. 2008). Verrill also makes mention of *Plexaura flavida* (Lamarck, 1815) (=*Muriceopsis flavida* [Lamarck, 1815]) in Bermuda, although he seems doubtful about the record; as he states "I have seen a few specimens from Bermuda, but have not found it myself" (Verrill 1907:305). This genus has not otherwise been reported from this region and no collections that include specimens from Bermuda are currently known.

After Verrill (1907), there was no summary of Bermudian octocorals for more than 50 years, until Frederick M. Bayer's comprehensive study of the western Atlantic in the 1960s. Bayer (1961) examined existing museum and personal collections and listed 18 species as present in Bermuda. Except for *M. muricata*, his list included all previously recorded species as well as new records of *Pseudoplexaura wagenaari* (Stiasny, 1941), *Muricea atlantica* (Kükenthal, 1919), *Eunicea succinea* (Pallas, 1766), *Eunicea fusca* Duchassaing and Michelotti, 1860, his newly-described *Eunicea clavigera* (previously recorded in Bermuda as *Gorgonia turgida* by Verrill [1869], see above) and a dubious record of *Eunicea laciniata* Duchassaing and Michelotti, 1860.

Several species have been added to records from Bermuda since Bayer's study, but these represent occasional reports for single taxa rather than comprehensive reviews. For instance, the encrusting form of *Briareum asbestinum* (Pallas, 1766) (=*B. polyanthes* [Duchassaing and Michelotti, 1860], *sensu* Bilewitch et al. [2010]) was first noted as being present in Bermuda in a report on novel biochemical metabolites by Grode et al. (1983). Despite being an obvious shallow-water species (Fig. 14.2), its restricted distribution was likely responsible for it being previously overlooked. *Briareum asbestinum* has only been recorded in Bermuda from several small islands at the northeastern end of the archipelago: Paget Is., Governor's Is., the south side of Higg's Is., the northeastern end of Smith's Is. and on St. David's Is. at the entrance to Smith's Sound (Hammond 2001; J.P. Bilewitch, unpubl. data).

The penultimate modern list of octocorals in Bermuda is found in the seminal description of Bermudian marine fauna and flora by Sterrer (1986). Aside from the previous records, the section on the Octocorallia (Cairns et al. 1986) adds *Plexaurella nutans* (Duchassaing and Michelotti, 1860) and *Muricea laxa* Verrill, 1864 to the known species in the islands, while omitting *E. succinea* and the questionable record of *E. laciniata*. If we include *E. succinea* and all other verifiable accounts, this would form a list of 20 shallow-water species known for Bermuda up to the 1980s.

New Records of Species Diversity

The most recent report of shallow-water octocorallian diversity in Bermuda has been produced through a combination of studies on museum collections by J.P. Bilewitch and extensive field surveys by L.P. Holland– both of which were completed in 2008. An unpublished report on these efforts (Bilewitch 2008) includes the taxa listed by Cairns et al. (1986) and adds five new species– *Eunicea knighti* Bayer, 1961, *Muricea pinnata* Bayer, 1961, *Muricea* cf. *waltonsmithi* Bayer, 1994, *Plexaurella grisea* Kunze, 1916 and *Leptogorgia* cf. *setacea* (Pallas, 1766) (Table 14.3).

The current list for Bermuda contains octocoral taxa for which there are reliable field records and for which identifiable specimens are held in the collections of BAMZ, USNM or YPM (Cairns et al. 1986; Bilewitch 2008; L.P. Holland, unpubl. data). All included species are considered valid but many require further taxonomic investigation and revision. In particular, *Eunicea knighti*, here reported as a new record for Bermuda, possesses overlapping character variation with *Eunicea calyculata* that can make intermediate forms difficult to identify (Bayer 1961). Likewise, *Muricea pinnata* and *Muricea waltonsmithi* intergrade with one another, and with *Muricea laxa* (Bayer 1994), again allowing only tenuous identifications for some specimens.

The identity of *Leptogorgia* cf. *setacea*, Bermuda's only shallow-water azooxanthellate octocoral, is deserving of further taxonomic investigation (Fig. 14.3). The only colonies found in Bermuda, from the Great Sound, are usually unattached, possess polyps in biserial rows, and have characteristic disk spindle sclerites (Bayer 1961). However, the colonies appear to lack anthocodial sclerites and the lateral branching exceeds that described for *L. setacea* (J.P. Bilewitch, unpubl. data). Additional taxonomic analysis might show that the Bermudian variety is a distinct, new species. This would make it the only known endemic octocoral species from Bermuda.

The extent of the distribution of the rarer species of shallow-water Octocorallia in Bermuda remains largely unknown and the recent addition of so many new octocoral species records for the region (Bilewitch 2008) presents the possibility of even more undocumented taxa in shallow reef or inshore waters. The fortuitous discovery of the *L.* cf. *setacea* population (K.A. Coates and S.A. Manuel, pers. obs. 2008) during systematic surveys of the marine benthic habitats of Bermuda has revealed a second species of octocoral, other than *B. asbestinum*, with a greatly restricted distribution in Bermuda. Both *B. asbestinum* and *L.* cf. *setacea* are known only from isolated localities, but other small, undocumented populations may exist. Both species occupy non-reef habitats that are not frequently explored by either scientific or recreational divers, so that these species are unlikely to be encountered. *Briareum asbestinum* is found on the subtidal rocky shores of a few northeastern islands and *L.* cf. *setacea* is found in the silty substrate of turbid, low-visibility waters in the Great Sound. Clearly, the diversity of some types of habitats is significantly underexplored compared to typical reef habitats, and these are deserving of comprehensive and structured studies.

Table 14.3 Current list of valid shallow-water octocoral species in Bermuda

Species
Briareidae
Briareum asbestinum (Pallas, 1766)
Gorgoniidae
Gorgonia ventalina Linnaeus, 1758
Leptogorgia cf. *setacea* (Pallas, 1766)
Antillogorgia acerosa (Pallas, 1766)
Antillogorgia americana (Gmelin, 1791)
Pterogorgia citrina (Esper, 1792)
Plexauridae
Eunicea calyculata (Ellis and Solander, 1786)
Eunicea clavigera Bayer, 1961
Eunicea fusca Duchassaing and Michelotti, 1860
Eunicea knighti Bayer, 1961
Eunicea tourneforti Milne Edwards and Haime, 1857
Eunicea tourneforti forma *atra* Verrill, 1901
Muricea atlantica (Kükenthal, 1919)
Muricea laxa Verrill, 1864
Muricea muricata (Pallas, 1766)
Muricea pinnata Bayer, 1961
Muricea cf. *waltonsmithi* Bayer, 1994
Plexaura flexuosa Lamouroux, 1821
Plexaura homomalla (Esper, 1792)
Plexaurella dichotoma (Esper, 1791)
Plexaurella grisea Kunze, 1916
Plexaurella nutans (Duchassaing and Michelotti, 1860)
Pseudoplexaura flagellosa (Houttuyn, 1772)
Pseudoplexaura porosa (Houttuyn, 1772)
Pseudoplexaura wagenaari (Stiasny, 1941)

Deeper-Water Octocorallia

History of Species Documentations

The documentation of deep-water octocoral records in Bermuda begins with the HMS Challenger Expedition, which reports *Sympodium armatum* Wright and Studer, 1889 (validity unknown) and *Acanella simplex* (Verrill, 1883) (=*Lepidisis simplex* [Verrill, 1883]) midway between Bermuda and Challenger Bank (Wright and Studer 1889:32, 273).

Verrill (1901c:53, 1907:317) was first to report a mesophotic species from the region– *Verrucella grandis* Verrill, 1901c (=*Ellisella grandis* [Verrill, 1901c]), collected from about 30 m on the northern fore-reef terrace. Over 30 years later, this

Fig. 14.3 Bermuda's only shallow-water azooxanthellate octocoral, *Leptogorgia* cf. *setacea*, was discovered in Bermuda's Great Sound in 2008. Scale equals 5 cm (Photo: LP Holland)

species was noted again (as *Scirpearia grandis* [Verrill, 1901c]) at 66 m in the same vicinity (Deichmann 1936).

The shallow-water study of Bayer (1961: Fig. 101) reports the deep-water genus *Scleracis* Kükenthal, 1919 for Bermuda. However, *Scleracis* was omitted in Cairns et al. (1986), which instead lists one pennatulacean, *Sclerobelemnon* cf. *theseus* Bayer, 1959, one nidaliid alcyonacean, *Nidalia occidentalis* Gray, 1835 and mentions three species of ellisellid but specifically names only *Ellisella barbadensis* (Duchassaing and Michelotti, 1864). Most recently, Cairns and Bayer (2003, 2004) describe *Narella alvinae* Cairns and Bayer, 2003 and add records of *Narella versluysi* (Hickson, 1909) and *Candidella imbricata* (Johnson, 1862) from Bermuda.

Current Studies of Species Diversity

A report to the Bermuda Zoological Society by Bilewitch (2008) lists the known species of deep-sea octocorals present in Bermuda (amended herein to include 33 species in total; Table 14.4). Except for Wright and Studer's (1889) questionable record of *Lepidisis simplex* (Verrill, 1883), all listed species are represented in one of the collections of BAMZ, YPM, or USNM. This expands upon the list of Cairns et al. (1986), which included only five deep-sea octocorals among the Pennatulacea, Nidaliidae and Ellisellidae. The most recent records belong to the Primnoidae (seven spp.), the Isididae (four spp.) and a chrysogorgiid– all calcaxonians. One species of clavulariid telestacean, *Telestula septentrionalis* Madsen, 1944 is also added based on two specimens in the USNM, collected from Muir Seamount to the northeast of Bermuda. Among the Holaxonia, seven records are added for the Paramuriceidae, one to the Keroididae, one to the Gorgoniidae and four to the Plexauridae.

Of the genera in Tables 14.3 and 14.4, only *Antillogorgia*, *Plexaurella*, *Muricea* and *Eunicea* are represented in both shallow and deep waters and *Plexaurella nutans* and *Antillogorgia acerosa* are the only species broadly recorded from 1 m to 55 m depth. Interestingly, Bayer (1961:129, 191) describes *Muricea elongata* Lamouroux, 1821 as an inshore species in the Caribbean whereas in Bermuda it has not been recorded from less than 45 m. This discrepancy could be due to taxonomic confusion or to the effects of localized environmental differences.

The specimen of the rare isidid *Chelidonisis aurantiaca* Studer, 1890, currently curated at BAMZ, is a noteworthy record. The species is represented by very few specimens from localities as disparate as Ireland and the Gulf of Mexico (Deichmann 1936; Bayer and Stefani 1987). Although Verrill mentions an unconfirmed record from Florida (see Bayer and Stefani [1987:984–985] for discussion), the BAMZ record is likely the first and only specimen from the Atlantic. Whether the Bermudian specimen represents the Caribbean subspecies *C. aurantiaca mexicana* (like most shallow-water octocoral taxa), the European subspecies *C. aurantiaca aurantiaca* or an intermediate of the two remains to be determined.

Table 14.4 Current list of valid deep-water octocoral species in Bermuda

Species
Order Alcyonacea
Suborder Alcyoniina
Nidaliidae
Nidalia occidentalis Gray, 1835
Nephtheidae
Gersemia rubiformis (Ehrenberg, 1834)
Suborder Calcaxonia
Ellisellidae
Ellisella atlantica (Toeplitz, 1929)
Ellisella barbadensis (Duchassaing and Michelotti, 1864)
Ellisella grandis (Verrill, 1901)
Nicella gracilis Cairns, 2007
Chrysogorgiidae
Chrysogorgia fewkesii Verrill, 1883
Primnoidae
Callogorgia gracilis (Milne Edwards and Haime, 1857)
Callogorgia verticillata (Pallas, 1766)
Candidella imbricata (Johnson, 1862)
Narella alvinae Cairns and Bayer, 2003
Narella versluysi (Hickson, 1909)
Narella bellissima (Kükenthal, 1915)
Thouarella (Euthouarella) grasshoffi Cairns, 2006
Isididae
Keratoisis flexibilis (Pourtalés, 1868)
Chelidonisis aurantiaca Studer, 1890
Lepidisis caryophyllia Verrill, 1883
Lepidisis simplex (Verrill, 1883)
Suborder Holaxonia
Gorgoniidae
Antillogorgia acerosa (Pallas, 1766)
Keroeididae
Thelogorgia vossi Bayer, 1991
Paramuriceidae
Bebryce parastellata Deichmann, 1936
Muriceides kükenthali (Broch, 1912)
Scleracis guadalupensis (Duchassaing and Michelotti, 1860)
Scleracis petrosa Deichmann, 1936
Thesea citrina Deichmann, 1936
Placogorgia cf. *intermedia* (Thomson, 1927)
Placogorgia tenuis (Verrill, 1883)
Plexauridae
Eunicea pinta Bayer and Deichmann, 1958
Lytreia plana? (Deichmann, 1936)
Muricea elongata Lamouroux, 1821
Plexaurella nutans (Duchassaing and Michelotti, 1860)
Order Pennatulacea
Suborder Sessiliflorae
Kophobelemnidae
Sclerobelemnon theseus Bayer, 1959
Order Telestacea
Clavulariidae
Telestula septentrionalis Madsen, 1944

Table 14.5 Current list of valid Antipatharia species in Bermuda

Species
Family Myriopathidae
Cupressopathes gracilis (Thomson and Simpson, 1905)
Tanacetipathes hirta (Gray, 1857)
Tanacetipathes tanacetum (Pourtalès, 1880)
Tanacetipathes thamnea (Warner, 1981)
Family Antipathidae
Antipathes atlantica Gray, 1857
Antipathes furcata Gray, 1857
Stichopathes lutkeni Brook, 1889
Stichopathes cf. *spiessi* Opresko and Genin, 1990
Family Schizopathidae
Parantipathes tetrasticha (Pourtalès, 1868)

Antipatharia: Deep and Deeper Water

Six families of Antipatharia or black corals are reported in the greater Caribbean Region (Opresko and Sanchez 2005). For Bermuda, nine species are known (exclusive of three specifically undetermined specimens: *Tanacetipathes* sp. BAMZ 2007.258.034 and USNM 77485 and *Stichopathes* sp. USNM 99863) from five genera and three families, Antipathidae, Myriopathidae and Schizopathidae (Table 14.5). These records are based on Cairns et al. (1986) and collections held at the BAMZ and the USNM, and include two new records for Bermuda. The new records, based on specimens collected during the Pacific Guardian Cruise in 2007 (identified by DM Opresko), are of *Stichopathes* cf. *spiessi* Opresko and Genin, 1990 from 784.9 m, BAMZ 2007.258.011, and *Parantipathes tetrasticha* (Pourtalès, 1868) from 778.4 m, BAMZ 2007.258.013. Bermuda was not included within the distributional region of the recent guide to Caribbean Antipatharia (Opresko and Sanchez 2005) but based on the authors' criteria the majority of Bermuda's species are shallow-water black corals (depth range: 50–72 m) with a few exceptions.

Conclusion

Knowledge of two aspects of biological diversity underlie any plans to conserve and manage ecosystems: what species occur and where they occur. From these basics, we develop hypotheses and research programs on what controls distribution of individual species and species groups – such research spans the gamut of major fields of biology and relies on input from physical and chemical oceanography and from palaeogeography, climatology and many other fields of science.

We are fortunate that the diversity of Bermuda's corals has been extensively studied by natural historians and taxonomic specialists since the mid- nineteenth century and there is a well-laid foundation for future studies. Unfortunately, and not for Bermuda alone, only précised versions of taxonomic foundations have been presented and consulted in many recent research programs. This leads to unnecessarily inaccurate representations of both what is present and where species occur and this restricts any understanding of what controls their presence.

In foundation studies for Bermuda, shallow water environments have received the majority of attention and it is clear that some taxa have been more comprehensively studied than others. Further field investigations and studies of pre-existing, well-curated collections can add substantially to our basic knowledge, the latter is exemplified here by the new records of Scleractinia, Octocorallia and Antipathiaria.

Competence in taxonomy is a limiting factor in many field studies, but good record keeping and well-curated specimens can overcome this barrier, as in the case of the new record for Bermuda of *Leptogorgia* cf *setacea*. New distribution and species records cannot be based on unsubstantiated field observations, and at some point taxonomic specialists are needed to interpret the many forms of data used to determine species identity, from molecules to morphology. Bermuda's coral reef environment is monitored by the Bermuda Government (Departments of Conservation Services [DCS] and Environmental Protection [DEP]) both internally and in collaboration with the NGO's, the Bermuda Zoological Society with DCS (Bermuda Reef Ecosystem Analysis and Monitoring) and the Bermuda Institute of Ocean Sciences (BIOS) with DEP (Marine Environmental Program). The current efforts, given better developed taxonomic protocols, have the potential to rapidly move forward understanding of diversity and functioning of Bermuda's coral reef environments and associated habitats.

The number of experienced and trained taxonomists has diminished in the past few decades. Although the advent of automated DNA sequencing in the 1980s revived research interests in systematics through a molecular biological approach, this has done little to cultivate understanding of proper taxonomic process (e.g., the ICZN Code) and morphology-based familiarity with taxa. Rather, molecular systematics has often attempted to circumvent problematic issues, such as morphological plasticity, rather than provide a better understanding of them. Although the genetic delineation of species boundaries represents a powerful tool for taxonomic classification, it should be used in conjunction with detailed morphological diagnoses of taxa, rather than attempting to obviate an intimate understanding of morphology and the processes that drive remarkable phenotypic variation.

Confusing Records of Scleractinia for Bermuda

Isophyllia rigida (Dana, 1848): A Complex History of Misidentification

Confusion about the presence of *Isophyllia rigida* in Bermuda has a long history, beginning with Verrill's original misidentification of some specimens of *Isophyllia* (referred to, at that time, as *Mussa*) in 1864. Subsequent reports of *I. rigida* were based on communication with or the literature of AE Verrill prior to his published recognition of this error in 1907. He (Verrill 1907:230) notes that, after viewing Dana's type of *I. rigida*, held at Yale University, he decided the species named in his 1864 paper was not *I. rigida* but a distinct species, *Mussa rosula* Verrill, 1907. This last species has since been synonymized with *I. sinuosa*. Laborel (1966) did not find *I. rigida* in Bermuda, and there are no other substantial, more recent, records. Perhaps recent species lists for Bermuda, that include *I. rigida* (e.g., Veron 2000), have relied on out-dated information and not on deposited specimens. *Isophyllia sinuosa* is well-documented and represented in collections for Bermuda.

Siderastrea siderea: A Question to Pursue

Currently considered rare, *Siderastrea siderea* was initially observed in Bermuda by Verrill (1900); however, Laborel (1966) did not find it. Verrill's specimens of *S. siderea* from Castle Harbor, Bermuda, are deposited at the Yale Peabody Museum (YPM 006787, YPM 008228). There are also three specimens from Bermuda, including a fossil specimen, deposited at the Smithsonian National Museum of Natural History (USNM 64392 [Pleistocene], USNM 1142515, USNM 16363) and one specimen (Fricke and Meischner 1985) at the Hessisches

(continued)

(continued)

Landesmuseum (HLM), Darmstadt, FRG (Federal Republic Germany) (HLM, Fri Be-6). Verrill (1901b:152) says that "Bermudian specimens that have been referred to the species *Siderastrea siderea* ...[as far as he has observed are not the typical form, and] ... may be an extreme variety of *Siderastrea radians*. The calices are not so large nor so deep as in the Florida form, nor are the septa so numerous (about 42–48)." Logan (1988) stated that virtually all Bermudian *Siderastrea* belonged to *radians*; however earlier, Dryer and Logan (1978) had reported *S. siderea* from northwest reefs in Castle Harbour. The presence of *S. siderea* is debatable (Laborel 1966; Logan 1988; W. Sterrer and S.R. Smith, pers. comm., 2008) and an in depth study, including fieldwork and observation of new material and deposited specimens is warranted. The large colonies of *S. siderea* common to Florida, the Bahamas and Caribbean are not known to occur in Bermuda.

Montastraea annularis and *Montastraea franksi*

Records of *M. annularis* in Bermuda predate or are based in literature that predates a recent revision of several species in the genus (Weil and Knowlton 1994), and almost certainly are incorrect. Early accounts of *M. annularis* are found in Verrill (1900, 1901a, b, 1907), Smith (1948), Squires (1958), and Cairns et al. (1986), among others. Weil and Knowlton (1994) redescribed *M. annularis* sensu stricto and resurrected and amended the descriptions of two older taxa, *Montastraea faveolata* (Ellis and Solander, 1786), and *M. franksi* (Gregory, 1895). Based on these new, and accepted, descriptions, it is recognized that *M. franksi* is common on Bermuda reefs, along with *M. cavernosa*, and also that it has probably been regularly misidentified as *M. annularis*. Nonetheless, to confirm that there is no objective basis for records of *M. annularis* in Bermuda, it is important to re-examine historical collections from Bermuda.

Acknowledgements Stephen Cairns and Dennis Opresko, National Museum of Natural History, Smithsonian Institution, assisted with the identification of specimens for new species records for Bermuda; Andrew Cabrinovic, Natural History Museum, United Kingdom and Lisa Greene, Bermuda Natural History Museum assisted with collection information; Alison Green, Bermuda Natural History Library and Michel Paon, Dalhousie University Libraries provided access to taxonomic resources; and Wolfgang Sterrer has worked for many years on documenting Bermuda's marine biodiversity. This is a contribution of the Bermuda Biodiversity Project BBP#201.

References

Appeltans W, Bouchet P, Boxshall GA, De Broyer C, de Voogd NJ, Gordon DP, Hoeksema BW, Horton T, Kennedy M, Mees J, Poore GCB, Read G, Stöhr S, Walter TC, Costello MJ. (eds) (2012) World register of marine species (WoRMS). http://www.marinespecies.org. Accessed 21 July 2012

Bayer FM (1961) The shallow-water Octocorallia of the West Indian Region. Martinus Nijhoff, The Hague

Bayer FM (1994) A new species of the gorgonacean genus *Muricea* (Coelenterata: Octocorallia) from the Caribbean Sea. Precious Corals Octocoral Res 3:23–27

Bayer FM, Stefani J (1987) New and previously known taxa of isidid octocorals (Coelenterata: Gorgonacea), partly from Antarctic waters. Proc Biol Soc Wash 100:937–991

Bilewitch JP (2008) Octocoral biodiversity in Bermuda. Report for the Bermuda Zoological Society, Bermuda, BAMZ #1915

Bilewitch JP, Coates KA, Currie DC, Trapido-Rosenthal HG (2010) Molecular and morphological variation supports monotypy of the octocoral *Briareum* Blainville, 1830 (Octocorallia: Alcyonacea) in the western Atlantic. Proc Biol Soc Wash 123:93–112

Brüggemann F (1877) Notes on stony corals in the collection of the British museum. III. A revision of the recent solitary Mussaceae. Ann Mag Nat Hist 20:300–313

Cairns SD (1979) The deep-water Scleractinia of the Caribbean and adjacent waters. Stud Fauna Curaçao 57:1–341

Cairns SD (2000) A revision of the shallow-water azooxanthellate scleractinia of the western Atlantic. Nat Hist Carib Reg 75:1–240

Cairns SD, Bayer FM (2003) Studies on western Atlantic Octocorallia (Coelenterata: Anthozoa). Part 3: the genus *Narella* Gray, 1870. Proc Biol Soc Wash 116:617–648

Cairns SD, Bayer FM (2004) Studies on western Atlantic Octocorallia (Coelenterata: Anthozoa). Part 5: the genera *Plumarella* Gray, 1870; *Acanthoprimnoa* n. gen.; and *Candidella* Bayer, 1954. Proc Biol Soc Wash 117:447–487

Cairns SD, Chapman RE (2001) Biogeographic affinities of the North Atlantic deep-water Scleractinia. In: Proceedings of the first international symposium on deep-sea corals, Ecology Action Centre and Nova Scotia Museum, Halifax, pp 30–57

Cairns SD, den Hartog JC, Arneson C, Rützler K (1986) Class Anthozoa (Corals, Anemones). In: Sterrer W (ed) Marine fauna and flora of Bermuda. A systematic guide to the identification of marine organisms. Wiley, New York, pp 159–194

Dana JD (1872) Corals and coral islands. Dodd and Mead, New York

Dawson JP (2002) Biogeography of azooxanthellate corals in the Caribbean and surrounding areas. Coral Reefs 21:27–40

de Pourtalès LF (1871) Deep sea corals, vol IV, Illust Cat Museum of Comparative Zoology. Printed for the Museum of Comparative Zoology, Cambridge, MA

Deichmann E (1936) The alcyonaria of the western part of the Atlantic Ocean. Mem Mus Comp Zool 53:1–317

Dryer S, Logan A (1978) Holocene reefs and sediments of Castle Harbour, Bermuda. J Mar Res 36:399–425

Frade PR, Reyes-Nivia MC, Faria J, Kaandorp JA, Luttikhiuzen PC, Bak RPM (2010) Semi-permeable species boundaries in the coral genus *Madracis*: introgression in a brooding coral system. Mol Phyl Evol 57:1072–1090

Fricke H, Meischner D (1985) Depth limits of Bermudan scleractinian corals: a submersible survey. Mar Biol 88:175–187

Grode SH, James TR, Cardellina JH II (1983) Brianthein Z, a new polyfunctional diterpene from the gorgonian *Briareum polyanthes*. Tetrahedron Lett 24:691–694

Hammond MP (2001) Range, distribution, and dispersion of a recent colonist, the gorgonian *Briareum polyanthes*, in St. Georges Harbour, Bermuda. Report for the Bermuda Aquarium Museum and Zoo, BAMZ #1102

Heilprin A (1888) Contributions to the natural history of the Bermuda Islands. Proc Acad Nat Sci Phil 40:302–328

Heilprin A (1889) The Bermuda Islands: a contribution to the physical history and zoology of the somers archipelago. Heilprin A, Philadelphia

Jones JM (1868) Contributions to the natural history of the Bermudas. Coralliaria. Trans Nova Scotia Inst Nat Sci 2(2):7–16

Laborel J (1966) Contribution à l'étude des Madreporaires des Bermuda (systématique et répartition). Bull Mus Nat Hist Nat 38:281–300

Locke JM (2009) *Madracis auretenra* (Scleractinia: Pocilloporidae) – testing our knowledge of systematics, biology and connectivity in the western North Atlantic. Ph.D. dissertation, University of Puerto Rico, Mayagüez

Locke JM, Weil E, Coates KA (2007) A newly documented species of *Madracis* (Scleractinia: Pocilloporidae) from the Caribbean. Proc Biol Soc Wash 120:214–226

Logan A (1988) Holocene reefs of Bermuda. Sedimenta IX. Comparative Sedimentology Laboratory, University of Miami, Miami Beach

Moseley HN (1881) Report on certain hydroid, alcyonarian, and madreporian corals procured during the voyage of H.M.S. Challenger. In: Thomson CW, Murray J (eds) Report on the scientific results of the voyage of the H.M.S. Challenger, vol 2. Royal Stationery Office, London

Opresko DM, Genin A (1990) A new species of antipatharian (Cnidaria: Anthozoa) from seamounts in the eastern North Pacific. Bull Mar Sci 46:301–310

Opresko DM, Sanchez JA (2005) Caribbean shallow-water black corals (Cnidaria: Anthozoa: Antipatharia). Carib J Sci 41:492–507

Quelch JJ (1886) Report on the reef corals collected by H.M.S. Challenger during the years 1873–76. In: Thomson CW, Murray J (eds) Report on the scientific results of the Voyage of the H.M.S. Challenger, vol 16. Royal Stationery Office, London

Rathburn R (1887) Annotated catalogue of the species of *Porites* and *Synaraea* in the United States National Museum, with a description of a new species of *Porites*. Proc US Nat Mus 10:354–366

Smith PGW (1948) Atlantic reef corals. A handbook of the common reef and shallow-water corals of Bermuda, Florida, the West Indies and Brazil. University of Miami Press, New York

Squires DF (1958) Stony corals from the vicinity of Bimini, Bahamas, British West Indies. Bull Am Mus Nat Hist 115:215–262

Sterrer WE (ed) (1986) Marine fauna and flora of Bermuda: a systematic guide to the identification of marine organisms. Wiley, New York

Sterrer WE (1998) How many species are there in Bermuda? Bull Mar Sci 62:809–840

Venn AA, Weber FK, Loram JE, Jones RJ (2009) Deep zooxanthellate corals at the high latitude Bermuda seamount. Coral Reefs 28:135

Veron JEN (2000) Corals of the world. Australian Institute of Marine Science, Townsville

Verrill AE (1864) List of the polyps and corals sent by the museum of comparative zoology to other institutions in exchange, with annotations. Bull Mus Comp Zool Harvard 1:29–60

Verrill AE (1869) Critical remarks on the halcyonoid polyps with descriptions of new species in the museum of Yale college. No. 4. Am J Sci Arts 48:419–429

Verrill AE (1900) Additions to the Anthozoa and Hydrozoa of the Bermudas. Trans Conn Acad Arts Sci 10:551–572

Verrill AE (1901a) Comparisons of the Bermuda, West Indian, and Brazilian coral faunae. Trans Conn Acad Arts Sci 11:169–206

Verrill AE (1901b) Variations on the nomenclature of Bermudian, West Indian, and Brazilian Reefs corals, with notes on various Indo-Pacific corals. Trans Conn Acad Arts Sci 11:63–168

Verrill AE (1901c) Additions to the fauna of the Bermudas from the Yale expedition of 1901, with notes on other species. Trans Conn Acad Arts Sci 11:15–62

Verrill AE (1907) The Bermuda Islands. Part IV. Geology and paleontology and Part V. An account of the coral reefs. Trans Conn Acad Arts Sci 12:45–348

Weil E, Knowlton N (1994) A multi-character analysis of the Caribbean coral *Montastraea annularis* (Ellis andSolander, 1786) and its two sibling species, *M. faveolata* (Ellis and Solander, 1786), and *M. franksi* (Gregory, 1895). Bull Mar Sci 54:151–175

Wells JW (1972) Some shallow water ahermatypic corals from Bermuda. Postilla 156:1–10

Wright EPW, Studer T (1889) Report on the Alcyonaria collected by H.M.S. Challenger during the years 1873–1876. In: Thomson CW, Murray J (eds) Report on the scientific results of the voyage of the H.M.S. Challenger, vol 31. Royal Stationery Office, London

Total Economic Value of Bermuda's Coral Reefs: A Summary

15

Samia Sarkis, Pieter J.H. van Beukering, Emily McKenzie, Luke Brander, Sebastiaan Hess, Tadzio Bervoets, Lois Looijenstijn-van der Putten, and Mark Roelfsema

Introduction

The economic valuation of ecosystem goods and services is increasingly used worldwide because it offers a useful means of including the concept of ecosystem value into policy discussions and decision-making (Van Beukering and Slootweg 2010). Currently, Bermuda supports one of the "healthiest" coral reef systems of the Wider Caribbean Region, with a coral cover ranging from 22% to 70%. Owing its presence to the proximity of the Gulf Stream, Bermuda's reef system is of global importance as the northernmost reef system in the Atlantic Ocean. The colder temperatures associated with its northerly latitude have mitigated, to a certain extent, climate change impacts such as "bleaching" events, especially that of 2005 which had a devastating effect on Caribbean reefs in general. Nonetheless, Bermuda's reefs are ranked worldwide at "high risk" by the World Resources Institute (Burke et al. 2008) attributed in great part to the island's high human population density within 20 km of the reefs. More specifically, immediate threats to Bermuda's coral system relate to increasing maritime traffic associated with the import of goods and to the changing tourism industry. Developments necessary to accommodate larger ships have potential direct and indirect impact on the reef system, leading to the loss of ecosystem goods and services provided by the reefs to Bermuda's community. An economic value can be attributed to both commercial and non-commercial goods and services, though non-commercial values are difficult to measure quantitatively and have the greatest uncertainty attached to them.

Bermuda's reef value involves six key ecosystem goods and services which can be quantified and compiled to provide a Total Economic Value (TEV) of Bermuda's reef (see Sarkis et al. 2010 for further detail). The greatest part of the TEV obtained for Bermuda's reefs refers to the more tangible direct and indirect use values. This study focused on these values in order to provide a TEV which is understood and widely accepted by the community and the government, and ultimately used in future policy and decision-making. The six key ecosystem goods and services used here are: (1) Coral reef-associated tourism, (2) Reef-associated fisheries, (3) Amenity or reef-associated surplus value on real estate, (4) Physical coastal protection, (5) Reef-associated recreational and cultural values, and (6) Research and education value. The approach used to quantify values for each are summarised in Fig. 15.1 including both data collection and valuation techniques.

The study area, estimated to be 400 km^2, is illustrated in Fig. 15.2, and does not include reefs on the outer edge of the North Lagoon.

Tourism Value

Tourism is a key industry in Bermuda. In 2007, over 660,000 people visited Bermuda (Department of Tourism 2007), and the importance of the coral reef ecosystem to the visitor's experience was investigated. Data was collected through available documentation and two surveys were developed for the study: (1) a reef-associated tourist operator survey providing revenue data, and (2) a tourist exit survey assessing the importance of coral reefs to the visitation experience. The economic analysis for estimating the reef-associated tourism value involves three methods:

(1) the travel cost method, resulting in the "consumer surplus", or the value of the coral reefs to tourist recreation from the visitor's perspective, (2) the net factor income method,

S. Sarkis
Department of Conservation Services, Government of Bermuda, 17 North Shore Road, Smith's Parish FL 03, Bermuda
e-mail: scsarkis@gov.bm

P.J.H. van Beukering • L. Brander • S. Hess • T. Bervoets
L. Looijenstijn-van der Putten • M. Roelfsema
Institute for Environmental Studies (IVM), University of Amsterdam, Amsterdam, The Netherlands
e-mail: pieter.van.beukering@vu.nl; lukebrander@gmail.com; sebastiaan.hess@ivm.vu.nl; manager@naturefoundationsxm.org

E. McKenzie
Joint Nature Conservation Committee, Peterborough, UK
e-mail: Emily.McKenzie@wwfus.org

C.R.C. Sheppard (ed.), *Coral Reefs of the United Kingdom Overseas Territories*, Coral Reefs of the World 4,
DOI 10.1007/978-94-007-5965-7_15, © Springer Science+Business Media Dordrecht 2013

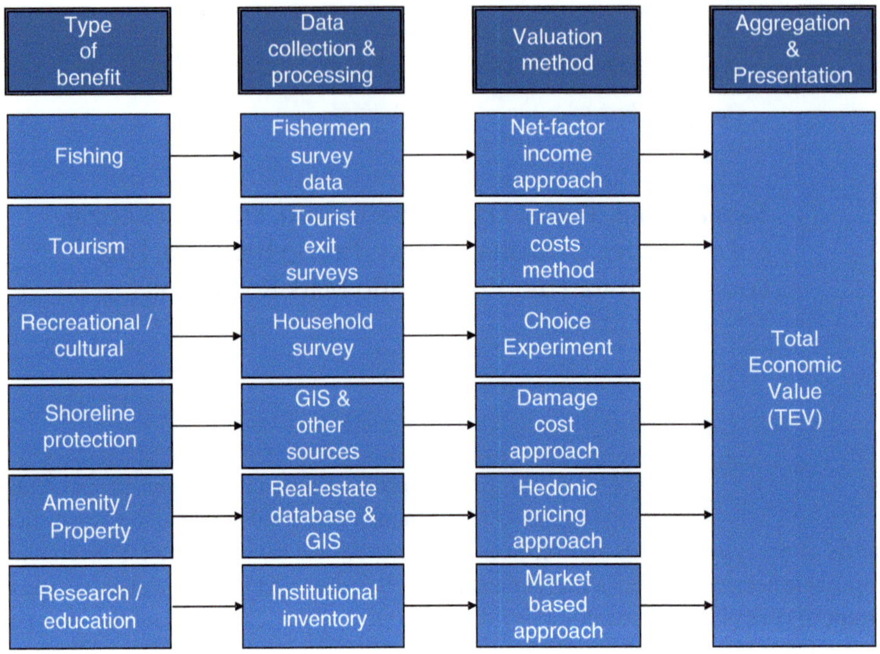

Fig. 15.1 Service and valuation techniques used in determining TEV for Bermuda's coral reefs

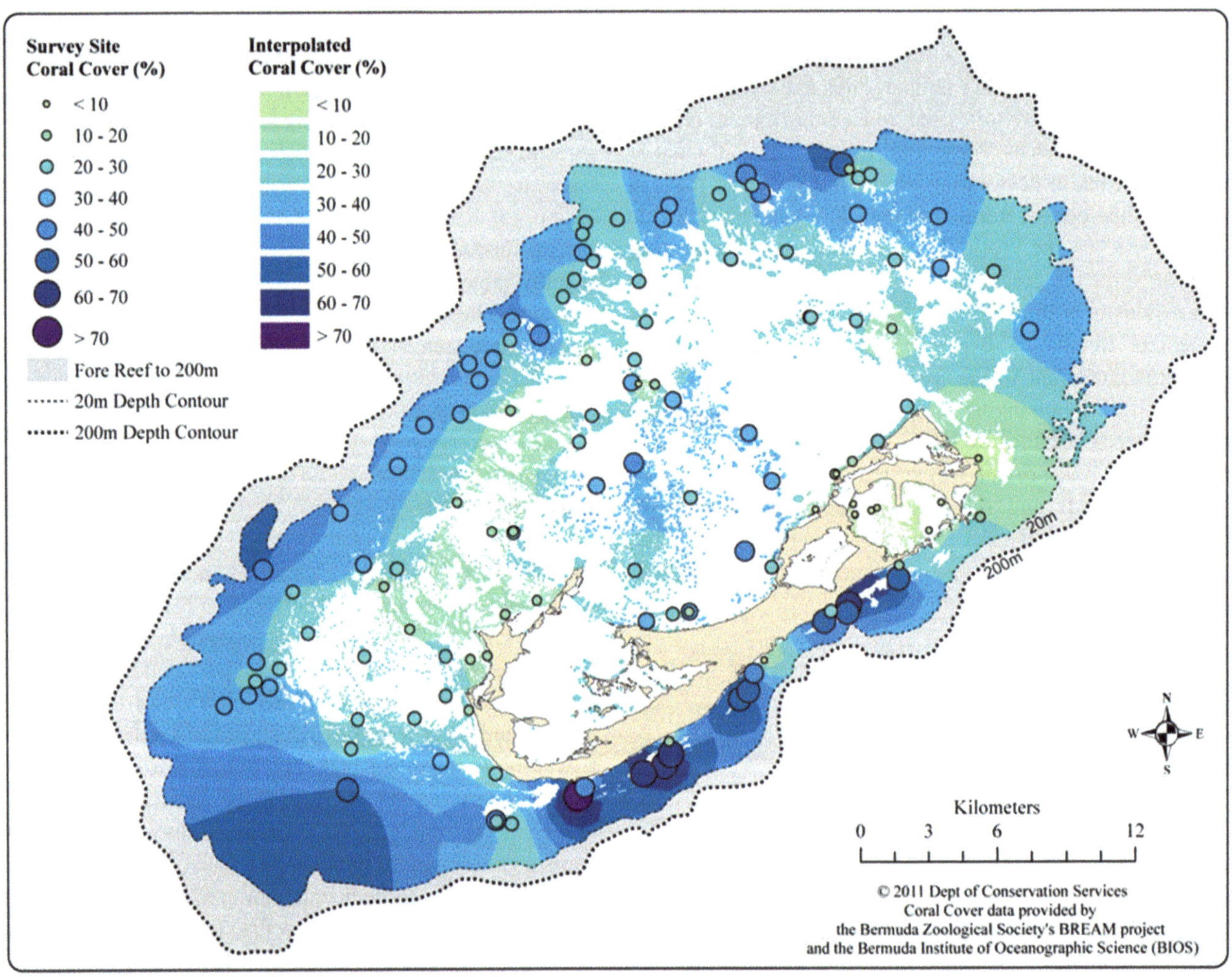

Fig. 15.2 Bermuda's known coral reef cover including hard and soft coral species (Taken from Sarkis et al. 2010)

Table 15.1 Motivation to visit Bermuda for both air and cruise tourists, expressed as a percentage of total interviewed. Significant difference between the two types of tourists is shown

	Air	*Cruise*	*Significant different*	Total
Business	4.5%	4.3%	No	4.4%
Snorkeling	7.3%	9.6%	Yes	8.6%
Diving	5.9%	5.5%	No	5.7%
Touring the reef	7.4%	9.1%	Yes	8.3%
Visiting the beach	16.1%	15.4%	No	15.8%
Shopping	11.1%	11.3%	No	11.2%
Sailing	6.2%	6.6%	No	6.4%
Sightseeing	13.7%	15.3%	Yes	14.5%
Playing golf	5.5%	5.4%	No	5.4%
Fishing	4.6%	5.2%	Yes	4.9%
Eating and drinking	16.0%	11.5%	Yes	13.7%
Other	1.5%	0.7%	Yes	1.1%
Total	100%	100%		100%
Total reef related	36.8%	39.7%		38.3%

resulting in the "producer surplus", or the value of coral reefs in the production of a marketed good generated through paid activities such as SCUBA diving and snorkeling, and (3) the contingent valuation method, providing a 'Willingness to Pay' value for coral reef conservation, an additional measure of "consumer surplus".

Current Bermuda Tourism Market

A total of 407 tourists were interviewed for this study, distributed evenly between cruise and air tourists. The cruise ship sector is substantial; in 2007, 53% of the total 663,767 visitors to the island, arrived by cruise ship. Survey results indicate that 38.3% of the tourists interviewed are motivated to visit Bermuda for a coral reef-associated reason, with snorkeling and touring the reef being the most popular activities (Table 15.1). Bermuda's "pristine" reefs are evidently well appreciated and 14% of interviewees confirmed they would not come to Bermuda should the coral reefs lose this quality. **This translates into a loss of 90,000 tourists per year if coral reef health declines**.

Operator Survey

Thirteen of the existing 40 reef-associated tour operators in Bermuda were interviewed, including all dive operators, a sample of glass bottom and rental boat operators, as well as charter boat operators. The reef-associated tourism gross revenue for Bermuda is estimated at $7.4 million (in 2007), with a profit margin estimated at 28%. Results concur with the tourist survey in that reef quality is important to the business, and that most visitors to Bermuda are sensitive to the health of the environment.

Consumer Surplus- Visitor's Perception of Coral Reef Value

Estimate of the reef-associated tourism value is based on the percentage of reef-associated recreation, stemming directly (such as diving) and indirectly (such as beach enjoyment) from the coral reef ecosystem. The consumer component of the coral-reef related tourism value of Bermuda's coral reefs in 2007 was calculated to be US$343 million, approximating US$190 million for the cruise ship sector and US$154 million for the air tourism sector.

Producer Surplus- Value of Marketed Goods Provided by Coral Reefs

Producer surplus is calculated from the expenditures of tourists visiting coral reefs, minus the cost of production. The producer surplus of reef-associated operators is based on the reported tourist expenditures for both air and cruise ship tourists. This is estimated at US$139 per tourist for air visitors and at US$55 per tourist for cruise ship passengers. Based on 663,767 visitors in 2007, the reef-related producer surplus value totals US$62.5 million in that year, the greatest part being attributed to the air tourism sector amounting to $42.9 million per year, more than twice the value of $19.5million per year contributed by the cruise ship tourism sector.

'Willingness to Pay' by Tourists

Results indicate that 68% of all tourists visiting Bermuda are – in principle – willing to pay, in addition to their current expenses, to fund activities to preserve Bermuda's coral reefs. The average cruise ship tourist is willing to pay US$28 per visit to Bermuda and the average airplane tourist is willing to pay markedly less, US$19. Extrapolating this information to a yearly basis, using the number of visitors recorded in 2007, a total sum of US$15.6 million – US$5.9 million from the air tourism sector, and US$9.7 million from the cruise ship sector- is potentially available for the conservation of Bermuda's coral reefs.

Total Tourism Value

The total tourism value of coral reefs in Bermuda is the sum of the **consumer surplus** (based on travel costs), and the **producer surplus** (based on tourist reef-associated

Table 15.2 Total tourism value and potential extra value (WTP) of Bermuda's coral reefs per year

Variable	Air	Cruise	Total
Consumer surplus	\$153.7 million	\$189.8 million	\$343.3 million
Producer surplus	\$42.9 million	\$19.5 million	\$62.4 million
Total tourism value	\$196.6 million	\$209.4 million	\$405.8 million
WTP for conservation	\$5.9 million	\$9.7 million	\$15.7 million

expenditures). All tourists add substantial reef related value to the island's economy; the total tourism value of Bermuda's coral reefs per visitor is US\$611 per tourist. This value is slightly lower for cruise ship tourists (US\$591 per visitor) than for air tourists (US\$635 per visitor). This difference is due in great part to higher on-island expenditures incurred by air visitors, resulting in a twofold producer surplus spent by this sector; air tourists add US\$43 million per year to the island's reef-related economy, compared to less than half that added by cruise ship tourists (US\$20 million per year) (Table 15.2). However, because of the higher volume of cruise ship visitors, the resulting reef-associated tourism value for cruise ship tourism (US\$197 million) is similar to that of air tourism (US\$209 million). In addition, appreciation of the reefs by cruise ship tourists is similar to that of airplane tourists, reflected in their '**Willingness to Pay**' for conservation (US\$9.7million for cruise ship and US\$5.9 million for air visitors). The **reef-associated tourism value** for Bermuda's coral reefs is **US\$406 million** per year (calculated in 2007). The '**Willingness to Pay**' extra for ensuring the preservation of reefs per year by all tourists is **US\$16 million**.

Fishery Value

Coral reefs are a crucial habitat for fish stock and in this way also provide important ecosystem services through the commercial and recreational fishery sector. The value of coral reef-associated fisheries encompasses both direct and indirect values. The direct value of coral reef-associated fisheries refers to the market value of the fish catch and the indirect value refers to the cultural and recreational importance of fishing in Bermuda. The direct value was obtained through existing data from the Marine Resources Section (GOB 2000) and from face-to-face interviews with fishermen. The indirect value was obtained through a survey designed and administered for the purpose of this study, where 400 households were interviewed face to face. The valuation focuses on fisheries that depend directly on coral reefs for at least one portion of their life cycle, and is referred to as "reef-associated". The fisheries value includes (1) commercial fisheries and (2) recreational fisheries. To avoid overestimation of the fishery value, a distinction is made between reef-associated and non reef-associated catch.

Commercial Fisheries

Based on existing records, 42% of the total commercial catch is considered reef-associated. Reef-associated catch for finfish ranges from 257,000 to 375,000 lbs., resulting in a value ranging from US\$1.6 million to US\$2.5 million; note that these data exclude the catches recorded for pelagic and sharks, which would increase fin fisheries value to a maximum of US\$5.1 million. The gross value of the whole reef-associated commercial fisheries including finfish and lobster species ranges from US\$2.5 million to US\$3.2 million, with a mean of US\$2.9 ± 0.3 million. Net values for the reef-catch were calculated by deducting estimated fishing costs. Total fishing costs were estimated at 40–80% of the gross value of the total catch; due to the small sample size ($n = 6$ fishermen), this should be considered as a very preliminary estimate. The **final net values of the commercial reef-associated catch**, for both finfish and lobster is calculated to be **US\$0.6 ± 0.06 million to US\$1.8 ± 0.2 million** based on a 20–60% profit margin.

Recreational Fisheries

To date, a recreational fishing license is required only for lobster diving in Bermuda; for this reason, there is no statistical record of recreational finfish catch. This study provides scope for a first assessment of recreational fisheries in Bermuda, based on responses of local residents in 400 households. Details of socio-demographics characteristics, such as ethnicity, income and household size, were compared with the 2004 Expenditure survey, conducted by the Department of Statistics (Government of Bermuda); results are similar, and details of these characteristics are given in the Recreational and Culture value section of this document.

Of the 400 households interviewed, 30% indicate that one or more household member fishes recreationally. This

Table 15.3 Recreational catch classification and share of catch

Fish type	Average share of total catch (%)	Share of fishermen not involved at all (%)	Share of fishermen with 90% or more (%)
Shallow reef fish (< 26 m)	72	12	55
Deep reef fish (> 26 m)	13	65	3
Deep Sea fish (e.g. tuna)	10	79	3
Lobster	1	95	0
Bait fish	4	90	3
Mussels	0	99	0

Table 15.4 Overview of commercial and recreational catch in Bermuda

	2000	2001	2002	2003	2004	2005	2006	2007
Commercial total finfish catch (× 1,000 lbs.)	856	640	832	727	804	834	784	886
Recreational total finfish catch (× 1,000 lbs.)	524	530	537	544	551	558	565	572
Total finfish catch (× 1,000 lbs.)	**1,380**	**1,170**	**1,369**	**1,271**	**1,355**	**1,392**	**1,349**	**1,458**
Commercial reef-ass. finfish catch (× 1,000 lbs.)[a]	362	327	329	257	282	357	375	349
Recreational reef-ass. finfish catch (× 1,000 lbs.)[a]	354	359	363	368	373	377	382	387
Total reef-ass. finfish catch (× 1,000 lbs.)[a]	**716**	**686**	**692**	**625**	**655**	**734**	**757**	**736**
Commercial lobster catch (× 1,000 individuals caught)	19.1	17.4	23.4	28.1	20.4	23.8	26.7	26.1
Recreational lobster catch (× 1,000 individuals caught)	2.7	2.8	2.8	2.8	2.9	2.9	2.9	3.0
Total lobster catch (× 1,000 individuals)	**21.9**	**20.1**	**26.2**	**30.9**	**23.3**	**26.7**	**29.7**	**29.0**
Reef-ass. commercial value low (× 1,000 USD)[a]	551	552	586	513	498	646	687	649
Reef-ass. commercial value high (×1,000 USD)[a]	1,654	1,657	1,759	1,540	1,494	1,938	2,061	1,947
Reef-ass. recreational value (× 1,000 USD)[a]	3,312	3,352	3,426	3,471	3,515	3,571	3,609	3,652

[a]Refers to reef-associated catch or value. Commercial value is net value

translates into a total of almost 16,000 recreational fishermen on Bermuda in 2007. Based on interview results, the motivation for fishing is foremost the strengthening of bonds with friends and family, and enjoyment, rather than fishing for food. Details on fishing periods, fishing methods and preferred sites were obtained (see Sarkis et al. 2010 for further details). On average, 72% of the catch is made up of shallow reef fish (Table 15.3). Deep "reef" fish (>26 m depth) and deep sea fish (e.g. Tuna) are targeted by a minority of recreational fishermen. Bait fishing is recorded as being only 4% of the recreational total catch. Lobsters and mussels are reported to be least targeted by the fishermen interviewed, and make up <1% of the recreational total catch.

Results indicate that the average reef finfish catch per fishing household, i.e. a household in which at least one of the members is a recreational fisherman, is 50±53 lbs per fishing household. The large standard deviation illustrates wide differences among fishermen's catch success, with a few fishermen catching much more than the main group. This total reef-associated finfish catch is estimated at 387,000 lb (in 2007), or 68% of the total (i.e. commercial and recreational) finfish catch. The lobster recreational catch ranges from 2,720 lobsters in 2000 to 2,973 in 2007 (with a total of 556 registered recreational lobster divers in 2007).

Based on the same market prices as those used for the commercial fishery value, the recreational reef-associated value is estimated to be US$3.5 million for finfish (excluding big game sport fishing), and US$ 0.1 million for lobster. This results in a **total recreational reef-associated fishery value of US$3.7 million** for 2007. There are no costs deducted to estimate the net value as this activity is done for enjoyment and not with a financial goal; for this reason, the recreational fishery value seems high in comparison to the values of the commercial sector. The reef-associated fishery is an important component of the total recreational fishery, comprising 79% of the total value in 2007.

Recreational fishermen caught 40% of the total finfish catch in weight in 2007. Taking into account the total reef-associated catch for finfish (i.e. both commercial and recreational), recreational fishermen are responsible for 53% of the total catch in weight (2007 data). Recreational lobster diving accounts for 9% of the total lobster catch (2007).

The **sum** of **the reef-associated commercial fishery** (both finfish and lobster) and of the **reef-associated recreational fishery** (finfish and lobster) result in a fishery ecosystem value estimated at **US$5 million** per year (Table 15.4).

Amenity Value

The environmental amenity of Bermuda's coral reefs potentially adds value to residential properties. The assessment of such a value is determined using the hedonic pricing method. The underlying question is whether the proximity of coral reefs, assessed by the view from the house, access to the beach, or pristine waterfront is perceived as an important attribute affecting the purchase price. Should a higher house price be attributed to a coral reef-related characteristic, the additional value is an estimate of the amenity value given to this environmental ecosystem by a homeowner. This study is one of few using this methodology; it is a complex analysis, requiring a large data set of house sales. A number of challenges and limitations were encountered in the valuation of this service, among which was the difficulty in accessing house sale data, and in establishing a coral reef attribute adequately reflecting the relationship between house prices and the coral reef ecosystem. Following a series of analyses, 'distance from house to beach', was accepted as the coral reef attribute best associated with house price. Beaches in Bermuda can be considered as a coral reef variable, because of their coralline origin, tightly linking their existence to the status of coral reefs.

The dataset included 593 residential buildings, with an equal share of condominiums and houses. The average price of a 2.6 bedroom/2.2 bathroom house was of $1.5million. Approximately 14% of houses bought were located on the waterfront, and on average residences were less than 1km away from beach or coral reef, and always less than 3 km; this proximity is due to the nature of Bermuda's coastline.

The amenity value is based on the estimated non-linear relationship between house prices and beach distance. The analysis reveals a quadratic relationship; unexpectedly, house prices are directly proportional to the distance from the beach up to 1.1 km with house price increasing with distance from the beach, with a reversal in trend beyond 1 km (Fig. 15.3).

The total amenity value is calculated as the difference between: (1) The total price of the houses sold in the dataset ($890 million) and (2) The extrapolated calculation of the house prices in a "deterioration" scenario- or should beaches disappear (US$ 728 million). This yields an amenity value of US$221,000 per house, and when extrapolated for all houses in Bermuda, a total amenity value of US$5.6 billion. Converting this value into equal annual amounts generates an **amenity value of coral reefs** in Bermuda of around **US$6.8 million** per year.

This leads to the conclusion that Bermudans implicitly enjoy the ecosystem services derived from coral reefs but because of its invisibility, they do not explicitly consider such an amenity when buying a residential property. Living close to a beach does not appear to be a determinant characteristic in the purchase of house in Bermuda, probably due to the island's narrowness (1.5 km at the widest point) and hence the natural proximity of residences to beaches. However, **should coral reefs and beaches become scarce** due to degradation of this valuable ecosystem, **the economic value in terms of a surplus on house prices is likely to become more apparent.**

Recreational and Cultural Value

Residents of Bermuda appear to place a high value on their coral reef resources, made apparent by the significant number of people using the island's marine environment for recreational purposes. Because most residents do not depend on the ocean for subsistence or livelihood, the relationship between Bermudans and the coral reefs can be described as predominantly recreational and cultural.

A large-scale resident survey was developed in order to have a better understanding of what connects Bermudan households to their reefs. A special valuation technique, choice modeling, was used to quantify recreational and cultural values related to

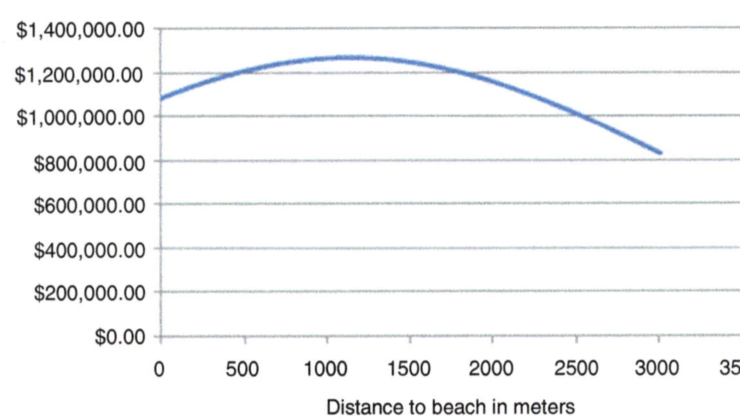

Fig. 15.3 The relationship between house price in Bermuda and distance to the beach ($n = 593$)

coral reefs. In total, 400 households collaborated, statistically representative of Bermuda's population.

The questionnaire includes seven sections: background of respondent; recreational use of reefs; environmental awareness; choice model; demographic characteristics; recreational fishing; diving and snorkeling. (Note that the recreational fishing section was added for the benefit of the Fishery value, and is not discussed within the context of the recreational and cultural value, but in the Fishery value section.) In the choice model section, each respondent was repeatedly asked to choose between complex, multi-attribute profiles describing various changes in Bermuda's coral reefs. The selection of coral reef attributes is specific to this case study and determined by consultation with focus groups and experts.

Choice Model Development

Three focus group discussions and one expert consultation were held. The three focus groups were: (1) Recreational fishers, comprising Bermudan residents who fish recreationally; (2) Snorkelers and scuba divers, comprising Bermudan residents who scuba dive and snorkel; and (3) Bermuda Residents, comprising both expatriates and Bermudans, who do not fish or scuba dive. The expert consultation was held with coral reef and fishery experts, from governmental departments (Conservation Services and Environmental Protection) and the Bermuda Institute for Ocean Science (BIOS), a local NGO.

Selection of attributes was based on the ability to determine the residents' use values for the coral reef ecosystem and enable the measurement of non-use values. Five attributes were identified: (1) recreational fishing, (2) coral diversity/fish diversity (or fish abundance), (3) recreational activities (scuba diving/snorkeling), (4) water quality (described as coral diversity, fish diversity, water clarity, and swimming restrictions), and (5) a payment vehicle (described as an environmental levy). The five attributes and their respective levels included in the design are summarised as follows:

1. Fish catch per trip: Percentage increase/decrease from the present catch. The levels are set at 20% higher catch, no change in catch, and 20% lower catch.
2. Quality of the Coral Reef: The variety and abundance of coral, reef fish and other creatures. The three levels for this attribute are poor, medium and high quality of the reef.
3. Swimming Restrictions: Number of closures during the swimming season (the summer months when the waters are warmest). The three levels for this attribute are set at 7 days, 4 days, and 0 days restrictions.
4. Water Clarity: Described as poor, medium or high clarity.
5. Payment vehicle: An environmental levy tied to the monthly electricity bill etc. The four levels for this attribute are set at 5 BMD, 10 BMD, 20 BMD, and 50 BMD per month.

Socio-Demographic Characteristics

A comparison of results obtained in this study was made with the 2004 Expenditure survey (Department of Statistics, Bermuda), indicating a similar ethnic composition (black = 59%, mixed = 8% and white = 27%), and household income. The majority (82%) of respondents were born in Bermuda. The level of educational attainment in the sample was normally distributed with 94% of the respondents having achieved, senior, technical or University level education. The average annual household income was US$124,900.

Reef-Related Activities

Swimming is by far the most popular marine-related recreational activity of the interviewed households. Beach picnic is also a popular leisure activity. While half of the respondents participate in water sports such as sailing, surfing, and boating, only 20% of the households go out snorkeling and/or diving. For the latter, seeing fish and coral species are the top two pleasures during the trip.

Environmental Awareness

The results of the survey indicate that Bermudans are concerned with the environment of their island. "Marine pollution" ranks a close second after the overarching concern of "Trash/littering and illegal dumping on the island"; the "Degradation of coral reefs" ranks 4th after "Increased development and lack of open space".

'Willingness to Pay' by Residents

Bermudan residents hold significant positive recreational and cultural values related to Bermuda's coral reefs and marine environment. Although there are issues associated with the payment of an environmental levy, residents are willing to pay to preserve their marine environment. Minimizing marine pollution, translated as the ability to swim anytime, anywhere was first and foremost, yielding an average 'Willingness to Pay' (WTP) of US$42 per month per household (Table 15.5). Second was maintaining coral reef quality (i.e. coral and fish diversity), resulting in an average WTP of US$32 per month per household. Third, water clarity (maintained by a healthy

Table 15.5 Average monthly household 'Willingness to Pay' (WTP) for environmental changes

Attribute	Change in good/service	Average monthly household WTP (USD)	Total monthly WTP (USD)
Coral reef quality	From poor to moderate	32.33	869,644
Swimming restrictions	From 4 to 0 days	42.17	1,134,360
Fish catch	From current to 20% higher	11.13	299,444
Visibility	From poor to moderate	27.42	737,699

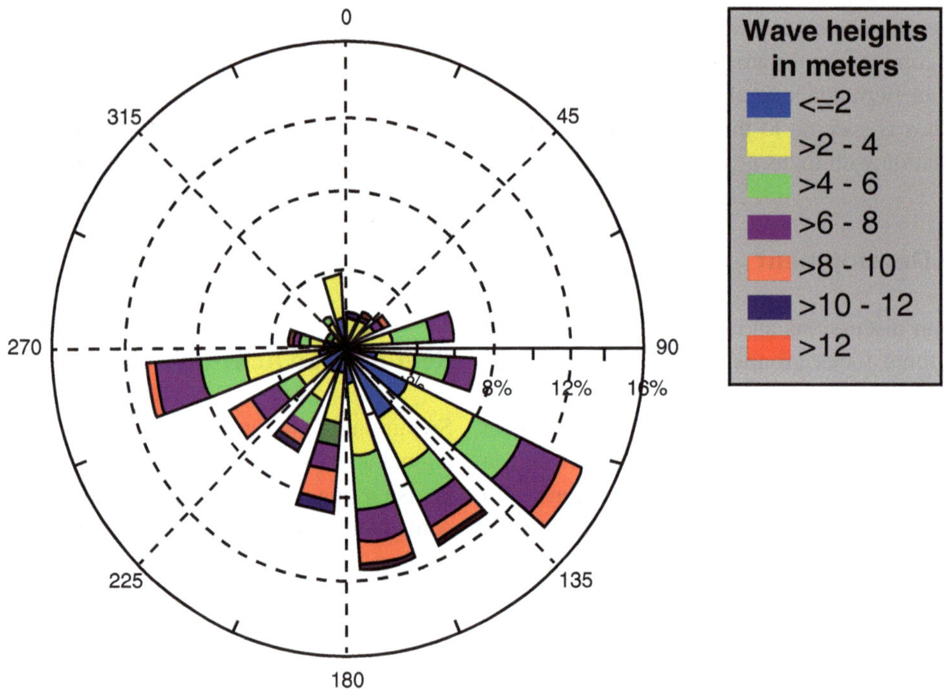

Fig. 15.4 Directional distribution of waves during storm events 1900–2003 (Taken from SWI (2004b))

coral reef system) was considered important and respondents were willing to pay on average US$27 per month to preserve or improve this attribute.

This implies that marine management policies resulting in improvements across all four environmental attributes (i.e. maintain/improve coral reef quality, avoid swimming restrictions, increase fish catch, and maintain/improve water clarity) would generate substantial benefits to the Bermudan population; more specifically, they would result in a welfare improvement equivalent to an increase in average monthly household income of around US$113. In aggregate terms, these improvements would be worth over **US$37 million** per year to the population of Bermuda, and considered to be the **total recreational and cultural value** of coral reefs to Bermuda.

Less than half of the respondents indicated that they would be willing to pay an environmental levy. This share is unusually high, compared to similar studies. The results of the choice experiment suggest that most Bermudans are actually willing to make clear trade-offs between levies and the non-monetary attributes. The contribution by residents for the preservation of the coral reef ecosystem enables the implementation of conservation and management measures.

Coastal Protection

Coral reefs absorb much of the incoming wave energy, functioning as natural breakwaters and helping to protect the shoreline from erosion and property damage. Awareness about the vital role of the rim and boiler reefs in protecting Bermuda's shoreline is raised in earlier coastal vulnerability assessment studies (Wallingford 1991; SWI 2004a, b). The current average number of storms passing Bermuda is calculated to be 11 every 10 years, estimated to increase to 13 storms every 10 years over the next 50 years (SWI 2004a). This excludes subtropical storms (Guishard et al. 2007).

Due to the typical trajectory of tropical storms in the Atlantic basin the majority of waves from storms and hurricanes come from the southeast affecting the relatively unprotected properties on the South shore (Fig. 15.4).

Despite the recognition of the importance of coral reefs to coastal protection, the economic value of this ecosystem service has not yet been estimated. Here the "avoided damages" approach is used to value this service. This involves the estimation of potential damage (and associated economic losses) to the Bermuda coastal area from a given storm event, with and without the presence of a reef.

The current study is one of the few examples of the valuation of coastal protection services provided by coral reef ecosystems. Not all of the required parameters were available for Bermuda at the time of analyses. The value for coastal protection in this case was obtained by combining local information about the island's coastal profile, the storm regime for Bermuda, historic information on property damage caused by storms, flood zones susceptible for high waves, coral reef locations, shoreline stability and the role of coral reefs, and property values for land areas.

Property damage information is based on reports for Hurricane Fabian, a Category 3 (bordering Category 4) storm, hitting Bermuda directly in 2003. This results in an average damage share of 27.5% – in other words, for a storm category 3 or 4, damage to property can be as high as a quarter of the property value –. Given the lack of reporting, damage in this study relates only to properties and excludes infrastructures (such as roads). It is recommended that further modeling and data gathering are conducted to improve on the existing calculations, as this value is based on numerous simplifying assumptions, and underestimates the true value.

The economic value of the **coastal protection** function of coral reefs in Bermuda is determined at **US$266 million** per year.

Research and Education Value

The research and education value of Bermuda's coral reefs is simply based on budgets of both governmental and non-governmental institutions in Bermuda. Only research and education budgets relating to the coral reef ecosystem are incorporated within this ecosystem service. Available data did not include monies invested in management and/or enforcement of coral reef-associated resources. The sum of all **research and education** activities associated with coral reefs in Bermuda amounts to **US$2.3 million** in 2007.

Total Economic Value

The value of the sum of compatible uses of the above goods and services constitutes the TEV of coral reef ecosystems. It is worth noting that although TEV is known as 'Total' Economic Value, this analysis has not included all goods and services provided by Bermuda's coral reefs and that some aspects of coral reefs may be 'invaluable' i.e. they have intrinsic value, beyond any benefits provided to people. Hence, the TEV estimated here is likely to under-estimate the true 'total' value of Bermuda's coral reefs.

The estimation of the various ecosystem service values involves a large number of assumptions that simplify the underlying dynamics and complexities. Therefore, lower and upper bound estimates are determined for each ecosystem service, recognizing the uncertainty surrounding the economic analysis. The basis for this range differs for each value category. The ranges estimated for each ecosystem service are presented in Table 15.6. The average annual value of the coral reef ecosystem amounts to **$722 million**. This high number certainly suggests that this ecosystem is highly valuable and worth conserving from an ecological, social and economic perspective. Lower and upper bound estimates were determined for each ecosystem service recognizing the uncertainty surrounding the economic analyses, and result in a TEV ranging from **$488 million to $1.1 billion per year**. Further study could allow for the reduction of uncertainties and thus the narrowing of the value range. N.B. the values are annual values, based on 2007 data and prices.

The contribution of ecosystem services to this value are, in order of importance: (1) Tourism (56% of TEV), (2) Coastal Protection (37%), (3) Recreational and Cultural (5%), (4) Amenity (1%), (5) Fishery (0.7%), and (6) Research and Education (0.3%) (Fig. 15.5).

The TEV of Bermuda's coral reefs depends on the ecological integrity of the coral reefs and socio-economic conditions. Degradation of the reefs is likely to lead to a loss of

Table 15.6 Upper and lower bound estimates of the annual benefits of coral reefs in Bermuda (2007US$ million/year)

Ecosystem service	Lower bound	Average	Upper bound
Tourism	324.7	405.9	487.1
Coastal protection	133.9	265.9	531.8
Recreation & Cultural	17.2	36.5	66.0
Fishery	4.3	4.9	5.6
Amenity	5.5	6.8	8.2
Biodiversity research	2.1	2.3	2.5
Total annual economic value	**487.7**	**722.4**	**1,101.2**

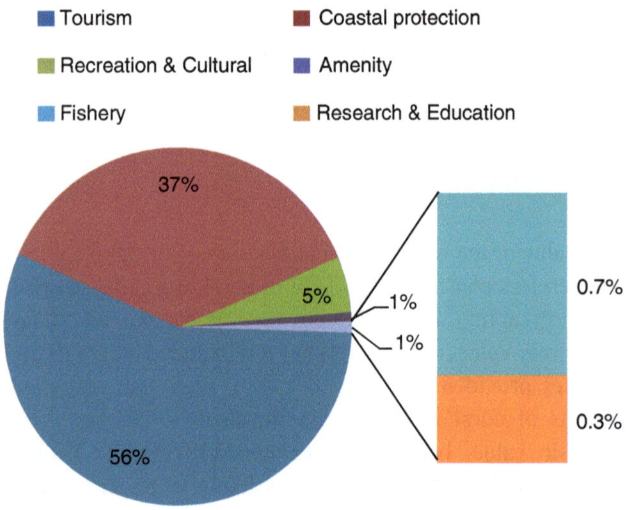

Fig. 15.5 Composition of ecosystem services valued for Bermuda's coral reefs

ecosystem service provision and a reduction in TEV. Using a discount rate of 4% for a 25-year period, it is evident that preserving the coral reefs (or delaying their degradation) in Bermuda pays off in economic terms. To place the TEV of coral reefs in context of the economy of Bermuda: in 2007, the Gross Domestic Product (GDP) of Bermuda amounted to US$5.85 billion (Government of Bermuda 2008). Based on this, the **TEV of coral reefs constitutes 12% of Bermuda's GDP**.

Conclusions

This first environmental economic valuation for Bermuda has paved the way for an alternative approach to conservation of natural resources. The findings will generate awareness among the general public of the valuable ecosystem services provided by Bermuda's coral ecosystem. Environmental valuation also provides a tool for government policy and decision-makers, and local businesses, to integrate the value of the coral reef ecosystem into marine public policy and decision-making, and business strategies. Four major recommendations were made within the scope of this study, identifying the gaps in current legislation, as well as the possibility of generating revenue for conservation of the coral reef system. The latter was triggered by the analysis on the importance of reefs to the general community and tourists. The value of coral reefs was evident by their willingness to trade off monies for preservation of this natural resource. Specific policy recommendations derived are:

Recommendation 1: Prioritize potential policy interventions in an economically sound manner, through improved legislation, integration of strategic environmental assessments (SEA) for major developments, incorporate TEV in extended cost-benefit analyses for marine developments, and establish damage compensation procedures for marine vessel groundings.

Recommendation 2: Make use of the cultural importance residents place on marine ecosystems to improve coral reef management,

Recommendation 3: Actively involve the tourism industry in the development of sustainable coral reef management, through the establishment of a vehicle for enabling community support for environmental conservation, thus allowing the use of funds currently put into the marine environment for other socio-economic needs.

Recommendation 4: Balance consumptive and non-consumptive uses of coral reefs by strategizing spatial management and protecting critical marine areas, through the careful delineation of Marine Protected Areas (MPAs), and prioritizing strong enforcement and protection of these zones by engaging boat and dive operators.

In future work, economic analysis could be effectively used to evaluate the feasibility of the potential measures recommended above. In addition, increased funding earmarked for coral reef ecosystem sustainability would allow for the assessment of research and management needs, and their implementation. This is required to ensure the continued provision of valuable ecosystem services to Bermuda's community. Some examples are:

- Monitoring and early detection of natural/human-induced changes;
- Enhancing enforcement capacity on the Bermuda platform;
- Developing and implementing mitigation measures of foreseen changes – i.e. due to climate change and/or coastal development;
- Researching coral restoration and growth, connectivity between fish productivity and coral reef habitats;
- Predicting wave impact on Bermuda's coastline and identifying flood zones- including collecting wave information during storms and hurricanes.
- Gaining a better understanding of coastal erosion parameters required for mitigation measures of natural and human induced erosion processes.

It is hoped that results on the TEV of Bermuda's coral reefs will encourage and facilitate marine policies that ensure the sustainability of these northernmost coral reef system in the world. A policy brief presented to the Bermuda Cabinet in September 2011, resulted in the approval of all four recommendations, and has initiated their implementation.

References

Burke L, Greenhalgh S, Prager D, Cooper E (2008) Coastal capital – economic valuation of coral reefs in Tobago and St. Lucia. WRI report. Washington, DC

Department of Tourism (2007) Visitor profile 2007. Government of Bermuda, Bermuda

GOB (2000) Marine resources and the fishing industry in Bermuda: a discussion paper. The Government of Bermuda, Ministry of the Environment, Bermuda

GOB (2008) Department of Statistics, Government of Bermuda, Bermuda

Guishard MP, Nelson EA, Evans JL, Hart RE, O'Connell DG (2007) Bermuda subtropical storms. Meteorol Atmos Phys 97(1–4):253–293

Sarkis S, van Beukering PJH, McKenzie E (eds) (2010) Total economic value of Bermuda's coral reefs: valuation of ecosystem services. Technical report, Department of Conservation Services, Government of Bermuda, 2010, 199pp

SWI, Smith Warner International (2004a) Coastal protection and development planning guidelines for Bermuda. Government of Bermuda, Ministry of the Environment. 47

SWI, Smith Warner International (2004b). Bermuda coastal erosion vulnerability assessment; final report. Government of Bermuda, Ministry of the Environment, 148pp

Van Beukering PJH, Slootweg R (2010) Valuation of ecosystem services: lesson from influential cases (Chapter 9). In: Slootweg R, Rajvanshi A, Mathur VB, Kolhoff A (eds) Biodiversity in environmental impact assessment. Cambridge University Press, Cambridge, pp 287–327

Wallingford HR (1991) Dymchurch and Pett Sea defences, extreme waves and water levels. Wallingford report EX 2312, Sept 1991, 54pp

Ascension Island's Hermatypic but Non-Reef Building Corals

16

Robert Irving

Introduction

Ascension Island lies approximately 8° south of the equator at 07° 57′S, 14° 22′W, 145 km west of the centre line of the mid-Atlantic Ridge. The nearest land to it is the similarly diminutive island of St Helena, some 1,130 km to the south-east, whilst the continental land masses of West Africa and Brazil are approximately 1,500 km and 2,400 km distant respectively. Ascension, therefore, has the peculiar distinction of being one of the most remote islands in the world.

The island is triangular in shape, being 14 km from east to west and 11.2 km from north to south (Fig. 16.1a, b). The coastline is about 45 km long and very rugged (Fig. 16.2), enclosing a land area of approximately 90 km^2 (35 $mile^2$). Much of the island's surface is pock-marked by volcanic peaks, craters and vast flows of lava (known locally as 'clinker'), solidified into extremely hard rock formations. In the sublittoral, rock surfaces are dominated by encrusting coralline algae and not, as one might expect, by corals. There are no coral reefs at all at Ascension and just five species of scleractinian coral have been recorded.

A Brief History of the Area

The island was first discovered by the Portuguese explorer Joao da Nova in 1501 and named Conception. However, this name was never publicised and it was re-named Ascension two years later by Alphonse D'Albuquerque (Packer 1983). To seafarers over the following 300 years it became known as a place where ships could take on supplies of turtle meat, as did Capt. James Cook in 1775 on HMS Resolution during his second world voyage of discovery. When, in 1815, Napoleon Bonaparte was imprisoned on St Helena, a small British garrison was stationed on Ascension in order to prevent the French from mounting a possible rescue attempt and the island has been continuously inhabited ever since. Around 1900, the first submarine telegraph cable was laid from Cape Town via St Helena, Ascension, the Cape Verde islands and on to the UK, establishing the presence of Cable and Wireless Ltd. on the island. In 1942 the US Army built an airstrip in the south-west of the island (known as Wideawake), followed by the establishment of a NASA tracking station in the 1960s. The BBC installed radio communications equipment and aerials to be able to relay World Service broadcasts from London to Africa and South America at around the same time. By 1980, the island could boast of having the greatest number of satellite dishes per unit area in the world! Ascension also acted as an important staging post for the transport of troops and equipment to and from the Falklands during the conflict in 1982. The RAF continues to have a base on the island to support its regular flights to the Falklands.

History of Research

Few marine biological studies had taken place at Ascension prior to the 1970s. Some small collections of echinoderms (obtained from dredged material) were made during the late 1800s and from visits made by Antarctic research vessels during the first half of the twentieth century. In 1971, Dr R.B. Manning of the US National Museum of Natural History (Smithsonian Institution) visited the island and made extensive collections of invertebrates, particularly of crustaceans, molluscs and echinoderms (Rosewater 1975). A second Smithsonian expedition was undertaken in 1976, with collections being made of shallow water marine invertebrates (Pawson 1978). In 1979, Drs J. Price and D. John from London's Natural History Museum investigated near-shore macroalgal communities (Price and John 1980). A few years later in 1985, a UK diving expedition (of which the author was a co-leader) set about cataloguing and photographing as many of the island's sublittoral habitats and invertebrates as

R. Irving (✉)
Sea-Scope Marine Environmental Consultants,
Devon, England, UK
e-mail: Robert@sea-scope.co.uk

C.R.C. Sheppard (ed.), *Coral Reefs of the United Kingdom Overseas Territories*, Coral Reefs of the World 4,
DOI 10.1007/978-94-007-5965-7_16, © Springer Science+Business Media Dordrecht 2013

Fig. 16.1 (**a**) Satellite view of Ascension Island. Satellite Image (c) Space Imaging. (**b**) Map of Ascension showing coastal locations mentioned in the text and the five near-shore sectors (as identified by Irving 1989), determined by predominant sublittoral habitat types. Illustration: J. Taylor)

Fig. 16.2 Dramatic yet stark coastal scenery off Ascension's south coast (Photo: R.A. Irving)

possible during a five week visit (Irving 1989). Subsequent studies have catalogued the larger Crustacea (Manning and Chace 1990) and collections of Porifera are being studied at present (pers. comm., C. Goodwin and R. van Soest).

At the time of writing, a major UK diving expedition to the island is underway, organised by the UK's Shallow Marine Surveys Group. The expedition aims to add to what is already known of the benthic habitats, species and marine biogeography, to produce a field guide to marine invertebrates, algae and fish; to report on the status of marine endemics; to report on the potential impacts of climate change; and to assist the Ascension Island Government in developing future biodiversity strategies for their shallow marine environment (http://www.smsg-falklands.org).

The Island Today

Ascension Island is part of the British Overseas Territory of St Helena, Ascension and Tristan da Cunha. In December 2011 its population numbered 873 (FCO website 2012). All three islands are overseen by a Governor (based in St Helena), although Ascension has its own island-based Administrator. Since April 2002 the island's revenue has derived from locally raised taxes, import duties, the sale of philatelic stamps and (since October 2010) the sale of fishing licences. The UK Government remains responsible for Ascension Island's external relations, defence and internal security.

Ascension is an important breeding site for the green turtle *Chelonia mydas* (between 6,000 and 15,000 nests are made each year) and for various species of sea bird, notably the sooty tern or wideawake *Onychoprion fuscatus* and the Ascension frigatebird *Fregata aquila*. In 2001, the British Government gave the Royal Society for the Protection of Birds (RSPB) £500,000, to rid Ascension Island of feral cats that had decimated the once huge seabird population, one of the world's most important breeding colonies. By early 2003 four species of seabird had already re-colonised the mainland (from their refuge on Boatswain Bird Island) as breeding species and predation on the sooty tern population had been eliminated. However, since the eradication of feral cats, the rat population has increased, as have rabbits, both of which are now subject to control programmes. A small Conservation Department comprising three core staff was created in 2001 and a National Park at Green Mountain was established in 2005.

Geological Background and Physical Parameters

Ascension is volcanic in origin, with at least sixty vents or ash cones of varying size. The last eruption is thought to have taken place about 600 years ago. Much of the coastal margin consists of basaltic or trachytic lava flows, with Green Mountain (roughly in the centre of the island) being principally formed of the latter. In geological terms it is a relatively young island: radioactive dating suggests a maximum age of 1.5×10^6 y B.P. (Bell et al. 1972), but many of the lower-lying lava flows are known to be younger (0.22–0.47×10^6 y B.P.). By comparison, St Helena has an age in the region of 20×10^6 y B.P.

For the most part, the island is influenced by the westward-flowing South Equatorial Current, which normally flows close to the surface. However, the island is also affected from time to time by the usually deeper-flowing (and colder) Southern Equatorial Counter-Current. The interaction of these two currents gives rise to localised turbulence, gyres and eddies which combine with upwellings in near-shore waters caused by the seabed topography, giving rise to considerable mixing of the water types.

Although the prevailing wind direction is from the south-east, the whole coastline is exposed to severe wave action at some time during the year (particularly from January to March). The least wave-exposed sites appear to be at English Bay in the north and north of Porpoise Point in the north-east. The tidal range is small (0.9 m) with tides being more or less regularly semi-diurnal (Price and John 1980). Seawater temperatures have a fairly narrow range, being slightly warmer (24–26°C) from February to May and cooler (22–24 °C) from August to November. The clarity of the coastal waters is generally very good (vertical visibility typically being 25–40 m), though the situation can change dramatically after one of the sudden though infrequent rainstorms which affects the island from time to time (Irving 1989). The results of sediment smothering after such a storm were apparent in rockpools in South West Bay in October 1985 when several dead *Favia gravida* colonies were observed (Taylor and Irving 1986).

Seabed Types

Near-shore areas of the seabed can be roughly divided into five main sectors. These are described below in clockwise order around the coast, starting in the north-east (Fig. 16.1b).

Sector 1: English Bay to North-East Bay

This is an area of massive outcrops of bedrock. Terraces and underwater cliffs are frequent, often with small caves or underhangs. The bedrock outcrops are separated by areas of coarse sand, occasionally with patches of mearl.

Sector 2: Hummock Point to Crystal Bay

This is the most exposed sector of the island, characterised by steep to vertical bedrock close inshore. Some localised shelter from the wind is provided behind Boatswain Bird Island and Rocks. Seabed profiles extending out from the coast range from wave-cut platforms, cliffs and screes to the vertical face of South-East Head, which drops to 38 m before meeting sand and mearl. At Hummock Point, the sand (unusually volcanic rather than maerl in origin) slopes away very steeply from about 24 m. (The boundary between sectors 2 and 3 can only be estimated, as few dives have been undertaken in this area on account of the frequent heavy swell).

Sector 3: Crystal Bay to South-West Bay

This area is dominated by boulder slopes composed of massive boulders of a minimum diameter of 3 m. Areas of scoured, outcropping bedrock have been recorded from two sites, with very few loose boulders. In many cases, the bedrock is undercut or pocked, providing daytime shelter for echinoids.

Sector 4: South-west Bay to Pyramid Point

The shallow shoals of the west coast have smaller and less prominent bedrock outcrops than other sectors. This is the least exposed part of the open coastline, especially the central area around Georgetown. However, the seasonal rollers which strike the island from the west each year make this still an exposed coastline in any UK understanding of the term. Mearl is abundant in the bays along the west coast.

Sector 5: Pyramid Point to White Rock Cove

This stretch of coast is bounded by low cliffs, many with breeding noddies and boobies. Sublittorally, there are steep boulder slopes extending to 24–28 m depth, where maerl is present overlying coarse maerl-sand. In places the boulders are very massive, whilst elsewhere smaller boulders are fused together with coralline algae. A wave-cut platform is typically present above the boulder slope, of varying width. At 8–10 m below sea level, the vertical walls of the platform are much undercut with caves and arches. This shallow area is both scenic and (relatively) biologically rich.

None of the above-mentioned sectors could be described as being fundamentally different from any of the others, any apparent differences being dependent upon a different mix of habitat types and associated species, rather than the presence or absence of either. In addition to these gross sector distinctions, certain locations stand out within them as being different from the surrounding seabed. These generally result from a combination of local topography and conditions. Most notable of these are Porpoise Point (probably the most sheltered site around the island) and English Bay (Taylor and Irving 1986).

Table. 16.1 Characteristic sublittoral species assemblages (After Irving 1989)

Predominant habitat type	Characteristic species assemblage	Locations (octants) where recorded
A. Bedrock		
(1) Shallow, wave-affected (particularly in the SE)	The zoanthid *Palythoa* sp. (Fig. 16.5) on upward-facing surfaces. Encrusting sponges prominent. Bivalve mollusc *Lopha* common.	Boatswain Bird Rock (E) Spire Rock (E)
(2) Shallow, vertical faces	A diversity of sponges and ascidians, usually in small patches, with encrusting lithothamnia. The echinoid *Diadema antillarum ascensionis* is likely to be present in holes.	Klinka Bay (N) Pyramid Point (NW)
(3) Underhangs, under boulders and shallow caves	The encrusting coral *Madracis decactis* (Fig. 16.3), solitary corals, *Astrangia solitaria* (Fig. 16.4) and *Rhizosmilia* sp. cf. *gerdae*, black coral *Antipathes* sp., and the zoanthid *Parazoanthus swiftii* (Fig. 16.5). Often with squirrelfish *Holocentrus* spp. present in or close to caves.	Widespread around all coasts. This community has the greatest invertebrate species diversity and richness.
B. Boulder slopes	Upper surfaces dominated by encrusting and columnar lithothamnia. The encrusting corals *Siderastrea radians* and *Madracis decactis* common as various colour morphs. Moray eels (especially *Lycodontis moringa* and *Muraena pavonina*) common.	South West Bay (SW) Lady's Loo (N) Archer Point (N)
C. Maerl, as		
(1) Branched 'hedgehogs'	With small filamentous blue-green algae and much coarse sand being visible.	Porpoise Point (NE) North East Point (NE)
(2) Cobbles	In beds, 5+ cobbles deep, with little visible sand. Occasionally with the fan shell *Pinna* and the conch *Strombus*.	Portland Point (SW) Pyramid Point (NW) Pillar Point (SE)

Fig. 16.3 The encrusting coral *Madracis decactis* (Photo: R.A. Irving)

Biological Characteristics

Sublittoral Community Types

Ascension has clear waters, barren rock (though frequently covered with the encrusting coralline algae lithothamnia) and large numbers of fishes, particularly of the triggerfish *Melichthys niger* (known as the blackfish). The sublittoral communities around the island are generally indistinct, although characterised by a low species diversity and abundance, particularly of sessile invertebrates. These impoverished communities are typically present in small patches, giving rise to a complex mosaic of types. However, certain species assemblages appear to be fairly consistent (Table 16.1).

Coral Fauna

There are no coral reefs at Ascension and in this respect the island resembles tropical West Africa (Price and John 1980). The coral fauna itself is extremely limited. Those scleractinian

Table 16.2 Species of Scleractinia (stony corals) and Milleporidae (fire 'corals') found during the Operation Origin expedition in 1985

Species	Description	Habitat notes	Location (octant)	Depth (bsl)	Zoogeographical notes
Madracis decactis (Fig. 16.3)	Thin, encrusting coral. Various-coloured polyps, from green to purple (when viewed under water).	On upper surfaces of bedrock is relatively shallow water.	White Rock (N) Pyramid Point (NW)	10–13 m	Caribbean & NE coast of Brazil. Also W. Africa and Cape Verde islands.
Favia gravida	Faviid coral. Grey-green coloured polyps.	Present in *Typhlatya* pools.	Shelley Beach (S)	0 m	Known from Brazil, West Africa and St Helena
Siderastrea radians	Large polyp coral.	Present on upper surface of bedrock, with algae growing around perimeter	Catherine Point (W)	16 m	Amphi-Atlantic but not off Brazil
Astrangia solitaria (Fig. 16.4)	Cup corals (brown/white)	Present on the undersides of bedrock and boulders.	Pyramid Point (NW) Pillar Point (SE) Archer Point (N) Spire Rock (E)	10–26 m	Western Atlantic, from Bermuda across the Caribbean to NE Brazil
Rhizosmilia sp. cf. *gerdae*	Cup coral (boulder)	Present under overhangs and on underside of cobble.	Spire Rock (E) Powerhouse Cove (N)	10–15 m	Insular Straits of Florida
Millepora alcicornis	Fire coral	In small patches on open areas of bedrock		10–15 m	Caribbean, Florida, Brazil and Cape Verde Islands
Millepora complanata	Fire coral	In small patches on open areas of bedrock		8–12 m	Caribbean, Florida, Bahamas and Venezuela

Fig. 16.4 The starfish *Linckia guildingii* with the brown coral *Astrangia solitaria* (Photo: R.A. Irving)

Fig. 16.5 Shallow bedrock covered with the zoanthid anemone *Palythoa* sp (Photo: J. Taylor)

corals which are present (Table 16.2, Figs. 16.3 and 16.4) are either ahermatypic encrusting forms (just three species were distinguished by the author from his visit in 1985) or are solitary/cup coral forms. Two species of fire coral (Hydrozoa: Milleporidae) were also recorded. The colonial relative *Palythoa* is also present in large sheets (Fig. 16.5).

Other Sublittoral Taxa

Ascension's sublittoral fauna is noticeably impoverished. The blackfish *Melichthys niger* (Fig. 16.6) deserves special mention on account of its surprising abundance, its cosmopolitan diet and its voracious grazing ability. There can be little doubt that the large numbers of these fishes have a significant effect on the local ecology. This species in particular, and probably the sea urchins *Echinometra lucunter polypora* in the sublittoral fringe and *Diadema antillarum ascensionis* in slightly deeper water, manage to keep in check the possible colonisation of rock surfaces by settling larvae. During daylight hours, when the blackfish rule supreme, the *Diadema* urchins hide away in holes and crevices, but under cover of darkness large numbers emerge to graze the miniscule epibionts which may remain on rocky substrata. 70% of the diet of blackfish seems to consist of algae (Randall and Klausewitz 1973; Price and John 1980). Only blue-green algae appear to be avoided by their grazing behaviour, and those macroalgal and sessile invertebrate species growing in places which are inaccessible to blackfish.

Fig. 16.6 The blackfish *Melichthys niger* is surprisingly colourful when seen close up (Photo: J. Taylor)

Biogeographic Background

With its isolated position in the middle of the Atlantic, Ascension lies a very long way from potential source areas of recruitment along the coastal fringes of both West Africa and Brazil, its nearest continental land masses. This fact alone has restricted the number of species that have reached the island. The majority of invertebrate species (including corals) have their centres of distribution centred on the Caribbean and/or the coast of Brazil with very few associated with West African faunas. This is true for most fishes: 30% of the 71 species of near-shore fishes are only recorded from the western and central Atlantic, whereas just 7% are recorded from the eastern and central Atlantic (Lubbock 1980).

Endemic Species

The number of endemic species which have been reported from Ascension is relatively low, reflecting the overall paucity of species diversity around the island. Two species of caridean shrimp are known to be endemic to Ascension, confined to isolated intertidal pools on the island's south coast, *Procaris ascensionis* and *Typhlatya rogersi*. *Procaris ascensionis* was described as being one of the most primitive living members of the Caridea yet discovered; and *Typhlatya rogersi* was described as the first species of that genus to have been collected outside the Gulf of Mexico and the Caribbean and the first known from salt water (Chace and Manning 1972).

Rosewater (1975) recorded a total of 89 species of molluscs from the island of which 6% occur at both Ascension and St Helena. For echinoderms 4 of the 25 species recorded (16%) are restricted to both Ascension and St Helena and just one species, the sea cucumber *Holothuria manningi*, is endemic to Ascension (Pawson 1978).

Despite the apparent abundance of fish life, the number of recorded near-shore fishes recorded from Ascension (71) is low for a tropical island (Lubbock 1980), attributed to the island's isolation and to low habitat diversity. 16% of these species are believed to be endemic to Ascension (Lubbock 1980) including the resplendent angelfish *Centropyge resplendens*; Apollo damselfish *Chromis* sp.; Lubbock's yellowtail damselfish *Stegastes lubbocki*; Ascension goby *Priolepis ascensionis*; white hawkfish *Amblycirrhitus earnshawi*; marmalade razorfish *Xyrichtys blanchardi*; seabream *Diplodus ascensionis*; and the Ascension wrasse *Thalassoma ascensionis*. A further 17% are known to be endemic to both Ascension Island and to St Helena. These are relatively high levels of speciation, a fact which can largely be assigned to the islands' isolation.

Conclusion

The low diversity of Ascension's sublittoral fauna appears to be largely the result of the island's isolation. Those sessile invertebrates which have managed to reach the island have only survived by avoiding the attention of the grazers, particularly the urchins (e.g. *Diadema antillarum ascensionis* and *Echinometra lacunter*) and the grazing fishes (e.g. the sergeant-major *Abudefduf saxitilis* and the ubiquitous blackfish *Melichthyes niger*). Typically this means these sessile species have become established in crevices, small caves or on the underside of overhangs. Besides the widespread cover of encrusting coralline algae ('lithothamnia') on most open rock surfaces, extensive sheets of zoanthid anemones *Palythoa* spp. appear to have filled the ecological niches where hermatypic corals might be expected to be, presumably a result of being either sufficiently distasteful or indigestible to would-be grazers. This lack of hermatypic corals also decreases the number of available habitats one might expect to find in the shallow sublittoral of a tropical island, and consequently the diversity of species too.

References

Bell JD, Atkins FB, Baker PE, Smith DGW (1972) Notes on the petrology and age of Ascension Island, South Atlantic. Eos 53:168

Chace FA Jr, Manning RB (1972) Two new Caridean shrimps, one representing a new family, from marine pools on Ascension Island (Crustacea: Decapoda: Natantia). Smithson Contrib Zool 131:18, 11 figures

FCO website (2012) http://www.fco.gov.uk/en/travel-and-living-abroad/travel-advice-by-country/country-profile/south-america/ascension-Island/

Irving RA (1989) A preliminary investigation of the sublittoral habitats and communities of Ascension Island, South Atlantic. Progress Underw Sci 13:65–78

Lubbock R (1980) The shore fishes of Ascension Island. J Fish Biol 17:283–303

Manning RB, Chace FA Jr (1990) Decapod and stomatopod Crustacea from Ascension Island, South Atlantic Ocean. Smithson contrib Zool 503:91

Packer JE (1983) A concise guide to Ascension Island, South Atlantic, 3rd edn. Ascension Island, Georgetown

Pawson DL (1978) The echinoderm fauna of Ascension Island, South Atlantic Ocean. Smithson Contrib Mar Sci 2:31

Price JH, John DM (1980) Ascension Island, South Atlantic: a survey of inshore benthic macro-organisms, communities and interactions. Aquat Bot 9:251–278

Randall JE, Klausewitz W (1973) A review of the triggerfish genus *Melichthys niger*, with a description of a new species from the Indian Ocean. Senckenberg biol 54:57–69

Rosewater J (1975) An annotated list of the marine mollusks of Ascension Island, South Atlantic Ocean. Smithson Contrib Zool 189:1–41

Taylor J, Irving RA (1986) Operation Origin expedition report. Unpublished report of a diving expedition to Ascension Island

British Indian Ocean Territory (the Chagos Archipelago): Setting, Connections and the Marine Protected Area

17

Charles R.C. Sheppard, Brian W. Bowen, Allen C. Chen, Matthew T. Craig, Jeff Eble, Nancy Fitzsimmons, Chai-hsia Gan, Michelle R. Gaither, Matthew Gollock, Shashank Keshavmurthy, Heather Koldewey, Jeanne A. Mortimer, David Obura, Miriam Pfeiffer, Alex D. Rogers, Anne L.S. Sheppard, Catherine Vogler, Gert Wörheide, Ming-Che Yang, and Chris Yesson

Geographical and Historical Setting

The British Indian Ocean Territory lies at the southern end of the Laccadive-Chagos ridge, in the centre of the tropical Indian Ocean. The area is geographically known as the Chagos archipelago, and has five islanded atolls and several other atolls and banks which are awash or completely submerged (Fig. 17.1 and Table 17.1). Its central feature is the 150 by 100 km Great Chagos Bank, the World's largest atoll in terms of area. This is mostly submerged, but there are eight islands on its western and northern rim. This is surrounded by the smaller atolls: Peros Banhos and Salomon to

C.R.C. Sheppard (✉) • A.L.S. Sheppard
School of Life Sciences, University of Warwick,
Gibbit Hill Campus, CV4 7AL, Coventry, UK
e-mail: Charles.Sheppard@warwick.ac.uk;
a.sheppard@warwick.ac.uk

B.W. Bowen
Hawai'i Institute of Marine Biology, University of Hawaii,
P.O. Box 1346, Kane'ohe, HI, 96744, USA
e-mail: bbowen@hawaii.edu

A.C. Chen • C.-h. Gan • S. Keshavmurthy • M. C. Yang
Biodiversity Research Centre, Academia Sinica,
128 Academia Road, Nankang, Taipei, 115, Taiwan
e-mail: acropora.chen@gmail.com; chgan@gate.sinica.edu.tw;
shashank@gate.sinica.edu.tw; jaymcyang@gmail.com

M.T. Craig
Department of Marine Sciences and Environmental Studies,
University of San Diego, San Diego, USA
e-mail: matthewcraig4@gmail.com

J. Eble
Center for Insect Science, University of Arizona,
1140 E. South Campus Dr, Tucson, AZ. 85721, USA
e-mail: eble@email.arizona.edu

N. Fitzsimmons
Faculty of Applied Science, University of Canberra,
ACT 2601, Canberra, Australia
e-mail: nancy.Fitzsimmons@canberra.edu.au

M.R. Gaither
Ichthyology Section, California Academy of Sciences,
San Francisco, CA, 94118, USA
e mail: michellergaither@gmail.com

M. Gollock • H. Koldewey • C. Yesson
Zoological Society of London, Regents Park, NW1 4RY, London, UK
e-mail: matthew.gollock@zsl.org; Heather.koldewey@zsl.org;
Chris.Yesson@ioz.ac.uk

J.A. Mortimer
Department of Biology, University of Florida, Gainesville, FL, USA
e-mail: Mortimer@ufl.edu

D. Obura
CORDIO East Africa, #9 Kibaki Flats, Kenyatta Beach,
Bamburi Beach, P.O.BOX 10135, Mombasa, 80101, Kenya
e-mail: cordioea@gmail.net

M. Pfeiffer
RWTH Aachen University, Templergraben 55, 52056, Aachen,
Germany
e-mail: pfeiffer@geol.rwth-aachen.de

A.D. Rogers
Department of Zoology, University of Oxford, The Tinbergen
Building, South Parks Road, Oxford, OX1 3PS, UK
e-mail: alex.rogers@zoo.ox.ac.uk

C. Vogler • G. Wörheide
Department für Geo- und Umweltwissenschaften Paläontologie &
Geobiologie, Ludwig-Maximilians-Universität,
Richard-Wagner-Str. 10, Munich, 80333, Germany
e-mail: catherine.vogler@gmail.com; woerheide@lmu.de

C.R.C. Sheppard (ed.), *Coral Reefs of the United Kingdom Overseas Territories*, Coral Reefs of the World 4,
DOI 10.1007/978-94-007-5965-7_17, © Springer Science+Business Media Dordrecht 2013

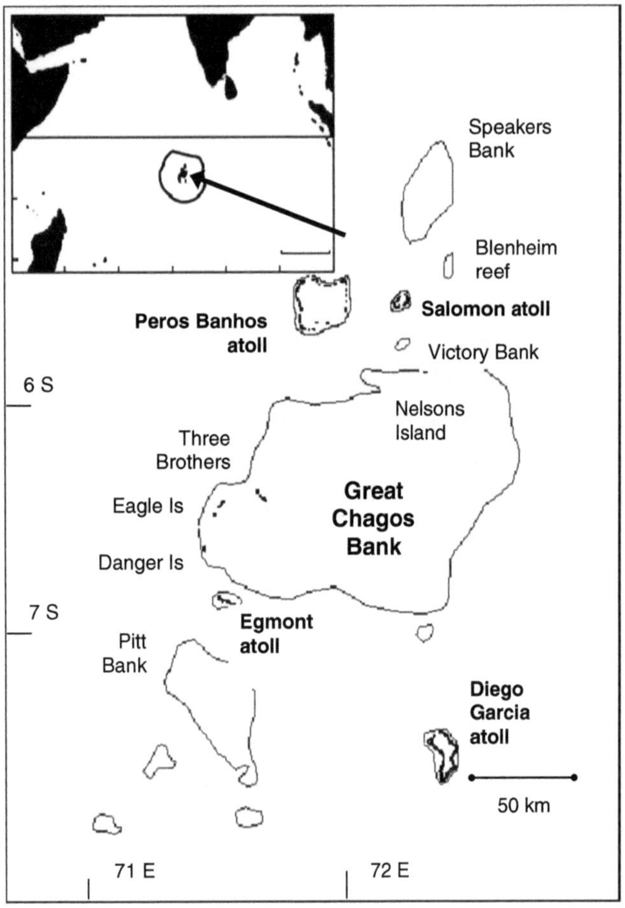

Fig. 17.1 The Chagos Archipelago. Inset shows location and MPA boundary (*circular shape* with flattened northern border). Main map: the five atolls with land are shown in *bold*, the islands on Great Chagos Bank and submerged reefs and atolls are not *bold*. All are located in the central area of the MPA

the north, Egmont (on some charts the 'Six Islands') to the Southwest, and Diego Garcia to the south. Among these lie many submerged reefs, the whole complex covering about 250 by 400 km. The submerged structures are one of the most notable features of the central Indian Ocean.

There are no records to suggest whether the widespread seafaring civilisations of 500–2,000 years ago discovered Chagos, and in this respect it differs from the relatively nearby Maldives. The islands' first documented discovery was by the Portuguese in the sixteenth Century but they were not inhabited for the next 200 years until, in the late eighteenth Century, the larger ones were farmed for coconuts or copra. The supply of copra, which exceeded about 0.5 million litres annually in the nineteenth Century (Moresby 1884), caused the atolls to be known as the Oil Islands. This lasted more than a century until the 1950s when the plantations on Egmont and the Great Chagos Bank were abandoned. The other plantations continued until the early 1970s, when all remaining atolls were evacuated, the archipelago became the British Indian Ocean Territory, and a military facility was built on Diego Garcia. Today all atolls except Diego Garcia are currently uninhabited, the latter supporting a communications, naval and air facility of a few thousand personnel.

The atolls and reefs attracted some distinguished early scientific attention. All its reefs were mapped in detail in 1837 by Moresby (1884), which permitted Darwin to incorporate them extensively in his exposition of coral reef formation, though Darwin did not land there. Bourne (1888) visited its islands and interpreted its rock strata as evidence against reef formation by subsidence, and Gardiner (1936) described several parts of Chagos which he had visited 30 years earlier. Then, few visits by scientists were undertaken until Stoddart and Taylor (1971) visited Diego Garcia and made observations of the atoll and collections of several groups of organisms from the island and reef flat. The first detailed, deeper reef ecology studies commenced in the 1970s when all the uninhabited atolls and several submerged banks were examined (Bellamy 1979; Sheppard and Wells 1988). No more scientific visits occurred until a visit in 1996 (Sheppard and Seaward 1999).

Table 17.1 Physical characteristics of the major atolls and banks. N.B. Several small banks with atoll cross-section also exist, whose shallowest rim sections are <10 m, but which have not been studied (from Sheppard and Seaward 1999)

Atoll	Atoll area (km^2)	Land area (ha)	Number of islands	Lagoon greatest depth (m)	Lagoon mean depth (m)	Shallowest depth (if no islands) (m)	% Rim enclosure by islands or reef flats
Atolls							
Diego Garcia	<200	2,733	4	31	10	–	95
Salomon	38	311	11	33	25	–	85
Peros Banhos	463	920	24	80	38	–	60
Great Chagos bank	18,000	445	8	~30	90	–	<5
Egmont	40	~300	2–3	26	12	–	35
Major submerged atolls							
Blenheim reef	40	0	0	17	8	awash	60
Victory bank	16	0	0	33	25	5	–
Speakers bank	680	0	0	35	44	7	–
Pitt bank	~1,200	0	0	44	35	<10	–

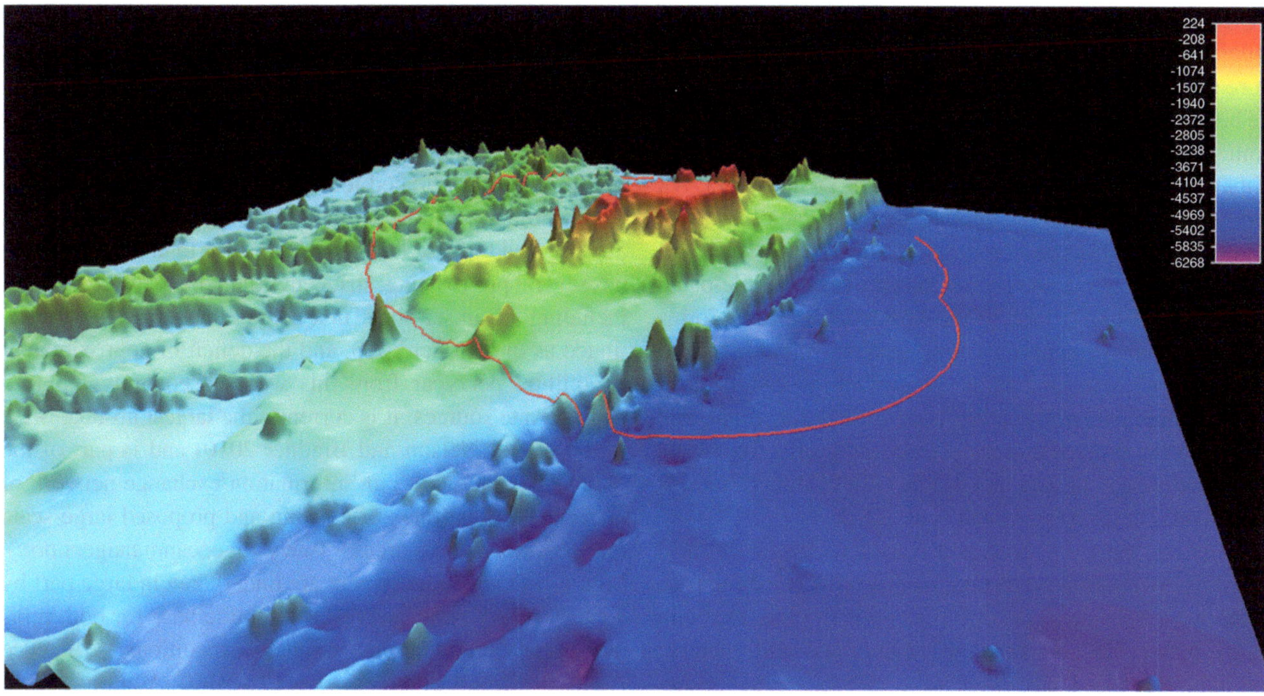

Fig. 17.2 Bathymetric interpretation of the Chagos MPA and surrounding sea bed. Thin *red line* marks the MPA boundary (Map: GEBCO/NOC) Scale bar in m

Chagos contains the largest expanse of totally unexploited reef in the Indian Ocean as well as some of the richest. With the exception of part of Diego Garcia, Chagos' reefs have been almost completely undisturbed for at least 30 years.

Geological and Physical Background

Chagos is the southernmost part of the Chagos – Laccadive ridge, formed when the Indian tectonic plate migrated northwards towards Asia. The entire chain was created by hot-spot activity since the late Cretaceous, and the trace of this can be followed from the Deccan traps in western India and from there southwards down the chains of atolls to Chagos (Eisenhauer et al. 1999). The archipelago is a group of limestone caps a few hundred metres to a few kilometres thick, resting on volcanic rock (Francis and Shor 1966), and there are numerous pinnacles, seamounts and knolls on the western side of the Territory with an abyssal plain to the east (Fig. 17.2).

Most of the present islands are those of typical atolls, located on the atoll rim, with elevations of no more than 1 or 3 m. Even the Great Chagos Bank itself has the same atoll structure of lagoon area and passes, though it is mostly submerged. Only two areas in Chagos have raised reefs; these are found in southern Peros Banhos and the adjacent, northwestern part of the Great Chagos Bank, both sites containing islands with small, uplifted and vertical cliffs about 6 m above high tide. One of the main features of the Chagos reefs is the number of submerged banks and 'drowned' atolls (see Table 17.1), the latter exhibiting typical atoll-like cross sections including 'lagoons' and passes cut through the submerged atoll rims. The atolls are classical islanded atolls, while Blenheim is an atoll of similar size but which is emerged at low tide only, though reports show that in the late 1700s it supported three vegetated islands. Other structures are atoll-shaped but submerged to at least 5 m.

All reefs visited have profusely growing corals, so the reason why some support islands, some are awash and others are drowned to 5 m or more is not known. Blenheim reef, for example, is a typical atoll in most respects except for its present lack of islands, and indeed its wave resistant algal ridges are the best developed in the Archipelago and possibly in the entire ocean. Other submerged reefs are crescent shaped (e.g. Benares, Colvocoresses) which may represent fragments of older atolls. On some large deeper reefs, such as Speakers Bank, corals are profuse, and thriving seagrasses have been found in parts of Speakers Bank and the Great Chagos Bank, indicating high benthic productivity. On the largest system of all, the Great Chagos Bank, complex patterns of reefs exist with several ring shaped structures within the main atoll rim, which attest to a complex past history of growth and erosion. Depths of over 1 km separate each atoll or reef.

Recent Research and the BIOT Marine Protected Area

Reasons behind the substantial recent research in Chagos are firstly, the understanding that unexploited reef systems like Chagos are increasingly scarce and so these reefs provide a valuable reference site, and secondly, increasing understanding that there is a need to afford full protection to more very large areas in the world. Coral reefs in particular are becoming increasingly affected by overfishing, pollution from agriculture and industry, shoreline construction and climate change (Wilkinson 2008; Burke et al. 2011). All coral reefs are highly vulnerable to these factors, which reduce their biomass and productivity, and consequences of reef deterioration may be greater than previously anticipated (Mora et al. 2011). In most of the world, conventional forms of marine management are failing to arrest the decline, so many marine science bodies, conservation organisations and international conventions have called for more and larger marine protected areas (MPAs) with effective levels of protection (e.g. Nelson and Bradner 2010; United Nations 2002; Wood et al. 2008; Veitch et al. 2012).

However, the term 'protected' varies widely, with most affording only partial protection, many allowing fishing (one of the most ecosystem distorting activities) while many, in reality, lack protection completely due to poor enforcement and implementation. The latter are commonly called 'paper parks' and, regrettably, these are the large majority given human pressures and poor in-country capacity in most places. For coral reefs, only 6% are effectively managed, 21% are ineffectively managed, and 73% lie outside any MPA (Burke et al. 2011). Reasons for MPA failures range from being declared to meet 'targets' which are inadequately resourced, to being simply overwhelmed by close proximity to human populations.

Because of this, the Pew Ocean Legacy Program recently included Chagos as one of five areas selected for protection, and promoted efforts to convince the UK Government to declare it a no-take MPA to the 200 NM boundary (Nelson and Bradner 2010). Part of this process was the creation of the Chagos Environment Network (CEN), a group of several leading UK science bodies and NGOs, whose aim was to ensure that Chagos' globally important natural environment would be conserved as a unique and valuable resource for present and future generations. In 2010, CEN responded to the UK Government's Consultation, saying that only designation as a no-take MPA "… guarantees full protection for the ecosystems and species of the Chagos Archipelago and its surrounding reefs, lagoons and waters. Only [this] provides the complete protection needed to underpin the Chagos Protected Area's value as an important global reference site for a wide range of scientific ecological, oceanographic and climate studies, as well as its continued benefits to humans into the future" (CEN 2010).

BIOT has its own administration, located in the Foreign and Commonwealth Office in London. The senior UK official in the archipelago is the commanding military officer, who is also the British Representative of the Commissioner. The UK Foreign Secretary announced the creation of BIOT as a no-take MPA, instructing the Commissioner to declare it as such in April 2010. The Commissioner, in Proclamation Number 1 of 2010, designated it such "in the name of the Queen". Existing environmental laws are currently being revised and consolidated to accommodate this status. Diego Garcia atoll is excluded from the MPA to its 3 NM boundary, and the area thus excluded is <1% of the total area though it has several pre-existing, strict environmental laws of its own, including a Ramsar site. Tuna fishing licences were discontinued as of October 2010, and the deficit of approximately $1 million per year from this was subsequently replaced from private sources. It is currently the largest no-take MPA in the world (Nelson and Bradner 2010) and is part of the 'Big Ocean Network', an information exchange network of managers and partners of existing and proposed large-scale marine managed areas (www.bigoceanmanagers.org/). Monitoring and enforcement are undertaken in large part by a patrol vessel which serves as a mobile base for both military purposes and civilian research expeditions. A Science Advisory Group provides advice to BIOT Administration, and a scientific review (Sheppard et al. 2012a) and the document 'Conservation and management in the British Indian Ocean Territory' which details scientific needs have also recently been released (Sheppard et al. 2012b).

Under the 2001 BIOT Environmental Charter, the UK Government facilitates the extension of the UKs ratification of multilateral environment agreements of benefit to the BIOT and which the BIOT has the capacity to implement. CITES and Convention on Migratory Species (CMS) have been extended to the territory, but the Convention on Biological Biodiversity (CBD) has not. The rationale is the current inability to fulfil all of the Convention requirements in Chagos, for practical reasons. But, as per the World Heritage Convention, the area is treated by the UK Government with no less strict regard, subject only to defence requirements, and in the case of CBD, the capacity to implement. Chagos harbours 76 threatened species (IUCN Red List) including Hawksbill turtle, Red foot booby, silky shark, Coconut crab, and Bigeye tuna, providing an internationally important refuge and reference site. This Ocean Legacy MPA will protect entire ecosystems rather than species in isolation, including deep-sea, pelagic, reef and small island systems including migratory species (cetaceans, sharks, turtles, birds) and those vulnerable to poaching and trade (sharks, turtles, sea cucumbers). While there is no international trade in CITES-listed species from Chagos, this emphasises its value as a reference site for comparison with exploited sites, particularly for corals, giant clams, cetaceans, marine turtles and sea cucumbers. This Convention is also relevant in Chagos for several bird species, notably boobies.

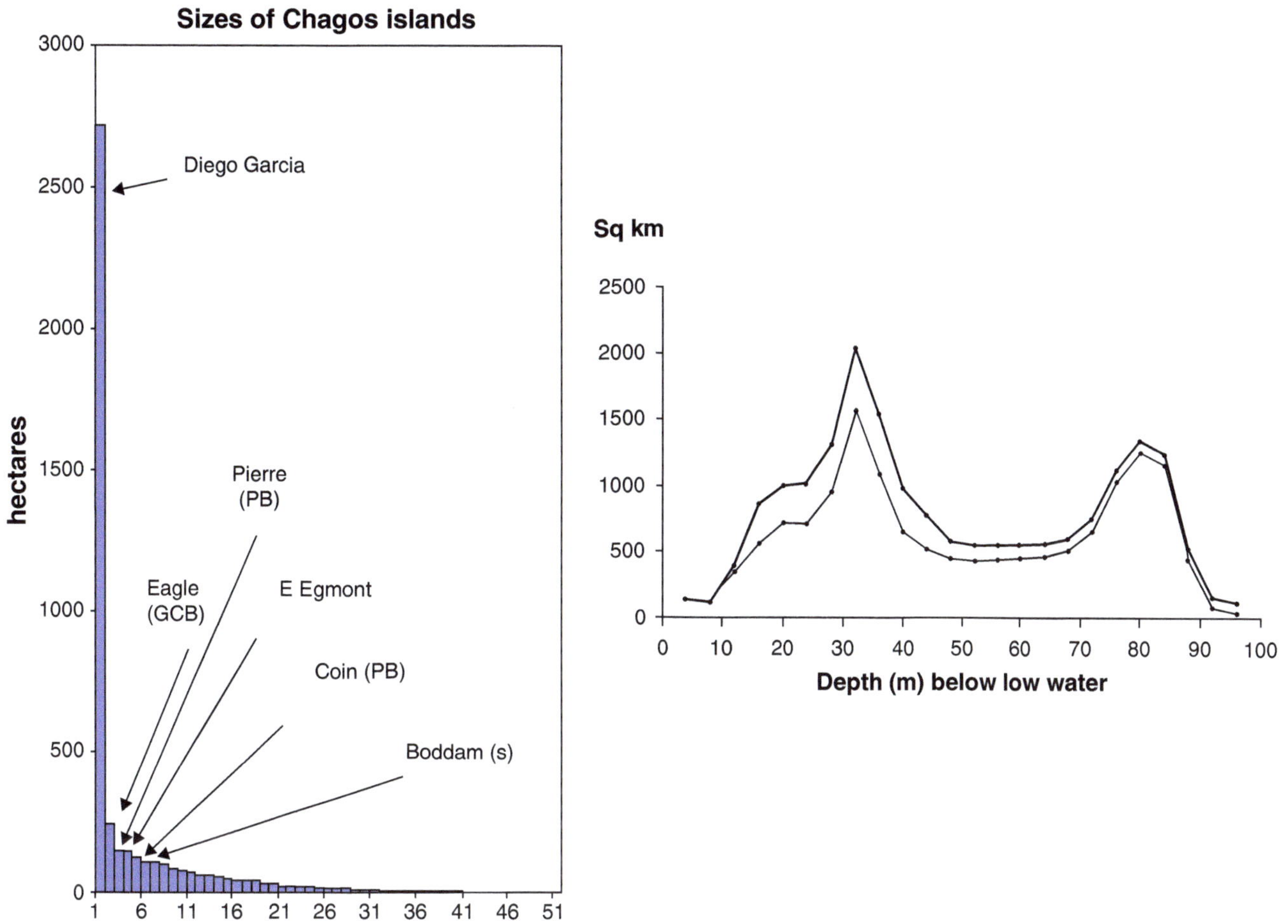

Fig. 17.3 (*Left*): Island areas in the Chagos Archipelago (scale in ha). The largest islands are named. *Right*: area of submerged substrate in the archipelago (scale in km^2) (From Dumbraveanu and Sheppard 1999)

Island and Reef Areas

Figure 17.3 (left) shows the areas of islands and reefs. There are over 50 small islands (Fig. 17.4), the number varying to some degree with tidal height and shifts of sand banks. In contrast to this, sublittoral substrate in the photic zone is calculated to be approximately 60,000 km^2 (Fig. 17.3, right) (Dumbraveanu and Sheppard 1999). The proportion of this which is actively growing reef remains uncertain because >95% of the territory still has never been studied, though some areas are apparently eroding and others support sand and/or large seagrass beds. There is enormous opportunity for new discoveries: in 2010 an expedition discovered many hectares of seagrass for example. Other parts of Chagos have been mapped using bathymetric or satellite data-based modelling (e.g. Purkis et al. 2008; Yesson et al. 2011, see Fig. 17.2).

This chapter summarises some of the recent scientific context of the British Indian Ocean Territory (BIOT) that is not covered in more detail in subsequent chapters. The Territory is, uniquely for the UK Territories, entirely built by coral reefs. Work carried out there over the past couple of decades has demonstrated the outstanding ecological value of the region, which has led to the creation of the world's largest marine protected area of about 640,000 km^2.

Biological Connections of Chagos in the Indian Ocean

Substantial values of Chagos are likely to be its role as a biological crossroad or stepping stone connecting different parts of the Indian Ocean, in possibly being a source of larvae to the west, and a reservoir of unexploited reef species for the western Indian Ocean. Currents and proximity of suitable substrata are key to these roles. To the east of Chagos, there is no shallow water until the Cocos-Keeling islands 2,750 km to the east, with Indonesia another 1,000 km further on. To the west, distances to shallow reefs are much less: 1,700 km to the Seychelles and only 1,050 km to the commonly overlooked Saya de Malha submerged banks at the northern end

Fig. 17.4 Three small islands on the Three Brothers group on the western Great Chagos Bank. These are, confusingly, three islands of the four that comprise the group. Note the huge reef surrounding the nearest; there are no reefs around the middle island (which has probably uplifted 2–3 m since being mapped as being only a shoal in the mid 1800s); while the furthest supports more typical reef flats all around (Photo Chris Davies)

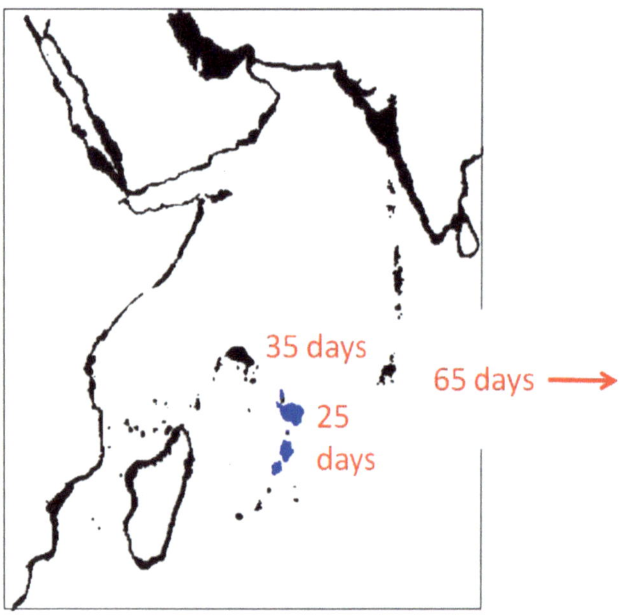

Fig. 17.5 Inert particle travel time (based on surface currents) between Chagos and adjacent reef systems ('Shoals of Capricorn' submerged reefs in blue)

of the 'Shoals of Capricorn' between the Seychelles and the Mascarenes. Figure 17.5 shows these, along with estimates of travel times for inert particles, given existing currents. Currents passing across Chagos flow towards Southeast Asia from approximately January to April, and towards the western Indian Ocean for the rest of the year, with fluctuations (Couper 1987). At 0.5 m s^{-1} (Bonjean and Lagerloef 2002) planktonic larvae from reef species would need 65 days to reach shallow habitat in the east, but only 35 and 25 days to reach the Seychelles and Saya de Malha reef systems respectively, well within the pelagic larval duration of many reef organisms. Due to its location, Chagos may be an important 'stepping-stone' for marine organisms in the Indian Ocean.

Fifteen years ago, mapping methods in which geographical distances were replaced by similarities of coral presences, showed that Chagos does appear to function as an East–West stepping stone for corals (Sheppard 1999). Using more recent synonymies, Chagos shows 82% of corals in common with the west and 88% in common with the east, and higher values are found with reef fishes (Fig. 17.6). (Such figures are strongly influenced by total numbers of species in each site which are, of course, highest in the East.) Extensive sampling (Obura 2012) has shown the coral fauna of Chagos to be more similar to the western Indian Ocean and its islands, including northern Madagascar, than it is to the much nearer Maldive archipelago, Sri Lanka or India located to its north (Fig. 17.7). Obura (2012) has shown that the previously accepted biogeographic zones (Fig. 17.7 left) need to be adjusted to encompass this biological connectivity (Fig. 17.7

right). This potentially reflects patterns of connectivity as well as the influence of habitat area on species diversity. The coral fauna of the eastern Indian Ocean is also more affiliated with the West Pacific/Coral Triangle fauna than it is to the central and western Indian Ocean fauna (Obura 2012).

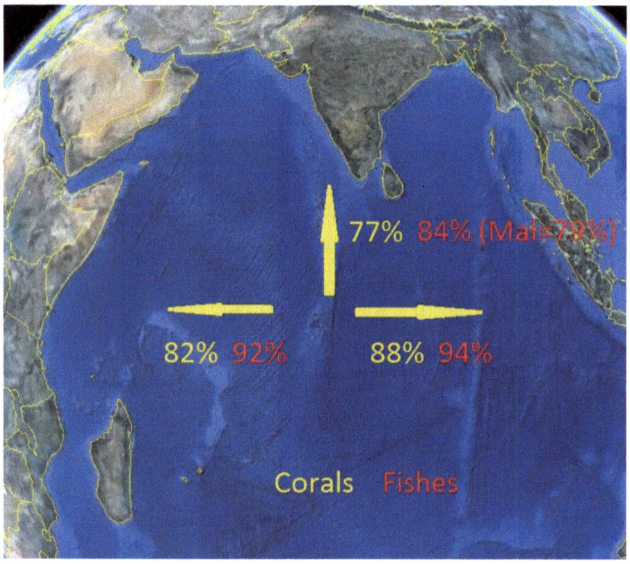

Fig. 17.6 Similarity coefficients of corals (*yellow*) and fishes (*red*) between Chagos reefs and with the west, north and eastern parts of the Indian Ocean

Preliminary examination of the coral *Platygyra daedalea* with five microsatellite loci, including samples from Chagos (Macdonald et al. submitted), revealed that, while the Chagos population had the lowest allelic diversity among the sites studied, it proved to be a source of genetic diversity for this species.

Another study has shown recent colonization of a fish species from the east, consistent with this stepping stone function, especially with reefs in the southern part of the group (Craig 2008).

Genetic programmes to examine connections between Chagos and other Indian Ocean reef sites have been initiated recently for numerous species, including about 24 reef fish species and several invertebrates including corals. For hawksbill turtles (*Eretmochelys imbricata*), genetic linkages were demonstrated for nesting females and foraging juveniles between Chagos and Seychelles, but no linkages were demonstrated with hawksbill rookeries of Western Australia (Mortimer and Broderick 1999; Mortimer et al. 2002). In the wider Indian Ocean, Vargas et al. (2010) identified nine genetic groupings, with hawksbill nesting in Chagos and Seychelles forming a single grouping distinct from those in the Arabian Gulf and from easterly sites including Western Australia, (Vargas et al. 2010 and Vargas et al. in prep). A similar pattern of connectivity was observed in foraging hawksbills in Chagos which derive mostly from rookeries in Chagos and Seychelles, though these rookeries were found to

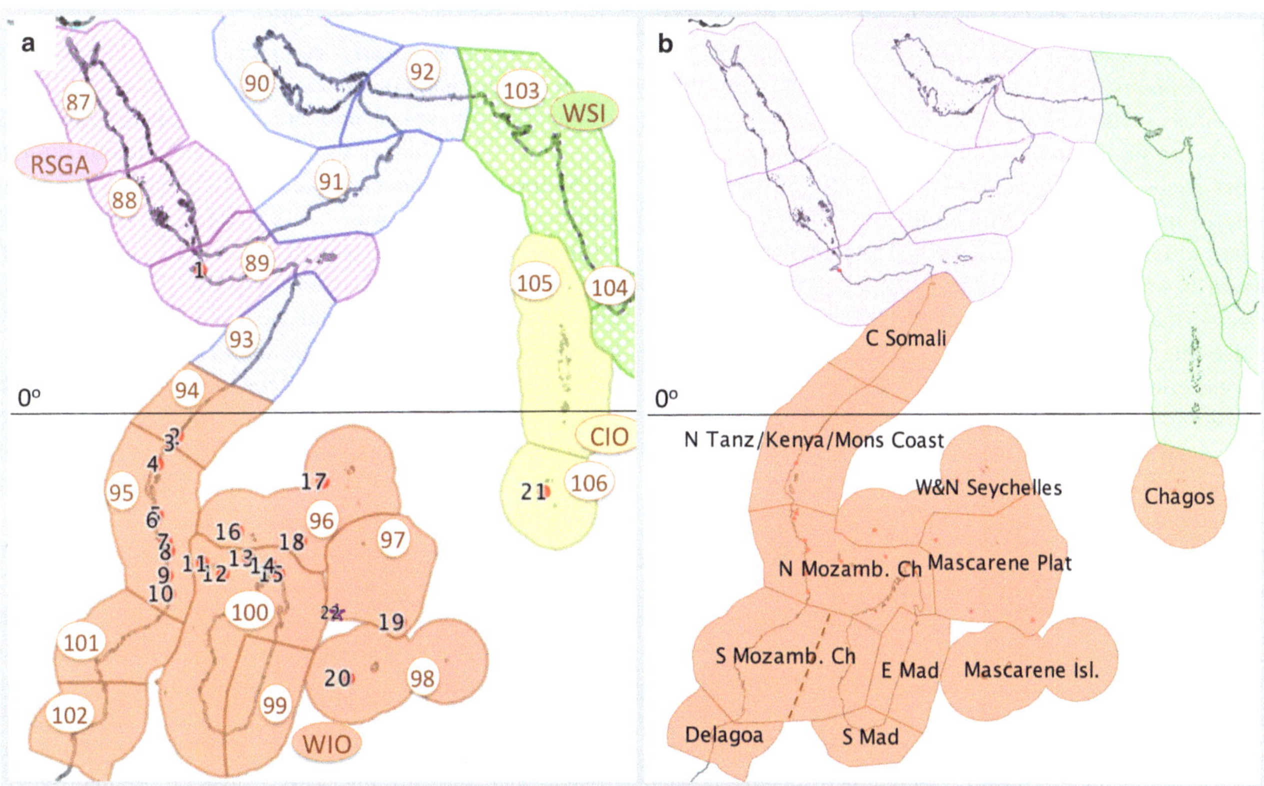

Fig. 17.7 *Left*: Ecoregions and provinces of the Western Indo-Pacific Realm – previous definitions. *Right*: revised ecoregions according to coral distributions. The boundaries correspond to the EEZs of each location (For details see Obura 2012)

also contribute substantially to foraging aggregations in Cocos Keeling (FitzSimmons 2010 unpublished report; Hahn et al. in prep). While most mtDNA lineages found in the Chagos and Seychelles were not observed elsewhere, some uncommon lineages were identical to those found in Iran, Oman, and Australia, supporting the stepping stone model.

The crown-of-thorns starfish, an important coral predator, was previously believed to be a single species, *Acanthaster planci*, but Vogler et al. (2008) have shown that the species includes four highly differentiated lineages with restricted distributions, which together form a species complex. Two of these lineages are found in the Indian Ocean, and data indicate that crown-of-thorns starfish from Chagos belong to the Southern Indian Ocean lineage (Fig. 17.8a). A more detailed phylogeographic study (Vogler et al. 2012) revealed high gene flow among the geographically distant populations in the southern Indian Ocean lineage, including between Chagos and sites in the eastern and western Indian Ocean. *Acanthaster* larvae can extend their developmental period to 7 weeks in marginal food regimes (Lucas 1982). Although the occurrence of facultative teleplanic larva remains to be confirmed (Birkeland and Lucas 1990), the low productivity found over most of the Southern Indian Ocean (<130 gC.m^{-2} day^{-1}; Reid et al. 2006) could result in extended larval durations there too, which would explain the observed high connectivity and low levels of genetic structure in the Southern Indian Ocean lineage, despite long geographic distances.

Added support for Chagos as a bridge between eastern and western Indian Ocean can be found in reef fishes (Eble et al. 2011; Gaither et al. 2010). For the peacock hind (*Cephalopholis argus*, Fig. 17.8b) and brown surgeonfish (*Acanthurus nigrofuscus*) (Fig. 17.8c), patterns of genetic linkage are similar to those observed in the hawskbill turtle, with Chagos showing greater genetic affinity with sites in the western Indian Ocean than with the east. Though, for both fishes, differences in affinity are marginal and appear to be driven by the relatively recent introduction of Pacific lineages to the eastern Indian Ocean (Eble et al. 2011).

Despite being geographically part of the Indian Ocean, the eastern Indian Ocean locations at Cocos Keeling and Christmas Islands, and Western Australia are more closely affiliated with the Pacific ichthyofauna, with only 5% of species at Cocos-Keeling being exclusively of Indian Ocean origin (Allen and Smith-Vaniz 1994). The latter islands are considered to be a part of the Indo-Polynesian Province stretching from the eastern Indian Ocean to Easter Island (Briggs and Bowen 2012). However, Cocos-Keeling and Christmas islands are a known region of overlap between Indian and Pacific faunas (Gaither et al. 2011), as indicated by the presence of a hybrid zone with a westward limit at Cocos-Keeling (Hobbs et al. 2009). Exceptions to this pattern are found among species with highly dispersive larvae, including the bluestriped snapper (*Lutjanus kasmira*; Gaither et al. 2010), trumpetfish (Bowen et al. 2001) and two moray eels (Genus *Gymnothorax*, Reece et al. 2010) which freely intermix across all their Indo-Pacific range. For the majority of less dispersive reef fishes, Chagos may act as a bridge between western Indian Ocean and Pacific populations.

The coconut crab, *Birgus latro*, is terrestrial though females lay eggs in the sea. Mitochondrial genetic work has compared Chagos with sites in the Seychelles and East Africa, and showed the Chagos population was significantly differentiated ($p < 0.05$) from Seychelles and East African populations (Tables 17.2, 17.3, and 17.4). Asymmetric gene flow, favouring migration from East Africa to Seychelles, and Seychelles to Chagos, comes from estimates of direction and mean number of migrants per generation between regions. The rate of immigration to Chagos from the west was measured at about 5 effective females per generation (breeding commences after about 5 years), using a measured mean effective female population size in the study of about 3,000, or about 0.1–0.2% per generation (N.B. this is not the counted population of individuals which is orders of magnitude greater). Thus for this species, Chagos receives more larvae from the west than flow from Chagos to the west (Fig. 17.8d). This predominantly eastward dispersal may result from the timing of egg release which partly coincides with the period of eastward current flow, and there is a high level of genetic connectivity. Additionally, a strong genetic connectivity among three sites was also observed through population structure analysis.

The pattern is clearly complex: earlier fish surveys (Winterbottom and Anderson 1997) in Chagos found two distinct assemblages: a northern portion sharing affinities with the eastern Indian Ocean and the southern portion (including Diego Garcia) more closely aligned with faunal assemblages further west. Taken together, these results confirm that Chagos is part of the western Indian Ocean province as described by Briggs (1974), though with respect to fishes, Briggs and Bowen (2012) additionally acknowledge affinities with the Indo-Polynesian province. Interestingly, Chagos shows less connectivity with the much closer Maldives to the North in some groups, which may be a function of the predominantly East–West currents. While there are recent reports of localized larval recruitment in predominately small-range fishes, these are countered by studies that show high genetic connectivity across large oceanic distances (Eble et al. 2011).

Thus early results from the genetic and distribution data indicate that Chagos is an important biogeographic crossroad between the eastern and western Indian Ocean. The so-far limited molecular data are generally concordant with biogeographic patterns indicating greater connectivity between Chagos and sites to the west (Saya de Malha Banks,

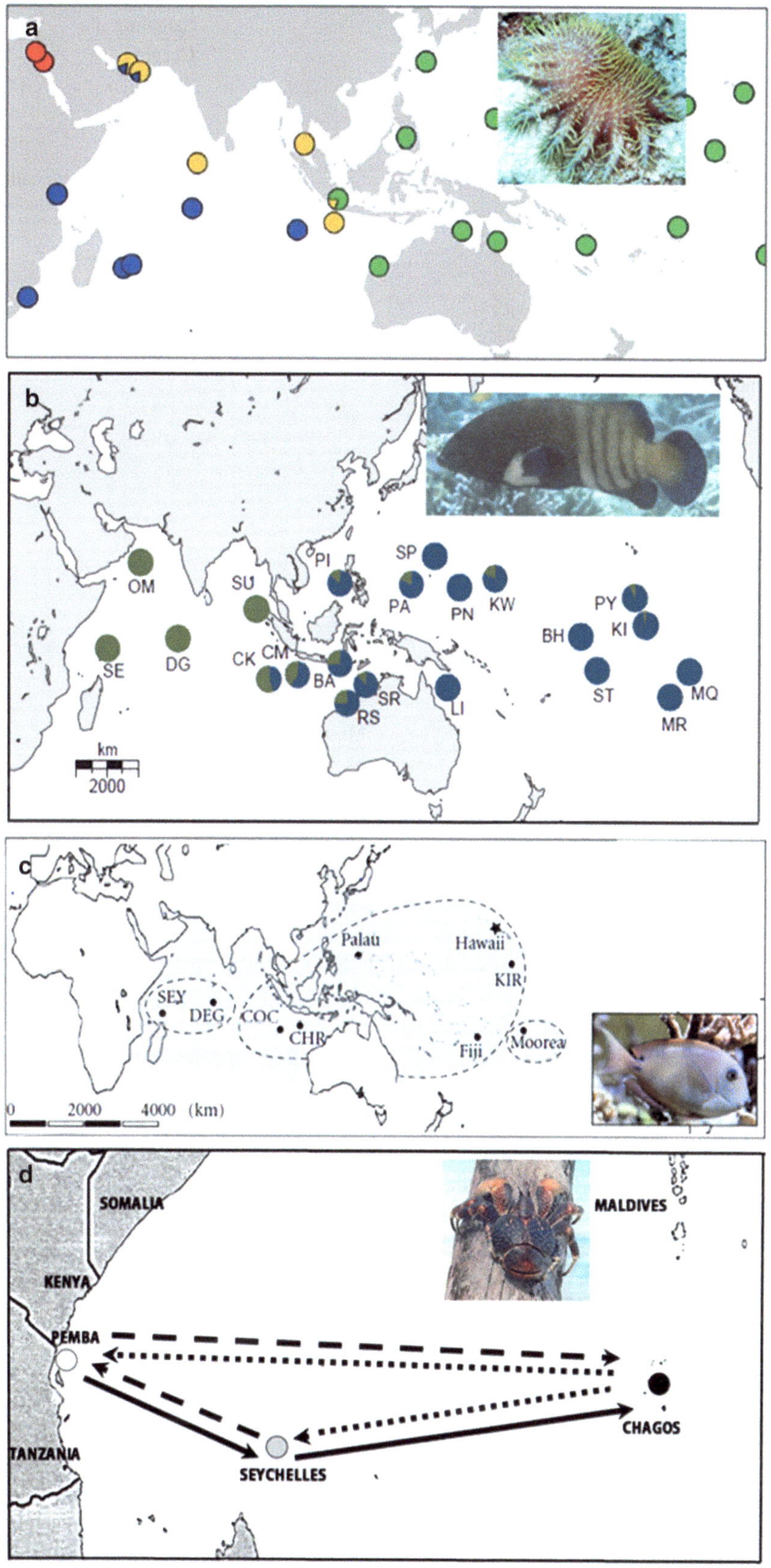

Fig. 17.8 (**a**) Crown of Thorns genetic groupings. (**b**): Peacock Hind (*Cephalopholis argus*). (**c**): Brown Surgeonfish (*Acanthurus nigrofuscus*). (**d**) Coconut crab (*Birgus latro*). Color coding for the Crown of Thorns (Vogler et al. 2008 and Vogler et al. 2012) and Peacock Hind (Gaither et al. 2011) indicate distinct genetic lineages. *Dashed lines* for the Brown Surgeonfish (Eble et al. 2011) indicate genetically independent populations (Photo credit: www.aquaportail.com. Image 12 b and 12c reprinted from Gaither et al. (2011) and Eble et al. (2011) with permission from the authors). For (**d**) solidity of *arrow lines* represents relative amounts of gene flow, so that for this terrestrial crab flow is mainly eastwards during the Equatorial Counter Current flow

Table 17.2 Coconut crabs: Molecular diversity of 600 bp mtDNA COX1 region sampling from Chagos, Seychelles and Misali populations

Sampling site	*n*	*Nh*	*S*	*h*	π	θ
Chagos	12	8	12	0.848	0.0067	4.97
Seychelles	18	16	18	0.987	0.0071	5.23
Misali	6	6	5	1	0.0056	4
Total	36	18	16	0.922	0.0076	4.65

n sample size, *Nh* number of haplotypes, *S* number of variable sites, *h* haplotypic diversity, π nucleotide diversity, θ Theta per sequence

Table 17.3 Pairwise F_{ST} values of population differentiation in coconut crab sampling in Western Indian Ocean

	Chagos	Seychelles	Misali
Chagos			
Seychelles	0.422*		
Misali	0.407*	0.017	

By Arlequin 3.1
$*p<0.05$

Table 17.4 Effective population size and mean of migrants per generation between populations of coconut crab calculated with Bayesian based on mtDNA COX1 region

		To		
	$N_{ef}(10^3)$	Chagos	Seychelles	Misali
From				
Chagos	2.96 (0–17.77)		0.4 (0–20.5)	0.59 (0–4.65)
Seychelles	38.04 (6.92–60.24)	3.35 (0–31.5)		1.6 (0–10.39)
Misali	3.9 (0–9.34)	1.29 (0–7.99)	32.34 (6.7–86)	

95% credible intervals from 100,000 drawn from Markov chains are indicated in brackets. Female effective population size $Nef=\theta/\mu$, μ is mutation rate per site per generation. For example, mutation rate of universal COX1 in terrestrial Jamaica land crab is 1.66% per site per million years – Schubart et al. 1998

Seychelles, and East Africa) than with those to the east (Cocos Keeling, Christmas Island, and west Australia). However, patterns of differentiation and migration vary considerably between species, which may reflect species level differences in dispersal ability, reproductive strategy, competitive ability, or habitat requirements. In some species, large distances may be successfully bridged by the existence of so-called 'teleplanic larvae' which afford greatly expanded larval durations (Scheltema 1988). In other species, like some corals, larvae are competent for up to 105 days (Wilson and Harrison 1998). This contrasts with reef fishes which exhibit an average larval duration of about 1 month (though it varies enormously; Brothers and Thresher 1985; Sale 2002).

Although it is probable that Chagos is an important stepping stone in the western Indian Ocean, the rate at which this happens for most groups is still not known. But in fact, the number of migrants needed to maintain genetic coherence between populations is small (Slatkin 1977, 1982). As noted by Hellberg (2007), for management purposes we need to know whether or not connections are made *en masse* every several thousand generations, or whether connections occur at demographically relevant intervals.

Nonetheless, patterns of connectivity as they exist today highlight the importance of the Chagos as a biogeographic stepping stone between the eastern and western Indian Ocean, although many questions remain unresolved. If Chagos is mainly a net recipient of larvae then its rich and relatively undamaged state affords it a very high conservation value as refugia. If Chagos exports biological diversity to the over-exploited sites to the west and east, then the reefs of Chagos, and of the MPA, would have even greater value.

Pelagic Fishing and Fisheries

Fisheries provided most of the income for Chagos until the MPA was created, with the last licences expiring on 31st October 2010. The main fisheries were longline and purse seine for tuna, and a smaller Mauritian inshore fishery also existed. There is a recreational fishery in Diego Garcia, which operates within the 3 NM limit surrounding the atoll, which is relatively small, taking (in 2008) 25.2 tonnes of tuna and tuna-like species (76% of the catch) the remainder being reef-associated species (Mees et al. 2009 and see Chap. 19).

Fig. 17.9 Shark poaching boats arrested in Diego Garcia (Photo A and C Sheppard)

The longline fishery in Chagos was active year-round, mainly under Taiwanese and Japanese flagged vessels targeting large pelagic species, including yellowfin tuna (*Thunnus albacares*), bigeye tuna (*Thunnus obesus*), swordfish (*Xiphias gladius*), striped marlin (*Tetrapturus audax*) and Indo-Pacific sailfish (*Istiophorus platypterus*), with annual catches ranging from 371 to 1,366 tonnes between 2005 and 2010 (Koldewey et al. 2010). The purse-seine fishery targeted yellowfin- predominantly juveniles – and skipjack tuna (*Katsuwonus pelamis*) and was highly seasonal, operating between November and March with a peak usually in December and January (Mees et al. 2009). Log book records show that catches, mainly by Spanish and French flagged vessels, were highly variable, ranging from <100 to ~24,000 tonnes annually (Koldewey et al. 2010).

The Mauritian inshore fishery targeted demersal species, principally snappers, emperors and groupers, and logbook records indicated that the catches were between 200 and 300 tonnes per year for the period 1991–1997, decreasing to between 100 and 150 tonnes from 2004 (Mees et al. 2008). There have been no fishing activities from Mauritius in the last few years prior to MPA designation.

Fisheries suffered from poor documentation of by-catch and from illegal fishing. By-catch was inadequately recorded through a logbook system supported by very limited observer coverage – mean coverage was 1.24% per season for longline fishing and 5.56% for purse-seine fishing (Koldewey et al. 2010). Even with this uncertainty, the by-catch in the Chagos was clearly substantial, particularly for sharks, rays and billfish (Pearce 1996; Roberts 2007; Graham et al. 2010; Koldewey et al. 2010). Shark and sea cucumber poaching (Fig. 17.9) are covered in later chapters.

Illegal fishing remains a management issue following the implementation of the MPA and enforcement will be key to its effectiveness, as for most MPAs globally. The size and location of Chagos as an MPA is particularly important in these respects as the western Indian Ocean has some of the most exploited, poorly understood and badly protected and managed coastal and pelagic fisheries in the world (Kimani et al. 2009; van der Elst et al. 2005), while overall catches from them continue to dramatically increase (FAO 2010). The Regional Fisheries Management Organisation (RFMO) that oversees the region containing Chagos is included within the Indian Ocean Tuna Commission (IOTC). Unfortunately, the IOTC, as with many tuna RFMOs, is widely recognised as having numerous legal and technical weaknesses Anonymous 2009 Cullis-Suzuki and Pauly 2010). Tuna in the Indian Ocean are considered to be close to the maximum sustainable yield (bigeye) or overexploited (yellowfin) and even skipjack, which is generally considered a highly productive and resilient species, has been highlighted for close monitoring (IOTC 2010). This also needs to be considered in a global context where, in 2011, all species of tunas, bonitos, mackerels, swordfish and marlins were assessed for the IUCN Red List of Threatened Species, with 11% of the world's 61 species documented to be at serious risk of extinction (Collette et al. 2011). Moreover, the IUCN classed a third of oceanic shark species, species that are regularly caught as by-catch in pelagic fisheries, in a threatened category (Camhi et al. 2009). Illegal, unreported and unregulated

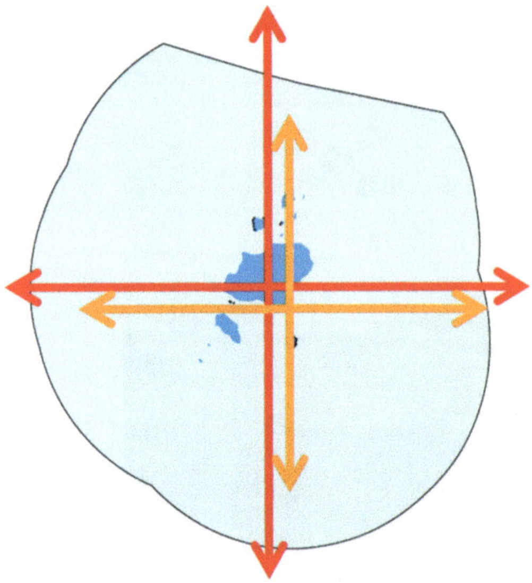

Fig. 17.10 The median lifetime ranges of skipjack (*red*) and yellowfin tuna (*yellow*), superimposed on a map of the Chagos MPA (Ranges from Pacific: Sibert and Hampton 2003)

fishing is not a trivial component of the catch and adds substantial uncertainty into population assessments (Ahrens 2010).

There is increasing evidence that large MPAs like Chagos can benefit pelagic species that have the potential to exhibit highly mobile behaviours (reviewed in Game et al. 2009; Koldewey et al. 2010). In reality, the phrase 'highly migratory' which is frequently used to describe tuna and oceanic sharks, often has little biological meaning, with studies of tuna mobility demonstrating they would benefit from national-level closures (Sibert and Hampton 2003). Pelagic fish demonstrate considerable stability and persistence, and predictability of some habitat features does occur within the pelagic realm (Alpine 2005; Baum et al. 2003; Etnoyer et al. 2004; Hyrenbach et al. 2000; Worm et al. 2003). Migratory predators like tuna do not move randomly, but associate with certain environmental and/or physical features (Hughes et al. 2010; Itano and Holland 2000; Morato et al. 2010a; Schaefer and Fuller 2010) – indeed tuna fisheries have been shown to benefit from such aggregations (Morato et al. 2010b), meaning that positive, measurable reserve effects on pelagic populations exist (Baum et al. 2003; Hyrenbach et al. 2002; Jensen et al. 2010; Roberts and Sargant 2002; Worm et al. 2003, 2005). Several studies have shown that migratory species can benefit from no-take marine reserves (Beare et al. 2010; Jensen et al. 2010; Palumbi 2004; Polunin and Roberts 1993). Ranges of skipjack and yellowfin tuna have not been measured in the Indian Ocean, but if their ranges in a Pacific archipelago (Sibert and Hampton 2003) are superimposed onto the Indian Ocean, the BIOT MPA would encompas the median lifetime range of these two key species (Fig. 17.10). If these ranges do apply to the Indian Ocean for these species, it is likely that the Territory does provide considerable scope for their conservation. Studies are now underway to test this in Chagos using a combination of pelagic video monitoring and tagging (Zoological Society of London and University of Western Australia).

Pelagic MPAs such as Chagos are thus an important tool in marine conservation management (Game et al. 2009) and are rapidly becoming a reality (Pala 2009), although the challenges relating to their implementation may be both costly and logistically challenging (Kaplan et al. 2010). Large MPAs are considered essential to protect species such as large pelagic fish and marine mammals (Wood et al. 2008) as well as offsetting the concentration of fishing effort outside them (Walters 2000) and maintaining ecological value (Nelson and Bradner 2010). Their importance for top predators has been highlighted by the most comprehensive, decade-long, open ocean tagging study in the Pacific that clearly demonstrated that top predators – including whales, seals, tuna, sharks, seabirds, turtles – exploit their environment in predictable ways, providing the foundation for spatial management of large marine ecosystems (Block et al. 2011). Extending to 200 NM, the Chagos MPA offers an extremely valuable opportunity to understand the effects of large-scale protection on pelagic, migratory species, both within the MPA and within a regional context.

Deep-Water Ecosystems

Seamounts

Yesson et al. (2011) determined that 86 seamounts (conical topographic rises of >1,000 m elevation) and 243 knolls (conical topographic rises of elevation 200–999 m) are predicted to occur within the Chagos MPA (Fig. 17.11). Chagos thus could contain more than 10% of all Indian Ocean seamounts and so the area is regionally important for these features as well. Given that globally only 506 seamounts and 606 knolls lie in protected areas (Yesson et al. 2011, based on the world database of protected areas 2009), this means that the Chagos MPA increased the world's protection of seamounts by 17% and knolls by 40%. Previous emphasis has been on shallow-water ecosystems, but protection of its seamounts is also important, especially considering their high biodiversity, often representing entirely unique ecosystems (Clark et al. 2006). Although the geology of some of the seamounts and ridges in the Indian Ocean has been explored, including the Chagos-Laccadives Ridge, seamount fauna is poorly known (Demopoulos et al. 2003; Rogers et al. 2007). Some data on fish exist, mainly resulting from exploratory or commercial fishing, but no specific information relates to the Chagos-Laccadive Ridge. Recent modelling studies based

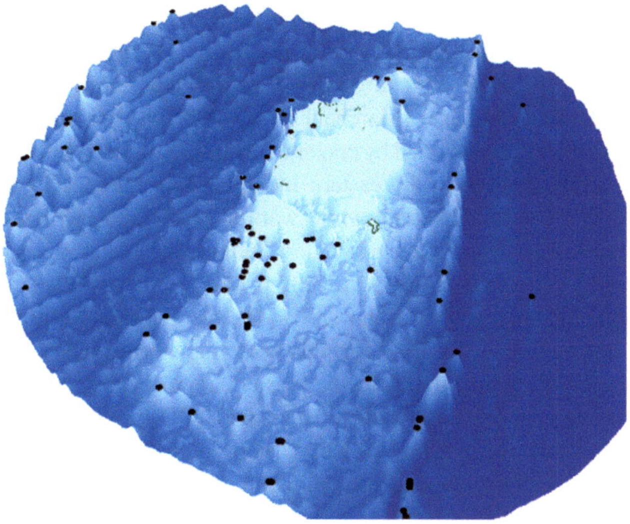

Fig. 17.11 Seamounts of the Chagos MPA as identified in Yesson et al. (2011). Bathymetry data from shuttle radar topography mission 30 arc-sec grid. (http://www2.jpl.nasa.gov/srtm/)

on 30-arc sec satellite bathymetry data indicate that the Indian Ocean hosts fewer seamounts than the Atlantic and Pacific Oceans (Yesson et al. 2011), and many are associated with ridges or originate at ridges.

Deep Water Fishing

The Indian Ocean suffers increasing pressure from deep-sea fishing that threatens these seamounts and other benthic habitats. The fact that deep-water fishing or trawling has never been documented in Chagos makes it particularly important. Deep-sea fishing in the Indian Ocean was mostly undertaken by distant-water fleets, particularly from the USSR. These fisheries targeted redbait (*Emmelichthys nitidus*) and rubyfish (*Plagiogeneion rubiginosus*) with catches peaking about 1980 and then decreasing to the mid 1980s (Clark et al. 2007). Fishing then switched to alfonsino (*Beryx splendens*) in the 1990s as new seamounts were exploited. Some exploratory trawling was also carried out on the Madagascar Ridge and South West Indian Ridge by French vessels in the 1970s and 1980s, particularly targeting Walter's Shoals and Sapmer Bank (Collette and Parin 1991). In the late 1990s, a new fishery developed on the South West Indian Ocean Ridge with trawlers targeting deep-water species such as orange roughy (*Hoplostethus atlanticus*), black cardinal fish (*Epigonus telescopus*), southern boarfish (*Pseudopentaceros richardsoni*), oreo (Oreosomatidae) and alfonsino (Clark et al. 2007). This fishery rapidly expanded, with estimated catches of orange roughy being approximately 10,000 t, until that fishery collapsed. Fishing then shifted to the Madagascar Plateau, Mozambique Ridge and Mid-Indian Ocean Ridge, targeting alfonsino and rubyfish (Clark et al. 2007). Most of these areas therefore have likely been significantly impacted by deep-sea bottom fisheries.

Deep-sea fishing in most of the Indian Ocean is continuing and showing signs of increasing its geographic spread, mainly targeting orange roughy and alfonsino. Recent fishing has also taken place on the Broken Ridge (eastern Indian Ocean), 90 East Ridge, possibly the Central Indian Ridge, the Mozambique Ridge and Plateau and Walter's Shoal (western Indian Ocean), where a deep-water fishery for lobster (*Palinurus barbarae*) has developed (Bensch et al. 2008). The banks around Mauritius and high seas portions of the Saya da Malha Bank have been targeted by fisheries for *Lutjanus* spp., and lethrinid fish (SWIOFC 2009), and there are also reports of unregulated gillnet fishing in the Southern Indian Ocean such as at Walter's Shoal, which target sharks (Shotton 2006). Currently, there is little or no information available on impacts of deep-sea fishing on high seas areas of the Indian Ocean on populations of target or by-catch species, or on seabed ecosystems. Reporting of data are complicated by issues of commercial confidentiality in fisheries where individual stocks may be located across a wide area (e.g. the South West Indian Ridge) and, up until June 2012, with the entering into force of the South Indian Ocean Fisheries Agreement (SIOFA), there was no adequate regional fisheries management organisation. At present, new fisheries are developing in the region with no apparent assessment of resource size or appropriate exploitation levels, and with no estimate of impacts on vulnerable marine ecosystems.

Global modelling studies are currently evaluating habitat suitability for deep-sea Scleractinia and Octocorallia, at 30-arc sec resolution (Davies and Guinotte 2011; Yesson et al. 2012) which indicate that suitable habitat for these organisms are likely to exist on deep slopes and seamounts within the Chagos MPA. Given the lack of a history of deep-sea fishing in the region around the Chagos Archipelago, it is likely that associated communities of deep invertebrates and fish are still largely intact, unlike on most other ridges in the Indian Ocean which have been subject to recent, continuing or expanding fisheries.

Considering the paucity of research on equatorial seamounts, the Chagos region is particularly important for deep-water ecosystem conservation too, both at a regional and global level. It also provides a unique opportunity to investigate the energetic links between production associated with shallow-water coral reefs and deep-water ecosystems.

Climate and Long Term Environmental Monitoring Programmes in BIOT

Work is increasingly needed on climate change projections, yet the Indian Ocean forms a very large gap in many global monitoring programmes. Numerous temperature loggers have been deployed on these reefs at 5, 15 and 25 m depth on several lagoonal and seaward facing reefs. These have shown upward movement of thermoclines with periods of 1–4 days, coinciding partly with times of the warmest seasonal temperatures (Sheppard 2009). This may have important consequences in terms of reducing stress on corals (next chapter). Furthermore, the temperature records report the possible intrusion of internal waves onto the platform, which is relevant to issues of nutrient cycling, though this key aspect is another which remains to be researched.

Climate observations are extremely sparse. Further, instrumental SST data are associated with errors that are as large as the SST anomalies across a range of time scales (Annamalai et al. 1999). There is currently one weather station at Diego Garcia, with much good data but with several gaps. 'Conventional' infrared satellite measurements of SST are not adequate to capture important SST perturbations that occur in the rainy season when convective activity and cloud cover is highest (Vialard et al. 2009), but from the late 1990s better quality SST measurement became possible to reveal the role of Chagos with respect to air-sea interactions. Therefore, a continuous monitoring programme of important climatic and oceanographic parameters at Chagos is needed for a better understanding of air-sea interactions that are crucial in global climate issues.

Chagos is situated in a key region of climate variability. It lies at the eastern margin of the 'Seychelles-Chagos thermocline ridge' (Hermes and Reason 2008), along which the thermocline rises close to the surface and upwelling of cold water (Vialard et al. 2009). In contrast to most upwelling regions, surface water of the Seychelles-Chagos ridge is extremely warm. In austral summer, the main rainy season, sea surface temperature (SST) varies only between 28.5°C and 30°C, a range in which the atmosphere is very sensitive to small oceanic changes (Timm et al. 2005). The high SST combined with the shallow thermocline makes this a region with very strong air-sea interactions.

The Seychelles-Chagos region has a distinct oceanic and atmospheric variability at multiple time scales, each with significant climatic consequences. It has the strongest intraseasonal SST variance in the Indo-Pacific warm pool, because the shallow thermocline is very responsive to atmospheric fluxes (Vialard et al. 2008, 2009). SST cooling of 1°C–1.5°C, lasting for 1–2 months, may occur during austral summer, followed by a short lag and sharp increase in atmospheric convective activity (Vialard et al. 2009) associated with the Madden-Julian-Oscillation (MJO) (Madden and Julian 1994). The MJO has a time scale of 30–80 days; it explains much of the variance of tropical convection, and modulates cyclonic activity.

On interannual time scales, the Seychelles-Chagos ridge is affected by the El Nino-Southern Oscillation (ENSO) and the Indian Ocean Dipole (IOD) (Webster et al. 1999; Saji et al. 1999). ENSO events lead to a displacement of the West Pacific warm pool and affect the Indian Ocean via the so-called atmospheric bridge. Anomalous subsidence over Indonesia during El Nino years leads to a cooling over the maritime continent and a basin-scale warming of the Indian Ocean, particularly the western part. The IOD is a coupled ocean–atmosphere instability centred in the tropical Indian Ocean that affects the climate of the countries that surround the Indian Ocean basin (Marchant et al. 2007). A positive IOD period is characterized by cooler than normal water and below average rainfall in the eastern Indian Ocean, Indonesia and over parts of Australia, while warmer water and increased rainfall is observed in the western Indian Ocean and equatorial East Africa. A negative IOD period is characterized by warmer than normal water and above average rainfall in the eastern Indian Ocean sector and cooler than normal water and below average rainfall in the western Indian Ocean. Some studies suggest that positive (negative) IOD events may be triggered by El Niño (La Niña), while others find that IOD events occur independently although they may overlap with El Niño/La Niña events, e.g. the strong 1998 ENSO (Saji and Yamagata 2003). The IOD has two 'centres of action': one off the coast of Sumatra, in the eastern Indian Ocean sector, and one at the Chagos Archipelago (Saji and Yamagata 2003).

Reliable instrumental records of the IOD are limited to the past 50 years. On decadal and longer time scales, information on climate variability at Chagos has been gained from geochemical proxies archived in long-lived corals (Pfeiffer et al. 2004, 2006, 2009; Timm et al. 2005). These data indicate that changes in the mean climate may influence the impact of large-scale climate phenomena at Chagos, possibly affecting climate in other countries surrounding the Indian Ocean. For example, a shift towards higher mean SSTs occurred in 1975 (Timm et al. 2005; Pfeiffer et al. 2006, 2009), after which small SST changes induced by ENSO caused much larger precipitation anomalies (Timm et al. 2005; Pfeiffer et al. 2006). Similar changes in rainfall variability have been found in South Africa (Richard et al. 2000), Sri Lanka and southern India (Zubair and Ropelewski 2006).

The long-term variability of the IOD has also been investigated with coral proxies from the Seychelles and Indonesia (Abram et al. 2008). The coral reconstruction suggests an increase in the strength and frequency of IOD events during the twentieth century, which may also influence the distribution of rainfall in the tropical Indian Ocean. The work highlights the importance of the Chagos Archipelago for climate variability research in the Indian Ocean sector and beyond, and emphasises the need for high-quality in-situ data recording of climatic parameters at Chagos.

Conclusions

Chagos is unquestionably a highly valuable biological asset in an ocean where most reefs show significant and continuing decline in health. The reefs are in exceptionally good condition (see next chapters on the corals and fishes) and are important in terms of biodiversity and productivity, and for their function as a biogeographic crossroads in the central Indian

Ocean. Its protection is of great importance given global pressures from overfishing and other activities resulting from human population growth. As such, Chagos joins a small handful of large, protected sites in the world, and is the only one in the Indian Ocean. The importance of the BIOT MPA was reinforced by a recent report across marine science disciplines suggesting that the world's ocean is at high risk of entering a phase of disturbance unprecedented in human history (Rogers and Laffoley 2011). Indeed, many more effective and very large MPAs need to be created if international goals for protecting the oceans are to be met, although the number of locations where this will be possible is diminishing.

The biological and conservation value of the Chagos therefore is evident. Nevertheless, the formation and maintenance of the MPA has been met with opposition from various sources, including the oceanic fishing industry, the government of Mauritius that claims the territory, and from some Chagossians who were removed approximately 40 years ago and some of their representatives. While the decision to remove the inhabitants at that time was based on politics and defence concerns rather than for conservation, the present excellent condition of this large area has been an unplanned consequence of the subsequent lack of exploitation. However, the MPA was created 'without prejudice' to any future resettlement, and if resettlement does occur then management must be adequate to avoid the problems evident from similar experience around the world.

The high value of relatively undisturbed areas encompassing a range of functionally linked ecosystems is becoming increasingly recognised at the same time that their number is diminishing world-wide (http://www.globaloceanlegacy.org/). While there are many 'managed' reef sites elsewhere in the tropical oceans, almost all are themselves in a poor condition compared to Chagos and a few Pacific sites. The 'shifting baseline syndrome' (Pauly 1995) applies to marine management and in many places has led to situations where decisions are taken that are highly disadvantageous to both the natural systems and to the people dependant on them. Most coral reefs remain unprotected or protected in name only (Burke et al. 2011). The huge scale of the interconnected network of atolls and banks in Chagos, and its effective governance, are likely to become increasingly important both directly and as a scientific reference site in the Indian Ocean. BIOT has, at present and the foreseeable future, a governance which will enable this protection to persist. Priorities for the region now are effective management, so that the benefits of a well protected MPA are likely to extend into the future including area far beyond the boundaries of the Chagos MPA.

Acknowledgements Some of this material was published previously in Sheppard et al. (2012) and is reproduced with permission of Wiley & Sons.

References

Abram NJ, Gagan MK, Cole JE, Hantoro WS, Mudelsee M (2008) Recent intensification of tropical climate variability in the Indian Ocean. Nat Geosci 1:849–853

Ahrens RNM (2010) Global analysis of apparent trends in abundance and recruitment of large tunas and billfishes inferred from Japanese longline catch and effort data. Ph.D. thesis, University of British Columbia, 422pp

Allen GR, Smith-Vaniz WF (1994) Fishes of the Cocos (Keeling) Islands. Atoll Res Bull 412:1–21

Alpine J (2005) The application of closed area management in pelagic ecosystems: a viable strategy for longline fisheries? B.Sc. thesis, CSIRO Marine Research and School of Zoology, University of Tasmania, 68pp

Annamalai H, Slingo JM, Sperber KR, Hodges K (1999) The mean evolution and variability of the Asian summer monsoon: comparison of the ECMWF and NCEP/NCAR reanalysis. Mon Weather Rev 127:1157–1186

Anonymous (2009) Report of the IOTC performance review panel: January 2009. Indian Ocean Tuna Commission

Baum J, Myers R, Kehler D, Worm B, Harley S, Doherty P (2003) Collapse and conservation of shark populations in the Northwest Atlantic. Science 299:389–392

Beare D, Hölker F, Engelhard GH, McKenzie E, Reid DG (2010) An unintended experiment in fisheries science: a marine area protected by war results in Mexican waves in fish numbers-at-age. Naturwissenschaften 97:797–808

Bellamy DJ (1979) Half of paradise. Cassell, London, p 180

Bensch A, Gianni M, Grébroval D, Sanders JS, Hjort A (2008) Worldwide review of bottom fisheries in the high seas. FAO Fisheries and Aquaculture Technical Paper 522

Birkeland CE, Lucas JS (1990) *Acanthaster planci:* major management problem of coral reefs. CRC Press, Boca Raton

Block BA, Jonsen ID, Jorgensen SJ, Winship AJ, Shaffer SA, Bograd SJ, Hazen EL, Foley DG, Breed GA, Harrison AL, Ganong JE, Swithenbank A, Castleton M, Dewar H, Mate BR, Shillinger GL, Schaefer KM, Benson SR, Weise MJ, Henry RW, Costa DP (2011) Tracking apex marine predator movements in a dynamic ocean. Nature 475:86–90. doi:10.1038/nature10082

Bonjean F, Lagerloef GSE (2002) Diagnostic model and analysis of the surface currents in the tropical Pacific Ocean. J Phys Oceanogr 32:2938–2954

Bourne GC (1888) The atoll of Diego Garcia and the coral formations of the Indian Ocean. Proc Roy Soc 43:440–461

Bowen BW, Bass AL, Rocha LA, Grant WS, Robertson DR (2001) Phylogeography of the trumpetfishes (*Aulostomus*): ring species complex on a global scale. Evolution 55:1029–1039

Briggs JC (1974) Marine zoogeography. McGraw-Hill, New York

Briggs JC, Bowen BW (2012) A realignment of marine biogeographic provinces with particular reference to fish distributions. Journal of Biogeography 39:12–30

Brothers EB, Thresher RE (1985) Pelagic duration, dispersal and the distribution of Indo-Pacific coral reef fishes. In: Reaka ML (ed) The ecology of coral reefs, vol 3, No 1, Symposia series for undersea research. NOAA's Undersea research program. US Department of Commerce, Washington, DC, pp 53–69

Burke L, Reytar K, Spalding M, Perry A (2011) Reefs at risk revisited. World Resources Institute, Washington, DC, p 130

Camhi DV, Valenti SV, Fordham SV, Fowler SL, Gibson C (2009) The conservation status of pelagic sharks and rays: report of the IUCN shark specialist group pelagic shark red list workshop. IUCN Species Survival Commission Shark Specialist Group, Newbury

Chagos Environment Network (2010) Consultation on whether to establish a Marine protected area in the British Indian Ocean Territory. protectchagos.org/wp-content/uploads/CEN_Submission-Final-FINAL.pdf

Clark MR, Tittensor D, Rogers AD, Brewin P, Schlacher T, Rowden A, Stocks K, Consalvey M (2006) Seamounts, deep-sea corals and fisheries: vulnerability of deep-sea corals to fishing on seamounts beyond areas of national jurisdiction. UNEP/WCMC, Cambridge, UK

Clark MR, Vinnichenko VI, Gordon JDM, Beck-Bulat GZ, Kukharev NN, Kakora AF (2007) Large-scale distant-water trawl fisheries on seamounts. In: Pitcher TJ, Morato T, Hart PJB, Clark MR, Haggan N, Santos RS (eds) Seamounts: ecology, fisheries & conservation, vol 12, Fish and Aquatic Resources Series. Blackwell, Oxford, UK, pp 361–399

Collette BB, et al (2011) High value and long life – double jeopardy for tunas and billfishes. Science 333:291–292

Collette BB, Parin NV (1991) Shallow-water fishes of Walters Shoals, Madagascar Ridge. Bull Mar Sci 48:1–22

Couper A (1987) The Times Atlas of the Oceans. Van Nostrand Reinhold, New York, p 256

Craig MT (2008) The goldrim surgeonfish (*Acanthurus nigricans*; Acanthuridae) from Diego Garcia, Chagos Archipelago: first record for the central Indian Ocean. Zootaxa 1850:65–68

Cullis-Suzuki S, Pauly D (2010) Failing the high seas: a global evaluation of regional fisheries management organizations. Mar Policy 34:1–141

Davies AJ, Guinotte JM (2011) Global habitat suitability for framework-forming cold-water corals. PLoS One 6(4):e18483

Demopoulos AWJ, Smith CR, Tyler PA (2003) Ecology of the deep Indian Ocean floor. In: Tyler PA (ed) Ecosystems of the world volume 28: ecosystems of the deep Ocean. Elsevier, Amsterdam, 569pp

Dumbraveanu D, Sheppard CRC (1999) Areas substrate at different depths in the Chagos Archipelago. In: Sheppard CRC, Seaward MRD (eds) Ecology of the Chagos Archipelago. Linnean Society/ Westbury, London, p 44

Eble JA, Craig MT, Rocha L, Bowen BW (2011) Not all offspring stay close to home: trans-Pacific dispersal and gene flow in Indo-Pacific reef fishes. J of Mar Bio. Article ID 518516, 12 pages doi:10.1155/2011/518516

Eisenhauer A, Heiss G, Sheppard CRC, Dullo WC (1999) Reef and Island formation and late Holocene Sea-level Changes in the Chagos Islands. In: Sheppard CRC, Seaward MRD (eds) Ecology of the Chagos Archipelago, Special Issue of Linnean Society. Westbury for the Linnean Society of London, London, pp 21–33

Etnoyer P, Canny D, Mate B, Morgan L (2004) Persistent pelagic habitats in the Baja California to Bering Sea (BSB) ecoregion. Oceanography 17:90–101

FAO Fisheries and Aquaculture Department (2010) The State of World Fisheries and Aquaculture 2010 (SOFIA). Food and Agriculture Organisation of the United Nations, Rome, 218pp

FitzSimmons N (2010) Final report: population genetic studies in support of conservation and management of hawksbill turtles in the Indian Ocean. Marine turtle conservation fund award 98210-7-G126. Unpublished report to Multinational Species Conservation Fund, 24pp

Francis TJ, Shor GG (1966) Seismic refraction measurements in the northwest Indian Ocean. J Geophys Res 71:427–449

Gaither MR, Toonen RJ, Robertson DR, Plane S, Bowen BW (2010) Genetic evaluation of marine biogeographical barriers: perspectives from two widespread IndoPacific snappers (*Lutjanus kasmira* and *Lutjanus fulvus*). J Biogeogr 37:133–147

Gaither MR, Bowen BW, Bordenave TR, Rocha LA, Newman SJ, Gomez JA, van Herwerden L, Craig MT (2011) Phylogeography of the reef fish *Cephalopholis argus* (Epinephelidae) indicates Pleistocene isolation across the Indo-Pacific Barrier with contemporary overlap in the Coral Triangle. BMC Evol Biol 11(189):15

Game ET, Grantham HS, Hobday AJ, Pressey RL, Lombard AT, Beckley LE, Gjerde K, Bustamante R, Possingham HP, Richardson AJ (2009) Pelagic protected areas: the missing dimension in ocean conservation. Trends Ecol Evol 24:360–369

Gardiner JS (1936) The reefs of the western Indian Ocean. I. Chagos Archipelago. II The Mascarene region. Trans Linn Soc London 19(2):393–436

Graham NAJ, Spalding MD, Sheppard CRC (2010) Reef shark declines in remote atolls highlight the need for multi-faceted conservation action. Aquat Conserv Mar Freshw Ecosyst 20:543–548

Hahn A, Jensen MP, Broderick D, FitzSimmons NN, Bell I, Mortimer JA, Whiting S, Limpus CJ, Trott S (In prep) Stock composition of hawksbill turtle (*Eretmochelys imbricata*) feeding grounds in the Indo-Pacific region

Hellberg ME (2007) Footprints on the water: the genetic wake of dispersal among reefs. Coral Reefs 26:463–473

Hermes JC, Reason CJC (2008) Annual cycle of the South Indian Ocean (Seychelles-Chagos) thermocline ridge in a regional ocean model. J Geophys Res 113:10. doi:10.1029/2007JC004363

Hobbs JA, Frisch AJ, Allen GR, Herwerden LV (2009) Marine hybrid hotspot at Indo-Pacific biogeographic border. Biol Lett 5:258–261

Hughes TP, Graham NAJ, Jackson JBC, Mumby PJ, Steneck RS (2010) Rising to the challenge of sustaining coral reef resilience. Trends Ecol Evol 25:633–642

Hyrenbach K, Forney K, Dayton P (2000) Marine protected areas and ocean basin management. Aquat Conserv 10:437–458

Hyrenbach K, Fernandez P, Anderson D (2002) Oceanographic habitats of two sympatric North Pacific albatrosses during breeding season. Mar Ecol Prog Ser 233:283–301

IOTC (2010) Report of the fourteenth session of the Indian Ocean tuna commission, Busan, Korea, 1–5 March 2010. IOTC-2010-S14-R[E]

Itano DG, Holland KN (2000) Movement and vulnerability of bigeye (*Thunnus obesus*) and yellowfin tuna (*Thunnus albacares*) in relation to FADs and natural aggregation points. Aquat Living Resour 13:213–223

Jensen OP, Ortega-Garcia S, Martell SJD, Ahrens RNM, Domeier ML, Walters CJ, Kitchell JF (2010) Local management of a "highly migratory species": the effects of long-line closures and recreational catch-and-release for Baja California striped marlin fisheries. Prog Oceanogr 86:176–186

Kaplan DM, Chassot E, Gruss A, Fonteneau A (2010) Pelagic MPAs: the devil is in the details. Trends Ecol Evol 25:62–63

Kimani EN, Okemwa GM, Kazungu JM (2009) Fisheries in the Southwest Indian Ocean: trends and governance challenges. In: Laipson E, Pandya A (eds) The Indian Ocean: resource and governance challenges. The Henry L. Stimson Centre, Washington DC, p 107

Koldewey H, Curnick D, Harding S, Harrison L, Gollock M (2010) Potential benefits to fisheries and biodiversity of the Chagos Archipelago/British Indian Ocean Territory as a no-take marine reserve. Mar Pollut Bull 60:1906–1916

Lucas JS (1982) Quantitative studies of feeding and nutrition during larval development of the coral reef asteroid *Acanthaster planci* (L.). J Exp Mar Biol Ecol 65:173–193

Macdonald AHH, Lamb J, Schleyer MH (submitted) Population structure of *Platygyra daedalea* on the south-east African coastline

Madden RA, Julian PR (1994) Observations of the 40–50 day tropical oscillation—a review. Mon Weather Rev 122:814–837

Marchant R, Mumbi C, Behera S, Yamagata T (2007) The Indian Ocean dipole – the unsung driver of climatic variability in East Africa. Afr J Ecol 45:4–16

Mees C, Pearce J, Clarke J, Wilson O, Carroll A (2008) IOTC eleventh session of the scientific committee Mahé, Seychelles, 1–5 Dec 2008: IOTC-2008-SC-INF12

Mees C, Pearce J, Clarke J, Wilson O (2009) UK (Chagos/BIOT) national report. IOTC twelfth session of the scientific committee Mahé, Seychelles, 30 Nov-4 Dec 2009: IOTC-2009-SC-INF08

Mora C, Aburto-Oropeza O, Ayala Bocos A, Ayotte PM, Banks S et al (2011) Global human footprint on the linkage between biodiversity and ecosystem functioning in reef fishes. PLoS Biol 9(4):e1000606. doi:10.1371/journal.pbio.1000606

Morato T, Hoyle SD, Allain V, Nicol SJ (2010a) Tuna longline fishing around West and Central Pacific Seamounts. PLoS One 5(12):e14453. doi:10.1371/journal.pone.0014453

Morato T, Hoyle SD, Allain V, Nicol SJ (2010b) Seamounts are hotspots of pelagic biodiversity in the open ocean. Proc Natl Acad Sci USA 107:9707–9711

Moresby, Commander Indian Navy (1884) Untitled. Trans Bombay Geogr Soc 1:307–310

Mortimer JA, Broderick D (1999) Population genetic structure and developmental migrations of sea turtles in the Chagos Archipelago and adjacent regions inferred from mtDNA sequence variation. In: Sheppard CRC, Seaward MRD (eds) Ecology of the Chagos Archipelago. Westbury Publishing/Linnean Society, London, pp 184–194

Mortimer JA, Day M, Broderick D (2002) Sea turtle populations of the Chagos Archipelago, British Indian Ocean Territory. In: Mosier A, Foley A, Brost B (compilers). Proceedings of the twentieth annual symposium on Sea turtle biology and conservation. NOAA Technical Memorandum NMFSSEFSC-477. pp 47–49

Nelson J, Bradner H (2010) The case for establishing ecosystem-scale marine reserves. Mar Pollut Bull 60:635–637

Obura DO (2012) The diversity and biogeography of Western Indian Ocean reef-building corals. PLOS ONE. http://dx.plos.org/10.1371/journal.pone.0045013

Pala C (2009) Protecting the last great tuna stocks. Science 324:1133

Palumbi SR (2004) Marine reserves and ocean neighbourhoods: the spatial scale of marine populations and their management. Annu Rev Environ Resour 29:31–68

Pauly D (1995) Anecdotes and the shifting baseline syndrome of fisheries. Trends Ecol Evol 10:430

Pearce J (1996) A review of the British Indian Ocean Territory fisheries conservation and management zone tuna fishery 1991–1995. Marine Resources Assessment Group, London

Pfeiffer M, Dullo W-C, Eisenhauer A (2004) Variability of the intertropical convergence zone recorded in coral isotopic records from the central Indian Ocean (Chagos Archipelago). Quaternary Res 61:245–255. doi:10.1016/j.yqres.2004.02.009

Pfeiffer M, Timm O, Dullo W-C, Garbe-Schönberg D (2006) Paired Sr/Ca and delta^{18}O records from the Chagos Archipelago: late 20th century warming affects rainfall variability in the tropical Indian Ocean. Geology 34:1069–1072

Pfeiffer M, Dullo W-C, Zinke J, Garbe-Schönberg D (2009) Three monthly coral Sr/Ca records from the Chagos Archipelago covering the period of 1950 to 1995 A.D.: reproducibility and implications for quantitative reconstructions of sea surface temperature variations. Int J Earth Sci 98:53–66

Polunin NVC, Roberts CM (1993) Greater biomass and value of target coral-reef fishes in two small Caribbean marine reserves. Mar Ecol Prog Ser 100:167–176

Purkis SJ, Graham NAJ, Riegl BM (2008) Predictability of reef fish diversity and abundance using remote sensing data in Diego Garcia (Chagos Archipelago). Coral Reefs 27:167–178

Reece JS, Bowen BW, Joshi K, Goz V, Larson L (2010) Phylogeography of two moray eels indicates high dispersal throughout the Indo-Pacific. J Hered 101:391–402

Reid D, Lal K, Mackenzie-Dodds J, Kaligis F, Littlewood D, Williams S (2006) Comparative phylogeography and species boundaries in *Echinolittorina* snails in the central Indo-West Pacific. J Biogeogr 33:990–1006

Richard Y, Trzaska S, Roucou P, Rouault P (2000) Modification of the southern African rainfall variability/ENSO relationship since the late 1960s. Clim Dynam 16:883–895

Roberts J (2007) Preliminary characterization of bycatch by tuna longliners operating in the Chagos/BIOT area. Report for MRAG, London

Roberts RC, Sargant H (2002) Fishery benefits of fully protected marine reserves: why habitat and behaviour are important. Nat Resoure Model 15:487–507

Rogers AD, Laffoley D d'A (2011) International Earth system expert workshop on ocean stresses and impacts. Summary report. IPSO Oxford, 18pp. http://www.stateoftheocean.org/pdfs/1906_IPSO-LONG.pdf

Rogers AD, Baco A, Griffiths HJ, Hart T, Hall-Spencer JM (2007) Corals on seamounts. In: Pitcher TJ, Morato T, Hart PJB, Clark MR, Haggan N, Santos RS (eds) Seamounts: ecology, fisheries and conservation. Blackwell, Oxford, pp 141–169

Saji NH, Yamagata T (2003) Structure of SST and surface wind variability during Indian Ocean Dipole Mode events: COADS observations. J Climate 16:2735–2751. doi:10.1175/1520-0442(2003) 016

Saji HH, Goswami BN, Vinayachandran PN, Yamagata T (1999) A dipole mode in the tropical Indian Ocean. Nature 401:360–363

Sale PF (ed) (2002) Coral reef fishes. Dynamics and diversity in a complex ecosystem. Academic, San Diego, 549pp

Schaefer KM, Fuller DW (2010) Vertical movements, behavior, and habitat of bigeye tuna (*Thunnus obesus*) in the equatorial eastern Pacific Ocean, ascertained from archival tag data. Mar Biol 157:2625–2642

Scheltema RS (1988) Initial evidence for the transport of teleplanic larvae of benthic invertebrates across the East Pacific Barrier. Biol Bull 174:145–152

Schubart CD, Diesel R, Hedges SB (1998) Rapid evolution to terrestrial life in Jamaican crabs. Nature 393:363–365

Sheppard CRC (1999) Corals of Chagos, and the role of Chagos reefs in the Indian Ocean. In: Sheppard CRC, Seaward MRD (eds) Ecology of the Chagos Archipelago. Westbury Publishing/Linnean Society, London, pp 53–66

Sheppard CRC (2009) Large temperature plunges recorded by data loggers at different depths on an Indian Ocean atoll: comparison with satellite data and relevance to coral refuges. Coral Reefs 28:399–403

Sheppard CRC, Seaward MRD (1999) Ecology of the Chagos Archipelago, Special Issue of Linnean Society. Westbury for the Linnean Society of London, London, 350pp

Sheppard CRC, Wells SM (1988) Coral reefs of the World. Volume 2: Indian Ocean, Red Sea and Gulf. UNEP Regional Seas directories and bibliographies. IUCN/UNEP, Gland/Nairobi, p 389

Sheppard CRC, and 40 others (2012a) Reefs and islands of the Chagos Archipelago, Indian Ocean: why it is the world's largest no-take marine protected area. Aquatic Conservation: Mar Freshw Ecosyst 22:232–261

Sheppard CRC and 14 others (2012b) Conservation and management in British Indian Ocean Territory (Chagos Archipelago). Report for FCO, London, pp 28

Shotton R (2006) Managment of demersal fisheries resources of the Southern Indian Ocean, vol 1020, FAO Fisheries Circular. FAO, Rome, 90pp

Sibert J, Hampton J (2003) Mobility of tropical tunas and the implications for fisheries management. Mar Policy 27:87–95

Slatkin M (1977) Gene flow and genetic drift in a species subject to frequent local extinctions. Theor Popul Biol 12:253–262

Slatkin M (1982) Testing neutrality in subdivided populations. Genetics 100:533–545

Stoddart DR, Taylor DL (1971) Geology and ecology of Diego Garcia Atoll, Chagos Archipelago. Atoll Res Bull 149:237

SWIOFC (2009) South West Indian Ocean Fisheries Commission, report of the third session of the scientific committee. Mahe, Seychelles, 17–20 Dec 2007

Timm O, Pfeiffer M, Dullo W-C (2005) Nonstationary ENSO-precipitation teleconnection over the equatorial Indian Ocean documented in a coral from the Chagos Archipelago. Geophys Res Lett 32:4. doi:10.1029/2004GL021738, L02701

United Nations (2002) Report of the World summit on sustainable development, Johannesburg, South Africa, 26 Aug–4 Sept 2002. A/CONF.199/20. UN, NY, USA

van der Elst R, Everett B, Jiddawi N, Mwatha G, Afonso PS, Boulle D (2005) Fish, fishers and fisheries of the Western Indian Ocean: their diversity and status. A preliminary assessment. Philos Trans R Soc B 363:263–284

Vargas SM, Jensen MP, Broderick D, FitzSimmons NN, Mobaraki A, Mortimer JA, Prince R, Bell I, dos Santos FR (In prep) Phylogeography of the hawksbill turtle (*Eretmochelys imbricata*) in the Indo-Pacific region

Vargas SM, Jensen MP, Mobaraki A, Santos FR, Broderick D, Mortimer JA, Limpus C, Whiting S, FitzSimmons N (2010) Phylogeography of the Hawksbill turtles (*Eretmochelys imbricata*) from the Indo Pacific Region. Abstract, 29 April 2010. Symposium proceedings, (in press)

Veitch L, Dulvy NK, Koldewey H, Lieberman S, pauly D, Robberts CM, Rogers AD, Balie JE (2012) Avoiding empty ocean commitments at Rio+20. Science 336:1383–1385

Vialard J, Foltz GR, McPhaden MJ, Duvel JP, de Boyer Montegut C (2008) Strong Indian Ocean sea surface temperature signals associated with the Madden-Julian Oscillation in late 2007 and early 2008. Geophys Res Lett 35:L19608. doi:10.1029/2008GL035238

Vialard J, Duvel JP, McPhaden MJ, Bouruet-Aubertot P, Ward B, Key E et al (2009) Cirene: air-sea Interactions in the Seychelles-Chagos thermocline ridge region. Bull Am Meteorol Soc 90:61. doi:10.1175/2008BAMS2499.1

Vogler C, Benzie J, Lessios H, Barber P, Wörheide G (2008) A threat to coral reefs multiplied? Four species of crown-of-thorns starfish. Biol Lett 4:696–699. doi:10.1098/rsbl.2008.0454

Vogler C, Benzie J, Barber P, Erdmann MV, Ambariyanto, Sheppard C, Tenggardjaja K, Gérard K, Wörheide G (2012) Phylogeography of the crown-of-thorns starfish in the Indian Ocean. PLoS ONE 7:e43499

Walters C (2000) Impacts of dispersal, ecological interactions, and fishing effort dynamics on efficacy of marine protected areas: how large should protected areas be? Bull Mar Sci 66:745–757

Ward JR, Rypien KL, Bruno JF, Harvell CD, Jordan-Dahlgren E, Mullen KM, Rodriguez-Martinez RE, Sanchez J, Smith G (2006) Coral diversity and disease in Mexico. Dis Aquat Organ 69:23–31

Webster PJ, Moore A, Loschnigg J, Leban M (1999) Coupled ocean–atmosphere dynamics in the Indian Ocean during 1997–98. Nature 401:356–360

Wilkinson C (2008) Executive summary. In: Wilkinson C (ed) Status of coral reefs of the World: 2008. Global Coral Reef Monitoring Network and Reef and Rainforest Research Center, Townsville, pp 5–19

Wilson JR, Harrison PL (1998) Settlement-competency periods of larvae of three species of scleractinian corals. Mar Biol 131:339–345

Winterbottom R, Anderson RC (1997) A revised checklist of the epipelagic and shore fishes of the Chagos Archipelago, Central Indian Ocean. JLB Smith institute of ichthyology. Ichthyol Bull 66:1–28

Wood LJ, Fish L, Laughren J, Pauly D (2008) Assessing progress towards global marine protection targets: shortfalls in information and action. Oryx 42:340–351

Worm B, Lotze K, Myers R (2003) Predator diversity hotspots in the blue ocean. Proc Natl Acad Sci 100:9884–9888

Worm B, Sandow M, Oschlies A, Lotze HK, Myers RA (2005) Global patterns of predator diversity in the open oceans. Science 309:1365–1369

Yesson C, Clark MR, Taylor M, Rogers AD (2011) The global distribution of seamounts based on 30-second bathymetry data. Deep Sea Res Part I Oceanogr Res Pap 58:442–453. doi:10.1016/j.dsr.2011.02.004

Yesson C, Taylor M, Tittensor DP, Davies A, Guinotte J, Baco-Taylor A, Black J, Hall-Spencer J, Rogers AD (2012) Global distribution and habitat preferences of deep-sea octocorals. J Biogeogr 39:1278–1292

Zubair L, Ropelewski CF (2006) The strengthening relationship between ENSO and northeast monsoon rainfall over Sri Lanka and southern India. J Climate 19:1567–1575

Coral Reefs of the Chagos Archipelago, Indian Ocean

18

Charles R.C. Sheppard, Mebs Ateweberhan, Allen C. Chen, Alasdair Harris, Rachel Jones, Shashank Keshavmurthy, Carl Lundin, David Obura, Sam Purkis, Peter Raines, Bernhard Riegl, Michael H. Schleyer, Anne L.S. Sheppard, Jerker Tamelander, John R. Turner, Shakil Visram, and Sung-Yin Yang

Introduction

Over the period of BIOT's existence, there have been a dozen scientific visits to its atolls by more than 50 scientists, and double this number have become engaged in work on materials sent back. It is clear, during this period when coral reefs in most of the Indian Ocean have become seriously degraded, that the reefs of Chagos persist in an exceptionally good state. The results of these research expeditions led increasingly to calls to extend conservation in the area, and data to support this concept has come from over 200 papers arising from the research. This summarises current understanding of the reefs, but starts with an account of their condition through the major bleaching event of 1998.

Coral Cover and Changes due to Mortality Episodes

There were no quantitative studies of reef condition on Chagos reefs before the 1970s, although descriptive studies, notably Stoddart and Taylor (1971), described land and reef flats in Diego Garcia. From the 1970s, episodic visits enabled a series of coral cover measurements to be taken on reef slopes.

Coral cover declined slightly between the first survey in 1978 (Sheppard 1980) and the next in 1996 (Sheppard 1999a) (Fig. 18.1). This was mainly due to loss of shallow and mid depth branching species, particularly *Acropora palifera* and table corals including *A. cytherea*. Causes of this were speculated upon at the time (Sheppard 1999b) but only later, after

C.R.C. Sheppard (✉) • M. Ateweberhan • A.L.S. Sheppard
School of Life Sciences, University of Warwick,
Gibbit Hill Campus, CV4 7AL, Coventry, UK
e-mail: Charles.Sheppard@warwick.ac.uk; mateweberhan@wcs.org; a.sheppard@warwick.ac.uk

A. Harris
Omnibus Business Centre, 39-41 North Road, N7 9DP, London
e-mail: al@blueventures.org

A.C. Chen • S. Keshavmurthy • S. Visram • S.-Y. Yang
Biodiversity Research Centre, Academia Sinica,
128 Academia Road, Nankang, Taipei, 115, Taiwan
e-mail: acropora.chen@gmail.com; shashank@gate.sinica.edu.tw; shakilvisram@yahoo.co.uk; likefh@gmail.com

R. Jones
Zoological Society of London, Regents Park,
NW1 4RY, London, UK
e-mail: Rachel.jones@zsl.org

C. Lundin
IUCN Marine Programme, Rue Mauverney 28, Gland, 1196, Switzerland
e-mail: carl.lundin@iucn.org

D. Obura
CORDIO East Africa, #9 Kibaki Flats, Kenyatta Beach, Bamburi Beach, P.O.BOX 10135, Mombasa, 80101, Kenya
e-mail: dobura@cordioea.org

S. Purkis • B. Riegl
National Coral Reef Institute,
Nova Southeastern University, Oceanographic Center,
8000 North Ocean Drive, Dania Beach, FL, 33004, USA
e-mail: purkis@nova.edu; rieglb@nova.edu

P. Raines
71 Badminton Road, SW12 8BL, London, UK
e-mail: rainespeter@gmail.com

M.H. Schleyer
Oceanographic Research Institute, 10712, Marine Parade,
Durban, 4056, South Africa
e-mail: schleyer@ori.org.za

J. Tamelander
UNEP Division of Environmental Policy Implementation, UN,
Rajdamnern Nok Av, Bangkok, 10200, Thailand
e-mail: tamelander@un.org

J.R. Turner
School of Ocean Sciences, Bangor University, Menai Bridge,
Anglesey, LL59 5AB, UK
e-mail: j.turner@bangor.ac.uk

C.R.C. Sheppard (ed.), *Coral Reefs of the United Kingdom Overseas Territories*, Coral Reefs of the World 4,
DOI 10.1007/978-94-007-5965-7_18, © Springer Science+Business Media Dordrecht 2013

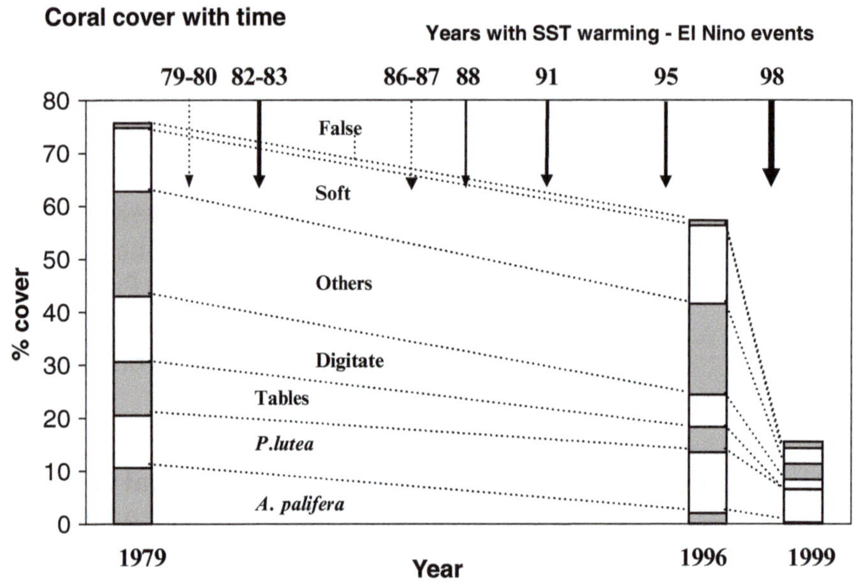

Fig. 18.1 Percent coral on reefs by the main coral types in 3 years, for the major live categories identifiable by snorkelling in 1999. *Arrows* along the top are dates and approximate relative severity (*arrow thickness*) of previous warming events (From Sheppard 1999b)

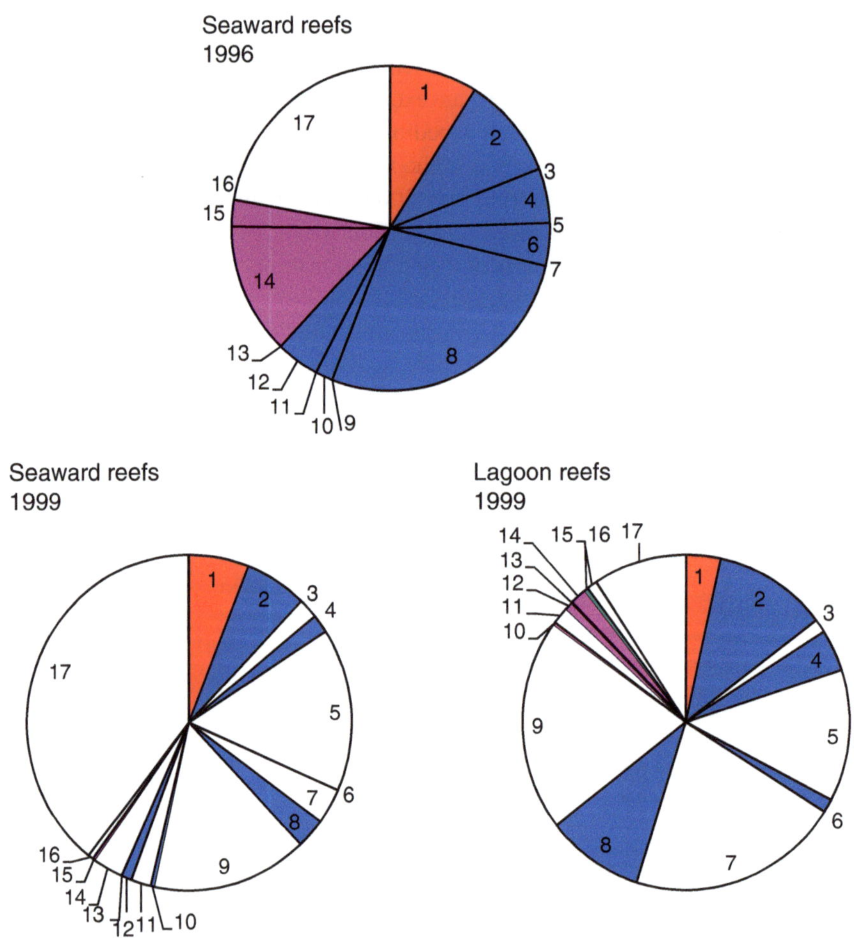

Fig. 18.2 Pie charts showing reef cover values before and after the 1998 warming event, from 1996 and seaward and algoonal reef slopes in 1999. Key: (*1*) Red algae, (*2*) *Porites* live, (*3*) *Porites* dead, (*4*) digitate coral live, (*5*) digitate coral dead, (*6*) table coral live, (*7*) table coral dead, (*8*) other corals live, (*9*) other corals dead, (*10*) *A. palifera* live, (*11*) *A palifera* dead, (*12*) *Heliopora* live, (*13*) Heliopora dead, (*14*) soft coral, (*15*) *Millepora* live, (*16*) *Millepora* dead, (*17*) bare substrate (From Sheppard 1999b)

much work globally and more surveys in Chagos, the coral cover decline was suggested to have been caused by several warming events. This modest decline occurred in many Indian Ocean island groups over this period (Ateweberhan et al. 2011) and in Kenya (Muthiga et al. 2008).

Severe warming in 1998 then caused massive mortality in shallow water on all Chagos reefs (Sheppard 1999b; Sheppard et al. 2002, Fig. 18.2) as it did throughout the Indian Ocean (Ateweberhan et al. 2011). Coral and soft coral mortality was almost total on several ocean-facing

Fig. 18.3 Typical ocean facing reef (10 m depth) in February 2001, 3 years after the 1998 mortality, showing almost no recovery and complete disintegration of coral skeletons

Fig. 18.4 Lagoon slope (10 m depth) in February 2001, showing survival of apparently healthy corals in the lagoons

reefs in shallow water to clearly defined depths, (Fig. 18.3) below which corals provided much higher cover. This killed zone extended deeper in southern atolls; in Diego Garcia for example the critical depth was >40 m depth, while in more northern atolls the killed zone extended to only about 10–15 m depth (Sheppard et al. 2002). Such variability was mirrored in the Indian Ocean as a whole (Sheppard 2006). Lagoon reefs of Chagos atolls were much less affected than ocean facing reefs (Figs. 18.2 and 18.4), with many retaining very good coral cover, including stands of *Acropora*. Post 1998, coral species diversity was greatest in deep lagoon areas.

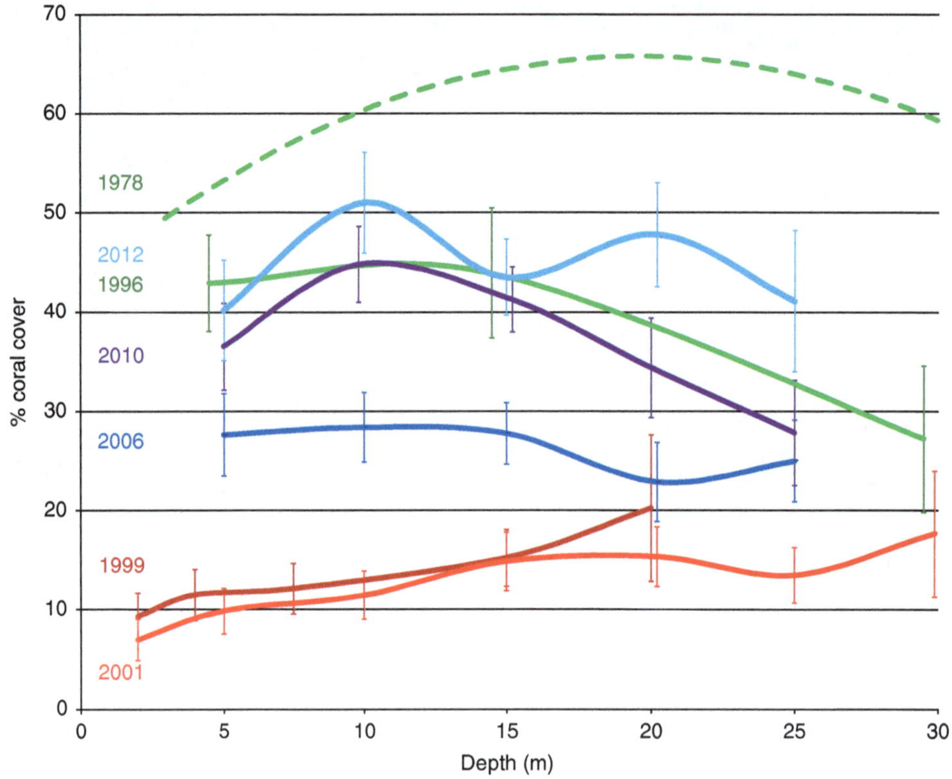

Fig. 18.5 Coral cover in depths to 25 m on ocean-facing slopes in different years (Sheppard 1980, 1999a, b; Sheppard et al. 2008, 2012, Sheppard unpublished data). Data is of all ocean-facing transects in this series measured on each date. *Bars* are error bars (Error data lost for 1978)

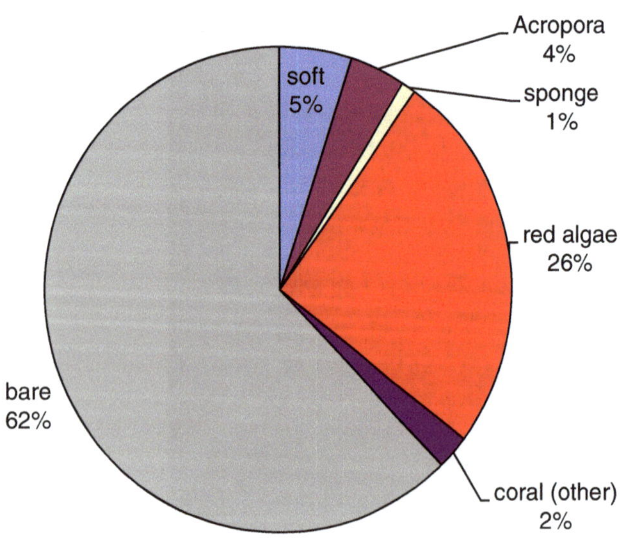

Fig. 18.6 Relative abundance of space occupancy of juvenile corals on 30 dead seaward slope *Acropora* tables between 8 and 15 m depth. The area marked 'bare' is in fact covered by films of microscopic and small filamentous algae (From Sheppard et al. 2002)

Coral Recovery

Several ocean-facing transects around the atolls have been monitored repeatedly from 1999 onwards (Fig. 18.5) and shows a pattern of excellent coral recovery which is illustrative of what can happen in the absence of other impacts. On these reefs, no increase of hard coral cover was seen for 3 years following the mortality, although by 2001 huge numbers of juveniles were present, at densities of up to 28 m^{-2}, the highest recorded globally at that time (Sheppard et al. 2002; Harris and Sheppard 2008). Coral juveniles in 2001 provided 6% cover on easily measured (but disintegrating) dead coral tables, with a further 5% cover provided by juvenile soft corals, indicating good recovery potential (Fig. 18.6). Significant increase in coral cover on even the most badly affected reefs became evident by 2006, especially in shallow water (Fig. 18.7) (Harris and Sheppard 2008; Sheppard et al. 2008). Cover by all major benthic categories in 2006 was good (Fig. 18.8) and was very similar in all atolls. Restoration of coral cover had fully occurred by 2012 (see curve for 2012 in Fig. 18.5).

Deeper, initial recovery was slower. Coral cover in 2011 reached values recorded in 1978 in a few transects in Diego Garcia, but most atolls were not surveyed during that year, but by 2012 cover values at most depths exceeded those seen immediately before the bleaching. The dip in this curve around 15 m depth is due to the loss and breakage of many formerly thriving table corals, attributable perhaps to severe storms that affected the area a few months before that set of measurements. A programme of juvenile (<15 mm colony diameter) counts has recently been commenced in Diego Garcia, and these show substantial recruitment at all depths

Fig. 18.7 Ocean facing reef slope (Isle Anglais, Salomon atoll) in 2006, showing very high cover of table corals (*Acropora cytherea*)

Fig. 18.8 Mean percentage cover values of life form and substrate categories pooled from all depths and all sites for each of four atolls, surveyed by video during 2006. *Bars* are error bars

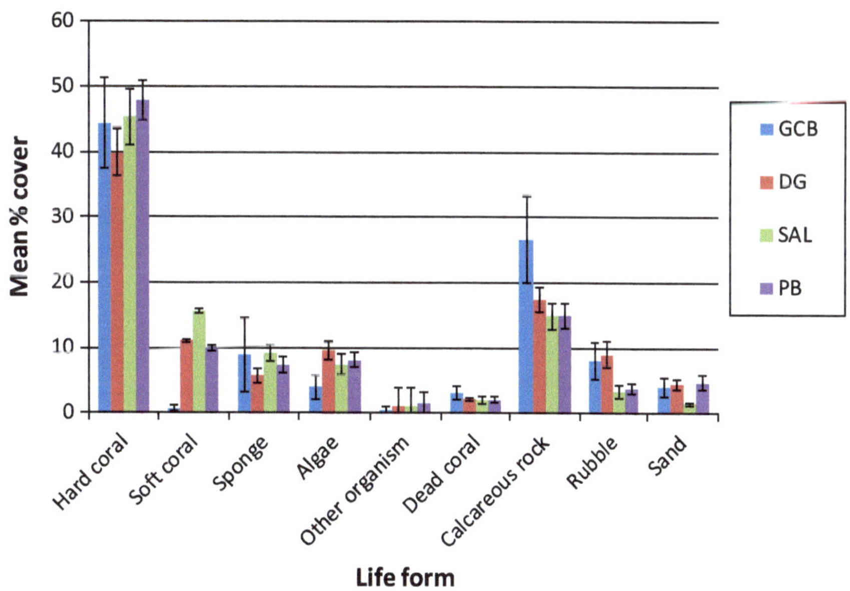

with greatest numbers in the mid photic depths (Fig. 18.9), with values as high or higher than have been recorded anywhere, though comparisons are difficult as different size thresholds defining the meaning of 'juvenile' have been used by different researchers in different places.

In 2006, extensive video surveys were taken and archived, showing mean percentage cover of several benthic categories (Bayley 2009) and this also serves as a detailed reference point (Fig. 18.10). These showed that in 2006 there were no significant differences in hard coral cover in pooled data between atolls, though significant differences existed at depth and site levels between ocean-facing reefs. Hard coral cover was significantly higher in lagoons (63.04 ± 3.19%) than ocean-facing slopes (39.69 ± 2.03%). Soft coral was higher on ocean-facing slopes (14.02 ± 1.58%) than in lagoons (2.65 ± 0.72%). Hard coral cover decreased between 6 and 25 m, but sponge and soft corals showed an increase with depth. Dead standing coral at most sites was much lower than a few years earlier at 3–13%, and rubble by this time did not change significantly at different depths.

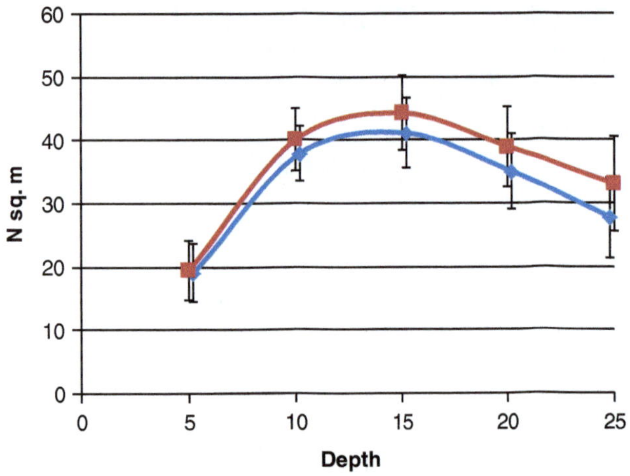

Fig. 18.9 Numbers of juvenile corals (<15 mm maximum diameter) at 5 m depth intervals on ocean facing reef slopes in Diego Garcia atoll in 2012. Bars are std errors. *Blue line* is 'Raw' data which is actual count taken from replicate 25 cm side quadrats, in areas of substrate chosen for having mostly 'bare space' i.e. without large coral colonies. *Red line* is 'adjusted', meaning adjusted for the proportion of space occupied by mature coral colonies which are therefore unavailable for occupation by juveniles

Structural complexity was reduced down to 15 m in the outer atolls, and to greater depths on the Great Chagos Bank and Diego Garcia atoll. By 2006 all shallow regions had developed sufficiently to form a canopy, with colonies competing with one another for space (O'Farrell 2007). Furthermore, deep lagoonal areas exhibited the highest numbers of small, juvenile coral colonies. Modelling studies indicate that such deep reef areas could be responsible for relatively rapid recolonization of denuded shallow reefs (Riegl and Piller 2003; Riegl et al. 2012).

The past decade has seen further coral bleaching events in Chagos, in 2003, 2004, 2005, and a mild event in 2010, but none were sufficient to cause mass mortality, though species specific coral mortality was recorded of many *Acropora cytherea* tables in 2010 (Pratchett et al. 2010). Given that warming episodes sufficient to kill corals are predicted to increase in frequency (Sheppard 2003; Hoegh-Guldberg et al. 2007) it is likely that intermittent interruptions to coral growth will continue. However, models based on recruit availability scaled to the present coral cover, suggest that

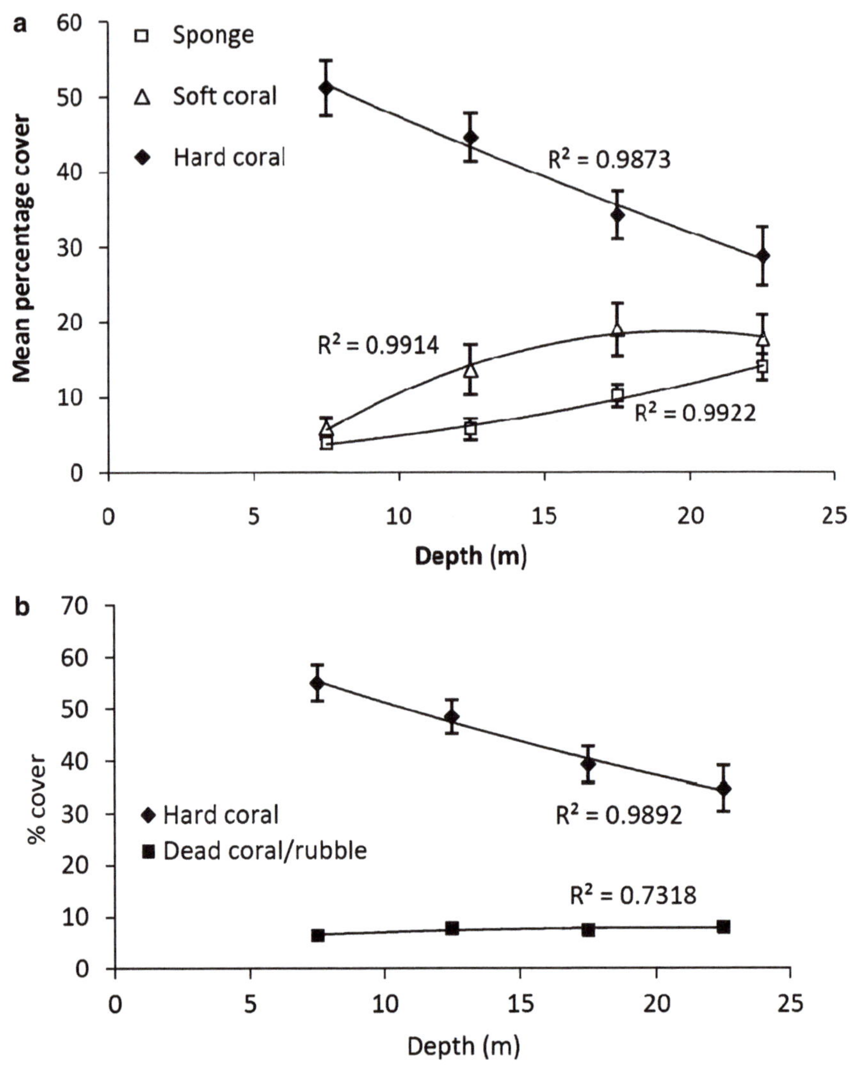

Fig. 18.10 *Top*: Mean percentage cover of hard coral, sponges and soft coral at depths on 16 seaward reef sites surveyed by video in 2006. (Error bars, 2nd order polynomial regression lines, with R^2 values). *Bottom*: Percentage cover of hard coral and rubble at depths for all 20 sites surveyed within Chagos by video in 2006 (Error bars, 2nd order polynomial regression lines, with R^2 values)

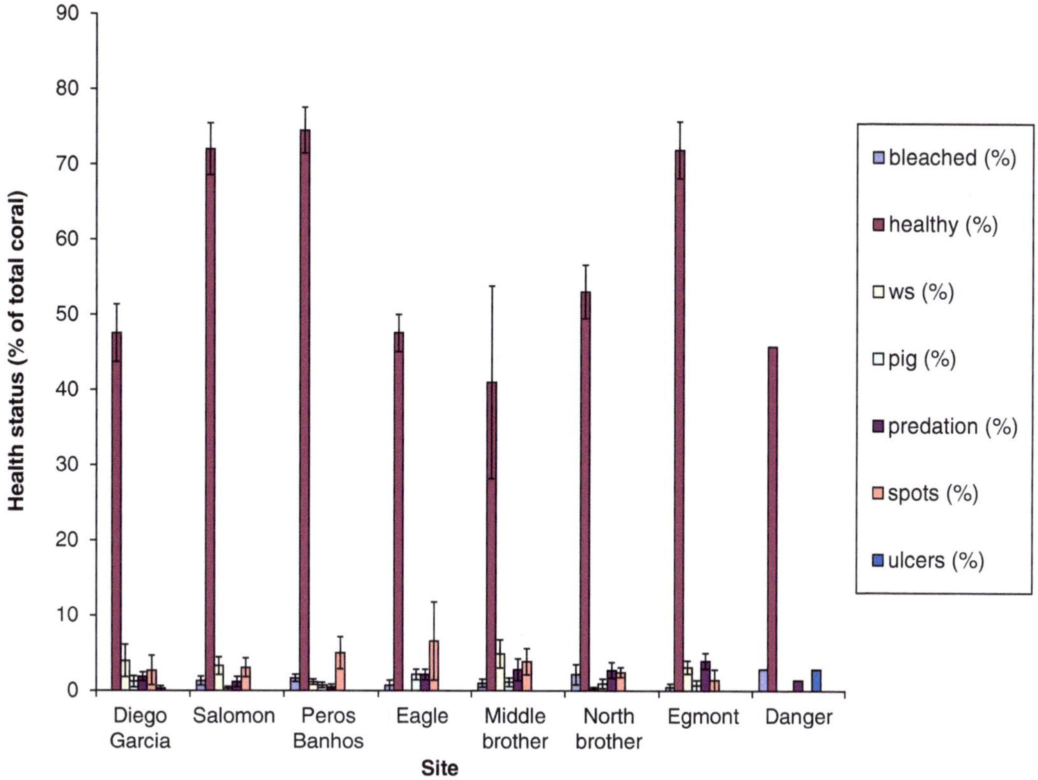

Fig. 18.11 Percent of adult colonies in different atolls showing disease or other adverse conditions

Chagos reefs will long be able to withstand recurring strong mortality events – even each decade - and still maintain high coral cover (Riegl et al. 2012). In the absence of direct human impacts, significant declines in cover are only expected if both larval supply decreases and coral mortality events increase in intensity and frequency.

Soft Corals

Pre- and post-1998 studies have also revealed some changes in the soft coral component. The principal octocorals before and after the 1998 ENSO shared many common taxa (Reinicke and van Ofwegen 1999; Schleyer and Benayahu 2010), but a few discontinuities in their biodiversity indicate subtle changes in more persistent genera (*Lobophytum, Sarcophyton*). Some fast-growing "fugitive" genera (e.g. *Cespitularia, Efflatounaria, Heteroxenia*) disappeared after the ENSO-related coral bleaching (Reinicke and van Ofwegen 1999; Schleyer and Benayahu 2010), suggesting that such transient fugitives might be eliminated from soft coral communities on isolated reef systems by bleaching disturbance of this nature. *Carijoa riseii*, a species often considered a fouling organism, and even an invasive in some places (Concepcion et al. 2010), was found in 2006. The observed post-ENSO recovery gives cause for hope for soft coral survival in the face of climate change. Worth noting is the condition of some northern reefs, facing southeast, which had previously been dominated by soft corals at around 5–10 m depth. These zooxanthellate soft corals died *en masse* in 1998 and, because they leave no skeleton, the appearance of these areas of reef was remarkable; vast expanses existed of apparently unoccupied, fairly smooth rock. These areas were devoid of any significant macroalgae too, indicative of both low nutrient levels and of the high densities of reef fish, including grazers (see next chapter), which persisted throughout this time. Turf algae was present, as always, but the overall and remarkable appearance was of fairly barren expanses of 'bare' substrate.

Coral Diseases

In 2006 a survey assessed corals along 37 transects at eight sites across the archipelago (Fig. 18.11). Overall prevalence of disease was 5.2%, which sits at the low end of the global spectrum where regional averages for 'white syndrome' alone are around 5% in parts of Australia, Palau and East Africa, 8% in the Philippines (Weil et al. 2002; Willis et al. 2004; Raymundo et al. 2005) up to around 13% at some sites in the Eastern Indian Ocean (Hobbs and Frisch 2010), and 20% in the Caribbean (Weil et al. 2006; Miller et al. 2009).

Temperature has been shown to be a key factor triggering diseases, with infection occurring rapidly at elevated temperatures (Ben-Haim and Rosenberg 2002; Bruno et al. 2007; Harvell et al. 2007). Thus the increasing frequency of raised temperature episodes gives cause for concern. Coral diseases often arise from changes to the normal, commensal relationship between the coral and the bacterial community in their mucus, skeleton and tissues (Rohwer and Kelley 2004; Lesser et al. 2007). Physiological stresses that cause corals to become overwhelmed by bacteria are often anthropogenic in origin, coming from sediment deposition, nutrient rise (Bruno et al. 2003; Kaczmarskay and Richardson 2011), or sea temperature rise (Harvell et al. 2007; Zvuloni et al. 2009). Other factors correlated with the likelihood of coral disease include geographical range and predator diversity (Diaz and Madin 2011), while a higher density of individuals also increases susceptibility (Willis et al. 2004; Bruno et al. 2007). Although remoteness from people is no guarantee of absence of disease, especially if temperature rises (Williams et al. 2007), mitigation of other human induced stress factors may reduce disease prevalence (Bruno et al. 2003; Harvell et al. 2007). At present, Chagos reefs have very low disease levels.

Marine Invasive Species

Marine Invasive Alien Species (IAS) are recognized as one of the most significant threats to global biodiversity (Wilcove et al. 1998; Bax et al. 2010) and documented IAS are commonly significantly underestimated. IAS pressure is driving global declines in species diversity, with the overall impact apparently increasing (McGeoch et al. 2010). Notably, over 100 introduced marine species covering 14 phyla are known from ports in the Pacific (Coles et al. 1999). Article 8 (h) of the Convention on Biological Diversity calls for prevention of introductions and control or eradication of alien species that threaten ecosystems, habitats or species, and the recently agreed Aichi Biodiversity Targets call for identifying pathways and putting in place by 2020 measures to prevent species introduction and establishment.

Ballast water and hull fouling provide the primary vectors for marine species introduction (Cohen and Carlton 1998; Ruiz et al. 2000; Hewitt et al. 2004). Navy and supply ships frequently arrive in Diego Garcia, mainly from the US, the Middle East and Singapore, but any ships, including recreational yachts, may carry hull fouling (e.g. Bax et al. 2002). Therefore pathways for species introductions to Chagos exist, as do preconditions for successful establishment (Tamelander et al. 2009). While most ships arrive to Chagos loaded, some may be empty and ballasted. Ballast water exchange occurs outside the lagoon and during mid-crossing in keeping with IMO ballast water management guidelines (IMO 2004).

A survey of non-native marine biota in Chagos was carried out in 2006 in all atolls (Tamelander et al. 2009) based on standard port survey methods (Hewitt and Martin 2001) but with a lower sensitivity. Hard and soft substrate benthic biota was sampled at 42 sites (19 sites in Diego Garcia, nine each on the Great Chagos Bank and at Peros Banhos, and five in the Salomon atoll). Twenty four phyla were represented in 2,672 samples, with four phyla (Bryozoa, Mollusca, Annelida and Porifera) each making up over 10% of the total number of specimens.

No non-native species were detected in the samples, the first time such a survey has not found species introduced as a result of human activities (Tamelander et al. 2009). This finding is testament to the ecological integrity of Chagos' marine ecosystems. Shallow marine habitats are believed to be particularly vulnerable to bioinvasions when degraded (Heywood 1995), but ecosystem health and high biodiversity confer higher resistance. Only 16% of marine ecoregions have no reported marine invasions, although the true figure may be lower due to under-reporting (Molnar et al. 2008).

Because controlling or eradicating a marine species once it is established is nearly impossible (e.g. Bax et al. 2002), management must focus on precautionary measures (Thresher and Kuris 2004; Carlton and Ruiz 2005). Successful prevention and management of IAS threats in Chagos is a prerequisite for effective management of the newly established MPA (Pomeroy et al. 2004; Tu 2009). Further, this needs to be devised in the broader context of climate change and the potentially greater risk of species spread and establishment that this may bring (Burgiel and Muir 2010; Bax et al. 2010).

Chagos Reef Condition in the Indian Ocean Context

Most Indian Ocean reef areas are heavily exploited and many have shown limited recovery following the 1998 bleaching disturbance (Wilkinson 2008; Harris 2010). Many reefs which declined catastrophically in 1998 and which also suffer from local impacts have not recovered significantly, or at all (Harris 2010). The 1998 bleaching event has the main determinant of coral cover change in the Indian Ocean since the 1970s (Ateweberhan et al. 2011), and the central regions which had some of the highest coral cover estimates prior to 1998 suffered the worst during the bleaching event. Subsequent recovery for most of these reefs now remain below average for the region, but of the central Indian Ocean reefs, recovery in Chagos is higher than elsewhere (Ateweberhan et al. 2011).

Globally, a third of reef-building corals are threatened with extinction (Carpenter et al. 2008) and today, in the Indian Ocean, only about a third of reefs may be attributed to a 'Low Threat Level' category (Wilkinson 2008). Chagos reefs fall within this minority group and contain a substantial proportion of reef area in very good condition. Reef area estimations are difficult, and have been subject to wide variation. Spalding et al. (2001) suggested the Indian Ocean has 32,000 km^2 of reefs (the Red Sea region and the Gulf region adding 17,400 and 4,200 km^2 more respectively), and, based on this, Chagos has 3,770 km^2 of reefs (Rajasuria et al. 2004) meaning Chagos comprises up to half of this ocean's reefs in a 'low threat level' category (Fig. 18.12).

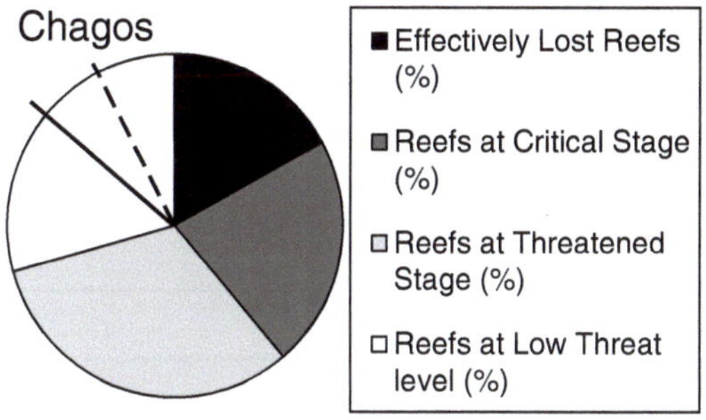

Fig. 18.12 Percent of reefs in different categories in the Indian Ocean. Categories are those from Wilkinson (2008). The probable proportion occupied by Chagos (solid line to vertical) is about half of the reefs in the 'best' category. From the *dashed line* to vertical is an alternative estimate of the proportion of Chagos reefs according to Spalding using slightly different categories (pers. comm.)

More recent calculations by Spalding (pers. comm. and see Burke et al. 2011) resulted in a revised estimate that Chagos provides 25% of reefs in the 'Low Threat' category. Even this "…25% of the region's low threat reefs is still an extraordinary proportion, and it is also worth stressing that in addition to this, these are by far the largest contiguous reef tracts considered to be under low threat" (Spalding pers. comm. 2011).

While both area values are very much less than the area of illuminated shallow limestone substrate which was calculated from detailed bathymetric plotting (see previous Chapter), the values used have the benefit that they were calculated consistently throughout the world, thus permitting comparisons; direct measurements based on bathymetry do not yet have a counterpart in most other countries. Far too many other parts of the Indian Ocean have shown very poor recovery by comparison (Tamelander and Rajasuria 2008; Wilkinson 2008; Burke et al. 2011).

Reasons for the good condition of Chagos reefs are likely to include remoteness from compounding human activities, but some additional factors may contribute. Strong light adapted 'Clade A' forms of symbiotic zooxanthellae have been identified in shallow corals in Chagos, occurring in approximately half of the shallow water *Acropora* colonies that were heavily affected by warming but which are now recovering strongly (Fig. 18.13). Also, an array of temperature

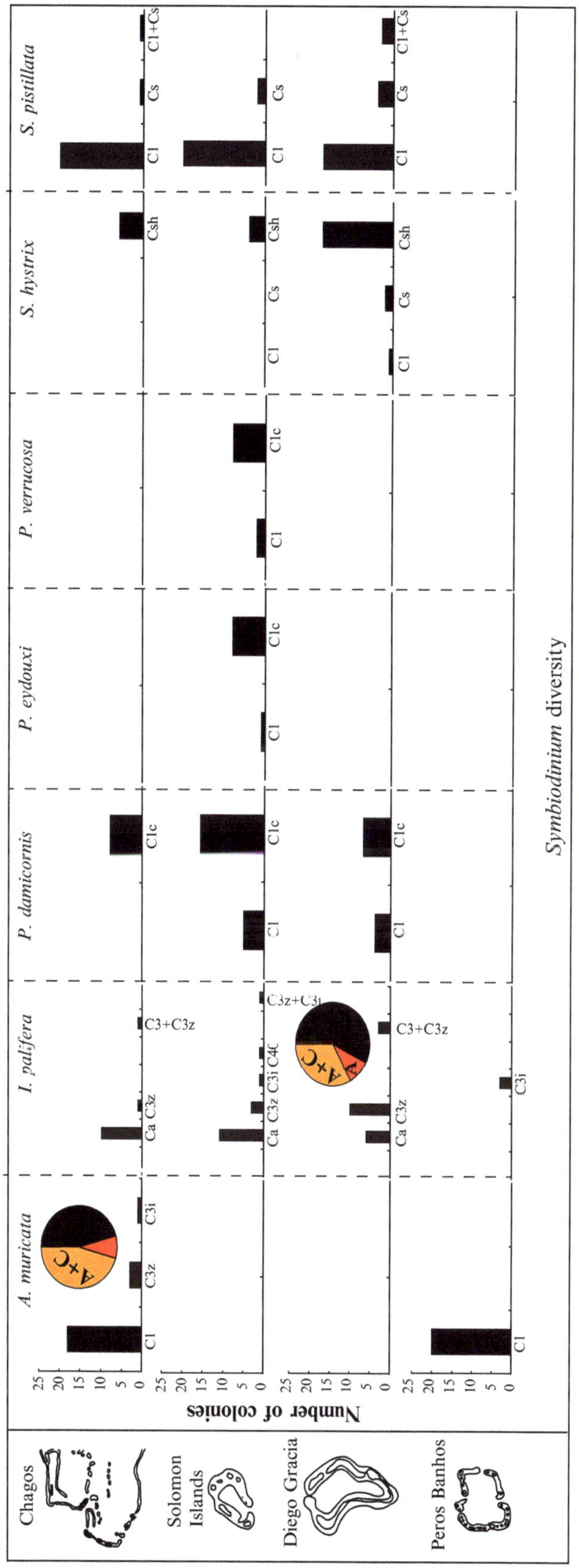

Fig. 18.13 Distribution of Symbiodinium Clade C, A and A+C in corals of Chagos (From Yang et al. 2012). Occurrences of *Symbiodinium* types in C (obtained from ITS2-DGGE band patterns) at four sites, Chagos, Solomon Islands, Diego Gracia and Peros Banhos in seven coral species; *Acropora muricata, Isopora palifera, Polillopora damicornis, Pocillopora eydouxi, Pocillopora verrucosa, Seriatopora hystrix* and *Stylophora pistillata* is shown using *black bars*. Blank spaces indicates absence of coral samples in those sites. Symbiodinium Clade A and Clade A+C were detected in samples from the colonies of *A. muricata* and *I. palifera* at Chagos and Diego Gracia respectively (shown as pie chart (Clade A+C – *Orange*, clade A – *Red*)). The proportion of Clade A was 9% (A. muricata) and 10% (*I. palifera*), while Clade A+C was 45% (*A. muricata*) and 33% *(I. palifera)* respectively in the Chagos and Diego Gracia

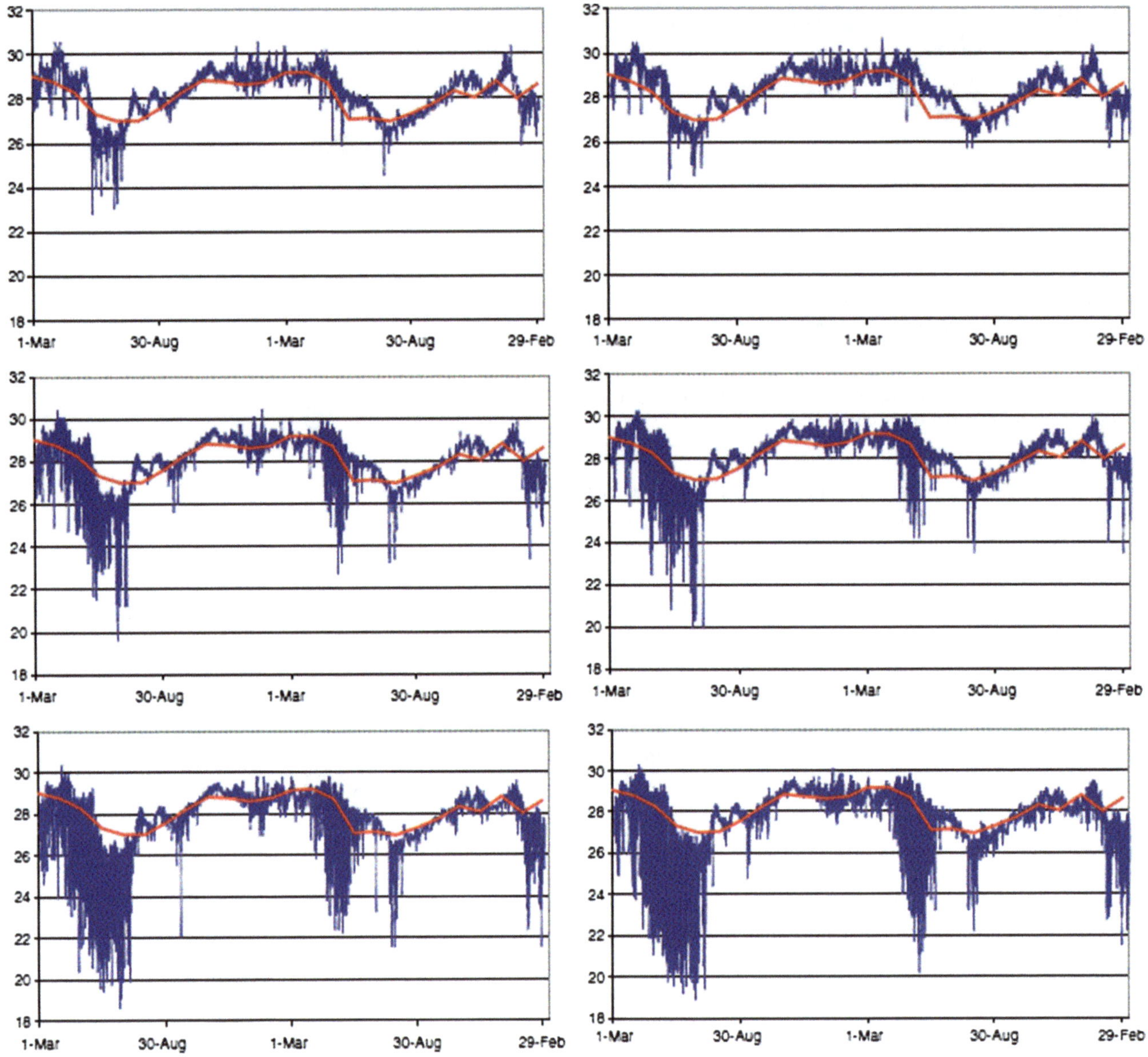

Fig. 18.14 Traces (*blue*) of the temperature data for (*left column*) northwest transect and (*right column*) northern transect. *Top row*: 5 m depth; *middle row*: 15 m depth and *bottom row*: 25 m depth. *Red line* is HadISST1 data (From Sheppard 2009)

sensors at different depths has identified regular incursions of deep, cool water that rise to cover reefs, including during the annual periods of greatest warming (Sheppard 2009) (Fig. 18.14).

But it is increasingly understood that direct human pressures are the main cause of reef degradation and this has often been underestimated in the past (e.g. Mora et al. 2011). Such activities impede recovery, and absence of herbivore extraction, pollution and sedimentation all increase reef resilience (Hughes et al. 2010). Most of Chagos has had no human population for about four decades. Diego Garcia' population imports all its requirements and for the last 15 years at least has had strong environmental management. Lack of human pressures is likely to be one major reason for the present good condition of these reefs.

In the Indian Ocean as a whole, direct human pressures can only increase further as human populations rise. The annual population growth rates may commonly be 2.5% in the region, especially on the coast where it is compounded by migration in some countries that have experienced wars or drought. As reefs degrade, the proportion of healthy reefs of the Indian Ocean contained in Chagos, already very high, continues to increase, so that a precautionary approach to their protection is merited.

Acknowledgements The authors thank the Administration of the British Indian Ocean Territory for permission to visit the area on various occasions, the military Commanders and personnel for much assistance on site, and to the Officers and crew of the BIOT Patrol Vessel *Pacific Marlin* for exceptional help on all visits to atolls away from Diego Garcia. The OTEP fund provided core funds for most visits, and all scientists involved received funding from numerous sources to carry out their own programmes of work in the archipelago. Some of this material was published previously in Sheppard et al. (2012) and is reproduced with permission of Wiley & Sons.

References

Ateweberhan M, McClanahan TR, Graham NAJ, Sheppard CRC (2011) Episodic heterogeneous decline and recovery of coral cover in the Indian Ocean. Coral Reefs 30:739–752

Bax N, Hayes K, Marshall A, Parry D, Thresher R (2002) Man-made marinas as sheltered islands for alien marine organisms: establishment and eradication of an alien invasive marine species. In: Vetich, Clout (eds) 2002. Turning the tide: the eradication of invasives pecies (proceedings of the international conference on eradication of island invasives). Occasional paper of the IUCN Species Survival Commission No. 27

Bax N, Williamson A, Aguero M, Gonzalez E, Geeves W (2010) Marine invasive alien species: a threat to global biodiversity. Mar Policy 27:313–323

Bayley D (2009) Video analysis of the reef structure and community composition of the Chagos archipelago, British Indian Ocean Territory. M.Sc. thesis, University of Wales, Bangor, 129p

Ben-Haim Y, Rosenberg E (2002) A novel *Vibrio* sp. pathogen of the coral *Pocillopora damicornis*. Mar Biol 141:47–55

Bruno JF, Peters LE, Harvell CD, Hettinger A (2003) Nutrient enrichment can increase the severity of coral diseases. Ecol Lett 6:1056–1061

Bruno JF, Selig ER, Casey KS, Page CA, Willis BL, Harvell CD, Sweatman H, Melendy AM (2007) Thermal stress and coral cover as drivers of coral disease outbreaks. Plos Biol 5:e124

Burgiel SW, Muir AA (2010) Invasive species, climate change and ecosystem- based adaptation: addressing multiple drivers of global change. Global Invasive Species Programme (GISP), Washington, DC/ Nairobi

Burke L, Reytar K, Spalding M, Perry A (2011) Reefs at risk revisited. World Resources Institute, Washington, DC, 130pp

Carlton JT, Ruiz GM (2005) Vector science and integrated vector management in bioinvasion ecology: conceptual frameworks. In: Mooney HA, Mack RN, McNeely JA et al (eds) Invasive alien species: a new synthesis. Island Press, Washington, DC

Carpenter K et al (2008) One-third of reef-building corals face elevated extinction risk from climate change and local impacts. Science 321:560–563

Cohen AN, Carlton JT (1998) Accelerating invasion rate in a highly invaded estuary. Science 279:55–58

Coles SL, DeFelice RC, Eldredge LG, Carlton JT (1999) Historical and recent introductions of non- indigenous marine species into Pearl Harbor, Oahu, Hawaiian Islands. Mar Biol 135:147–158

Concepcion G, Kahng SE, Crepeau M, Franklin EC, Coles S, Toonen RJ (2010) Resolving natural ranges and marine invasions in a globally distributed octocoral (genus *Carijoa*). Mar Ecol Prog Ser 401:113–127

Diaz M, Madin J (2011) Macroecological relationships between coral species' traits and disease potential. Coral Reefs 30:73–84

Harris AR (2010) Analysis of recovery patterns of Indian Ocean coral reefs through examination of scleractinian communities and populations. Ph.D. thesis, University of Warwick, 273pp

Harris A, Sheppard CRC (2008) Status and recovery of the coral reefs of the Chagos archipelago, British Indian Ocean Territory. In: Obura D, Souter D, Linden O (eds) Indian Ocean coral reefs. CORDIO, Kalmar University, Sweden, pp 61–69

Harvell D, Jordán-Dahlgren E, Merkel S, Rosenberg E, Raymundo L, Smith G, Weil E, Willis B (2007) Coral disease, environmental drivers, and the balance between coral and microbial associates. Oceanography 20:172–195

Hewitt CL, Martin RB, Centre for Research on Introduced Marine Pests (2001) Revised protocols for baseline port surveys for introduced marine species: survey design, sampling protocols and specimen handling, vol 22, CRIMP Technical Report. CSIRO Marine Research, Hobart

Hewitt CL, Campbell ML, Thresher RE, Martin RB, Boyd S, Cohen BF, Currie DR, Gomon MF, Keough MJ, Lewis JA, Lockett MM, Mays N, McArthur MA, O'Hara TD, Poore GCB, Ross DJ, Sotrey MJ, Watson JE, Wilson RS (2004) Introduced and cryptogenic species in Port Philip Bay, Victoria, Australia. Mar Biol 144:183–202

Heywood V (ed) (1995) Global biodiversity assessment. Cambridge University Press, Cambridge

Hobbs JPA, Frisch AJ (2010) Coral disease in the Indian Ocean: taxonomic susceptibility, spatial distribution and the role of host density on the prevalence of white syndrome. Dis Aquat Organ 89:1–8

Hoegh-Guldberg O, Mumby PJ, Hooten AJ, Steneck RS, Greenfield P, Gomez E, Harvell CD, Sale PF, Edwards AJ, Caldeira K, Knowlton N, Eakin CM, Iglesias-Prieto R, Muthiga N, Bradbury RH, Dubi A, Hatziolos ME (2007) Coral reefs under rapid climate change and ocean acidification. Science 318:1737–1742

Hughes TP, Graham NAJ, Jackson JBC, Mumby PJ, Steneck RS (2010) Rising to the challenge of sustaining coral reef resilience. Trends Ecol Evol 25:633–642

IMO (2004) International convention for the control and management of Ships' Ballast Water & Sediments. Adopted by consensus in London on Friday 13 Feb 2004, not in force (May 2011)

Kaczmarsky L, Richardson LL (2011) Do elevated nutrients and organic carbon on Philippine reefs increase the prevalence of coral disease? Coral Reefs 30:253–257

Lesser MP, Bythell John C, Gates Ruth D, Johnstone Ron W, Hoegh-Guldberg O (2007) Are infectious diseases really killing corals? Alternative interpretations of the experimental and ecological data. J Exp Mar Biol Ecol 346:36–44

McGeoch MA, Butchart Stuart HM, Dian S, Elrike M, Kleynhans Elizabeth J, Andy S, Janice C, Michael H (2010) Global indicators of biological invasion: species numbers, biodiversity impact and policy responses. Divers Distrib 16:95–108

Miller J, Muller E, Rogers C, Waara R, Atkinson A, Whelan KRT, Patterson M, Witcher B (2009) Coral disease following massive bleaching in 2005 causes 60% decline in coral cover on reefs in the US Virgin Islands. Coral Reefs 28:925–937

Molnar JL, Gamboa RL, Revenga C, Spalding MD (2008) Assessing the global threat of invasive species to marine biodiversity. Front Ecol Environ 6:485–492

Mora C, Aburto-Oropeza O, Ayala Bocos A, Ayotte PM, Banks S et al (2011) Global human footprint on the linkage between biodiversity and ecosystem functioning in reef fishes. PLoS Biol 9(4):e1000606. doi:10.1371/journal.pbio.1000606

Muthiga N, Costa A, Motta H, Muhando C, Mwaipopo R, Schleyer M (2008) Status of coral reefs in East Africa: Kenya, Tanzania, Mozambique and South Africa. In: Wilkinson C (ed) Status of coral reefs of the world: 2008. Global Coral Reef Monitoring Network and Reef and Rainforest Research Center, Townsville, pp 91–104

O'Farrell S (2007) Competition and aggression between scleractinian corals and macroalgae on post-disturbance reefs in the Chagos archipelago. M.Sc. thesis University of Wales, Bangor, 58p

Pomeroy RS, Parks JE, Watson IM (2004) How is your MPA doing? A guidebook of natural and social indicators for evaluating marine protected area management effectiveness. IUCN, Gland/Cambridge, UK,xvi + 216 pp

Pratchett MS, Graham NAJ, Sheppard CRC, Mayes B (2010) Are infestations of *Cymo melanodactylus* killing *Acropora cytherea* in the Chagos archipelago? Coral Reefs 29:941

Rajasuriya A, Zahir H, Venkataraman K, Islam Z, Tamelander J (2004) Status of coral reefs in South Asia: Bangladesh, Chagos, India, Maldives and Sri Lanka. In: Wilkinson C (ed) Status of coral reefs of the world: 2004, vol 1. Australian Institute of Marine Science, Townsville, pp 213–234, 301p

Raymundo LJ, Rosell KB, Kaczmarsky L, Reboton C (2005) Coral diseases on Philippine reefs: Genus *Porites* is a dominant host. Dis Aquat Organ 64:181–191

Reinicke GB, van Ofwegen LP (1999) Soft corals (Alcyonacea: Octocorallia) from shallow water in the Chagos archipelago: species assemblages and their distribution. In: Sheppard CRC, Seaward MRD (eds) Ecology of the Chagos archipelago, vol 2, Linnean Society Occasional Publications. Westbury for the Linnean Society of London, London, pp 67–90

Riegl B, Piller WE (2003) Possible refugia for reefs in times of environmental stress. Int J Earth Sci 92:520–531

Riegl B, Sheppard CRC, Purkis SJ (2012) Human impact on atolls leads to coral loss and community homogenisation: a modeling study. PLoS One 7:e36921

Rohwer F, Kelley S (2004) Culture-independent analyses of coral-associated microbes. In: Loya Y, Rosenberg E (eds) Coral health and disease. Springer, Berlin, pp 265–277

Ruiz GM, Rawlings TK, Dobbs FC, Drake LA, Mullady T, Huq A, Colwell RR (2000) Global spread of microorganisms by ships. Ballast water discharged from vessels harbours a cocktail of potential pathogens. Nature 408:49–50

Schleyer MH, Benayahu Y (2010) Pre- and post-1998 ENSO records of shallow-water octocorals (Alcyonacea) in the Chagos archipelago. Mar Pollut Bull 60:2197–2200

Sheppard CRC (1980) Coral cover, zonation and diversity on reef slopes of Chagos atolls, and population structures of the major species. Mar Ecol Prog Ser 2:193–205

Sheppard CRC (1999a) Changes in coral cover on reefs of Chagos over eighteen years. In: Sheppard CRC, Seaward MRD (eds) Ecology of the Chagos archipelago. Linnean Society/Westbury, London, p 91

Sheppard CRC (1999b) Coral decline and weather patterns over 20 years in the Chagos archipelago, central Indian Ocean. Ambio 28:472–478

Sheppard CRC (2003) Predicted recurrences of mass coral mortality in the Indian Ocean. Nature 425:294–297

Sheppard CRC (2006) Longer term impacts of climate change. In: Cote I, Reynolds J (eds) Coral reef conservation. Cambridge University Press, New York, pp 264–290

Sheppard CRC (2009) Large temperature plunges recorded by data loggers at different depths on an Indian Ocean atoll: comparison with satellite data and relevance to coral refuges. Coral Reefs 28:399–403

Sheppard CRC et al (2012) Reefs and islands of the Chagos archipelago, Indian Ocean: why it is the world's largest no-take marine protected area. Aquatic Conservation Mar Freshw Ecosyst 22:232–261

Sheppard CRC, Spalding MD, Bradshaw C, Wilson S (2002) Erosion vs recovery of coral reefs after 1998 El Nino: Chagos reefs, Indian Ocean. Ambio 31:40–48

Sheppard CRC, Harris A, Sheppard ALS (2008) Archipelago-wide coral recovery patterns since 1998 in the Chagos archipelago, central Indian Ocean. Mar Ecol Prog Ser 362:109–117

Spalding MD, Ravilious C, Green EP (2001) World atlas of coral reefs. University of California Press, Berkely, 424pp

Stoddart D R, Taylor DL (eds) (1971) Geology and ecology of Diego Garcia Atoll, Chagos Archipelago. Atoll Res Bull 149:237pp

Tamelander J, Rajasuria A (2008) Status of coral reefs in South Asia: Bangladesh, Chagos, India, Maldives and Sri Lanka. In: Wilkinson C (ed) Status of coral reefs of the world: 2008. Global Coral Reef Monitoring Network and Reef and Rainforest Research Center, Townsville, pp 119–130

Tamelander J, Campbell M, Lundin CG (2009) Detecting bioinvasions on small islands in the Indian Ocean – project completion report. IUCN Global Marine Programme 2009. IUCN, Gland

Thresher RE, Kuris AM (2004) Options for managing invasive marine species. Biol Invasions 6:295–300

Tu M (2009) Assessing and managing invasive species within protected areas. In: Ervin J (ed) Protected area quick guide series. The Nature Conservancy, Arlington, p 36

Weil E, Urreiztieta I, Garzon-Ferreira J (2002) Geographic variability in the incidence of coral and octocoral diseases in the wider Caribbean. In: Proceedings of the 9th international coral reef symposium, vol 2, Bali, pp 1231–1237

Weil E, Smith GW, Gil-Agudelo DL (2006) Status and progress in coral reef disease research. Dis Aquat Organ 60:1–7

Wilcove DS, Rothstein D, Dubow J, Phillips A, Losos E (1998) Quantifying threats to imperiled species in the United States. Bioscience 48:607–615

Wilkinson C (ed) (2008) Status of coral reefs of the world: 2008. Global Coral Reef Monitoring Network and Reef and Rainforest Research Center, Townsville, p 296

Williams GJ, Davy SK, Aeby GS (2007) Coral disease at Palmyra Atoll, a remote reef system in the Central Pacific. Coral Reefs 27:207

Willis B, Page C, Dinsdale E (2004) Coral diseases on the great barrier reef. In: Loya Y, Rosenberg E (eds) Coral health and disease. Springer, Berlin, pp 69–104

Yang SY, Keshavmurthy S, Obura D, Sheppard CRC, Visram S, Chen CA (2012) Diversity and distribution of *Symbiodinium* associated with seven common coral species in the Chagos archipelago, Central Indian Ocean. PlosOne 7(e35836):9

Zvuloni A, Artzy-Randrup Y, Stone L, Kramarsky-Winter E, Barkan R, Loya Y (2009) Spatio-temporal transmission patterns of black-band disease in a coral community. PlosONE 4:e4993

The Status of Coral Reef Fish Assemblages in the Chagos Archipelago, with Implications for Protected Area Management and Climate Change

19

Nicholas A.J. Graham, Morgan S. Pratchett, Tim R. McClanahan, and Shaun K. Wilson

Introduction

Coral reef fishes are conspicuous components of coral reef ecosystems, representing the most diverse vertebrate communities on Earth (Jones et al. 2002). Reef fishes have attracted a great deal of research and management attention because of their vulnerability to fishing, and because many fishes have important functional roles within coral reef ecosystems. These functions include herbivory which mediates the competition between corals and benthic algal communities (Bellwood et al. 2004), predation on invertebrates that modify habitat (McClanahan 2000), or piscivory that mediates competitive interactions among fish (Hixon 1991). Fishes also form some of the strongest ecosystem service links between coral reefs and human societies, providing benefits such as food and income through fisheries and tourism (Polunin and Roberts 1996, Williams and Polunin 2000).

The sensitivity of fishes to depletion through fishing activities has been documented in a range of studies with rapid depletion of fish biomass even with relatively light fishing pressure. The most vulnerable groups are often species with higher trophic levels, such as groupers and snappers (Russ and Alcala 1989; McClanahan 1994; Jennings and Polunin 1997; McClanahan et al. 2008). Additionally, there are significant changes in reef fish size resulting in ecosystems being dominated by small-bodied individuals and species (Dulvy et al. 2004; Graham et al. 2005; McClanahan and Omukoto 2011). Even recreational fishing can have substantial impacts on fish populations (Coleman et al. 2004). Given the known sensitivity of reef fish assemblages to fishing, much interest has developed around management techniques to redress declines and restore pristine ecosystems (Pitcher 2001; McClanahan 2011).

A great deal of management and research attempting to restore coral reef ecosystems has focused on no-take marine protected areas (or marine reserves), with most studies documenting increases in the abundance, biomass and sometimes the diversity of reef fish assemblages within well-enforced marine reserves (Russ et al. 2005; McClanahan et al. 2007a; Stockwell et al. 2009). However, considerable debate has questioned whether marine reserves, which are generally <10 km^2 (McClanahan et al. 2009), are capable of promoting biomass and trophic structures that foster ecological processes in a 'pristine' environment (Graham et al. 2011a). Indeed, recent studies in remote, uninhabited atolls of the NW Hawaiian Islands and northern Line Islands in the Pacific have documented reef fish biomass, dominated by top predators, that far exceeds the biomass values quantified for even relatively large and old marine reserves (Friedlander and DeMartini 2002; Sandin et al. 2008; DeMartini et al. 2008; Williams et al. 2011). The Chagos Archipelago is likely to be a similar reference site in the Indian Ocean, as it covers a very large area (with ~60,000 km^2 of potential reef area) and the northern atolls have been uninhabited since the early 1970s with very little fishing activity even before the current no-take areas status was declared in April 2010 (Sheppard et al. 2012). Extensive fish surveys have been conducted across much of the western Indian Ocean and these provide a useful context for evaluating the status of the Chagos (Graham et al. 2008; McClanahan et al. 2011).

Studies in remote Pacific locations have highlighted the high abundance and biomass of reef sharks (Sandin et al. 2008; Nadon et al. 2012). In Chagos, however, sharks are one of the few groups that have been substantially depleted through illegal fishing activities. The relative abundance of

N.A.J. Graham (✉) • M.S. Pratchett
ARC Centre of Excellence for Coral Reef Studies, James Cook University, QLD 4811 Townsville, Australia
e-mail: nick.graham@jcu.edu.au; Morgan.pratchett@jcu.edu.au

T.R. McClanahan
Wildlife Conservation Society, Marine Programs, 10460 Bronx, NY, USA

S.K. Wilson
Department of Environment and Conservation, Marine Science Program, 6151 Kensington, WA, Australia

C.R.C. Sheppard (ed.), *Coral Reefs of the United Kingdom Overseas Territories*, Coral Reefs of the World 4,
DOI 10.1007/978-94-007-5965-7_19, © Springer Science+Business Media Dordrecht 2013

reef sharks in the northern atolls of Chagos are estimated to have declined by ~90% from the 1970s to 2006 (Anderson et al. 1998; Spalding 2003; Graham et al. 2010). Consequently, the trophic structure of reef fish communities at Chagos may differ from other remote locations. These results emphasize the importance of monitoring reef shark populations into the future, as efforts to enforce the recently declared Chagos no-take marine protected area develop and take effect.

Reef fishes are also sensitive to disturbances to their coral reef habitat (Jones and Syms 1998). If reef corals die and the physical structure of the reef subsequently erodes, reef fish abundance and diversity can decline (reviewed by Wilson et al. 2006; Pratchett et al. 2008). Fish species specialized on corals for food (Pratchett et al. 2006; Graham 2007), habitat (Munday 2004, Wilson et al. 2008a, Pratchett et al. 2012) or settlement (Jones et al. 2004) typically show the largest declines in abundance. Body size also appears to be a key determinant of fish species vulnerability to habitat degradation, with smaller body size classes showing the greatest declines (Graham et al. 2006, 2007; Wilson et al. 2008b, 2010). Chagos, like much of the northern Indian Ocean, was badly impacted by the 1998 thermal anomaly event, with up to 80% of the corals dying on many reefs (Sheppard 1999). However, the reefs have shown a strong recovery and by 2010 coral cover had returned to pre-1998 levels (Ateweberhan et al. 2011; Sheppard et al. 2012). It is useful to understand how reef fish have responded to these substantial benthic changes in remote locations such as Chagos, where local anthropogenic impacts are minimal and to allow comparisons with other locations with substantial local human impacts.

In this chapter we assess the reef fish biomass and trophic structure at Chagos, comparing data within and among atolls (Salomon, Peros Banhos, Great Chagos Bank and Diego Garcia), with a particular focus on the effects of recreational fishing around Diego Garcia. These data are then put into the context of the wider western Indian Ocean, including smaller marine reserves embedded in fished coastlines. Trends in reef shark abundances at Chagos are updated from Graham et al. (2010) to include data from 1975 to 2012. Finally, we assess the impacts of habitat change in Chagos on reef fish assemblages, compared to other locations in the western Indian Ocean, with a particular focus on the importance of specialization in corallivorous fishes.

Reef Fish Biomass Among Atolls of Chagos

It is estimated that there are at least 784 species of fishes in Chagos (e.g. Winterbottom and Anderson 1999). Spalding (1999) compared the density and diversity of fishes at Chagos to the relatively nearby Seychelles in 1996, reporting slightly higher species richness per site in Chagos, but slightly lower species over smaller replicate count areas (Spalding 1999). Following the 1998 mass bleaching substantial declines in species richness and abundances of certain guilds (e.g. coral feeding fishes) were reported across the western Indian Ocean where coral cover declined, but the fish assemblages at Chagos displayed relative stability (Graham et al. 2008; Sheppard et al. 2012; see 'Habitat degradation effects on reef fishes in Chagos' below). All of this work indicated that reef fish assemblages in Chagos tolerated the 1998 disturbance well, but the biomass of these assemblages had not been quantified.

Larger bodied fish contribute disproportionately to standing biomass estimates of fish assemblages. However large fish are much more vulnerable to fishing (Dulvy et al. 2004; Olden et al. 2007; Graham et al. 2011b), and substantial declines in reef fish biomass have been detected on coral reefs associated with increasing fishing pressure or efficiency (McClanahan 1994; Jennings and Polunin 1996; Russ and Alcala 1989; Newman et al. 2006; Cinner et al. 2009, 2013). Fish biomass is thought to be important to the functioning of coral reefs, maintaining a range of ecosystem processes (McClanahan et al. 2011; Mora et al. 2011). It is therefore critical to have reliable estimates of reef fish assemblage biomass in the absence of fishing pressure. Small marine reserves can substantially enhance fish biomass compared to adjacent fished areas (e.g. Russ et al. 2005; McClanahan et al. 2007a), but large, remote and unfished locations in the Pacific far outstrip the biomass values recorded in even the most successful, well managed, small marine reserves (Friedlander and DeMartini 2002; Stevenson et al. 2007; Sandin et al. 2008). In Chagos, a small seasonal (June – August) Mauritian fishery operated in the archipelago from the early 1970s until 2010, and was under license since 1991 (Mees 1996). This fishery was relatively small, targeting grouper and snappers on banks and in deeper water and there was no uptake of licences in some years (Mees 1996, C. Mees, personal communication). It is therefore unlikely that this fishery affected the reef fish populations at the depth and habitats that we have quantified them here.

At Diego Garcia there is also a small recreational fishery associated with the US Navy base. Affects from this recreational fishery on local fish communities may be significant and needs to be assessed.

Reef fish biomass at Chagos was first quantified in the 2010 Chagos scientific expedition. During this trip 18 sites were surveyed using underwater visual census techniques on the outside of Salomon and Peron Banhos atolls and around the Three Brothers and Eagle Island on the Great Chagos Bank (Fig. 19.1). A further 6 sites were surveyed on the outside of Diego Garcia during the 2012 expedition using the same techniques (Fig. 19.1). At each site fish were surveyed within four 50 by 5 m belt transects, by the same observer (N. Graham), along the 7–9 m depth contour. All diurnally active, non-cryptic, reef associated fishes larger than 8 cm

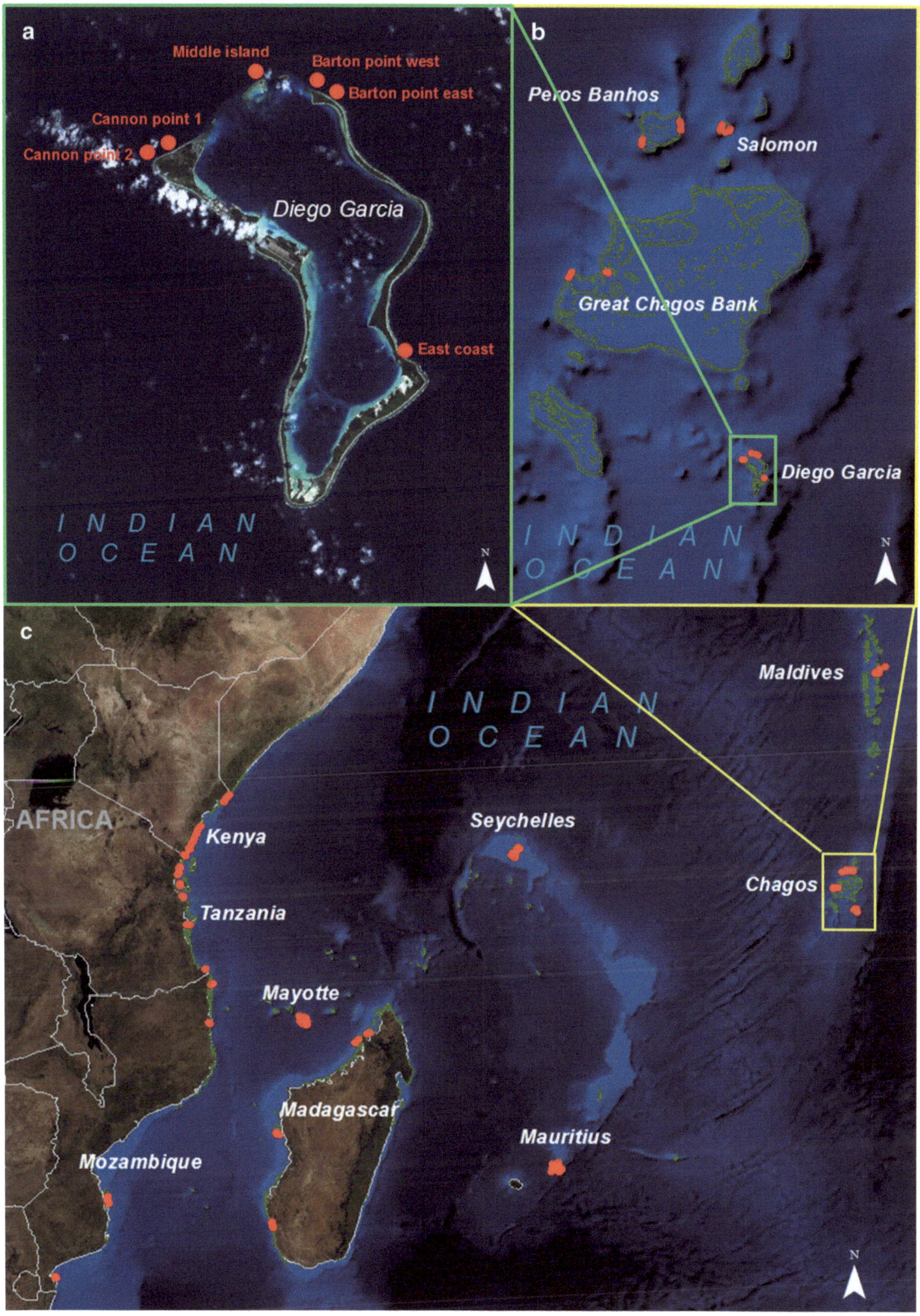

Fig. 19.1 Map of sites surveyed: (**a**) Diego Garcia, (**b**) Chagos archipelago and (**c**) The western Indian Ocean

were identified to species, counted and length estimated to the nearest centimeter. Larger more mobile species were surveyed on a first pass of the transect, while smaller fish, principally damselfishes, were surveyed within a 2 m belt on the return pass. Biomass was estimated for each counted individual using the length estimate and published length-weight relationships for that species or genus (Froese and Pauly 2012). Benthic cover, was quantified every 50 cm along each 50 m transect and structural complexity was estimated on a 6 point visual scale following Wilson et al. (2007).

Diego Garcia had similar coral cover to the Great Chagos Bank and Peros Banhos atoll sites, but lower cover than Salomon atoll (Table 19.1 and Fig. 19.2a). Sites at Salomon atoll had the highest coral cover, but cover did not differ significantly to Peros Banhos. Diego Garcia also had among the lowest structural complexity values, and was similar only to the Great Chagos Bank, while sites at Peros Banhos had the highest values (Table 19.1 and Fig. 19.2b).

In terms of total fish biomass, Diego Garcia had the lowest while Peros Banhos and the Great Chagos Bank had the highest values (Table 19.1 and Fig. 19.3a). Pairwise comparisons indicate that total fish biomass between Diego Garcia and Salomon atoll were similar, however the mean values at Diego Garcia are less than half those of Salomon. Shark biomass was not statistically different among atolls (Table 19.1 and Fig. 19.3b), although this may be due to high variation within atolls likely due to the relatively small sampling units for these large mobile organisms. Despite this variation, the Great Chagos Bank appears to have considerably higher shark biomass than the other three atolls. If the total fish biomass is broken down into families typically targeted in fisheries (e.g. groupers, snappers, and emperors) versus non-targeted fish families (e.g. damselfish, butterflyfish, and angelfish), significant variation among atolls is apparent for targeted biomass (Table 19.1 and Fig. 19.3c), but not non-targeted biomass (Fig. 19.3d). Targeted fish biomass was highest at Peros Banhos, intermediate on the Great Chagos Bank and lowest at Salomon followed by Diego Garcia (Fig. 19.3c).

Higher levels of reef structural complexity can have a positive influence on reef fish biomass, and is generally more important than coral cover for fish biomass (Cinner et al. 2009). If any effect of reef structural complexity on targeted fish biomass is controlled for with an analysis of covariance, the difference in biomass among atolls still comes out as highly significant, explaining 29% of the total variance in the data (structural complexity, $F=8.23$, $p<0.01$; atoll, $F=21.45$, $p<0.0001$). Structural complexity is therefore influencing reef fish biomass at Chagos, but the difference among atolls occurs independent of structural complexity, which can be noted from the patterns in structural complexity versus biomass among atolls (Figs. 19.2 and 19.3).

Table 19.1 One way ANOVA results of the differences among atolls of Chagos in hard coral cover (%), structural complexity, total fish biomass (kg/ha), untargeted fish biomass (kg/ha), targeted fish biomass (kg/ha) and shark biomass (kg/ha). $*p<0.05$, $**p<0.01$

	df	F	*p* value
Hard coral cover	(3,96)	6.467	<0.0001**
Structural complexity	(3,96)	13.432	<0.0001**
Total fish biomass	(3,96)	23.39	<0.0001**
Untargeted fish biomass	(3,96)	1.324	0.271
Targeted fish biomass	(3,96)	32.204	<0.0001**
Shark biomass	(3,96)	1.405	0.246

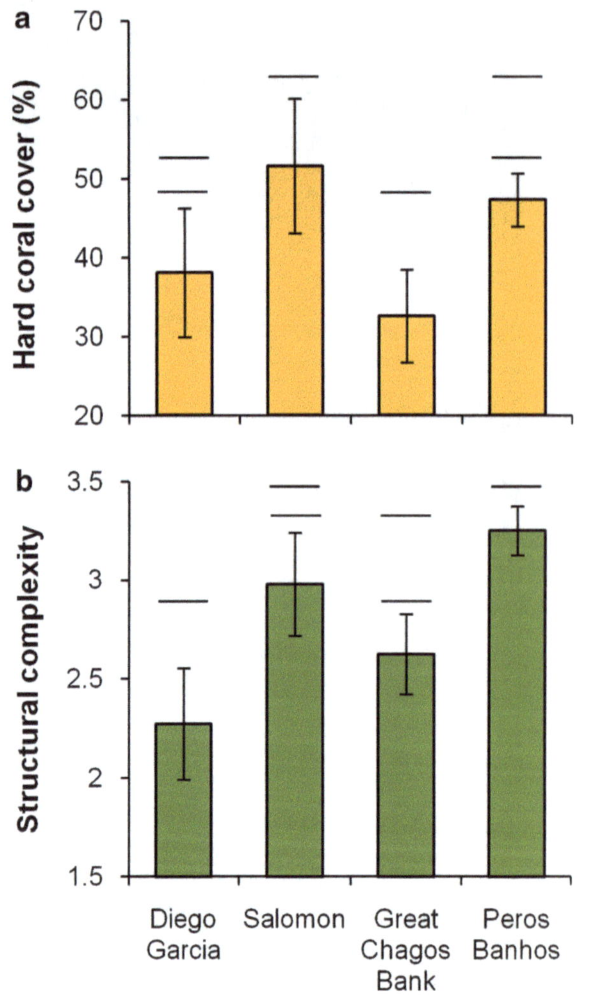

Fig. 19.2 Variation in (**a**) Hard coral cover (%) and (**b**) Structural complexity among atolls within the Chagos archipelago. Error bars represent one standard error of the mean. *Horizontal lines* show homogeneous subsets from post hoc comparisons using the Tukey test

Breaking down the fish biomass by trophic level and body size classes produces some illuminating patterns (based on species level information obtained from FishBase; Froese and Pauly 2012). The fish communities around the northern atolls appear to be represented by higher trophic level and larger sized fish than those around Diego Garcia (Fig. 19.4).

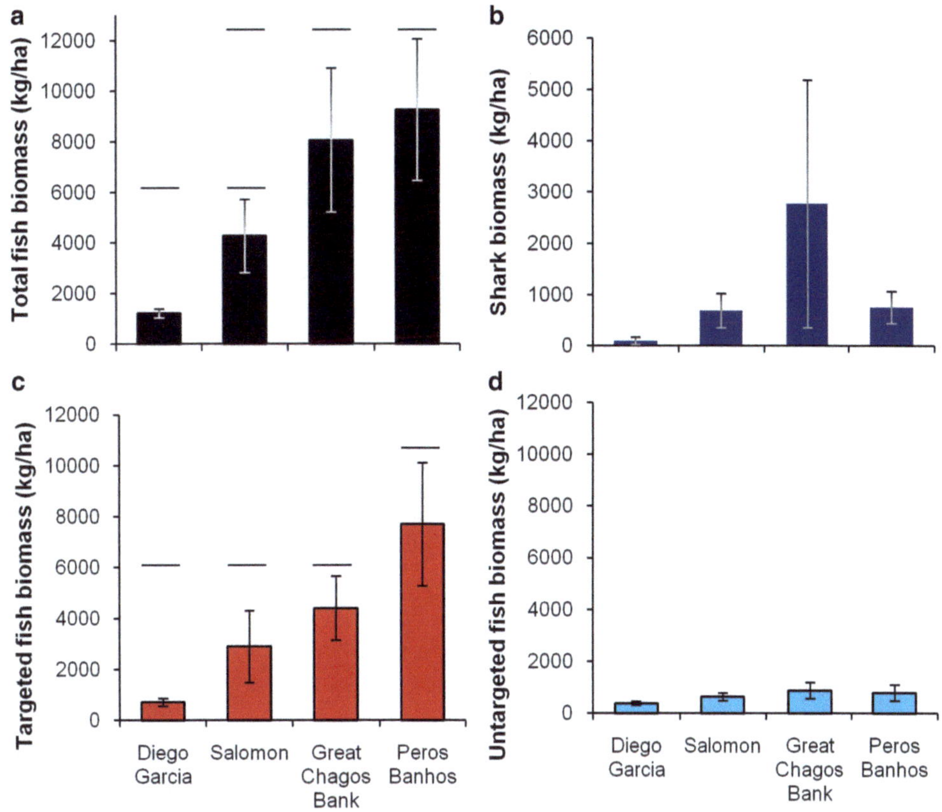

Fig. 19.3 Variation in (**a**) Total fish biomass, (**b**) Shark biomass, (**c**) Targeted fish biomass and (**d**) Untargeted fish biomass (kg/ha) among atolls in the Chagos archipelago. Error bars represent one standard error around the mean. *Horizontal lines* show homogeneous subsets from post hoc comparisons using the Tukey test

Only ~35% of the community around Diego Garcia is composed of fish with a trophic level greater than 3.5, while this portion of the community makes up ~60–70% of the community in the three northern atolls (Fig. 19.5a). Similarly, fish with a body size greater than 50 cm only make up ~ 25% of the community biomass around Diego Garcia, whereas they make up between 40 and 60% in the northern atolls (Fig. 19.5b). Large bodied fish are typically the most vulnerable to exploitation (Olden et al. 2007), and can decline substantially in response to fairly light fishing pressure (Dulvy et al. 2004; Wilson et al. 2010). Similarly, higher trophic level organisms are typically vulnerable to exploitation and are conspicuously missing in many food webs (Estes et al. 2011). These patterns, along with the biomass analyses (Fig. 19.3), indicate that the recreational fishery around Diego Garcia is reducing fish biomass and size compared to the relatively unfished northern atolls.

Community composition of fish varied considerably among sites and atolls (multidimensional scaling plot, MDS, Fig. 19.6). All the sites around Diego Garcia fall to the far left hand side of the ordination, while the other atolls, in particular Peros Banhos, spread out towards the right hand side of the ordination (Fig. 19.6a). The 10 most important fish families driving these patterns are shown as vectors (Fig. 19.6b), with almost all families, including all those that are potentially targets of fishing, increasing to the right side of the ordination. This indicates that in almost all cases the biomass of these important families is higher in the northern atolls than around Diego Garcia, potentially highlighting the influence of recreational fishing on the fish community biomass.

Focusing specifically on Diego Garcia provides some interesting patterns. Among the six sites around the atoll there were significant differences in hard coral cover (Table 19.2 and Fig. 19.7a) and structural complexity (Table 19.2 and Fig. 19.7b). There was a gradient from a coral cover of 8% and a relatively low structural complexity value of 1.5 at Cannon Point site 1 to a high coral cover of 59% at Barton Point west and a high structural complexity of 3.4 in Barton Point east. Fish biomass showed considerable variability among sites, but there was no significant difference when assessing total fish biomass (Fig. 19.7c). However, if only fish families known to be targeted in fisheries of the region (based on body size, and susceptibility to fishing gears) were assessed, biomass at Cannon Point site 2 was lower than at Middle island (Table 19.2 and Fig. 19.7d). Interestingly, the sites at Cannon Point are adjacent to the Navy base, and as such may experience the most recreational fishing pressure. Although both coral cover and structural complexity were

Fig. 19.4 High fish biomass at Chagos. (**a**) A large coral trout, *Plectropomus laevis*, (**b**) A school of paddletail snapper, *Lutjanus gibbus*, (**c**) A school of captain parrotfish, *Chlorurus enneacanthus*, (**d**) A mix of snapper species along the reef bottom, including red snapper, *Lutjanus bohar*, paddeltail snapper, *Lutjanus gibbus*, and midnight snapper, *Macolor niger* (Photo's: N Graham)

generally lower at the Cannon Point sites, regression analysis showed no significant relationship between benthic reef structure and total, targeted or untargeted fish biomass at Diego Garcia.

Comparison of Fish Biomass Across the Wider Indian Ocean

Overfishing on coral reefs is a pervasive problem, which is thought to have started centuries ago in some locations (Jackson et al. 2001; McClanahan and Omukoto 2011). Research indicates that the first effects of fishing are often the most severe in terms of biomass reduction, as slow-growing and large bodied fishes are vulnerable to fishing, and also contribute significantly to standing biomass (Reynolds et al. 2001). Given widespread depletion of fish assemblages in many countries, small marine reserves have become a dominant management tool attempting to reverse depletion trends and return fish biomass to high levels (Graham et al. 2011a). Putting the Chagos fish biomass in a broader geographic context produces an insightful view of the status of some of these smaller marine reserves that are embedded in fished landscapes. Broad spatial comparisons also gives biomass estimates in the northern atolls context and provides a clearer picture of just how heavily impacted the fish assemblages are around Diego Garcia. Fish biomass and coral cover has been collected in fished areas and marine reserves using comparable survey methods across 8 other western Indian Ocean countries (Fig. 19.1c). These data, collected principally by T McClanahan, N Graham and S Wilson, cover over 100 sites across the region and provide a basis for comparisons with fish and coral communities in Chagos. Some of the marine reserves in the region have reached asymptotes in fish biomass recovery (McClanahan et al. 2009), and prior to these surveys of Chagos, the relatively lightly fished Maldives was the site with the greatest recorded reef fish biomass (McClanahan 2011).

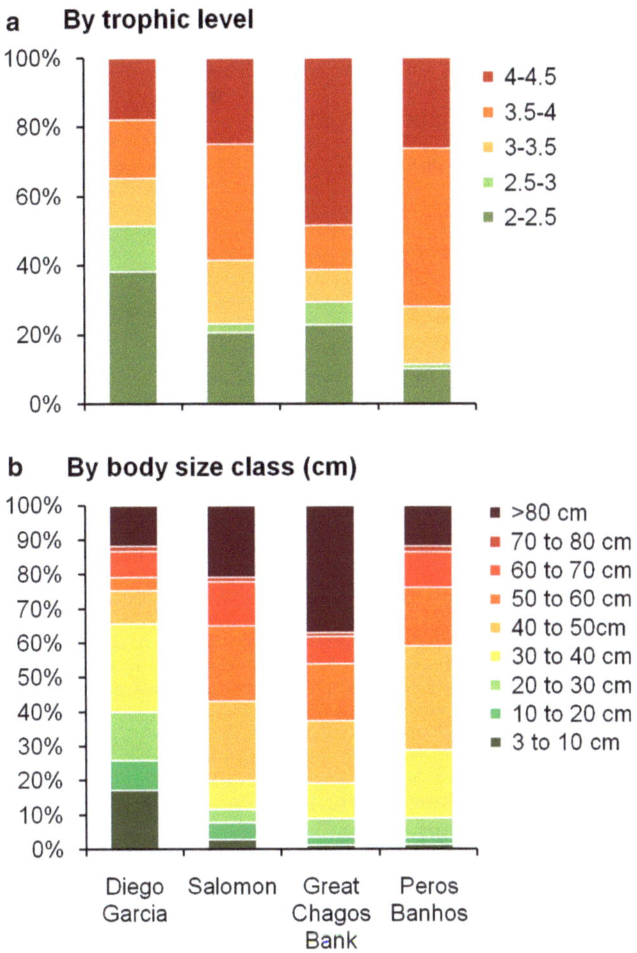

Fig. 19.5 Relative contributions of fish (**a**) Trophic levels and (**b**) Body size classes (cm) to the total fish biomass among atolls within the Chagos archipelago

Nevertheless, both historical records of fish catch from before the thirteenth century (McClanahan and Omukoto 2011) and evaluations of life history characteristics indicated that these reserves were not in pristine condition (McClanahan and Humphries 2012).

Hard coral cover varies substantially across the region (Table 19.3 and Fig. 19.8). The Seychelles have the lowest coral cover with an average of 14±3%, Diego Garcia falls within the middle group with average hard coral cover of 38±4%, while the northern atolls in Chagos have among the highest coral cover values in the region, with an average of 44±4% (Fig. 19.8). The western Indian Ocean was one of the regions most affected by the 1998 thermal anomaly; losing ~45% total coral cover (Ateweberhan et al. 2011). However, the impacts varied greatly around the region with extensive coral decline at the low latitude island nations (Seychelles, Chagos and Maldives) whilst impacts at higher latitude nations to the south, such as Mauritius, Tanzania, Mozambique and Madagascar, were minimal (McClanahan et al. 2007b; Graham et al. 2008; Ateweberhan et al. 2011).

Many of the patterns in coral cover reported here are a consequence of the 1998-bleaching event (Fig. 19.8). For example, the inner Seychelles lost >90% of its coral cover in 1998, and recovery of cover has been slow and variable among sites (Wilson et al. 2012). Similarly, both the Maldives and Kenya still have reduced coral cover at many sites in the aftermath of the 1998 bleaching event (McClanahan 2008; Ateweberhan et al. 2011). Given the substantial declines in coral cover in Chagos following 1998 (Sheppard 1999), the high cover reported here reflects substantial recovery not seen in other locations similarly impacted across the region. Indeed, the rapid recovery of hard coral cover in Chagos has been well documented (Sheppard et al. 2008; 2012) and contrasts markedly to many other locations in the region (Ateweberhan et al. 2011).

We do not have consistent structural complexity data across all of these locations but it is expected that structural complexity and coral cover will be positively correlated at this scale (Graham et al. 2008). Importantly, reef habitats in Chagos support much higher fish biomass than equivalent sites in the western Indian Ocean with comparable coral cover. This suggests human population density, distance to markets and community level economic development (which relate to fishing pressure and efficiency) are likely to be the key determinants of variation in reef fish biomass across the region (Cinner et al. 2009).

Total fish biomass in the three northern atolls of Chagos dwarf the biomass seen at all other reef sites across the wider region (Table 19.3 and Fig. 19.9a). Indeed, biomass values in the northern atolls are six times greater than those recorded from even the most successful small marine reserves in the region (McClanahan et al. 2009, 2011). A great deal of this difference is due to greater abundance of higher trophic level fishes and those with a larger overall body size in the northern atolls (Graham and McClanahan in press). The performance of the marine reserves across the region, in terms of building up fish biomass compared to fished areas, varies substantially (Fig. 19.9a). Some locations, such as Kenya, have relatively large (up to ~30 km^2) and old (up to ~40 years) marine reserves that are well enforced and compliance is high. These reserves recover from the effects of fishing in ~20 years, when estimates of fish biomass asymptote (McClanahan and Graham 2005), indicating they have reached their full biomass potential. Nevertheless, they continue to change in terms of the composition of the fish fauna towards slow-maturing and growing species (McClanahan and Humphries 2012).

Biomass in the northern atolls of Chagos is, therefore, much higher than even the most effective smaller reserves. Biomass differences are attributable to greater abundance of larger, higher trophic level fishes in the northern atolls. Many of these fish, such as the grey reef shark, have limited reef fidelity and large home ranges (Heupel et al. 2010).

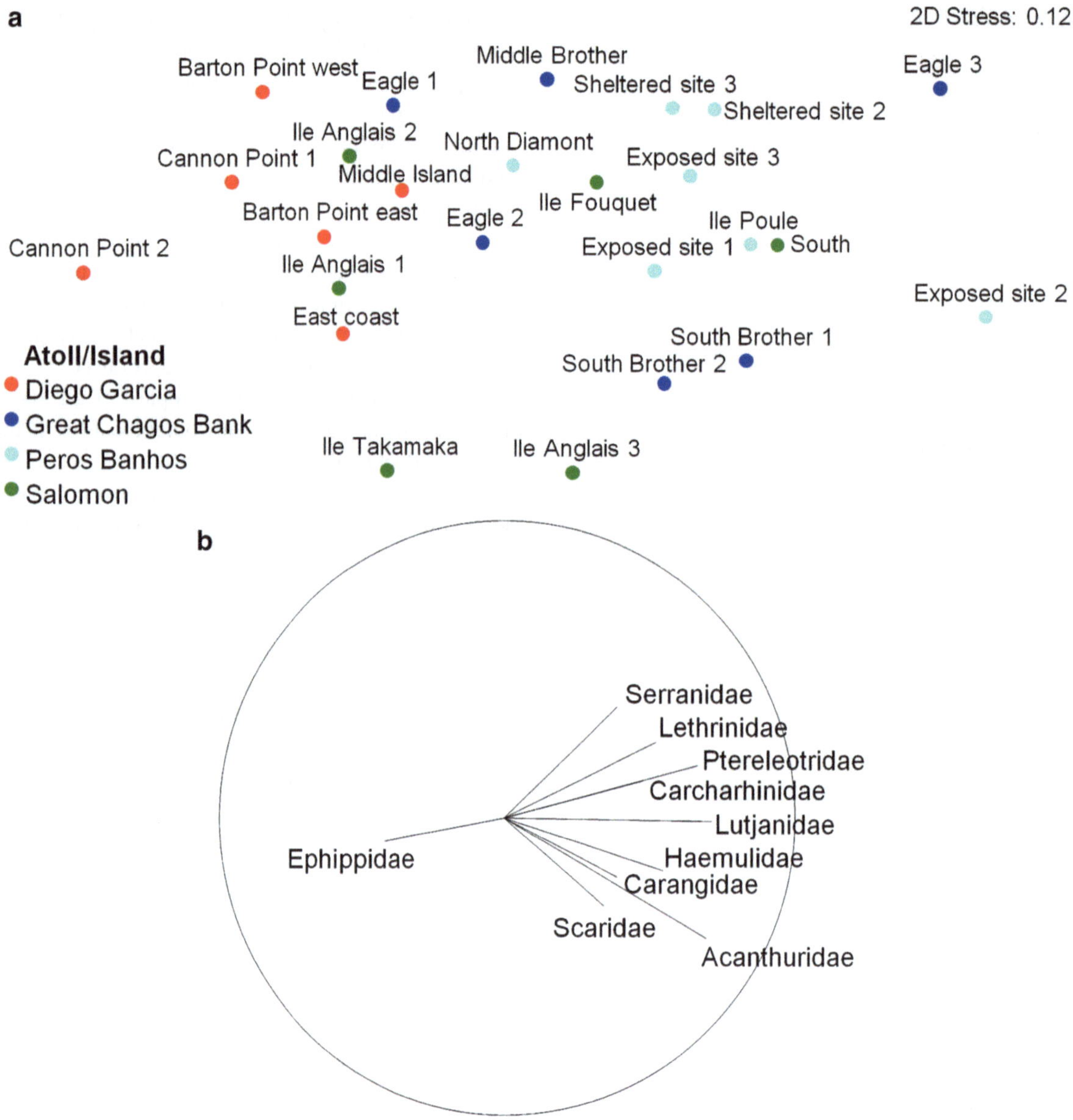

Fig. 19.6 Multidimensional scaling analysis of fish family biomass. (**a**) Plot of the spatial variation in fish family level biomass among sites in Chagos. Each site is coloured according to its corresponding atoll. (**b**) The relative contribution and direction of influence of the top ten fish families to the observed variation among sites. Only fish families with a correlation > 0.3 with either axis MDS1 or MDS2 were selected for representation in the vector plot

Hence, it is unlikely that small reserves embedded in fished landscapes are adequately protecting the top end of the food-web, with larger fish exposed to fishing pressure when they move outside park boundaries (Graham and McClanahan in press). The archaeological record of fish bones also support this contention of reduced trophic level and other life history characteristics in these marine reserves (McClanahan and Omukoto 2011).

Fish biomass at Diego Garcia is much more comparable to some of the other locations across the region (Fig. 19.9a, b). It should be noted however, that Diego Garcia is similar to the highest values reported across the region, for example the Maldives, Mayotte and Kenyan marine reserves (McClanahan et al. 2011). When the fish community is broken down by species targeted by fisheries, versus those that are not, the differences become more apparent. While the northern atolls have an exceptionally high targeted fish biomass (mean 4,930 kg/ha), Diego Garcia had values similar to the best performing marine reserves in the region (Fig. 19.9b). This indicates that although the recreational fishery around Diego Garcia does seem to be having an impact of the fish biomass compared to the northern atolls, the biomass in Diego Garcia is still very high, and is far from an overfished status.

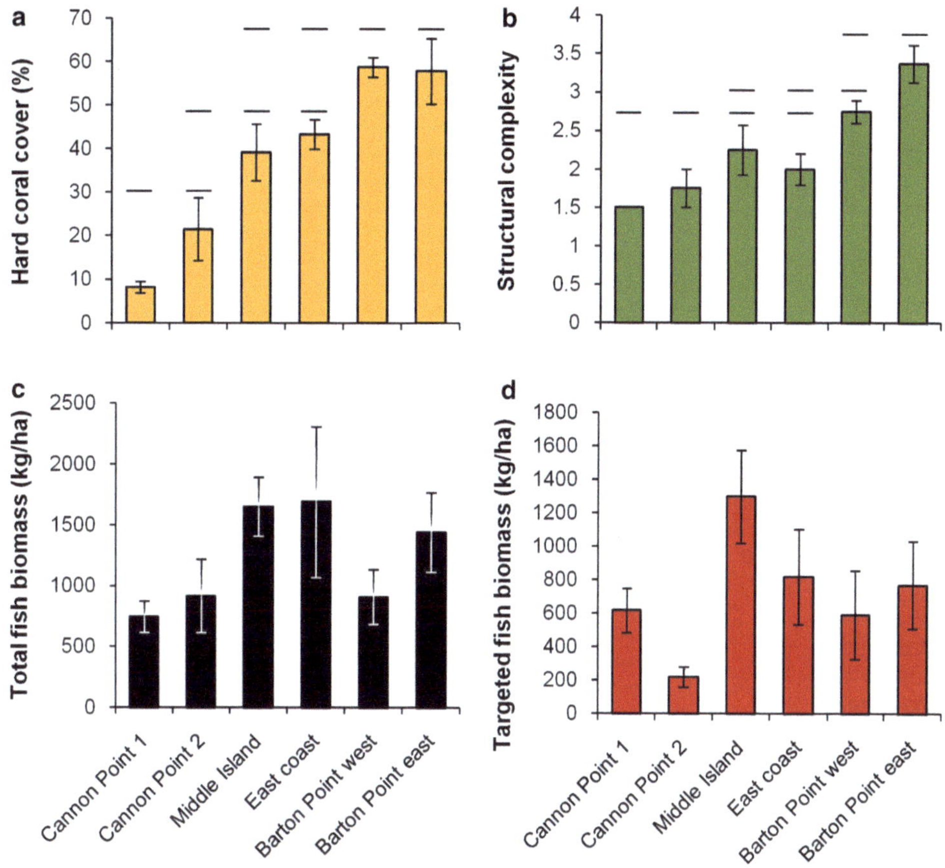

Fig. 19.7 Variation in (**a**) Hard coral cover (%), (**b**) Structural complexity, (**c**) Total fish biomass (kg/ha) and (**d**) Targeted fish biomass (kg/ha) among sites at Diego Garcia. Error bars represent one standard error of the mean. *Horizontal lines* at a same level show homogeneous subsets from post hoc comparisons using the Tukey test

Table 19.2 One way ANOVA results of the differences among the six sites around Diego Garcia in hard coral cover (%), structural complexity, total fish biomass (kg/ha) and targeted fish biomass (kg/ha). * $p<0.05$, ** $p<0.01$

	df	F	*p* value
Hard coral cover	(5,18)	14.477	< 0.0001**
Structural complexity	(5,18)	10.004	< 0.0001**
Total fish biomass	(5,18)	1.49	0.242
Targeted fish biomass	(5,18)	2.862	0.045*

Table 19.3 One way ANOVA results of the differences among surveyed countries in the Indian Ocean in hard coral cover (%), total fish biomass (kg/ha), targeted fish biomass (kg/ha) and untargeted fish biomass (kg/ha). *$p<0.05$, **$p<0.01$

	df	F	*p* value
Hard coral cover	(9, 159)	11.202	< 0.0001**
Total fish biomass	(15, 218)	6.395	< 0.0001**
Targeted fish biomass	(15, 218)	7.972	< 0.0001**
Untargeted fish biomass	(15, 218)	14.09	< 0.0001**

Various studies have shown that the first effects of fishing are often the most substantial in terms of reducing fish biomass, mainly because the slow-growing and large bodied fish that constitute most of biomass are the first to be caught and removed (e.g. Jennings and Polunin 1996, McClanahan and Omukoto 2011), and even recreational fishing can have substantial impacts on fish biomass (Coleman et al. 2004). The recreational fishery around Diego Garcia should be very carefully controlled and monitored to prevent any further declines in fish biomass in the future, and ideally management put in place to try to rebuild these fish stocks toward the status of the northern atolls. Diego Garcia is currently the only atoll in Chagos not included in the no-take marine protected area, however this does not mean more restrictive and targeted management, such as protecting a large portion of the atoll, or putting caps on fishing effort or restrictions of specific species, could not be achieved. It should be noted that some of the marine reserves in other countries across the region are not enhancing fish biomass compared

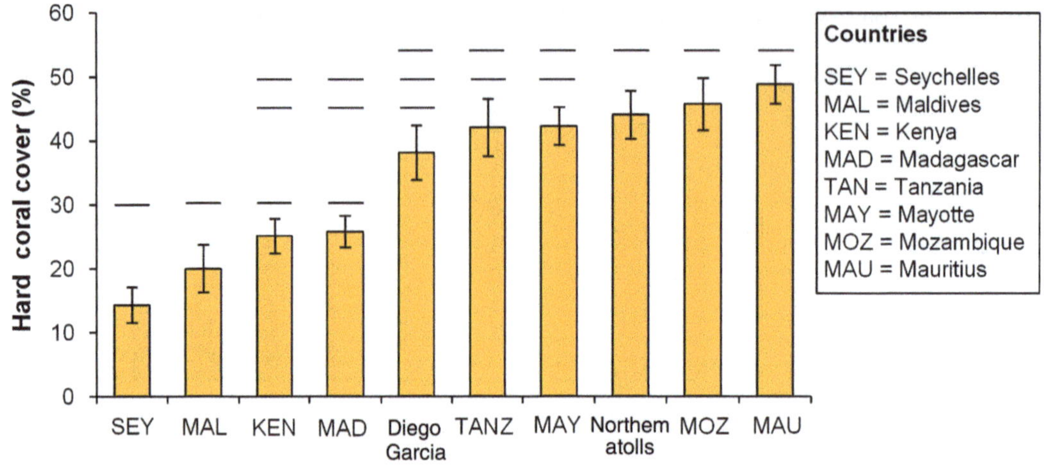

Fig. 19.8 Variation in hard coral cover (%) across nine countries in the western Indian Ocean. Diego Garcia and Chagos other atolls (Salomon, Peros Banhos and Great Chagos Bank) are presented separately. Error bars represent one standard error around the mean

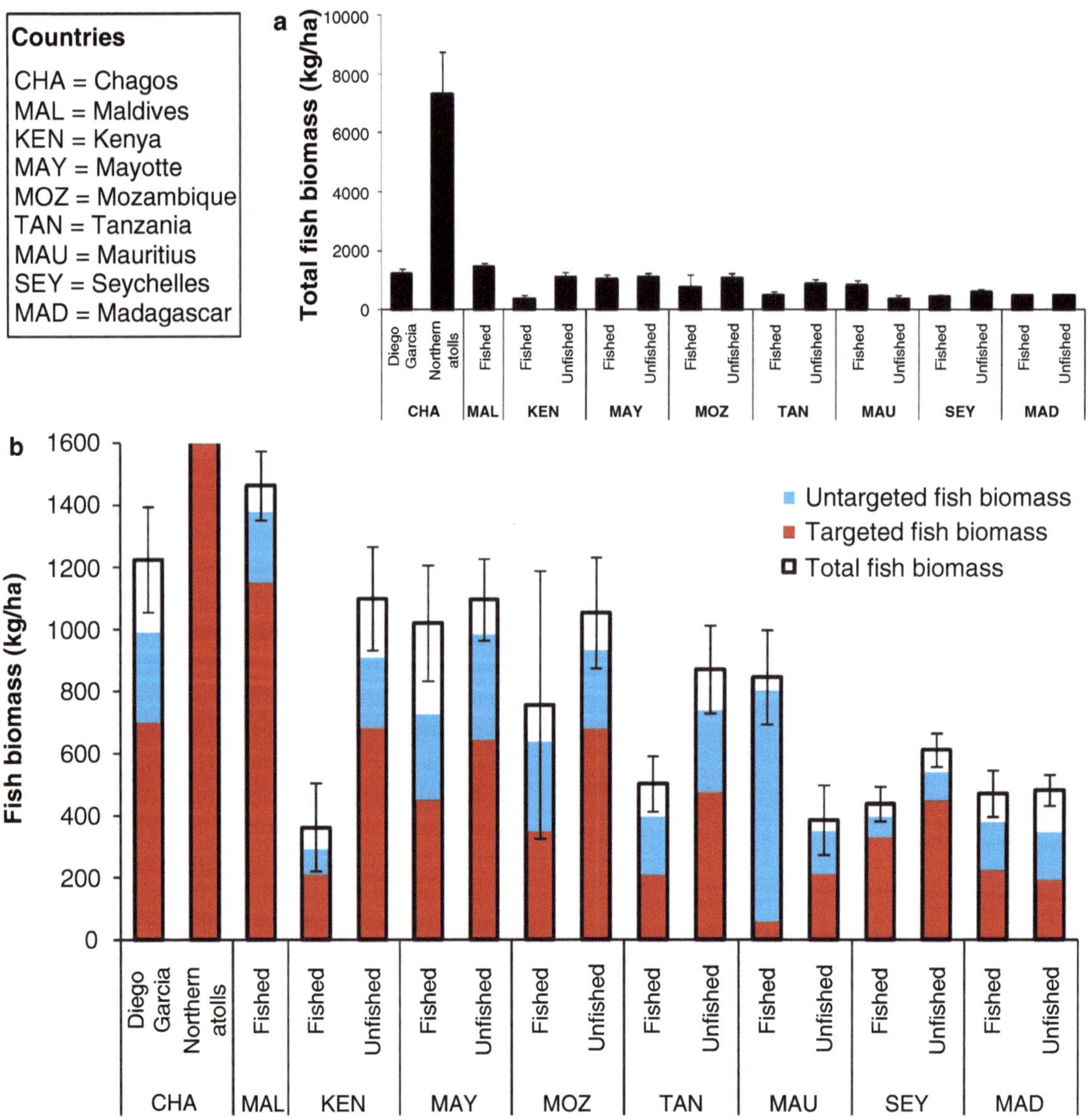

Fig. 19.9 Variation in reef fish biomass across nine countries in the western Indian Ocean including sites in fished and unfished areas. (**a**) Total fish biomass (kg/ha), (**b**) Total fish biomass (kg/ha) broken down by targeted and untargeted fish families. The open portion of the bars in panel (**b**) represent families that could not be easily assigned as targeted or untargeted. The y axis on panel (**b**) Have been cut off at 1,600 kg/ha to facilitate comparison between Diego Garcia and the other eight countries. Error bars represent 1 standard error around the mean

to fished areas in the same countries. This is likely due to a combination of some of the areas being small and not very old (McClanahan et al. 2009) and in some locations weak compliance by fishers and heavy fishing around the borders of the reserves (Pollnac et al. 2010; Daw et al. 2011).

It is clear that the biomass of reef fish in the northern atolls of Chagos is remarkably high, especially in the context of the wider western Indian Ocean region. The most likely explanation for this is the lack of fishing at a large scale in Chagos compared to the other countries. This may be thought of as an exploitation gap; the effect fishing has on reef fish assemblages compared to semi-pristine unfished locations. The key ecosystem processes fish provide (Bellwood et al. 2004; McClanahan et al. 2011; Mora et al. 2011) suggests that the high, relatively unexploited, biomass of fish in Chagos is key to the stability and health of the coral reef ecosystem in Chagos as a whole. Indeed, high abundance/biomass of the reef fish assemblage is a likely reason recovery of reefs in Chagos from the 1998 thermal anomaly was rapid compared to other reefs in the region. Locations such as Chagos are increasingly rare in the world's oceans and provide unique opportunities to understand what coral reef ecosystems should look like and how they function with minimal human impacts. They also represent some of the few areas likely to persist in a coral dominated and relatively intact state in the face of climate change.

Aside from Chagos, exceptional reef fish biomass and coral reef ecosystem condition has been documented from several locations in the Pacific Ocean. The uninhabited, and now protected, northwest Hawaiian Islands is one such example, where fish biomass, particularly the biomass of apex predators, is far greater than the biomass recorded around the populated main Hawaiian Islands (Friedlander and DeMaritini 2002). Similarly, the remote uninhabited northern Line Islands, in particular Palmyra and Kingman atolls, have exceptional reef fish biomass compared to the populated and fished southern Line Islands (Stevenson et al. 2007; DeMartini et al. 2008; Sandin et al. 2008; Williams et al. 2011). The biomass on these remote unfished Pacific locations is comparable to the biomass reported here for Chagos (for example biomass at Kingman atoll was reported as 5,500 kg/ha; Sandin et al. 2008). However, many of the locations in the Pacific report top predators, such as reef sharks, dominating the biomass of the assemblage (Friedlander and DeMartini 2002; Sandin et al. 2008). The contribution of fish trophic groups to the composition of fish biomass in Chagos is much more balanced (Graham and McClanahan in press). It must however be noted that reef sharks are one of the few groups of fish that have experienced some declines in Chagos, largely due to illegal fishing. It is therefore important to understand longer term trends in reef shark abundance and composition in Chagos.

Trends in Reef Shark Relative Abundances 1975–2012

Reef sharks are some of the most important apex predators in coral reef ecosystems, with evidence of their presence influencing trophic structure (DeMartini et al. 2008), prey fish demographics (Ruttenberg et al. 2011) and prey fish behaviour (Madin et al. 2010). Many species of sharks are vulnerable to over-exploitation due to their 'slow' life history traits, including late age at maturity, large body size, and low fecundity with a long gestation period (Reynolds et al. 2001). As such, reported declines in shark abundances has increased over the past two decades, in conjunction with increases in the trade for shark fin (Fong and Anderson 2002; Clarke et al. 2007; Dulvy et al. 2008). For example, spatial comparisons between remote coral reef locations or no-go preservation areas and more heavily used reef areas indicate substantial differences in shark abundances (Robbins et al. 2006, Stevenson et al. 2007; DeMartini et al. 2008; Sandin et al. 2008; Hisano et al. 2011).

The Chagos archipelago is not immune to the effects of shark fishing. Indeed, reef sharks are one of the only reef associated fish groups that have been heavily targeted (Sheppard et al. 2012). Aside from the small Mauritian reef fishery pre-2010, there was a licensed blue-water fishery in the territory largely for tuna, which stopped with the creation of the no-take marine protected area in 2010. Between 2006 and 2010 all licensed fishing vessels had to declare the quantities of shark fin bycatch and other products on board upon inspection by the Chagos fishery patrol vessel and in log books. Furthermore, shark finning was banned in 2006 and use of wire trace on longlines has been banned in the consolidated fisheries ordinance since 1999, with an associated penalty of £100,000 (Anon 2007). The majority of the fishing for reef associated sharks is thought to be by illegal vessels principally from Sri Lanka (Anderson et al. 1998). The number of Sri Lankan fishing vessels detected in the archipelago between 2002 and 2009 ranged from 3 to 26 per year, with number of arrests ranging from 1 to 8 (Graham et al. 2010). Vessels arrested usually have a hold full of sharks, retaining both the carcass and the fins.

The impacts of illegal fishing on the reef shark populations in Chagos has been well documented, with a 90% decline in the number of sharks observed per scientific dive between the 1970s and 1996 (Anderson et al. 1998), a small increase by 2001 (Spalding 2003) and a number similar to 1996 observed in 2006 (Graham et al. 2010). Grey reef sharks consistently made up the majority of the sharks observed across the entire sampling period (Fig. 19.10). Between 1996 and 2006 there was a small decline in the relative number of tawny nurse sharks and an increase in the proportion of silvertip sharks seen (Graham et al. 2010).

Fig. 19.10 A grey reef shark, *Carcharhinus amblyrhynchos*, which numerically dominate the reef shark assemblages of the Chagos Archipelago as observed by divers (Photo: N Graham)

We update earlier assessments of relative shark abundance by including data collected in the 2010 and 2012 Chagos expeditions. The earlier data (1975–1996) collected by Charles Anderson, Ron Crosby, Charles Sheppard and Mark Spalding was collated and compiled by Charles Anderson to investigate changes in shark abundance in the archipelago through time (Anderson et al. 1998). This was updated with data from 2001 collected by Mark Spalding (Spalding 2003), and data collected by Nick Graham in 2006 (Graham et al. 2010). Data presented here continue this time series and were collected by Nick Graham on the 2010 and 2012 expeditions. All observations were made on the outer reef slopes of Salomon, Peros Banhos and the Eagle and Brothers group of the Great Chagos Bank. Dives were typically ~1 h in duration and between 5 m and 25 m depth. From 1996 onwards the species of shark was also recorded. Although density cannot be estimated with these observational data, they are thought to provide reliable data on the relative abundance of sharks among years (Anderson et al. 1998).

The number of sharks seen per dive declined from 4.2 in the 1970s to 0.7 by 1996, increased in 2001 at 1.4 sharks per dive and declined again to 0.4 by 2006 (Anderson et al. 1998; Spalding 2003; Graham et al. 2010). Updating these data to include observations from 2010 and 2012 provides cause for some cautious optimism. The number of sharks seen per dive in 2010 rose to 1.5 and in 2012 an average of 1.0 shark was seen per dive (Fig. 19.11). These numbers are still fluctuating, and are a long way from the values recorded in the 1970s, however it is positive to see the very low numbers in 1996

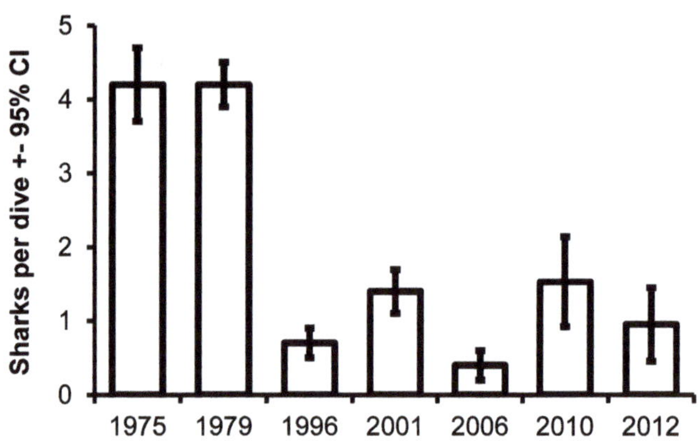

Fig. 19.11 Change in relative abundance of reef-associated sharks seen per scientific dive in the Chagos Archipelago, 1975–2012 (Updated from Anderson et al. 1998, Spalding 2003 and Graham et al. 2010)

and 2006 have not been repeated. It should be noted that the 95% confidence intervals around the 2010 and 2012 estimates are relatively large. Observational data of this kind are inherently variable, and in 2010 and 2012 up to 4 and 5 sharks were seen on individual dives respectively, while other dives recorded zero sharks. Continuing this time series will be important to see if these higher abundances are maintained or improved upon in the future.

Grey reef sharks continue to dominate the assemblage on the outer reef slopes in terms of proportional abundance (Fig. 19.12). The substantial contribution of silvertip sharks seen in 2006 (Graham et al. 2010), did not continue into

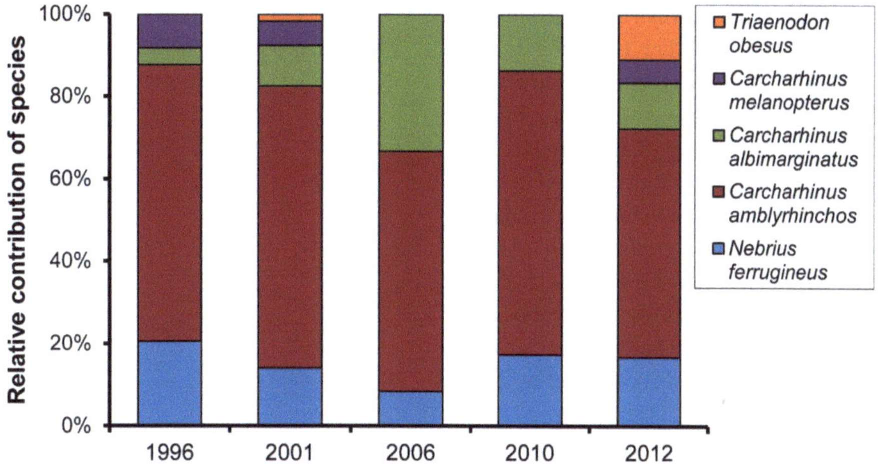

Fig. 19.12 Relative species-level contribution to shark assemblages seen per scientific dive in the Chagos Archipelago, 1996–2012. *Triaenodon obesus*: whitetip reef shark; *Carcharhinus melanopterus*: blacktip reef shark; *Carcharhinus albimarginatus*: silvertip shark; *Carcharhinus amblyrhynchos*: grey reef shark; *Nebrius ferrugineus*: tawny nurse shark (Updated from Spalding 2003 and Graham et al. 2010)

2010 and 2012, although the proportion did remain higher than in 1996 and 2001 (Fig. 19.12). Tawny nurse sharks were recorded in similar proportion to that seen in 1996 and 2001. Interestingly, blacktip reef sharks and, in particular, whitetip reef sharks that were absent from counts in 2006 and 2010, were recorded in 2012. It is positive to see all five species of shark being represented in the most recent surveys, again offering cautious optimism.

Illegal vessels with sharks on board are still being detected in Chagos (N Graham pers obs), and fishes with slow life histories such as sharks are typically depleted far more rapidly than they recover. It is therefore important that enforcement continues to develop and employ new technologies that will improve the protection of shark populations in the no-take area, and that this is augmented with other multi-faceted conservation measures (Graham et al. 2010). If the no-take area can be successfully protected, there is every reason to believe that given sufficient time shark populations in Chagos can recover to densities recorded in the 1970s. Given the exceedingly high fish biomass results given above, it will be interesting to see what added contribution to these values recovering shark populations make, or if they reduce the biomass of lower trophic levels through predation.

Habitat Degradation Effects on Reef Fishes in Chagos

One of the greatest threats to reef fish assemblages, other than fishing, is habitat degradation and loss. Substantial changes in coral cover have been documented through time in all coral reef regions of the world, with considerable ongoing declines in the Caribbean (e.g., Gardner et al. 2003) and Indo-west Pacific (e.g., Bruno and Selig 2007). Much of this change is associated with pulse disturbances, such as coral bleaching, crown-of-thorns starfish outbreaks, and tropical storms (Bruno and Selig 2007). Similarly, there was marked decline in coral cover across the Indian Ocean due to the 1998 thermal anomaly followed by some recovery (Ateweberhan et al. 2011). There is now substantial evidence that such habitat degradation has substantial negative effects on the abundance of many trophic groups of fish, on reef fish diversity, and can lead to local extinctions of highly specialized species (Wilson et al. 2006; Pratchett et al. 2008, 2011). Interestingly, the suite of species highly vulnerable to population declines following habitat degradation is different to the suite of species vulnerable to fishing (Wilson et al. 2008b, 2010; Graham et al. 2011b). While large bodied, often piscivorous species are typically most vulnerable to fishing, small bodied species that use live coral or the structure it provides for shelter, settlement or food are most vulnerable to habitat loss (Graham et al. 2011b). This is a concern, because both fishing and habitat degradation occur together on many of the world's coral reefs, effectively resulting in complementary effects that greatly extend the range of fishes facing significant population declines. We have ascertained above that fishing is having a negligible impact on reef fish communities in the northern atolls of Chagos. However, it is imperative to also assess if the reef fish assemblages are under threat from habitat change.

The greatest impact to coral cover and associated reef habitat in Chagos since scientific study of the reefs began was the 1998 thermal anomaly (Sheppard 1999). This thermal anomaly had severe, but variable, impacts on reefs across the western Indian Ocean. In order to assess the effects of this large scale habitat disturbance to reef fish assemblages, a large-scale study identified all the surveys of reef condition and associated fish assemblage structure from the Indian Ocean region conducted in the mid-1990s before the thermal anomaly and repeated these surveys using identical methods in 2005 and 2006 after the thermal anomaly. Analyses of these data showed that at locations where coral and associated structure had been lost, fish species richness, abundance of fish that feed on coral or plankton, and fish with a maximum body size less than 20 cm declined (Graham et al. 2008; MacNeil and Graham 2010). Chagos was interesting, because although coral mortality in 1998 had been up to 80% at many sites (Sheppard 1999), recovery of coral cover by 2006 was already substantial (Graham et al. 2008; Sheppard et al. 2008). Fish species richness showed very little change after the thermal anomaly in Chagos (Fig. 19.13a), although abundance of coral feeding fish (corallivores) declined in proportion with the amount of coral that was lost across the region (Graham et al. 2008). The extent of corallivore decline was however less severe than many other locations in the western Indian Ocean, largely because coral cover had recovered substantially in Chagos by the 2006 repeat surveys (Graham et al. 2009; Fig. 19.13b).

The corallivores are one of the most vulnerable fish groups to coral loss worldwide (Pratchett et al. 2008). Of particular interest is the degree of specialization among different species of butterflyfish. The family can be crudely broken into non-coral feeders, facultative coral feeders (i.e. those species that feed on coral, but will also feed on benthic algae and mobile invertebrates), and obligate corallivores (i.e. those species that feed exclusively on corals) (e.g. Pratchett 2005; Fig. 19.14). The extent to which abundances decline following coral mortality varies greatly among these classifications, with obligate specialists showing the greatest population losses (Pratchett et al. 2006; Graham 2007). If the measure of specialization is improved to include the number of coral species fed upon by different species of butterflyfishes, the importance of dietary specialization becomes even more evident. For example, at Lizard Island on the Great Barrier Reef, Australia, a gradient in specialization exists among 10 obligate coral feeding butterflyfish from a species that will only feed on a subset of less than 10 species of corals, to a more generalist feeder that will consume tissue from over 50 species of coral. Following a crown-of-thorns starfish outbreak that caused extensive loss of live coral cover at Lizard Island, the most specialized butterflyfish declined by ~80%, while the more generalist feeder only declined by ~20% (Pratchett 2001; Pratchett et al. 2008).

Interestingly, the extent of decline in corallivores in Chagos after the 1998 thermal anomaly was not related to degree of specialization (Graham 2007; Graham et al. 2009). This was perhaps due to the rapid recovery of coral on reefs in Chagos masking any effects. However, from a range of studies worldwide, it is clear that the degree of dietary specialization confers a significant disadvantage to coral feeding fishes in the face of coral loss and increasing habitat disturbances (Lawton et al. 2011). The obvious question, therefore, is what is the advantage of such dietary specialization. Noting that coral cover had recovered to pre-1998 levels by the 2010 Chagos expedition (Sheppard et al. 2012), M. Pratchett quantified the degree of feeding specialization of butterflyfish at the archipelago to compare to their patterns of numerical dominance in the assemblage. Interestingly, there was a very strong positive relationship between feeding specialization and abundance, with the most specialized being most abundant (Pratchett et al. in press). Such information from a unique and relatively undisturbed location such as Chagos provides a starting point to understanding why some species are so specialized. Specialization may have evolved to provide a competitive advantage, perhaps associated with the quality of the tissue of different corals (Graham 2007), during long geological periods of relatively stable conditions on coral reefs (Pandolfi et al. 2011). However, specialized feeding on corals is likely to confer a

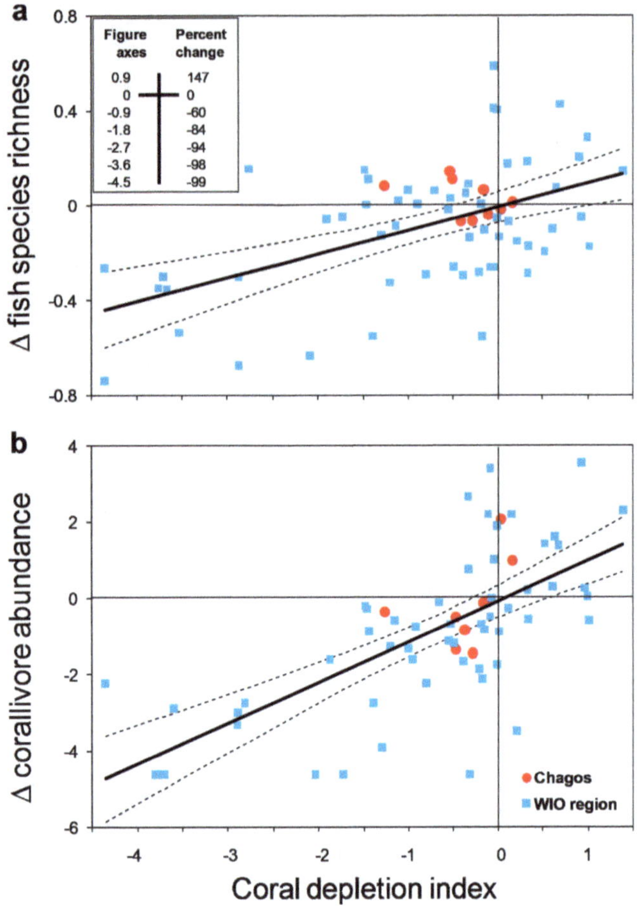

Fig. 19.13 Change in (**a**) Reef fish species richness and (**b**) The abundance of corallivores, across seven countries in the Indian Ocean before and after the 1998 coral mortality event. Chagos sites represented by red circles, and all other data as blue squares (Adapted from Graham et al. (2008))

Fig. 19.14 Butterflyfish from the Chagos Archipelago. (**a**) *Chaetodon meyeri* are obligate coral feeders, but are known to feed on at least 24 coral genera in Chagos (Pratchett et al. in press). (**b**) Two *Chaetodon auriga* and a *Chaetodon kleinii* from the Chagos Archipelago. Both species are facultative coral feeders, taking between 10% and 30% of their bites from corals in Chagos (Pratchett et al. in press) (Photo's: M Pratchett)

significant disadvantage as coral cover declines due to the effects of climate change and human impacts.

Dire predictions for the future of coral dominated ecosystems under global climate change suggest that if carbon emissions continue under a business as usual scenario, reefs may be functionally lost by the middle of the century (Sheppard 2003; Hoegh-Guldberg et al. 2007; Veron et al. 2009). However, it is also emerging that different species of reef building corals and fishes are differentially vulnerable to the effects of climate change and other disturbances (McClanahan et al. 2007b; Graham et al. 2011b; Pandolfi et al. 2011), and composition of coral reef communities may change substantially rather than be lost altogether. Coral reefs will also have a much better chance of bouncing back from disturbances and coping with a certain degree of climatic warming if local human derived stresses can be greatly reduced (Hughes et al. 2010). Chagos is an excellent example, being one of the few locations in the Indian Ocean that is largely free of human impact, and one of the few that has recovered rapidly from the 1998 thermal anomaly (Sheppard et al. 2012). It is highly likely that large remote locations, such as Chagos, will be some of the longest lasting and best hopes for the future of coral dominated ecosystems as our climate continues to change.

Conclusion

The rapid recovery of coral cover at Chagos following the 1998 thermal anomaly, the stability of the fish assemblages through this event and the exceptionally high reef fish biomass on the reefs all provide compelling evidence of the unique status of the coral reefs of Chagos. The archipelago provides an example of community structure that most reefs likely resembled in the past, and also demonstrates the considerable resilience of reef ecosystems that are largely devoid of chronic anthropogenic pressures. It also indicates the high sensitivity of the ecosystem to even light recreational fishing. Such a location is not only useful as a reference point to compare with other reefs and management strategies globally (Knowlton and Jackson 2008), but is also likely to act as a large reservoir of diversity as climate change and other human pressures interact and cause uncertain futures for coral reef ecosystems. Chagos is not immune to all human impacts, and the reduced fish biomass around Diego Garcia and shark abundances testament. The recent establishment of the no-take marine protected area around Chagos can facilitate a move towards improved enforcement, appropriate international policies and engagement with fishers and consumers to help reverse trends in reef shark abundance and promote recovery of their populations. Similarly, better monitoring and regulations of any continuing recreational fishery around Diego Garcia may help enhance the reef fish biomass of that atoll towards the status of the northern atolls.

Acknowledgements Funding for this work was provided by the British Foreign and Commonwealth Office, the Leverhulme Trust, Fisheries Society of the British Isles, Western Indian Ocean Marine Science Association and research fellowships from the Australian Research Council and Queensland Smart Futures Fund. The research could not have happened without the organization of Chagos expeditions by Charles Sheppard and Heather Koldewey. Many thanks to Bob Long for help in the field around Diego Garcia and to Cindy Huchery for help with figures and analyses. The Wildlife Conservation Society through grants from the Western Indian Ocean Marine Science Association supported much of the larger western Indian Ocean data collection.

References

Anderson C, Sheppard C, Spalding M, Crosby R (1998) Shortage of sharks at Chagos. Shark News 10:1–12

Anon (2007) The fisheries (Conservation and management) ordinance 2007. British Indian Ocean Territory Administration, London

Ateweberhan M, McClanahan TR, Graham NAJ, Sheppard C (2011) Episodic heterogeneous decline and recovery of coral cover in the Western Indian Ocean. Coral Reefs 30:739–752

Bellwood DR, Hughes TP, Folke C, Nyström M (2004) Confronting the coral reef crisis. Nature 429:827–833

Bruno JF, Selig ER (2007) Regional decline of coral cover in the Indo-Pacific: timing, extent, and subregional comparisons. PLoS One 2:e711

Cinner JE, McClanahan TR, Daw TM, Graham NAJ, Maina J, Wilson SK, Hughes TP (2009) Linking social and ecological systems to sustain coral reef fisheries. Curr Biol 19:206–212

Cinner JE, Graham NAJ, Huchery C, MacNeil MA (2013) Global impacts of local human population density and distance to markets on the condition of coral reef fisheries. Conserv Biol doi:10.1111/j.1523-1739.2012.01933.x

Clarke SC, Milner-Gulland EJ, Bjørndal T (2007) Perspective: social, economic and regulatory drivers of the shark fin trade. Mar Resour Econ 22:305–327

Coleman FC, Figueira WF, Ueland JS, Crowder LB (2004) The impact of United States recreational fisheries on marine fish population. Science 305:1958–1960

Daw T, Cinner JE, McClanahan TR, Graham NAJ, Wilson SK (2011) Design factors and socioeconomic variables associated with ecological responses to fishery closures in the western Indian Ocean. Coast Manage 39:412–424

DeMartini EE, Friedlander AM, Sandin SA, Sala E (2008) Differences in fish-assemblage structure between fished and unfished atolls in the northern Line Islands, central Pacific. Mar Ecol Prog Ser 365:199–215

Dulvy NK, Polunin NVC, Mill AC, Graham NAJ (2004) Size structural change in lightly exploited coral reef fish communities: evidence for week indirect effects. Can J Fish Aquat Sci 61:466–475

Dulvy NK, Baum JK, Clarke S, Compagno LJV, Cortes E, Domingo A, Fordham S, Flowler S, Francis MP, Gibson C, Martinez J, Musick JA, Soldo A, Stevens JD, Valenti S (2008) You can swim but can't hide: the global status and conservation of oceanic pelagic sharks and rays. Aquat Conserv Mar Freshwat Ecosyst 18:459–482

Estes JA, Terborgh J, Brashares JS, Power ME, Berger J, Bond WJ, Carpenter SR, Essington TE, Holt RD, Jackson JBC (2011) Trophic downgrading of planet earth. Science 333:301

Fong QSW, Anderson JL (2002) International shark fin markets and shark management: an integrated market preference–cohort analysis of the blacktip shark (*Carcharhinus limbatus*). Ecol Econ 40:117–130

Friedlander AM, DeMartini EE (2002) Contrasts in density, size, and biomass of reef fishes between the northwestern and the main Hawaiian Islands: the effects of fishing down apex predators. Mar Ecol Prog Ser 230:253–264

Froese R, Pauly D (2012) FishBase (www.fishbase.org). World wide web electronic publication

Gardner TA, Côté IM, Gill JA, Grant A, Watkinson AR (2003) Long-term region-wide declines in Caribbean corals. Science 301:958–960

Graham NAJ (2007) Ecological versatility and the decline of coral feeding fishes following climate driven coral mortality. Mar Biol 153:119–127

Graham NAJ, McClanahan TR (in press) The last call for marine wilderness? BioScience

Graham NAJ, Dulvy NK, Jennings S, Polunin NVC (2005) Size-spectra as indicators of the effects of fishing on coral reef fish assemblages. Coral Reefs 24:118–124

Graham NAJ, Wilson SK, Jennings S, Polunin NVC, Bijoux J, Robinson J (2006) Dynamic fragility of oceanic coral reef ecosystems. Proc Natl Acad Sci USA 103:8425–8429

Graham NAJ, Wilson SK, Jennings S, Polunin NVC, Robinson J, Bijoux JP, Daw TM (2007) Lag effects in the impacts of mass coral bleaching on coral reef fish, fisheries, and ecosystems. Conserv Biol 21:1291–1300

Graham NAJ, McClanahan TR, MacNeil MA, Wilson SK, Polunin NVC, Jennings S, Chabanet P, Clark S, Spalding MD, Letourneur Y, Bigot L, Galzin R, Öhman MC, Garpe KC, Edwards AJ, Sheppard CRC (2008) Climate warming, marine protected areas and the ocean-scale integrity of coral reef ecosystems. PLoS One 3:e3039

Graham NAJ, Wilson S, Pratchett M, Polunin N, Spalding M (2009) Coral mortality versus structural collapse as drivers of corallivorous butterflyfish decline. Biodivers Conserv 18:3325–3336

Graham NAJ, Spalding MD, Sheppard CRC (2010) Reef shark declines in remote atolls highlight the need for multi-faceted conservation action. Aquat Conserv Mar Freshwat Ecosyst 20:543–548
Graham NAJ, Ainsworth TD, Baird AH, Ban NC, Bay LK, Cinner JE, De Freitas DM, Diaz-Pulido G, Dornelas M, Dunn SR, Fidelman PIJ, Foret S, Good TC, Kool J, Mallela J, Penin L, Pratchett MS, Williamson DH (2011a) From microbes to people: tractable benefits of no-take areas for coral reefs. Oceanogr Mar Biol Annu Rev 49:105–136
Graham NAJ, Chabanet P, Evans RD, Jennings S, Letourneur Y, MacNeil MA, McClanahan TR, Ohman MC, Polunin NVC, Wilson SK (2011b) Extinction vulnerability of coral reef fishes. Ecol Lett 14:341–348
Heupel MR, Simpfendorfer CA, Fitzpatrick R (2010) Large-scale movement and fidelity of grey reef sharks. PLoS One 5:e9650
Hisano M, Connolly SR, Robbins WD (2011) Population growth rates of reef sharks with and without fishing on the Great Barrier Reef: robust estimation with multiple models. PLoS One 6:e25028
Hixon MA (1991) Predation as a process structuring coral reef fish communities. In: Sale PF (ed) The ecology of fishes on coral reefs. Academic Press, New York, pp 475–508
Hoegh-Guldberg O, Mumby PJ, Hooten AJ, Steneck RS, Greenfield P, Gomez E, Harvell CD, Sale PF, Edwards AJ, Caldeira K, Knowlton N, Eakin CM, Iglesias-Prieto R, Muthiga N, Bradbury RH, Dubi A, Hatziolos ME (2007) Coral reefs under rapid climate change and ocean acidification. Science 318:1737–1742
Hughes TP, Graham NAJ, Jackson JBC, Mumby PJ, Steneck RS (2010) Rising to the challenge of sustaining coral reef resilience. Trends Ecol Evol 25:633–642
Jackson JBC, Kirby MX, Berger WH, Botsford LW, Bourque BJ, Bradbury RH, Cooke R, Erlandson J, Estes JA, Hughes TP, Kidwell S, Lange CB, Lenihan HS, Pandolfi JM, Peterson CH, Steneck RS, Tegner MJ, Warner RR (2001) Historical overfishing and the recent collapse of coastal ecosystems. Science 293:629–637
Jennings S, Polunin NVC (1996) Effects of fishing effort and catch rate upon the structure and biomass of Fijian reef fish communities. J Appl Ecol 33:400–412
Jennings S, Polunin NVC (1997) Impacts of predator depletion by fishing on the biomass and diversity of non-target reef fish communities. Coral Reefs 16:71–82
Jones GP, Syms C (1998) Disturbance, habitat structure and the ecology of fishes on coral reefs. Aust J Ecol 23:287–297
Jones GP, Caley MJ, Munday PL (2002) Rarity in coral reef fish communities. In: Sale PF (ed) Coral reef fishes dynamics and diversity in a complex ecosystem. Academic, San Diego, pp 81–102
Jones GP, McCormick MI, Srinivasan M, Eagle JV (2004) Coral decline threatens fish biodiversity in marine reserves. Proc Natl Acad Sci USA 101:8251–8258
Knowlton N, Jackson JBC (2008) Shifting baselines, local impacts, and global change on coral reefs. PLoS Biol 6:e54
Lawton RJ, Messmer V, Pratchett MS, Bay LK (2011) High gene flow across large geographic scales reduces extinction risk for a highly specialised coral feeding butterflyfish. Mol Ecol 20:3584–3598
MacNeil MA, Graham NAJ (2010) Enabling regional management in a changing climate through Bayesian meta-analysis of a large-scale disturbance. Glob Ecol Biogeogr 19:412–421
Madin EMP, Gaines SD, Warner RR (2010) Field evidence for pervasive indirect effects of fishing on prey foraging behavior. Ecology 91:3563–3571
McClanahan TR (1994) Kenyan coral reef lagoon fish: effects of fishing, substrate complexity, and sea urchins. Coral Reefs 13:231–241
McClanahan TR (2000) Recovery of a coral reef keystone predator, Balistapus undulatus, in East African marine parks. Biol Conserv 94:191–198
McClanahan TR (2008) Response of the coral reef benthos and herbivory to fishery closure management and the 1998 ENSO disturbance. Oecologia 155:169–177
McClanahan TR (2011) Coral reef fish communities in management systems with unregulated fishing and small fisheries closures compared with lightly fished reefs – Maldives vs Kenya. Aquat Conserv Mar Freshwat Ecosyst 21:186–198
McClanahan TR, Graham NAJ (2005) Recovery trajectories of coral reef fish assemblages within Kenyan marine protected areas. Mar Ecol Prog Ser 294:241–248
McClanahan TR, Humphries A (2012) Differential and slow life history responses of fishes to coral reef closures. Mar Ecol Prog Ser 469:121–131
McClanahan TR, Omukoto JO (2011) Implications of modern versus historical fisheries catches for marine protected area goals. Conserv Biol 25:945–955
McClanahan TR, Graham NAJ, Calnan JM, MacNeil MA (2007a) Toward pristine biomass: reef fish recovery in coral reef marine protected areas in Kenya. Ecol Appl 17:1055–1067
McClanahan TR, Ateweberhan M, Sebastian CR, Graham NAJ, Wilson SK, Guillaume MMM, Bruggemann JH (2007b) Western Indian Ocean coral communities: bleaching responses and susceptibility to extinction. Mar Ecol Prog Ser 337:1–13
McClanahan TR, Hicks CC, Darling SE (2008) Malthusian overfishing and efforts to overcome it on Kenyan coral reefs. Ecol Appl 18: 1516–1529
McClanahan TR, Graham NAJ, Wilson SK, Letourneur Y, Fisher R (2009) Effects of fisheries closure size, age, and history of compliance on coral reef fish communities in the western Indian Ocean. Mar Ecol Prog Ser 396:99–109
McClanahan TR, Graham NAJ, MacNeil MA, Muthiga NA, Cinner JE, Bruggemann JH, Wilson SK (2011) Critical thresholds and tangible targets for ecosystem-based management of coral reef fisheries. Proc Natl Acad Sci USA 108:17230–17233
Mees CC (1996) The mauritian banks fishery: a review and spatial analysis. Technical report. Marine Resources Assessment Group London
Mora C, Aburto-Oropeza O, Ayala Bocos A, Ayotte PM, Banks S, Bauman AG, Beger M, Bessudo S, Booth DJ, Brokovich E, Brooks A, Chabanet P, Cinner JE, Cortés J, Cruz-Motta JJ, Cupul Magaña A, DeMartini EE, Edgar GJ, Feary DA, Ferse SCA, Friedlander AM, Gaston KJ, Gough C, Graham NAJ, Green A, Guzman H, Hardt M, Kulbicki M, Letourneur Y, López Pérez A, Loreau M, Loya Y, Martinez C, Mascareñas-Osorio I, Morove T, Nadon M-O, Nakamura Y, Paredes G, Polunin NVC, Pratchett MS, Reyes Bonilla H, Rivera F, Sala E, Sandin SA, Soler G, Stuart-Smith R, Tessier E, Tittensor DP, Tupper M, Usseglio P, Vigliola L, Wantiez L, Williams I, Wilson SK, Zapata FA (2011) Global human footprint on the linkage between biodiversity and ecosystem functioning in reef fishes. PLoS Biol 9:e1000606
Munday PL (2004) Habitat loss, resource specialization, and extinction on coral reefs. Glob Change Biol 2004:1–6
Nadon MO, Baum JK, Williams ID, McPherson JM, Zgliczynski BJ, Richards BL, Schroeder RE, Brainard RE (2012) Re-creating missing population baselines for Pacific reef sharks. Recreación de las Líneas de Base Poblacionales Faltantes para Tiburones de Arrecife en el Pacífico. Conserv Biol 26:493–503
Newman MJH, Paredes GA, Sala E, Jackson JBC (2006) Structure of Caribbean coral reef communities across large gradient of fish biomass. Ecol Lett 9:1216–1227
Olden JD, Hogan ZS, Zanden MJV (2007) Small fish, big fish, red fish, blue fish: size-biased extinction risk of the world's freshwater and marine fishes. Glob Ecol Biogeogr 16:694–701
Pandolfi JM, Connolly SR, Marshall DJ, Cohen AL (2011) Projecting coral reef futures under global warming and ocean acidification. Science 333:418–422
Pitcher TJ (2001) Fisheries managed to rebuild ecosystems? Reconstructing the past to salvage the future. Ecol Appl 11:601–617
Pollnac R, Christie P, Cinner JE, Dalton T, Daw TM, Forrester GE, Graham NAJ, McClanahan TR (2010) Marine reserves as linked social-ecological systems. Proc Natl Acad Sci USA 107:18262–18265

Polunin NVC, Robert CM (eds) (1996) Reef fisheries. Chapman & Hall, London

Pratchett MS (2001) Dynamics of outbreak populations of crown-of-thorns starfish (Acanthaster planci L.), and their effects on coral reef ecosystems. Ph.D. thesis, James Cook University

Pratchett MS (2005) Dietary overlap among coral-feeding butterflyfishes (Chaetontidae) at Lizard Island, northern Great Barrier Reef. Mar Biol 148:373–382

Pratchett MS, Graham NAJ, Cole AJ (in press) Specialist corallivores dominate butterflyfish assemblages in the Chagos Archipelago, central Indian Ocean. J Fish Biol

Pratchett MS, Wilson SK, Baird AH (2006) Declines in the abundance of Chaetodon butterflyfishes following extensive coral depletion. J Fish Biol 69:1–12

Pratchett MS, Munday PL, Wilson SK, Graham NAJ, Cinner JE, Bellwood DR, Jones GP, Polunin NVC, McClanahan TR (2008) Effects of climate-induced coral bleaching on coral-reef fishes: ecological and economic consequences. Oceanogr Mar Biol Annu Rev 46:251–296

Pratchett MS, Hoey AS, Wilson SK, Messmer V, Graham NAJ (2011) Changes in biodiversity and functioning of reef fish assemblages following coral bleaching and coral loss. Diversity 3:424–452

Pratchett MS, Coker DJ, Jones GP, Munday PL (2012) Specialization in habitat use by coral reef damselfishes and their susceptibility to habitat loss. Ecol Evol 2:2168–2180

Reynolds JD, Jennings S, Dulvy NK (2001) Life histories of fishes and population responses to exploitation. In: Reynolds JD, Mace GM, Redford KH, Robinson JG (eds) Conservation of exploited species. Cambridge University Press, Cambridge, pp 47–168

Robbins WD, Hisano M, Connolly SR, Choat JH (2006) Ongoing collapse of coral-reef shark population. Curr Biol 16:2314–2319

Russ GR, Alcala AC (1989) Effects of intense fishing pressure on an assemblage of coral reef fishes. Mar Ecol Prog Ser 56:13–27

Russ GR, Stockwell B, Alcala AC (2005) Inferring versus measuring rates of recovery in no-take marine reserves. Mar Ecol Prog Ser 292:1–12

Ruttenberg BI, Hamilton SL, Walsh SM, Donovan MK, Friedlander A, DeMartini E, Sala E, Sandin SA (2011) Predator-induced demographic shifts in coral reef fish assemblages. PLoS One 6:e21062

Sandin SA, Smith JE, DeMartini EE, Dinsdale EA, Donner SD, Friedlander AM, Konotchick T, Malay M, Maragos JE, Obura D, Pantos O, Paulay G, Richie M, Rohwer F, Schroeder RE, Walsh S, Jackson JBC, Knowlton N, Sala E (2008) Baselines and degradation of coral reefs in the Northern Line Islands. PLoS One 3:e1548

Sheppard CRC (1999) Coral decline and weather patterns over 20 years in the Chagos Archipelago, Central Indian Ocean. Ambio 28:472–478

Sheppard CRC (2003) Predicted recurrences of mass coral mortality in the Indian Ocean. Nature 425:294–297

Sheppard CRC, Harris A, Sheppard ALS (2008) Archipelago-wide coral recovery patterns since 1998 in the Chagos Archipelago, central Indian Ocean. Mar Ecol Prog Ser 362:109–117

Sheppard CRC, Ateweberhan M, Bowen BW, Carr P, Chen CA, Clubbe C, Craig MT, Ebinghaus R, Eble J, Fitzsimmons N, Gaither MR, Gan CH, Gollock M, Guzman N, Graham NAJ, Harris A, Jones R, Keshavmurthy S, Koldewey H, Lundin CG, Mortimer JA, Obura D, Pfeiffer M, Price ARG, Purkis S, Raines P, Readman JW, Riegl B, Rogers A, Schleyer M, Seaward MRD, Sheppard ALS, Tamelander J, Turner JR, Visram S, Vogler C, Vogt S, Wolschke H, Yang JMC, Yang SY, Yesson C (2012) Reefs and islands of the Chagos Archipelago, Indian Ocean: why it is the world's largest no-take marine protected area. Aquat Conserv Mar Freshwat Ecosyst 22:232–261

Spalding MD (1999) Biodiversity patterns in Chagos reef fishes. In: Sheppard CRC, Seaward MRD (eds) The ecology of the Chagos Archipelago. Westbury Publishing, Great Britain, pp 119–136

Spalding MD (2003) Partial recovery of sharks in the Chagos Achipelago. Shark News 15:12–13

Stevenson C, Katz LS, Micheli F, Block B, Heiman KW, Perle C, Weng K, Dunbar R, Witting J (2007) High apex predator biomass on remote Pacific Island. Coral Reefs 26:47–51

Stockwell B, Jadloc CRL, Abesamis RA, Alcala AC, Russ GR (2009) Trophic and benthic responses to no-take marine reserve protection in the Philippines. Mar Ecol Prog Ser 389:1–15

Veron JEN, Hoegh-Guldberg O, Lenton TM, Lough JM, Obura DO, Pearce-Kelly P, Sheppard CRC, Spalding M, Stafford-Smith MG, Rogers AD (2009) The coral reef crisis: the critical importance of 350 ppm CO2. Mar Pollut Bull 58:1428–1436

Williams ID, Polunin NVC (2000) Differences between protected and unprotected reefs of the western Caribbean in attributes preferred by dive tourists. Environ Conserv 27:382–391

Williams ID, Richards BL, Sandin SA, Baum JK, Schroeder RE, Nadon MO, Zgliczynski B, Craig P, McIlwain JL, Brainard RE (2011) Differences in reef fish assemblages between populated and remote reefs spanning multiple Archipelagos across the Central and Western Pacific. J Mar Biol 2011:1–14, Article ID 826234

Wilson SK, Graham NAJ, Pratchett MS, Jones GP, Polunin NVC (2006) Multiple disturbances and the global degradation of coral reefs: are reef fishes at risk or resilient? Glob Change Biol 12:2220–2234

Wilson SK, Graham NAJ, Polunin NVC (2007) Appraisal of visual assessments of habitat complexity and benthic composition on coral reefs. Mar Biol 151:1069–1076

Wilson SK, Burgess SC, Cheal AJ, Emslie M, Fisher R, Miller I, Polunin NVC, Sweatman HPA (2008a) Habitat utilization by coral reef fish: implications for specialists vs. generalists in a changing environment. J Anim Ecol 77:220–228

Wilson SK, Fisher R, Pratchett MS, Graham NAJ, Dulvy NK, Turner RA, Cakacaka A, Polunin NVC, Rushton SP (2008b) Exploitation and habitat degradation as agents of change within coral reef fish communities. Glob Change Biol 14:2796–2809

Wilson SK, Fisher R, Pratchet MS, Graham NAJ, Dulvy NK, Turner RA, Cakacaka A, Polunin NVC (2010) Habitat degradation and fishing effects on the size structure of coral reef fish communities. Ecol Appl 20:442–451

Wilson SK, Graham NAJ, Fisher R, Robinson JI, Nash K, Chong-Seng K, Polunin NVC, Aumeeruddy R, Quatre R (2012) The effect of macroalgal expansion and marine protected areas on coral recovery following a climatic disturbance. Conserv Biol 26:995–1004

Winterbottom R, Anderson RC (1999) Fishes of the Chagos Archipelago. In: Sheppard CRC, Anderson RC (eds) Ecology of the Chagos Archipelago. Westbury Publishing, Great Britain, pp 101–117

20 Coral Islands of the British Indian Ocean Territory (Chagos Archipelago)

Peter Carr, Jesse C. Hillman, Mark R.D. Seaward, Scott Vogt, and Charles R.C. Sheppard

Introduction

All islands of this territory are coral islands and cays. Of the *c*. 64,000 km^2 of reefs and shoals, less than 60 km^2 (<0.1%) is land, comprising some 47 islands grouped in the four atolls of Diego Garcia, Salomon Islands, Peros Banhos and the Egmont Islands, with eight more islands on the rim of the largest atoll in the world, the Great Chagos Bank.

The archipelago's first documented discovery was by the Portuguese in the sixteenth century, but it was over 200 years later that they were first successfully colonised by the French who established themselves on the largest land mass of Diego Garcia (following an unsuccessful attempt earlier in the same year by the British). As well as receiving lepers, this initial colonisation established coconut *Cocus nucifera* plantations that were farmed principally for the production of copra oil. In 1814, following the ceding by France of Mauritius and its lesser dependencies as part of the Treaty of Paris, the British assumed control of the "Oil Islands" and expanded the plantation regime to all accessible islands in the archipelago. Other short-lived and less successful economic ventures in the Chagos have included whaling, a coaling station, guano mining and native hardwood timber extraction (Scott 1961; Edis 1993).

P. Carr (✉) • C.R.C. Sheppard
School of Life Sciences, University of Warwick,
Gibbit Hill Campus, CV4 7AL, Coventry, UK
e-mail: petecarr1@hotmail.com; Charles.Sheppard@warwick.ac.uk

J.C. Hillman
Spring Cottage, Bluehill, St. Helena Island,
South Atlantic, STHL 1ZZ, UK
e-mail: hillman.jesse@gmail.com

M.R.D. Seaward
School of Life Sciences, University of Bradford,
Bradford, BD7 1DP, UK
e-mail: M.R.D.Seaward@Bradford.ac.uk

S. Vogt
Naval Facilities Engineering, Far East Command, Yokosuka, Japan
e-mail: Scott.Vogt@fe.navy.mil

The military have had a long association with the archipelago. The unsuccessful British attempt at colonisation in 1786 was accompanied by military personnel. Royal Navy hydrographers have been amongst those who have charted the archipelago's seas and mapped the islands, and the twentieth Century saw a British military presence based on Diego Garcia, particularly during World War Two, that included the Royal Air Force and Royal Marines (Edis 1993). Since an Exchange of Notes (EoN) in 1966 between the United Kingdom and the United States of America, all of the islands and seas have been given over to the defence needs of both Governments. This EoN resulted in the final closure of the coconut plantations, the clearing of all the islands of the remaining human populations, the creation of a military support facility on Diego Garcia and the naming of the Chagos Archipelago as the British Indian Ocean Territory (BIOT). The 1966 EoN is due for review in 2016 and, if neither country contests the agreement, the Territory will remain dedicated to defence purposes for a further 25 years (http://www.fco.gov.uk).

Since man first colonised the islands, the impact on the fragile environment has been catastrophic, as was the case on oceanic islands world-wide. In the Chagos this has been principally due to deforestation to make way for coconut plantations, (Bourne 1971) and the introduction of invasive plants (e.g. Topp and Sheppard 1999) and animals (Symens 1999; Hilton and Cuthbert 2010). The two historic anthropogenic factors that still traumatically impact nearly every island are the introduction of Black Rats *Rattus rattus* and the lack of management of the former coconut plantations – the unmanaged coconut groves have become dark, dank, inhospitable monocultures, offering little in the way of biodiversity (Carr 2011a). In the Chagos it is estimated that historically well over two thirds of the archipelago land mass, and all except remote and treacherous to land on islands, is now unmanaged, near-sterile "coconut chaos" (Fig. 20.1).

However, despite the ecological ravages wrought by man's presence over the past centuries, some of the islands remain rat and "coconut chaos" free, and are havens for

C.R.C. Sheppard (ed.), *Coral Reefs of the United Kingdom Overseas Territories*, Coral Reefs of the World 4,
DOI 10.1007/978-94-007-5965-7_20, © Springer Science+Business Media Dordrecht 2013

Fig. 20.1 "Coconut chaos." A term coined for the state of the former coconut plantations after over 40 years of neglect. These dark, dank, inhospitable habitats offer little to biodiversity and occupy over two thirds of the terrestrial land mass in the Marine Protected Area

Fig. 20.2 Despite the ecological ravages wrought by man's presence over the past centuries, some of the islands remain havens for breeding seabirds and other wildlife, as this Brown Booby *Sula leucogaster* colony on North Brother, one of the largest in the Indian Ocean shows

breeding seabirds and other wildlife (Fig. 20.2). Eighteen species of seabird are now known to breed in the Chagos (Carr 2011a) with five in internationally important numbers (BirdLife International 2004; Carr 2006). Two species of endangered sea turtle, the Hawksbill *Eretmochelys imbricata* and Green *Chelonia mydas* still breed on undisturbed beaches despite the persecution they have historically received (Mortimer and Day 1999), as does the unique Coconut Crab *Birgus latro* (Barnett et al. 1999), a species listed as Data Deficient on the IUCN Threatened Species List. These remaining vestiges of once flourishing populations have earned protection status for certain areas of the archipelago.

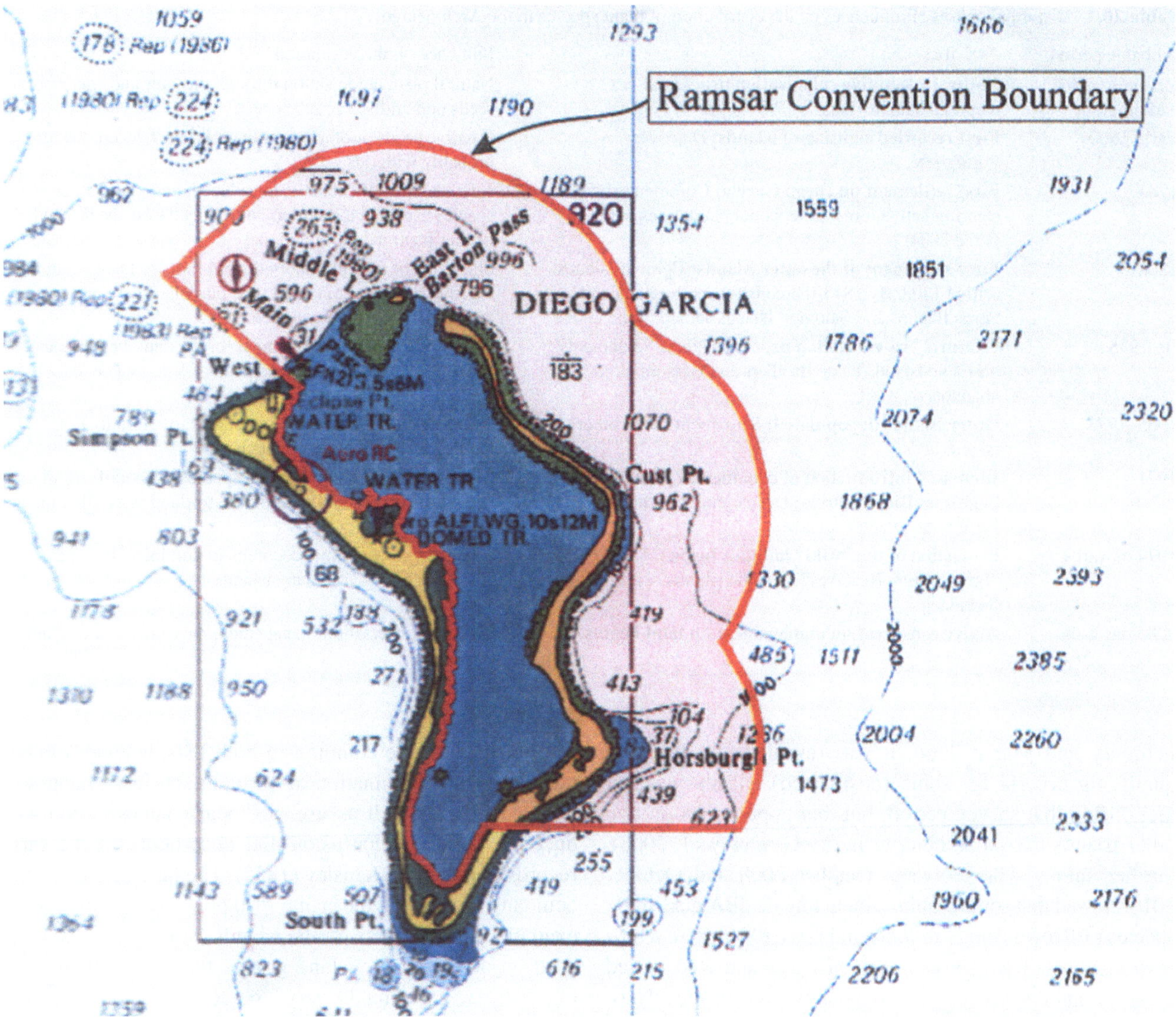

Fig. 20.3 Map of the Ramsar site on the atoll Diego Garcia

Sites of Global Importance and Protected Areas

There are presently three categories of sites of global importance in these coral islands: a single IUCN Category 1 No-Take Marine Protected Area (MPA); a Ramsar site and ten IUCN Important Bird Areas (IBAs), with a further two proposed. The Ramsar site, in Diego Garcia, designated in July 2001 covers 35,424 ha (Fig. 20.3). It extends over the eastern, uninhabited, arm of the island, the entire lagoon (with the unexplained exception of the southern barachois which is of particular importance with regard to juvenile Hawksbill turtles, though this area does have a separate protection), the three islets in the mouth of the lagoon and seas ocean-side of the eastern arm out to three nautical miles. 'The site qualifies for the [Ramsar] List under 7 of the 8 Ramsar Criteria and is a particularly good example of a relatively unpolluted coral reef system in a near-natural state, of special value for maintaining the genetic and ecological diversity of the region, especially its marine life' (www.ramsar.org accessed 27 June 2011). Contributing factors to the Ramsar designation were breeding populations of the two endangered turtles and Coconut Crab, which fulfil Criteria 2 and, from an avian perspective, the eastern arm and three islets which regularly hold more than 20,000 waterbirds, thus fulfilling Criteria 5 of the Ramsar requirements. The most abundant waterbird here is the Red-footed Booby *Sula sula* and its' numbers (over 3,500 pairs) also qualify this site as an IUCN qualified Important Bird Area (IBA).

At present there are ten designated IBAs (BirdLife International 2004; Carr 2006) and two proposed IBAs (McGowan et al. 2008), though the true situation concerning IBAs may not be clear-cut. In a census of the seabird populations in

Table 20.1 Historical phases of influences on the distribution of plant species in the Archipelago

Probable period	Activities	Influence on the vegetation
Up until 1500 and on-going	**Natural vegetative colonisation processes:** No recorded human visits.	Natural plant colonisation through bird, air and water-borne seeds and "rafts"
Early 1500s	**First recorded sighting of islands:** Discovery by Portuguese.	Possible seeds or other plant material introduced, together probably with rats
1784	**First settlement on Diego Garcia:** Commencement of clearing native forest for coconut plantations begins (Vine 2004)	New varieties of coconuts introduced, and many other plant species useful to the settlers; introduced bird species and further mammal introductions;
1808	**First settlement of the outer islands:** Egmont Islands settled 1808. By 1813, Three Brothers, Eagle Island, Peros Banhos and Salomon Islands settled (Vine 2004)	Shiploads of topsoil imported from India to Diego Garcia, together with the contained seed burden of herb and grass species from the sub-continent
By 1935	**"Natural" re-vegetation process begins:** Settlements on Eagle Island, Three Brothers and Egmont atoll abandoned	After 120 years of management for coconut, vegetation starts to return to unmanaged state of 'coconut chaos'.
1965–1973	**Outer islands depopulated:** Settlement on all other islands ends.	After 200 years the vegetation begins to return to a unmanaged state of 'coconut chaos'.
1971	**Increased introduction of ornamental plant species begins on Diego Garcia:** Development of the US Facility on Diego Garcia	Increased introduction of plant species to beautify "downtown" area; unintentional introduction of other plants and vertebrates.
1974 onwards	**Protection of the "wild" habitats begins:** Creation of "Strict Nature Reserves" on some islands, with landing forbidden	Benign neglect permits coconuts to dominate, leading to "Coconut Chaos" on many islands
2006 onwards	**Active conservation management on land begins**	Positive conversation management measures of the islands commences

February and March 2006, it was revealed that the IBA qualifying criteria for some (designated) islands was not met during that survey period, but that populations on two other islands *did* meet the criteria (McGowan et al. 2008). Further survey work between October 2008 and October 2010 showed that some seabirds, including the IBA qualifying species of Brown *Anous stolidus* and Lesser Noddy *A. tenuirostris*, Sooty Tern *Onychoprion fuscatus* and Red-footed Booby, do not necessarily breed annually, synchronously, or on the same islands each season.

All of the IBAs are also Strict Nature Reserves (SNRs) or, in the case of Barton Point and the three associated islets in the mouth of Diego Garcia lagoon, a Conservation (Restricted) Area. This affords them a degree of protection, both through BIOT Law and effective enforcement of the Law.

Vegetation and Plant Colonisation

The terrestrial vegetation on all islands has gone through a number of phases (Table 20.1). Historical accounts give some indication of past vegetation of the islands, and many mention their wooded appearance from the sea and the dominant presence of the coconut palm. They also mention the continued presence of very large trees and associated forest on some islands (Hemsley 1855; Bourne 1888; Willis and Gardiner 1931). Hemsley (1855) lists 43 species considered to be native, which compares well with Hamilton and Topp's (2009) list of 42 "native" species, although only 25 species can be shown to be common to both lists, leaving a maximum possible combination of 60 plant species. Changes in nomenclature as well as use of "local" names have introduced a degree of confusion and uncertainty in the early records as noted by Hemsley (1855). The latter, for example, bemoans the use of the name "Mapou" (or "Mapan") for example for large trees on the islands – this Creole name is used on other oceanic island groups for a number of large tree species, including the very large Rose Tree *Barringtonia asiatica*.

A great deal of information on the distribution of plant species throughout the Archipelago was amassed by Topp (e.g. Hamilton and Topp 2009; Topp and Sheppard 1999). He also wrote of vegetation assemblages and composition on the numerous islands he visited, and Hamilton and Topp (2009) list 234 plant species from all islands and included information as to whether they consider a species to have been introduced or to be naturally-occurring. Removing all those species from the list that occur on only one island – Diego Garcia – as likely to have been introduced by man, (especially since the islands' military development phase since 1971), also leaves a core list of 42 "native" species – two of which are marine sea-grasses.

These native species vary in occurrence from two species found on only one island through to three species found on over 40 islands. Twenty seven native plant species are recorded on over 11 islands. The presence of a species on more than one island can be an indication of the length of time that species has had to spread through the Archipelago,

Table 20.2 Number of vegetation types found on Eagle island in 2006. The data was obtained from 2,821 30×30 m squares established for the purpose of laying out rodent poison bait stations

Vegetation type	30×30 m squares	% Total surveyed	Area covered (ha)
Palm forest	1,365	48.4	122.9
Scaevola thicket	543	19.2	48.9
Open clearing	342	12.1	30.8
Palm swamp	203	7.2	18.3
Hernandia woodland	192	6.8	17.3
Mangrove	125	4.4	11.3
Indian almond	16	0.6	1.4
Hibiscus thicket	11	0.4	1.0
Guettarda grove	10	0.4	0.9
Beach	7	0.2	0.6
Hernandia forest	5	0.2	0.5
Ficus grove	1	0	0.1
Takamaka	1	0	0.1
Total	**2,821**	**100**	**253.9**

as well as of the rate of species spreading between islands. The species might also of course have been secondarily spread through the area by man, either purposefully if useful, or unintentionally if it has a dispersal mechanism favoured by human activity.

A species-island area curve (Fig. 20.4) shows a relationship of increasing numbers of native species with island size, and of total species. However most islands are small, between 1 and 100 ha, with only four islands between 100 and 1,000 ha, and one, Diego Garcia, being over 1,000 ha. The larger islands had the main coconut plantations and were therefore most influenced by man, including the dispersal and use of beneficial native plant species.

Vegetation Assemblages

Due to the limited topography, mean island height is no more than a couple of metres, with only occasional elevations of dunes found more than 6 m above sea surface, and few vegetation assemblages other than coconut palm forest and thicket can still be discerned. There are several relict patch habitats and species, as well as examples of those few species able to counteract the coconut dominance. These relict habitats have tended to take the form of forest and woodland patches with associated shrub and herb species. However, in the absence of earlier detailed accounts of the pre-human vegetation it is difficult to judge as to how representative of the original situation these may be, though Table 20.2 shows habitats identified in Eagle Island which show the communities that can be supported wherever suitable conditions exist (Hillman 2007). Coconut palms were found in all habitats, but dominated two – palm forest and palm swamp – to a total of 55.6% of the land surface. Dense *Scaevola* thicket and open clearing were common also. Small areas at the north centre of the island comprised recent low sand dune formation, with a surface binding of mosses and other non vascular plants. The only two other native species to dominate some areas were *Hernandia sonora* that formed woodland towards the north of the island and forest with large mature trees in the south where plantation activities were limited, and groves of *Guettarda speciosa;* both species seemed to be able to hold their own against coconut invasion.

Vegetation Dynamics

Limited past data indicate the dynamics of the vegetation and the rate at which change can occur. At the time of a 1975 expedition to this island an extensive *Typha domingensis* bulrush swamp area along the western seaboard was recorded. By 2006 no *Typha* plants were found (although the species is still known from Diego Garcia). The area in 2006 held a senescent *Lumnitzera racemosa*, small-leaved or black mangrove swamp, along with an area of coconut palm swamp of even-aged trees, which at the time had not started fruiting. It is surmised that either a tidal wave or flood event had floated a lot of coconuts into the area which took root, the incoming seawater having killed off the *Typha* that had previously survived on the fresh-water lens. At the same time, developing fruits of the grey mangrove *Avicennia marina* were recorded on Eagle Island beaches but the plant at the time was unrecorded for the archipelago. In 2010 another *Lumnitzera* mangrove swamp area was discovered on Moresby Island in the Peros Banhos atoll. This is in addition to the *Pemphis acidula* mangroves on Ile Anglais and Ile du Passe in the Salomons. It is unknown whether these are recent arrival, or have remained previously unrecorded.

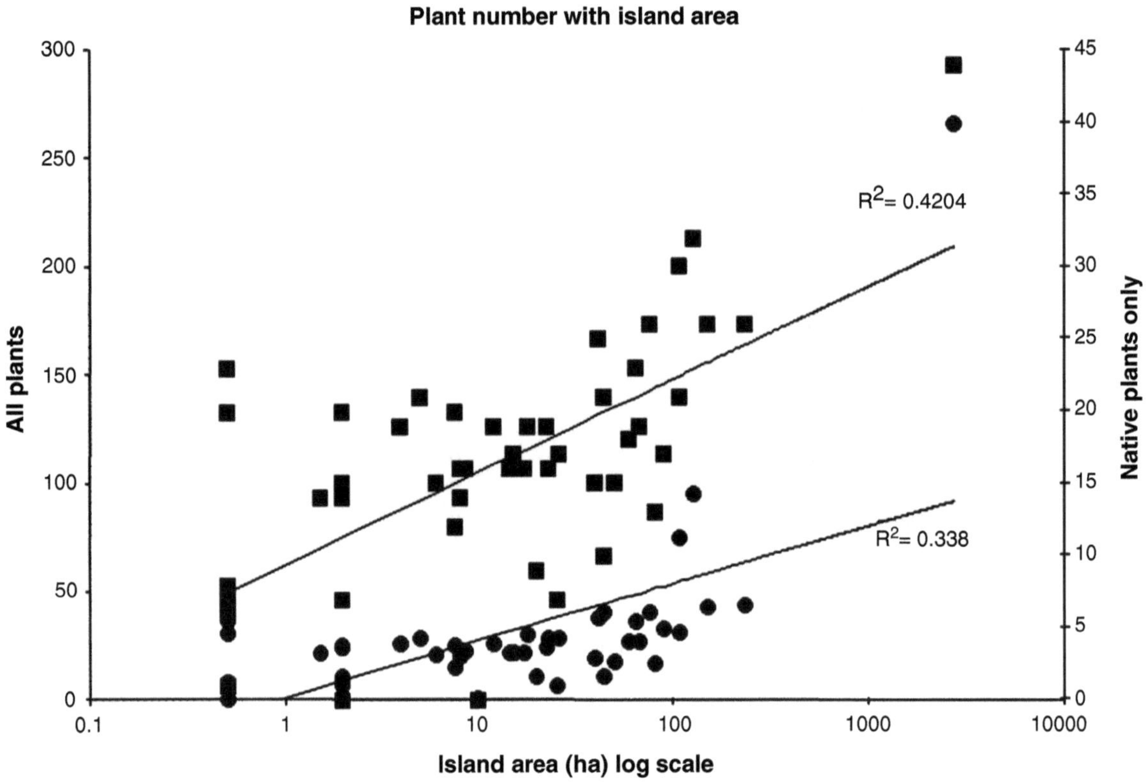

Fig. 20.4 Number of plants on each island against island area. Squares and top regression line are data for all plants, *dots* and *bottom* regression line are data for native species only. *Left* y axis is numbers of all plants, *right* y axis is numbers of native species only (Data from Topp and Sheppard (1999))

Another example of vegetation dynamics involving anthropogenic influence is seen on East Island in Diego Garcia atoll, the island most heavily influenced by man. Historical notes record:

- Bourne (1888) noted that *Pisonia* existed generally where there was exposed coral rubble, and that there were no coconuts or tall trees on East Island.
- Stoddart (1971) made no record of *Pisonia grandis* on East Island, but did note coconut thicket on the western half of this island. The shrub *Scaevola toccada* covered half the island, and *Tournefortia (Argusia) argentea* was occasional within the *Scaveola. Hernandia* formed a woodland of massive trees in the western half of the island.
- Topp (1988) noted only three *Pisonia grandis* trees on the island prior to 1988.
- Hillman (2009) recorded that much of the island was dominated by *Pisonia grandis* trees, with grassy "lawns" beneath. *Scaveola* was limited and patchy, *Tournefortia* comprised a strong woodland fringe at the upper NE beach crest, and coconuts were restricted to two small patches in the NW and SW around the coaling station ruins. *Hernandia* was limited to a single large tree.

These observations show how the natural *Pisonia* forest became altered over a century to coconut dominated habitat and, after 40 years of no management, changed back to a *Pisonia* dominated habitat.

Lower Plants, Fungi and Cyanobacteria

Although the diversity of algae, mosses, liverworts, cyanobacteria, fungi and lichens is relatively low on Chagos, these provide high cover and are ecologically important (Seaward 1999; Seaward and Aptroot 2000; Watling and Seaward 2004; Seaward et al. 2006). The cyanobacteria extensively clothe eroded corals, thin sandy soils and tree trunks, and fix nitrogen directly and thereby make this vitally important element, which is absent from the poorly developed soils, available to the other plants. The mosses and liverworts are important in stabilizing soils, and are often the primary stages in the succession of higher vegetation in exposed areas. In the wooded areas, the mosses and liverworts clothe the bark of living trees; the nature of this epiphytic flora changes with age and as the trunks die, different mosses and liverworts support the decomposition process undertaken mainly by fungi. Lichens too colonize all available tree bark surfaces, as well as being found on the living leaves of various tree species, the long-lived evergreen nature of the leaves allowing these slow growing organisms to establish themselves.

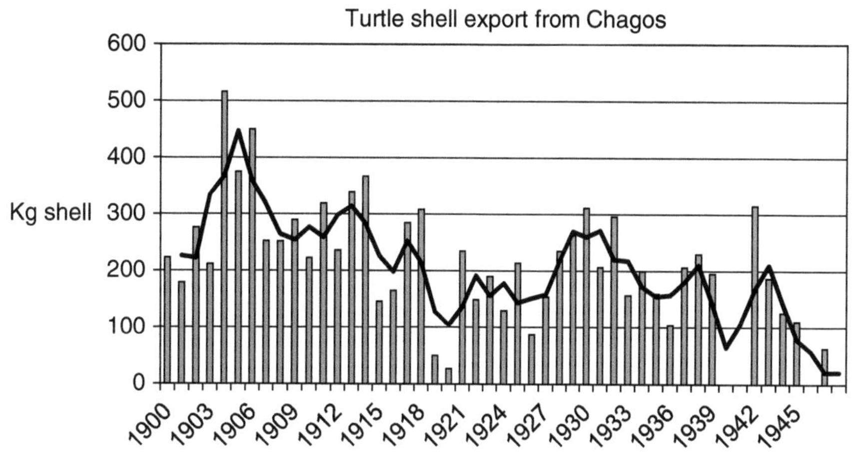

Fig. 20.5 Statistics documenting kilograms of raw shell exported annually from Chagos into Mauritius during the period between 1904 and 1945. Data are from "The *Blue* Book for the Colony of Mauritius". Two kilos of shell are approximately equivalent to one adult turtle (Mortimer and Day 1999)

Many of these organisms are particularly sensitive to environmental changes brought about by natural and human disturbances, such as climatic change and rising seawater levels. Furthermore, lichens are remarkable bioaccumulators of elements (such as heavy metals), radionuclides and compounds (such as PCBs) and therefore able to detect range dispersal of pollutants (even in low concentrations). No doubt these organisms will prove invaluable, as they have done elsewhere, in biomonitoring programmes.

Due to the relatively short history of the current terrestrial habitats of the Chagos Archipelago, as would be expected, there are no endemic species of these groups; for example, a new liverwort taxon, *Cololejeunea planissima* var. *chagosensis*, known only from Chagos (Seaward et al. 2006), will undoubtedly occur elsewhere, but there is always the possibility that it has disappeared (or will disappear) from its original site(s). However, due to the geographical setting of the islands their establishment is remarkable, not only in terms of the distances travelled by the spores and propagules, but also in terms of their origin – there are interesting affinities of the new floras to those of south-east Asia, India or Africa, as yet to be fully determined. Our knowledge of these groups on other Indian Ocean islands (as well as the land masses fringing the ocean) has increased very considerably in recent years due to a greater interest in tropical and/or island floras.

Terrestrial Fauna

Being remote, geologically young and true oceanic islands, the terrestrial fauna of the Chagos is naturally impoverished. As far back as the late nineteenth and early twentieth Century, scientific expeditions were visiting the area (e.g. Bourne 1888; Gadow and Gardiner 1907). Prior to the creation of a military facility on Diego Garcia scientific surveys for this atoll were published (e.g. Stoddart and Taylor 1971). Following the creation of the military facility, scientific research has been undertaken slightly more regularly and a synopsis of some of this work is found in Ecology of the Chagos Archipelago (Sheppard and Seaward 1999). A more recent publication by Sheppard et al. (2012) covers several scientific disciplines including a brief summary of terrestrial understanding.

There are no endemic bird species, no native mammals and two native reptiles. From the very limited invertebrate recording there appear to be one endemic moth, one endemic sub-species of moth and two endemic sub-species of butterflies (Barnett and Emms 1999). However, the archipelago remains under- recorded in this respect and several groups of organisms have not been thoroughly researched, arthropods and several invertebrate orders being examples.

Red-Listed Vertebrates

Two key terrestrial (non-avian) vertebrate species are the sea turtles on the IUCN Red List of Threatened Species, one being assessed as Critically Endangered (Hawksbill) and the other as Endangered (Green). Populations of both species are of global significance given their Red List status (Mortimer and Day 1999) and both are now protected, as is much of their breeding habitat. However, during plantation times they were heavily exploited; Hawksbill shell was exported in substantial quantity until their numbers declined (Fig. 20.5), and Greens were used for food. The numbers used for food remains unknown, although turtle holding pens in at least two atolls suggest consumption was common. Furthermore, throughout the Indian Ocean consumption of turtle eggs was commonplace and a likely source of further attrition in Chagos also. Their survival and continued recovery following two centuries of exploitation would be enhanced by long-term monitoring, by turtle awareness campaigns on Diego Garcia, better discouragement of poaching and eradication programmes of invasive mammals (Mortimer and Day 1999).

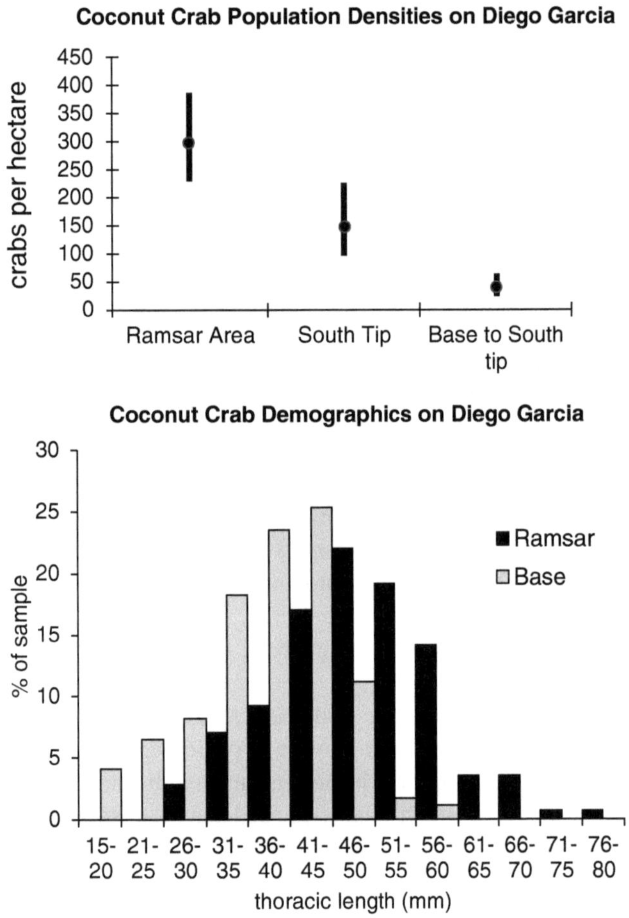

Fig. 20.6 Coconut crab densities and sizes in Diego Garcia. *Top*: numbers in three locations (mean and 95% C.I.). *Bottom*: Crab size structure in Ramsar site and in an inhabited area

The world's largest arthropod, the Coconut Crab is also on the IUCN Red List, designated as Data Deficient. This species has been extirpated world-wide from most areas with significant human populations (likely through harvesting), including mainland Australia and Madagascar. Island wide crab surveys were performed on Diego Garcia during October, 2010 (Vogt 2011). Utilizing DISTANCE methodology (Buckland et al. 2001), 20, 300 m transects were surveyed for Coconut Crabs. The highest densities were recorded in the Minni Minni conservation area with an overall density of 298 crabs per hectare, Within Minni Minni the highest density transect had 489 crabs per hectare. This is one of the highest, if not the highest, densities recorded for this species. Densities decline the closer one gets to the inhabited areas, with 147 per hectare on the southern tip of the island, to 39 per hectare on the western side of the island including the military base (Fig. 20.6 top). Crab sizes were also larger in the Minni Minni area (Fig. 20.6 bottom) It is thought that the high Minni Minni population density exists because of the lack of harvesting (Vogt 2011) and it is possible this density is mirrored in the undisturbed northern atolls. Coconut Crabs have a slow growth rate and a very low juvenile recruitment due to an oceanic larval stage (Brown and Fielder 1991). Because of this, populations are easily over-harvested and slow to recover.

Avifauna

Historically the bird populations have received most attention. There have been general accounts (e.g. Bourne 1886; Loustau-Lalanne 1962; Hutson 1975), some species specific analysis (e.g. Ripley 1969; Benson 1970) and one broad-ranging analysis summarising the ornithological knowledge of the Chagos at that time (Bourne 1971). Carr (2011a) gives the most up to date information on the species present throughout the Chagos and their numbers. There has been no fixed long-term monitoring, though thorough censuses of the internationally important breeding seabirds occurred in February/March 1996 (Symens 1999); March 2006 (McGowan et al. 2008), February 2010 and February 2012 (Carr in prep.), all as part of organised scientific research expeditions.

The avifauna of the Chagos naturally divides itself in to three groups: introduced exotics; seabirds; and landbirds, which include regular migrants, vagrants and natural colonists.

Introduced exotics are Madagascar Red Fody *Foudia madagascariensis*; Common Myna *Acridotheres tristis*; Madagascar Turtle-dove *Nesoenas picturata*; Zebra Dove *Geopelia striata*; Cattle Egret *Bubulcus ibis* and Red Junglefowl *Gallus gallus* (Feral Chicken). With the exception of the fody, which is found throughout the archipelago, none have sustained viable populations away from Diego Garcia and man. Feral chickens are an interesting conundrum; populations do exist on islands never permanently inhabited by man e.g. Nelsons Island, and it is presumed these isolated populations were released as a food source by either former plantation workers or latterly by illegal poachers and that their numbers are sustained by reintroductions (Carr 2011b).

There is evidence of failed introductions throughout the archipelago. House Sparrow *Passer domesticus* was still present in Peros Banhos and the Salomons in 1960 (Loustau-Lalanne 1962), but despite extensive searches between 2008 and 2010, this species was not recorded and appears extinct (Carr 2011a). Interestingly, the sparrow was never introduced to Diego Garcia. Common Myna, another obligate associate of man was, and still thrives there, though it too has not been found on the other atolls where it was once recorded (Carr 2011a). Other failed bird introductions include Grey Francolin *Francolinus pondicerianus*, (Loustau-Lalanne 1962), a guineafowl *Numididae* species thought to be Helmeted *Numida meleagris* (Lever 1987) and a bulbul (Bourne 1971), thought to be the Réunion Bulbul *Hypsipetes borbonicus*.

Seabirds are the archetypal avifauna of the Chagos. Up until 1971, 22 species of seabirds were recorded of which 14 have since been proven to breed. In 1996, the breeding species increased to 17 after the first comprehensive seabird survey with the additions of Masked Booby *Sula dactylatra*, Roseate *Sterna dougallii* and Little Tern *Sterna albifrons* (Symens 1999). An eighteenth breeding seabird species was added when Red-tailed Tropicbird *Phaethon rubricauda* was recorded as nesting on Diego Garcia (Guzman 2003).

There is evidence that at least Red-footed Booby is increasing in both numbers and range, this is likely to be due to the reduction of anthropogenic interference. Diego Garcia had very few pairs breeding on the mainland in 1984–6 (Topp 2003), but in May 2005, there were 4,370 breeding pairs there (Carr 2011a). Similarly, in 1996, Red-footed Booby was not found breeding in the Salomon Islands (Symens 1999). In 2006, 323 pairs were breeding on two islands (McGowan et al. 2008); by 2009, this had increased to four islands (Carr 2011a). It is likely that this species is re-colonising areas where it once bred, now that man has vacated.

Sooty Tern, the most numerous breeding species in the Chagos, has been proven to switch breeding islands during different breeding cycles. Without conclusive evidence from tracking, it appears there are "island clusters" that breeding populations use, and these are centred upon the Three Brothers, and groups of islands in eastern Peros Banhos (Parasol and Longue; the Bois Mangues and, the Coquillages). It has also been shown that this species is a less than annual breeder in the Chagos. Table 20.3 details breeding records for the Three Brothers and the two Coquillages for years 2009, 2010 and 2012. The "island-hopping" breeding strategy is not unique to the Chagos; where it occurs in the Seychelles it has been attributed to infestations of an argasid tick *Ornithodoros capensis* (Feare 1976). It has been suggested that tick infestation is also the cause for island desertions in the Chagos following the finding of dying chicks on Ile Longue in 2009 that were heavily infested with ticks (Carr in prep.).

The shifting of breeding colonies of Sooty Tern impacts upon the designation of Important Bird Areas in that those islands surveyed in "non-breeding" years could potentially lose their IBA status. This has led to the proposal that "island clusters" are designated as IBAs, not specific islands. Table 20.4 gives the proposed designations of IBAs in the Chagos (Carr 2011b).

Of the non-breeding seabirds, by 1971, 22 species were reported as having definitely occurred (Bourne 1971); by 2009 this figure had increased to 37 (Carr 2011a). This increase is presumed to be a result of greater recording effort, better optical equipment and the quality of field-guides available rather than a genuine increase in the number of seabird species present or passing through.

Due to its isolation and relatively young geological age, the Chagos has an impoverished land bird community. Further, due to a lack of varied habitat and altitude, and little permanent water or extensive wetlands, estuaries or mudflats, the area does not support large numbers of associated birds, e.g. ducks, geese, storks, herons and waders (Carr 2011a). Bourne (1971) speculated of BIOT that "....it seems possible

Table 20.3 Breeding (≥10,000 pairs) Sooty Tern *Onychoprion fuscata* records from selected "island clusters" of the Chagos Archipelago. The "island-hopping" by this breeding species brings challenges when classifying IUCN Important Bird Areas

Month/Year	Island cluster				
	Three Brothers			The Coquillages	
	South	Middle	North	Petite	Grand
JUL 2009	X		X	X	
FEB 2010		X		X	
FEB 2012		X			X

Table 20.4 Proposed "island cluster" Important Bird Areas in the Chagos Archipelago. The four newly proposed IBAs would replace the ten designated and two proposed island-specific IBAS

Criteria	Name	Key species
A4ii, A4iii	Eastern Diego Garcia Island Group	Red-footed Booby
	Includes eastern Diego Garcia from the Plantation Gate to Barton Point (19 miles) and East, Middle and West Island	
A4i, A4ii and A4iii	Western Great Chagos Bank Island Group	Audubon's Shearwater
	Includes Danger, Sea Cow, Eagle, the Three Brothers and Resurgent	Red-footed Booby
		Lesser Noddy
		Common Noddy
		Sooty Tern
A4i, A4ii and A4iii	Nelson's Island	Red-footed Booby
		Lesser Noddy
		Common Noddy
A4i, A4ii and A4iii	Eastern Peros Banhos Island Group	Red-footed Booby
	Includes all islands from Ile du Passe through to Ile Vache Marine inclusive	Lesser Noddy
		Common Noddy
		Sooty Tern

that there was once quite a rich landbird community, comparable to that of Christmas Island further east." This is certainly not true of the present avifauna. The three landbirds that are successful natural colonists are Striated Heron *Butorides striata*; White-breasted Waterhen *Amaurornis phoenicurus*; and Common Moorhen *Gallinula chloropus*. The heron has likely been on all of the atolls for centuries; the latter two are recent colonists to Diego Garcia. Both species now breed throughout the year on Diego Garcia with the waterhen population in 2012 being estimated at 75–100 pairs and the moorhen at 20–25 pairs.

With the recent increase in ornithological recording on Diego Garcia, it has been possible to assess which species are vagrants and which are regular migrants, albeit in small numbers. Some such as Whimbrel *Numenius phaeopus*, Ruddy Turnstone *Arenaria interpres* and Curlew Sandpiper *Calidris ferruginea* turn up all across the archipelago in their hundreds, if not in low thousands. Sanderling *Calidris alba*, Bar-tailed Godwit *Limosa lapponica*, Common Greenshank *Tringa nebularia*, Wood *Tringa glareola* and Common Sandpiper *Actitis hypoleucos* also appear to turn up annually on any of the atolls but in much smaller numbers. It is still unknown whether the final group of waders, those with very few records, such as Common Ringed Plover *Charadrius hiaticula*, Marsh Sandpiper *Tringa stagnatilis* and Common Redshank *T. totanus* are annual visitors in extremely low numbers or merely occasional overshooting migrants. Of the non-waders, Garganey *Anas querquedula* appears annually, at least on Diego Garcia, as does Oriental Pratincole *Glareola maldivarum* (Carr 2011a).

Insects

Two surveys have been undertaken that recorded the insect species present. The first in 1975 concentrated purely on Diego Garcia (Hutson 1981), the second in 1996 visited 24 islands and identified 96 species (Barnett and Emms 1999). Three endemic sub-species of Lepidoptera have been claimed (Meadow Argus *Junonia villida chagoensis*; Common Eggfly *Hypolimnus bolina euphonoides* and a hummingbird hawk-moth *Macroglossum corythus oceanicum*) and a single species of moth is thought to be endemic, *Sticoptera hironsi*. In addition to these two surveys, post the creation of the military facility on Diego Garcia, limited records have been kept of the insects present, particularly those deemed as a pest to man. Most recorded insects are widespread across the Indo-Australian tropics and include commensals of man and agricultural pests (Barnett and Emms 1999), likely indicating introduction by man rather than natural colonisation.

Terrestrial Challenges

The ecological integrity of every island has been affected to some degree by man, either through the introduction of alien species, both accidentally and deliberately, or by the introduced coconut mono-cultures. The latter areas now have very little to offer for biodiversity.

Of the introductions, mammals (Bruner 1995; Barnett and Emms 1998; Symens 1999); plants (Topp and Sheppard 1999) and the birds (Lever 1987) have been recorded throughout the archipelago. To a much lesser degree, introduced insects (Barnett and Emms 1999) and reptiles and amphibians (Carr 2008) have been recorded. It is the introduced mammals that are responsible for the greatest environmental damage. Over the ages these have involved domestic cattle *Bos sp.*, horse *Equus ferus*, donkeys *Equus africanus*, sheep *Ovis aries*, domestic pig *Sus scrofa domestica* and dogs *Canis lupus* (Scott 1961) and, Black Rats *Rattus rattus* and feral/domestic cats *Felis catus*, which have had and are still having the greatest detrimental impact upon the fragile oceanic island ecosystems of the archipelago (Hilton and Cuthbert 2010). Along with direct predation and persecution by man, the introduced mammals were responsible for the demise of the vast seabird colonies that once existed on Diego Garcia (Bourne 1886) and on other islands where guano deposits indicate they once existed (Bourne 1971).

Globally the impact of invasive plants on oceanic islands has been well-studied (e.g. Brockie et al. 1988; Caujapé-Castells et al. 2009). Whilst higher plants have been surveyed throughout the Chagos (Topp and Sheppard 1999; Clubbe 2010), only Diego Garcia has had any assessment made of the impact of alien plants, with the giant sensitive plant *Mimosa invisa*, dwarf sensitive plant *Mimosa pudica* and ironwood trees *Casuarina equistifolia* having been identified as invasive and are therefore controlled. The native parasitic vine *Cassytha filiformis* is also controlled on Diego Garcia in specific shoreline areas where it is reducing plant cover in areas where coastal erosion is a concern. Other alien plant species of concern on Diego Garcia are: *Leucaena leucocephala*; *Laurentia longiflora* and *Rivina humilis* (DG INRMP 2005). The ornamental mayflower tree *Tabebuia pallida* and *Pipturus argenteus* should be added to this list, the former now totally dominates the island for 1.5 km south of East Point Plantation and the latter is colonising and dominating any disturbed open area on the eastern arm of Diego Garcia.

Of the introduced birds, House Crow *Corvus splendens* has the potential to become invasive and ecologically damaging, though as it has an obligate association with humans and therefore it is unlikely to colonise the uninhabited northern atolls. This pest species was first recorded in the Chagos when a single bird was found present on Diego Garcia from at least May 2002 (Guzman 2003). This, or another bird, was joined by a second in at least October 2008 (Carr 2011a). The long-term fate of these two birds has yet to be revealed. It is noteworthy and of concern that the only known arrivals of pairs of House Crows that have *failed* to breed have been in southern Chile, where the pair died during their first boreal winter, and in Australia where they were quickly trapped and shot (Nyári et al. 2006).

Of the six introduced reptiles known to occur, Cane Toad *Bufo marinus* and the agamid lizard *Calotes versicolor* are restricted to Diego Garcia: both of these species were first recorded in the late twentieth century and their ecological impact is unknown. Two geckos occur on Diego Garcia: the House Gecko *Hemidactylus frenatus* and Mourning Gecko *Lepidodactylus lugubris*, the former also being found on islands in both the Salomons and Peros Banhos and neither being deemed invasive (Barnett and Emms 1998). There were formerly two introduced species of terrapin, Indian Pond Terrapin *Melanochelys trijuga thermails* and East African Black Mud Turtle *Pelusios subniger* both appearing to be now extinct, the last scientific records being in 1885 and 1905 respectively. The ecological impact of the terrapins is again unknown.

Too little is known of the invertebrates to make any meaningful comment on what species constitute a threat to the island ecosystems of the Chagos. Historically the Coconut Rhinoceros Beetle *Oryctes rhinocerus* has been deemed a pest though with the termination of coconut farming this is no longer the case. On Diego Garcia, the island with the most introduced invertebrates, there are several species that are deemed and treated as pests, with the majority appearing to be recent (post-1971) introductions. Aside from health and nuisance issues, the impact of these invertebrate "pests" on ecosystems remains unknown, though any release of these species on to the more pristine islands north of Diego Garcia should be considered an undesirable state.

The future

Global events also present a challenge to the low lying islands of the Chagos. Climate change and the related sea level rising are possibly the greatest concern for the long-term future of the islands. Several inter-related aspects are important, summarised in Sheppard et al. (2012) and Chap. 20. Firstly warming of the oceans has killed corals on previous occasions, and with coral mortality comes damage to the 'natural breakwater' around the islands. Of longer term but similar concern is ocean acidification which will have similar consequences to the corals and especially to the strong algal ridges which are a key part of these shallow reefs. Secondly, there is increasing concern about sea level rise (see Chaps. 17 and 18). The latest predictions for global sea level rise have increased earlier IPCC estimates substantially, such that global average rises are now predicted to be up to 2 m by the end of the Century, with obvious consequence to Chagos islands. These factors both cause on going coastal erosion.

Despite some adverse forecasts of the future of all coral islands, islands of Chagos are perhaps those most likely to withstand global changes and human pressures the longest while their fringing reefs remain undamaged. With this in mind, with the creation of the Marine Protected Area in 2010 and its associated tiny fraction of land, the possibility of rectifying some of the historic environmental wrongs now clearly exists. Plans exist for attempts at rat eradication and restoration of the vegetation of selected islands. In 2009, reforestation of some of the former coconut plantations on the eastern arm of Diego Garcia was started and is showing promising, indeed spectacular, results. Under the management of the MPA, it is not beyond the bounds of reality to believe that some form of restoration//rehabilitation of the remainder of the islands is now a distinct possibility.

Acknowledgements Some of this material was published previously in Sheppard et al. (2012) and is reproduced with permission of Wiley & Sons.

References

Barnett LK, Emms C (1998) An annotated checklist of the Chagos Archipelago terrestrial fauna (omitting birds) recorded during the 1996 Friends of the Chagos expedition. Phelsuma 6:41 52

Barnett LK, Emms C (1999) The insects of the Chagos Archipelago. In: Sheppard CRC, Seaward MRD (eds) Ecology of the Chagos. Linnean Society Occasional Publications 2. Westbury Publishing, Otley, W. Yorkshire, pp 241–256

Barnett LK, Emms C, Clarke D (1999) The coconut or robber crab (Birgus latro) in the Chagos Archipelago and its captive culture at London Zoo. In: Sheppard CRC, Seaward MRD (eds) Ecology of the Chagos. Linnean Society Occasional Publications 2. Westbury Publishing, Otley, W. Yorkshire, pp 273–284

Benson CW (1970) The systematic status of *Streptopelia picturata* on Diego Garcia. Bull Br Orn Club 90:32–35

BirdLife International (2004) Important Bird Areas in Asia: key sites for conservation, BirdLife Conservation Series No. 13, Cambridge, UK, pp 43–44

Bourne GC (1886) General observations on the fauna of Diego Garcia. Proc Zool Soc Lond 1886:331–334

Bourne GC (1888) The atoll of Diego Garcia and the coral formations of the Indian Ocean. Proc Roy Soc London 43(1887–1888):440–461

Bourne WRP (1971) The birds of the Chagos group, Indian Ocean. Atoll Res Bull 149:175–207

Brockie RE, Loope LL, Usher MB, Hamann O (1988) Biological invasions of island nature reserves. Biol Conserv 44:9–36

Brown IW, Fielder DR (1991) The coconut crab: aspects of the biology and ecology of *Birgus latro* in the Republic of Vanuatu. ACIAR Monograph No. 8, 136 pp. Australian Centre for International Agricultural Research, Canberra

Bruner PL (1995) Avifaunal and feral mammal survey of Diego Garcia, Chagos Archipelago, British Indian Ocean Territory. Diego Garcia

Integrated Natural Resources Management Plan. Diego Garcia: US State Department of Defence. Appendix F1
Buckland ST, Anderson DR, Burnham KP, Laake JL, Borchers DL, Thomas L (2001) Introduction to distance sampling, estimating abundance of biological populations. Oxford University Press, Oxford, 432pp
Carr P (2006) British Indian Ocean Territory. In: Sanders SM (ed) Important Bird Areas in the United Kingdom Overseas Territories. RSPB, Sandy, pp 37–55
Carr P (2008) Diego survey III – expedition report. Sea Swallow 57:16–62
Carr P (2011a) Birds of the British Indian Ocean Territory. Published by Pisces Publications for the RSPB, Sandy
Carr P (2011b) British Indian Ocean Territory: Important Bird Areas. Br Birds 104:642–659
Caujapé-Castells J, Tye A, Crawford DJ, Santos-Guerra A, Sakai A, Beaver K, Lobin W, Florens FBV, Moura M, Jadim R, Gómes I, Kueffer C (2009) Conservation of oceanic island floras: present and future global challenges. Perspectives in Plant Ecology, Evolution and Systematics (2010). doi:10.1016/j.ppees.2009.10.001
Clubbe C (2010) Botanical Explorations of the Chagos Archipelago. Chagos News 36:8–10
Diego Garcia Integrated Natural Resources Management Plan (2005) Internal report to the US DoD
Edis R (1993) Peak of Limuria: the story of Diego Garcia and the Chagos Archipelago. Chagos Conservation Trust, London
Feare CJ (1976) Desertion and abnormal development in a colony of sooty terns *Sterna fuscata* infested by virus-infected ticks. Ibis 118:112–115
Gadow H, Gardiner JS (1907) The Percy Sladen Expedition Trust to the Indian Ocean. Aves, with some notes on the distribution of the land-birds of the Seychelles. Trans Linn Soc Lond Zool II 12:103–110
Guzman N (2003) Wildlife on Diego Garcia. Chagos News 22:5–7
Hamilton M, Topp J (2009) British Indian ocean territory plant species checklist. Kew. Manuscript 19 August 2009. 20pp.
Hemsley WB (1855) Report on the vegetation of Diego Garcia. Herbarium Royal Gardens Kew Trans Linn Soc xxii:45
Hillman JC (2007) Report on the state of the environment of Eagle Island, Chagos Archipelago. Chagos Ecological Restoration Project, Fauna and Flora International, Cambridge, UK
Hillman JC (2009) Report to the Chagos Conservation Trust on accompanying the Australian National University Archaeological Expedition to Diego Garcia, Nov-Dec 2009
Hilton GM, Cuthbert RJ (2010) The catastrophic impact of invasive mammalian predators on birds of the UK overseas territories: a review and synthesis. Ibis 152:443–458
Hutson AM (1975) Observations on the birds of Diego Garcia, Chagos Archipelago, with notes on other vertebrate. Atoll Res Bull 175:1–25
Hutson AM (1981) A preliminary checklist of the insects of Diego Garcia Atoll, Chagos Archipelago. Atoll Res Bull 243:1–30
Lever SC (1987) Naturalized birds of the World. Longman Scientific and Technical, Harlow
Loustau-Lalanne P (1962) The birds of the Chagos Archipelago, Indian Ocean. Ibis 104:67–73
McGowan A, Broderick AC, Godley BJ (2008) Seabird populations of the Chagos Archipelago: an evaluation of Important Bird Area sites. Oryx 42:424–429
Mortimer JA, Day M (1999) Sea turtle populations and habitats in the Chagos Archipelago. In: Sheppard CRC, Seaward MRD (eds) Ecology of the Chagos. Linnean Society Occasional Publications 2. Westbury Publishing, Otley, W. Yorkshire, pp 159–176
Nyári Á, Ryall C, Townsend Peterson A (2006) Global invasive potential of the house crow *Corvus splendens* based upon ecological niche modelling. J Avian Biol 37:306–311
Ripley SD (1969) Comments on the little green heron of the Chagos Archipelago. Ibis 111:101–102
Scott R (1961) Limuria. The lesser dependencies of Mauritius. Oxford University Press, London
Seaward MRD (1999) Cryptogamic flora of the Chagos Archipelago. In: Sheppard CRC, Seaward MRD (eds) Ecology of the Chagos Archipelago. Linnean Society Occasional Publications 2. Westbury Press, London, pp 207–224
Seaward MRD, Aptroot A (2000) The lichen flora of the Chagos Archipelago, including a comparison with other island and coastal tropical floras. Trop Bryol 18:185–198
Seaward MRD, Ellis LT, Pócs T, Wigginton MJ (2006) Bryophyte flora of the Chagos Archipelago. J Bryol 28:11–19
Sheppard CRC, Seaward MRD (eds) (1999) Ecology of the Chagos. Linnean Society Occasional Publications 2. Westbury Publishing, Otley, W. Yorkshire, 350pp
Sheppard CRC, Ateweberhan M, Bowen BW, Carr P, Chen CA, Clubbe C, Craig MT, Ebinghaus R, Eble J, Fitzsimmons H, Gaither MR, Gan C-H, Gollock M, Guzman A, Graham NAJ, Harris A, Jones R, Keshavmurthy S, Koldewey H, Lundini CG, Mortimer JA, Obura D, Pfeiffer M, Price ARG, Purkis S, Raines P, Readman JW, Riegl B, Rogers A, Schleyer M, Seaward MRD, Sheppard ALS, Tamelander J, Turner JR, Visram S, Vogler C, Vogt S, Yang JM-C, Yang S-Y, Yesson C (2012) Reefs and islands of the Chagos Archipelago, Indian Ocean: why it is the world's largest no-take marine protected area. Aquat Conserv Mar Freshwat Ecosyst 22:232–261. doi:10.1002/aqc.1248, Published online in Wiley Online Library (wileyonlinelibrary.com)
Stoddart DR (1971) Land vegetation of Diego Garcia. Atoll Res Bull 149:127–142
Stoddart DR, Taylor J (1971) Geology and ecology of Diego Garcia atoll, Chagos Archipelago. Atoll Res Bull 149:1–237
Symens P (1999) Breeding seabirds of the Chagos Archipelago. In: Sheppard CRC, Seaward MRD (eds) Ecology of the Chagos. Linnean Society Occasional Publications 2. Westbury Publishing, Otley, W. Yorkshire, pp 257–272
Topp JMW (1988) An annotated checklist of the flora of Diego Garcia, British Ocean Territory. Atoll Res Bull 313:1–21
Topp JMW (2003) Editor's note. In: Guzman N (2003) Wildlife on Diego Garcia. Chagos News 22:5–7
Topp JMW, Sheppard CRC (1999) Higher plants of the Chagos Archipelago. In: Sheppard CRC, Seaward MRD (eds) Ecology of the Chagos Archipelago. Linnean Society Occasional Publications 2. Westbury Publishing, Otley, W. Yorkshire, pp 225–240
Vine D (2004) Exile in the Indian Ocean: documenting the injuries of involuntary displacement. Doctoral Program in Anthropology, Graduate Center, City University of New York
Vogt S (2011) Population densities and demographics of the Coconut Crab, *Birgus latro*, on Diego Garcia, British Indian Ocean Territory. A report to NAVFACECFE, Yokuska
Watling R, Seaward MRD (2004) Some fungi of Indian Ocean Islands. Bot J Scotland 56:65–84
Willis JC, Stanley Gardiner J (1931) Flora of the Chagos Archipelago. Trans Linn Soc 19(2):301–306

Contaminants, Pollution and Potential Anthropogenic Impacts in Chagos/BIOT

21

James W. Readman, Francis DeLuna, Ralf Ebinghaus, AntenorNestor Guzman, Andrew R.G. Price, Emily E. Readman, Anne L.S. Sheppard, Victoria A. Sleight, Renate Sturm, Richard C. Thompson, Andrew Tonkin, Hendrik Wolschke, Robyn J. Wright, and Charles R.C. Sheppard

Introduction

This chapter reviews available data on a broad range of chemical contaminants and pollutants within the Chagos Archipelago. These have been measured during several past projects. It provides an evaluation of levels of pollution, and compares these levels with concentrations recorded in other regions of the world, showing that contamination levels are amongst the lowest in the world so far recorded. Whilst much of the data relate to information published in the open literature, this chapter includes details of the extensive pollution monitoring under the 'Final Governing Standards' (FGS) in the atoll Diego Garcia which contains a military facility. A description of what is routinely measured and the frequency of measurements is provided.

To summarise the data presented, hydrocarbon analyses reveal compounds primarily of a natural origin with negligible evidence of contamination from petroleum or combustion origins. Tar balls, however, have been reported on several beaches in the Archipelago. Analyses of faecal steroids provide negligible evidence of sewage contamination. The large group of compounds known as 'persistent organic pollutants' (POPs), including PCBs and pesticides, were generally below analytical detection limits, as were polyfluorinated compounds, brominated, chlorinated and organo-phosphorous flame retardants, fluorinated tensides, and surfactants (PFOS) .

Investigations into antifouling biocides and herbicides in Diego Garcia also revealed negligible contamination. Metal concentrations, too, are very low, although some elevated copper concentrations were recorded in 1996, probably originating from the historical fungicide treatments used in coconut agriculture.

Levels of most contaminants are typically comparable to those recorded in environments perceived to be pristine, for example, the Antarctic.

Concerning the extensive regulatory monitoring in Diego Garcia, which includes analyses by accredited US laboratories of over one hundred metals and organic contaminants, results generally reveal concentrations that are below the limits of detection, which is in agreement with the open literature surveys. These legislated assessments are designed to ensure both environmental and human health preservation. Whilst many detection limits are higher than those of the independent surveys, they generally confirm the pristine nature of the Archipelago.

Beach surveys have, however, revealed a surprisingly high number of pieces of debris throughout the Archipelago. Investigations indicate that these are mainly plastics of South East Asian origin. However, the number of litter pieces in Diego Garcia was less than in the other atolls, with reductions being attributed to sporadic beach clean-up events.

J.W. Readman (✉) • E.E. Readman
Plymouth Marine Laboratory, PL1 3DH, Plymouth, UK
e-mail: jwre@pml.ac.uk

F. DeLuna • A. Guzman
NAVFACFE PWD DG Environmental,
PSC 466 Box 5, FPO AP 96595-0005, Olongapo, USA
e-mail: Francis.DeLuna.RP@fe.navy.mil;
AntenorNestor.Guz.RP@fe.navy.mil

R. Ebinghaus • R. Sturm • H. Wolschke
Institute of Coastal Research, Department for Environmental Chemistry, Helmholtz-Zentrum Geesthacht, Max-Planck-Str. 1, 21502 Geesthacht, Germany
e-mail: ralf.ebinghaus@hzg.de; Renate.Sturm@hzg.de;
hendrik.wolschke@hzg.de

A.R.G. Price
Environment Department, York University, YO1 5DD, York, UK

School of Life Sciences, University of Warwick,
Gibbet Hill Campus, CV4 7AL, Coventry, UK
e-mail: andrew.price@warwick.ac.uk

A.L.S. Sheppard • C.R.C. Sheppard
School of Life Sciences, University of Warwick,
Gibbet Hill Campus, CV4 7AL, Coventry, UK
e-mail: a.sheppard@warwick.ac.uk; Charles.Sheppard@warwick.ac.uk

V.A. Sleight • R.C. Thompson • A. Tonkin • R.J. Wright
School of Marine Science and Engineering, Plymouth University,
PL4 8AA, Plymouth, UK
e-mail: R.C.Thompson@plymouth.ac.uk

C.R.C. Sheppard (ed.), *Coral Reefs of the United Kingdom Overseas Territories*, Coral Reefs of the World 4,
DOI 10.1007/978-94-007-5965-7_21, © Springer Science+Business Media Dordrecht 2013

Fig. 21.1 Great Chagos Bank, eastern side of Eagle Island. Collecting sediment samples at low tide for subsequent analysis of PFOS and related compounds. The top approx 1 cm of sediment was removed and placed in dichloromethane pre-cleaned aluminium containers. Samples were prepared for return by wrapping immediately in dichloromethane pre-cleaned aluminium foil and then freezing on immediate return to the ship (within 1 h)

Microplastic contamination in the Chagos Archipelago is shown to be both widespread and relatively high compared to other locations on a global scale. There were significantly more pieces at uninhabited sites compared to the inhabited Diego Garcia. The reason for this spatial pattern is not clear but indicate the potential for microplastics to accumulate in remote locations.

Holothurian (sea cucumber) poaching has been another significant environmental pressure on the coral reefs of Chagos and is included in this review, particularly in view of the reported ecological benefits of the group to reef health and resilience.

Concentrations of Potential Pollutants

The features of the archipelago described in earlier chapters show the geographical isolation and ecological significance of the Chagos archipelago. These features render Chagos of special interest with respect to 'baseline' measurements and the threat of any potential effects of pollutants (Sheppard et al. 2012). Prior to a 1996 scientific expedition, however, there were negligible data in the open literature concerning the extent of contamination of the archipelago (Readman et al. 1999; Everaarts et al. 1999). Sediment samples collected from Ile Diamant (Peros Banhos), Takamaka (Salomon Islands), Middle Brother (Great Chagos Bank) and Diego Garcia during the 1996 event were analysed for hydrocarbons, steroids, organochlorines and toxic metals at the IAEA Marine environment laboratory. Concurrent analytical intercomparison exercises to certify reference materials afforded stringent quality assurance. Subsequent expeditions have further investigated persistent organic pollutants (POPs) and contaminants potentially linked to the military base on Diego Garcia. Indeed, as mentioned previously, associated with this base is regular monitoring and, for the first time, we provide details of this monitoring programme.

In addition to the chemical contaminants, results from beach litter and tar ball surveys together with microplastic assessments are provided.

For the open literature surveys, sediment samples for analyses were collected from beaches at low tides (Figs. 21.1–21.2), prepared appropriately and, if freezing was required, frozen within one hour. In the case of water samples, these were extracted by concentration using relevant cartridges within the necessary time periods. The cartridges were then frozen and were maintained frozen throughout transportation to the analytical laboratories.

Oil and Combustion Products

Oils contain a complex mixture of hydrocarbons and combustion processes result in the formation of toxic polycyclic

Fig. 21.2 Sampling sites in northern atolls. (**a**): Great Chagos Bank – Middle Brother beach. A sample was collected nearby. (**b**): PerosBanhos – Ile du Coin beach

aromatic hydrocarbons. Analyses of the hydrocarbons in the sediment samples collected in 1996 (Readman et al. 1999) revealed a dominance of short-chain n-alkanes of odd carbon number, such as n-C_{17}, n-C_{15}, and short-chain alkenes such as heptadecene, nonadecene and acyclic C_{25} alkenes. These compounds are natural/biogenic and appear to have an exclusive autochthonous source, such as planktonic plants. There was negligible evidence of contamination from petroleum. For example, total alkane (Σn-C_{14-34}) concentrations were less than 0.1 $\mu g.g^{-1}$ dry sediment at all stations, including Diego Garcia. To place this into perspective, total alkanes in Antarctic sediments have been reported to vary from 0.3 to 85.1 $\mu g.g^{-1}$ dry sediment (Cripps 1994). An unresolved complex mixture (UCM) in the chromatograms resulting from analyses by gas chromatography is often taken as a measure of chronic oil contamination. It was only measurable in one sample that came from the inhabited Diego Garcia island at 1.9 μg.g^{-1} dry sediment. This compares, for example, to 0.3–4.8 μg.g^{-1} dry sediment for Crete, Greece and 10–104 μg.g^{-1} dry sediment for UK estuaries (Readman et al. 2002). Concerning polycyclic aromatic hydrocarbons, total concentrations did not exceed 20 ng.g^{-1} dry sediment with compositional patterns reflecting both pyrolitic and petrogenic sources. Again, for comparison, total concentrations ranging between 8 and 280 ng.g^{-1} dry sediment have been reported for the Antarctic (Cripps 1994).

Although oil slicks have not been reported in the Chagos region, tar balls were observed at eight beaches/islands throughout the Archipelago in 1996, at three in 2006 and were not evident in 2010. This decrease may reflect improved international ship ballast cleaning measures over that time period throughout the Indian Ocean (Price, unpublished data). No tar balls have been reported in Diego Garcia.

Sewage

The sediments collected in 1996 in Diego Garcia were also analysed for steroids (Readman et al. 1999). No evidence of sewage contamination, as would have been demonstrated by the presence of the faecal steroid coprostanol (5ß-cholestan-3ß-ol), was observed. Indeed, coprostanol was below the limit of detection (less than 1 ng.g^{-1} dry sediment) at all stations, including Diego Garcia. Desmethyl sterols [cholesterol (cholest-5-en-3ß-ol), campesterol (24α-methylcholest-5-en-3ß-ol), stigmasterol (24-ethylcholesta-5,22-dien-3ß-ol) and β-sitosterol (24α-ethylcholest-5-en-3ß-ol)] dominated all sediments examined. These are natural, and compositional ratios between the sterols are consistent with the input of organic matter derived primarily from planktonic or benthic algal sources, with a small terrestrial component. Details relating to sewage treatment in the populated Diego Garcia island are provided in a later section.

Organochlorines: Persistant Organic Pollutants (POPs)

These compounds gained notoriety through Rachel Carson's book "Silent Spring", published in 1962. This exposed the hazards associated with the pesticide DDT which can biomagnify and contaminate food chains, harming animals, particularly at the higher trophic levels, including humankind. POPs are organic (usually halogenated) compounds or mixtures that share four characteristics; high toxicity; persistence; potential for bioaccumulation; and ability for long-range transport. Examples include the pesticides lindane, chlordane, heptachlor, aldrin, dieldrin and endrin, and industrial polychlorinated biphenyls (PCBs) used in transformers and electrical components. In response to concerns relating to the protection of human health and that of the environment, the United Nations Stockholm Convention on POPs was adopted in 2001 and, following appropriate notification, became binding international law for those participating governments in 2004. The Convention seeks the elimination, or restriction of production and use, of all intentionally produced POPs (i.e. industrial chemicals and pesticides).

Sedimentary polychlorinated biphenyls (PCBs) and organochlorine pesticides in Chagos have been investigated (Readman et al. 1999; Table 21.1). Only some PCB congeners, lindane and DDMU (a metabolite of pp-DDT) were above the detection limits of the analytical technique (between 2 and 18 pg.g^{-1} dry weight, depending on the compound). Total PCB concentrations (160–250 pg.g^{-1} dry sediment) are much lower than those reported for deep and remote sediments, such as the Sargasso Sea (600 pg.g^{-1}, Fowler 1990) and the Mediterranean basin (2,000 pg.g^{-1}, Tolosa et al. 1995). The PCB congener distribution was dominated by penta and hexa-substituted species, approximately reflecting the Aroclor 1254 commercial mixture. The predominance of the lower chlorinated PCB congeners and lindane suggest atmospheric deposition as the main route of introduction for organochlorine compounds into the sediments. The pristine nature of Chagos with respect to organochlorine compounds is also reported by Everaarts et al. (1999).

Flame Retardants and Polyfluorinated Compounds

Whilst organochlorine POPs are of particular concern, other persistent compounds are emerging that also exhibit global ubiquity. Of these, flame retardants and polyfluorinated compounds (PFCs)[e.g. perfluorooctanesulfonate (PFOS), a synthetic fluorosurfactant used in abundance for many years as a fabric protector/stain repellent (Scotchgard®)] are important and have warranted inclusion in the Stockholm Convention. In addition, some can be linked closely with aviation, and hence to Diego Garcia. To investigate this issue, in 2010 coastal sediment samples were collected for analyses from Diego Garcia, (the inhabited atoll) and from selected uninhabited atolls and islands: Salomon Atoll, Ile Boddam; Peros Banhos, Ile du Coin; Great Chagos Bank, Middle Brother Lagoon; Great Chagos Bank, Eagle Island; Great Chagos, Bank Danger Island; and Egmont Atoll (see Fig. 17.1 in Chap. 17).

For the flame retardants, sediment samples were Soxhlet-extracted with dichloromethane. For quality control, the

Table 21.1 Concentrations of organochlorine compounds and lipid contents in sediment extracts (Taken from Readman et al. 1999)

Site		Diamante	Takamaka	Middle brother		
Atoll		Peros Banhos	Salomons	Chagos bank	Diego Garcia	
Dry/Wet ratio	–	0.81	0.76	0.80	0.80	0.82
Lipid contents	$mg.g^{-1}$	0.009	0.043	0.005	0.012	0.012
HCB	$pg.g^{-1}$	<2	<2	<2	<2	<2
Lindane	$pg.g^{-1}$	22	29	17	14	23
pp′ DDE	$pg.g^{-1}$	<10	<10	<10	<10	<10
pp′ DDD	$pg.g^{-1}$	<18	<18	<18	<18	<18
pp′ DDT	$pg.g^{-1}$	<17	<17	<17	<17	<17
DDMU	$pg.g^{-1}$	10	18	7	6	6
op DDE	$pg.g^{-1}$	<5	<5	<5	<5	<5
op DDD	$pg.g^{-1}$	<8	<8	<8	<8	<8
Heptachlor	$pg.g^{-1}$	<3	<3	<3	<3	<3
Aldrin	$pg.g^{-1}$	<5	<5	<5	<5	<5
Dieldrin	$pg.g^{-1}$	<6	<6	<6	<6	41
Endrin	$pg.g^{-1}$	<8	12	<8	<8	<8
Arochlor 1254	$pg.g^{-1}$	200	210	160	250	160
Arochlor 1260	$pg.g^{-1}$	<100	<100	<100	<100	<100
PCB 44	$pg.g^{-1}$	<9	<9	<9	<9	<9
PCB 52	$pg.g^{-1}$	28	28	24	22	24
PCB 87	$pg.g^{-1}$	<9	<9	<9	11	<9
PCB 101	$pg.g^{-1}$	18	15	15	26	15
PCB 105	$pg.g^{-1}$	<10	<10	<10	<10	<10
PCB 118	$pg.g^{-1}$	16	12	10	28	<6
PCB 128	$pg.g^{-1}$	<13	<13	<13	<13	<13
PCB 138	$pg.g^{-1}$	<13	<13	<13	<13	<13
PCB 149	$pg.g^{-1}$	12	9	7	14	<7
PCB 153	$pg.g^{-1}$	21	17	16	21	19
PCB 170	$pg.g^{-1}$	<21	<21	<21	<21	<21
PCB 180	$pg.g^{-1}$	<16	<16	<16	<16	<16
PCB 187	$pg.g^{-1}$	<12	<12	<12	<12	<12
PCB 200	$pg.g^{-1}$	<15	<15	<15	<15	<15

"<" indicates that the peak is less than three times the blank results

samples were spiked with a standard solution containing 10 mass labelled internal standards prior to extraction. The extracts were concentrated by rotary evaporation and purified/fractionated on a silica gel column. The brominated and chlorinated flame retardants were eluted using hexane followed by the organophosphorus compounds using dichloromethane/acetone. Analyses were performed using gas chromatography – mass spectrometry (GC-MS) in selective ion monitoring mode. For the determination and quantification of the brominated and chlorinated flame retardants, chemical ionisation was used.

For the polyfluorinated compounds, extraction and analyses were conducted according to International Council for the Exploration of the Sea (ICES) guidelines. Sediments, spiked with an internal standard mix of 50 mass labelled PFCs, were ultrasonically extracted with methanol. Extracts were cleaned using SupelcleanENVICarb cartridges. Concentrations of the PFCs in the samples were determined using high performance liquid chromatography with tandem mass spectrometer interfaced with an electrospray ionisation source in negative-ion mode (HPLC-(−)ESI-MS/MS). Appropriate blanks, recovery experiments and replicates were simultaneously processed. Results have been reported (Wolschke et al. 2011).

Of the brominated, chlorinated and organo-phosphorus flame retardants analysed, only Dechlorane Plus® (a polychlorinated flame retardant) was recorded above the limits of quantification. This compound occurred in eight of the 20 sediment samples analysed, albeit at low concentrations (≤ 38.4 $pg.g^{-1}$ dry sediment). Seven of the eight samples were from Diego Garcia with its associated military base, so elevated concentrations may not be surprising. The eighth sample was from Salomon Atoll, Ile Boddam, adjacent to a jetty/yacht anchorage. For comparison, concentrations of the

Table 21.2 Brominated and chlorinated flame retardants investigated and corresponding LOQs

Acronym	Name	CAS-No.	LOQ [pg/g dw]
BDE-28	2,4,4′-Tribromodiphenyl ether	41318-75-6	5
BDE-47	2,2′,4,4′-Tetrabromodiphenyl ether	5436-43-1	5
BDE-66	2,3′,4,4′-Tetrabromodiphenyl ether	84303-45-7	5
BDE-85	2,2′,3,4,4′-Pentabromodiphenylether		5
BDE-99	2,2′,4,4′,5-Pentabromodiphenylether	60348-60-9	5
BDE-100	2,2′,4,4′,6-Pentabromodiphenylether	189084-64-8	5
BDE-153	2,2′,4,4′,5,5′-Hexabromodiphenylether	68631-49-2	5
BDE-154	2,2′,4,4′,5,6′-Hexabromodiphenylether	207122-15-4	5
BDE-183	2,2′,3,4,4′,5′,6-Heptabromodiphenylether	207122-16-5	5
BDE-209	Decabromodiphenylether	1163-19-5	25
DPTE	2,3-dibromopropyl-2,4,6-tribromophenyl ether	35109-60-5	5
EHTBB	2-ethyl-1-hexyl 2,3,4,5-tetrabromobenzoate	183658-27-7	5
HBB	Hexabromobenzene	87-82-1	5
OBIND	Octabromotrimethylphenylindane		10
PBEB	Pentabromoethylbenzene	85-22-3	5
PBT	Pentabromotoluene	87-83-2	5
s-DP	syn-Dechlorane Plus ®	13560-89-9	1
a-DP	anti-Dechlorane Plus ®	13560-89-9	1
aCl11DP	Cl11 Dechlorane Plus ®		5
aCl10DP	Cl10 Dechlorane Plus ®		5
Dec 602	Dechlorane 602	31107-44-5	5
Dec 603	Dechlorane 603	13560-92-4	5
Dec 604	Dechlorane 604	34571-16-9	5

Table 21.3 Organophosphorous flame retardants investigated and corresponding LOQs

Acronym	Name	CAS number	Chemical formula	LOQ [pg/g dw]
TCEP	Tris(2-chloroethyl) phosphate	115-96-8	$C_6H_{12}Cl_3O_4P$	50
TCPP	Tris(1-chloro-2-propyl)phosphate	13674-84-5	$C_9H_{18}Cl_3O_4P$	250
TDCP	Tri (dichloroisopropyl) phosphate	13674-87-8	$C_9H_{15}Cl_6O_4P$	250
TMP	Trimethylphosphate	512-56-1	$C_3H_9O_4P$	250
TEP	Triethylphosphate	78-40-0	$C_6H_{15}O_4P$	10
TiPrP	Triisopropylphosphate	513-02-0	$C_9H_{21}O_4P$	10
TPrP	Tripropylphosphate	513-08-6	$C_9H_{21}O_4P$	10
TiBP	Triisobutylphosphate	126-71-6	$C_{12}H_{27}O_4P$	10
TBEP	Tris(2-butoxyethyl) phosphate	78-51-3	$C_{18}H_{39}O_7P$	50
TEHP	Tris(2-ethylhexyl) phosphate	78-42-2	$C_{24}H_{51}O_4P$	50
TPhP	Triphenylphosphate	115-86-6	$C_{18}H_{15}O_4P$	25
TCP	Tricresylphosphate	1330-78-5	$C_{21}H_{21}O_4P$	100

compound recorded in the Great Lakes (USA and Canada) range from 14 to 4,390 pg.g^{-1} dry sediment (Shen et al. 2010). A listing of the other flame retardants that were screened for, together with their analytical quantification limits are provided in Tables 21.2 and 21.3.

The polyfluorinated compounds (including fluorinated tensides and surfactants) are used for a variety of purposes, including surfactant coatings for clothing and carpets, as well as in paints, paper products, fire fighting foams, insecticide formulations and as processing aids for production of polytetrafluoroethylene (Teflon®). A comprehensive overview about poly- and perfluorinated compounds (PFCs) in the aquatic environment is given in a review by Ahrens (2011). Ionic polyfluorinated compounds were only detected in one of the twenty sediment samples. PFOS, PFHxS and PFOA, at low concentrations (2.4, 0.028 & 0.105 ng.g^{-1} dry weight, respectively), were recorded in a sample from Diego Garcia adjacent to a landfill and burn pit site. For comparison, ∑PFC concentrations of<LOQ to 85 ng.g^{-1} dry weight have been reported for Arctic Lakes in Canada (Stock et al. 2007).

Fig. 21.3 Sampling sites in Diego Garcia for antifouling analyses (From Guitart et al. 2007)

Herbicides and Antifouling Agents

Antifouling biocides on boats and ships provide a threat at very low concentrations, especially to the algal symbionts of corals. In 2006, replicate water samples were taken from fourteen coastal locations focussed around the Diego Garcia lagoon (Fig. 21.3) but also including oceanic reference sites. Following on-site extraction, the samples were analysed by GC-MS for the popular antifouling booster biocides Irgarol®1051, chlorothalonil, dichlorofluanid and Sea Nine 211®, together with triazine herbicides (atrazine, simazine and ametryn) (Guitart et al. 2007). Results revealed negligible contamination, with levels generally below the limit of detection (<1 ng L^{-1}). Only in two harbour samples (Stations 2 and 1) was an antifoulant (Irgarol®1051) detected, at concentrations of 8 and 2 ng L^{-1} respectively. With respect to the antifouling agents and herbicides analysed, it was considered that they pose no chemical threat to the coral communities. Further investigations were, however, suggested by Guitart et al. (2007) into which antifouling products/herbicides are used in the region.

Toxic Metals

Everaarts et al. (1999) analysed surface sediments and biota collected during the 1996 expedition. They reported that, in the sediments, concentrations of copper, zinc, cadmium, lead, chromium and nickel were exceptionally low. Analyses of the biota revealed that lead and chromium (non-essential elements) levels were also very low. Concentrations of cadmium in macrobenthic invertebrates were similar to those found in open ocean areas (5–32 $\mu g.g^{-1}$ dry weight). Copper and zinc concentrations were, however, elevated in hermit crabs and clams. The elevated copper concentrations probably originate from the historical fungicide treatments previously used in coconut agriculture. In bivalves and sea stars from Salomon atoll, nickel was also detected at elevated levels.

Contamination by Solid Waste

Shoreline Debris

Despite their near pristine status in terms of chemical contaminants, Chagos beaches accumulate surprisingly high densities of solid debris. Observations were made in 1996, 2006, and 2010 at 20 sites in the outer atolls, and one in Diego Garcia as part of rapid environmental assessments (Price 1999; Price and Harris 2009; Price et al., in prep.). Median levels of the number of litter pieces were high (score 4) in all years; this corresponds to 1,000–9,999 items (geometric mean 3,162) per terrestrial portion of a site inspection quadrat, i.e. 500 m (along the beach) to 250 m 'inland' from the shore. Items were mainly macroplastics, polystyrene (Styrofoam) and rope, much being lost fishing gear or debris discarded from ships, most commonly of south-east Asian origin. Levels in Diego Garcia in all years were two orders of magnitude less than in other atolls, reflecting periodic clean-up events in that inhabited atoll. The method did not determine size categories or weight; most items were a few cm in size or less, but several northern islands, which are uninhabited, appear to collect substantial volumes of larger flotsam. Similar numbers are found in remote Pacific atolls (Price and Harris 2009) where ocean current gyres are the main transport vector. Driftwood and lost timber from ships was low on beaches in all years, but decreased over time from 1996 to 2006, attributed to use for fuel by illegal fishing camps on the islands during this period of increasing fishing pressure (Price and Harris 2009; Price et al. 2010). While these are unsightly, they have the potential also to impede nesting turtles in some areas.

Microplastics

Plastic debris now contaminates marine habitats from the poles to the equator (Thompson et al. 2009; Cole et al. 2011). Whilst most attention has addressed debris items that are visible to the naked eye, attention is increasingly being focussed towards smaller particles termed microplastics (Thompson et al. 2004; Browne et al. 2011; Andrady 2011). Small fragments such as these have the potential to be ingested by a wide range of organisms (Browne et al. 2008; Andrady 2011). They can also accumulate and transport hydrophobic pollutants such as PAHs, PCBs, DDT, PBDEs (Teuten et al. 2007; Hirai et al. 2011). Thompson et al. (2004) and Browne et al. (2011) report widespread contamination of shorelines and the water column with microscopic plastics including brightly coloured granular and fibrous fragments.

Quantitative sampling for microplastics was undertaken using sediment collected from the low water mark at 20 sites in Chagos during 2010. Six samples were from remote, uninhabited atolls (Salomon, Peros Banhos, Great Chagos Bank and Egmont Atoll) while fourteen were from Diego Garcia. The method to extract microplastics followed that of Thompson et al. (2004). Sediment (50ml) was added to a saturated solution of NaCl, shaken vigorously for 30 s then allowed to settle for two minutes. The supernatant was then separated by filtration. Three sequential extractions were made for each sample. Particles were removed from filter papers and identified spectroscopically using a Bruker IFS 66 equipped with a Hyperion 1,000 IR microscope MCT detector and Specac DC 2 diamond compression cell. Identity was confirmed if samples achieved >70% match with reference spectra. Spectra with a match to reference of 60–70 % were examined individually and only included in the data where key features of synthetic polymers were apparent. Blank control samples were run at regular intervals to check for procedural contamination.

Synthetic polymers were found at all 20 sites (Table 21.4) including nylon, polyethylene, polyester, polypropylene and rayon, with an average size of 1.5 mm ± 1.6 mm (mean ± 1SD; range 30 μm – 4 mm). There was no significant difference in particle size between Diego Garcia and the northern atolls. The abundance of synthetic pieces was 4.55 ± 2.74 (mean ± 1SD) fragments per 50ml of sediment and was toward the upper end of the range reported from other locations worldwide (Browne et al. 2010, report a range of 0.4–6 pieces per 50 ml of sediment). Hence it is apparent that microplastic contamination in the Chagos Archipelago is both widespread and relatively high compared to other locations. There were some differences in relative abundance among sites with significantly more pieces at uninhabited sites compared to the inhabited military facility at Diego Garcia ($t_{2\text{-tailed}} = 2.46$, d.f. = 17, $P = 0.025$; Fig. 21.4) and the greatest number of synthetic pieces at the uninhabited Great Chagos Bank Eagle Island site. The reason for this spatial pattern is not clear but the results clearly indicate the potential for microplastics to accumulate in remote locations (Hirai et al. 2011; Browne et al. 2011).

Radioactive Contamination

A report (Ministry of Defence 2008) analysed sediments at 21 sites in Diego Lagoon. From the report: "Gamma spectroscopy analysis of underwater sediment did not detect any trace of cobalt-60. The measured gross gamma dose rates within the intertidal zone are indistinguishable from background levels at any point. Gamma spectroscopy analysis of sediment samples taken from the intertidal zones did not detect any trace of cobalt-60. It was therefore concluded that there has been no radiological hazard to any inhabitant of Diego Garcia as a result of the operation of nuclear powered submarines."

Table. 21.4 Number of synthetic fragments (i.e. microplastics) at each site, uninhabited sites indicated by *shading*

Sample Location	Number of synthetic fragments (per 50ml sediment)
Salomon Atoll	4
Peros Banhos Atoll	6
Great Chagos Bank Middle Brother	6
Great Chagos Bank Eagle Island	13
Great Chagos Bank Danger Island	3
Egmont Atoll	5
Diego Garcia	2
Diego Garcia (seaward east of atoll)	3
Diego Garcia (back of Barachois)	2
Diego Garcia (beach rock at edge of lagoon)	3
Diego Garcia (atoll seaward)	5
Diego Garcia (turtle cove)	2
Diego Garcia (turtle cove Barachois)	6
Diego Garcia (southern tip of atoll)	5
Diego Garcia (off landfill site)	1
Diego Garcia (end of runway)	3
Diego Garcia (entrance to small boat harbour)	9
Diego Garcia (lagoon beside accommodation blocks)	5
Diego Garcia (nearby pipe running into sea)	5
Diego Garcia (yacht club)	3

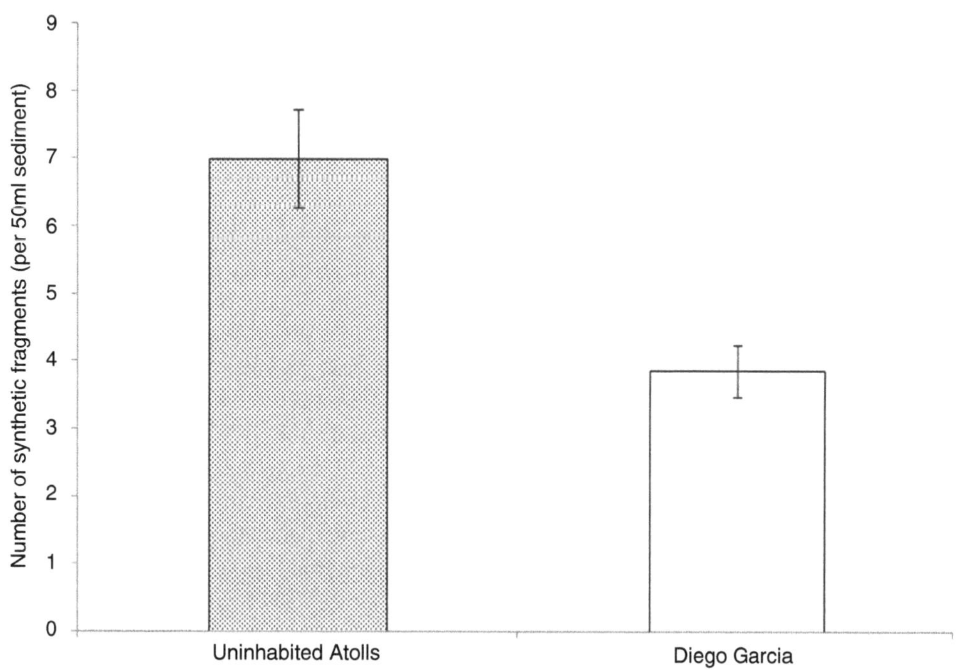

Fig. 21.4 Number of synthetic fragments (microplastics) extracted from intertidal sediment collected at sites in the Chagos Archipelago (Mean ± SE, Uninhabited Atoll N = 6, Diego Garcia N = 13)

In addition, several dosimeters are deployed around the atoll, especially in the vicinity of the port area. These are sent to the USA for analysis, and results are not available on the island.

Mandatory Monitoring (Diego Garcia)

As previously noted, Diego Garcia is the only inhabited atoll in Chagos and has a US Navy Support Facility on the western arm and some much smaller facilities extending to the far south of the western arm of the atoll. The population is commonly between 3,000 and 5,000 persons. The eastern arm is not inhabited but, like the rest of the atoll, it used to have coconut plantations throughout; now it is 'coconut chaos' (Chap. 20) and is part of a Strict Nature Reserve into which access requires permission. The lagoon is used for pre-positioning large ships, commonly numbering between six and a dozen.

As part of the US Navy operating procedures, the document *Final Governing Standards* prescribes practices of work and covers matters of environmental importance or consequences, particularly relating to pollution control.

Standard operating practices aim to reduce or eliminate pollutants and other environmental problems. Over the last 15 years intensive efforts have been made to reduce causes of pollution, and to treat waste appropriately according to US Navy regulations and which are compatible with and required by UK and USA standards. As appropriate, waste is either treated on island, or else is exported to waste treatment plants overseas for processing or recycling.

As part of these processes, analytical laboratories exist in Diego Garcia which carry out many analyses directly, while analyses for many other substances are sent to accredited laboratories overseas. In total, well over one hundred potential pollutants are analysed at regular intervals depending on substance. Analyses may be daily (even hourly in a few cases) to multi annually, with most being routinely analysed at weekly or monthly intervals. The Appendix tabulates the substances analysed in fresh water, drinking water, in leachates from various 'sumps' designed to collect runoff percolating through waste facilities, and from several other points of origin or discharge. In addition, monitoring is done of seawater in the lagoon (to monitor the anchored ships). Thousands of such analyses are performed annually (excluding a far greater number of analyses of the water supply, later section). The vast majority of measurements report values below detectable or reportable limits; when a reportable limit is observed, remedial action is taken. As part of the programme, for each sampling event, any or all of three types of blanks are used: trip blanks for detecting contamination during field handling, shipment, or in the laboratory; equipment blanks to determine the effectiveness of field decontamination procedures; and field blanks to determine potential contamination of water during field procedures. Much of the large volume of data generated is planned for electronic recording in future, but mostly exists on paper only at present.

In the past 15 years, reduction of causes and sources of pollution has been seen as being essential, and this course is now pursued strongly. For example, Fig. 21.5 shows a broad overview of the reduction in pesticide use over the last decade, resulting from focussed use and application, and the use of alternative methods. In addition, the entire process is subject to internal and external audits.

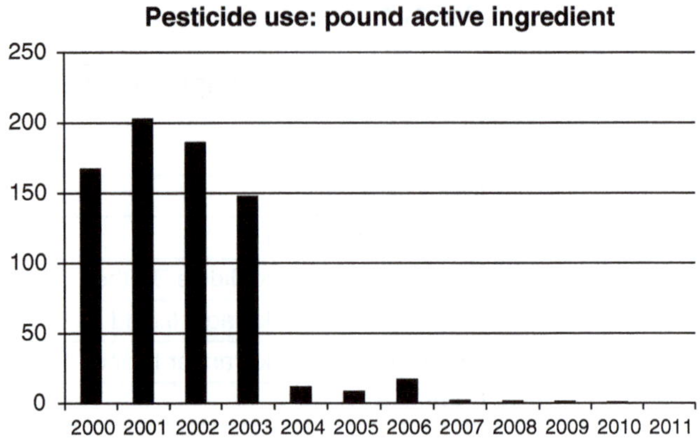

Fig. 21.5 Pesticide use in Diego Garcia. Units (Pounds) are those used in Diego Garcia. Values in 2010 and 2011 are 1.23 and 0.538 respectively

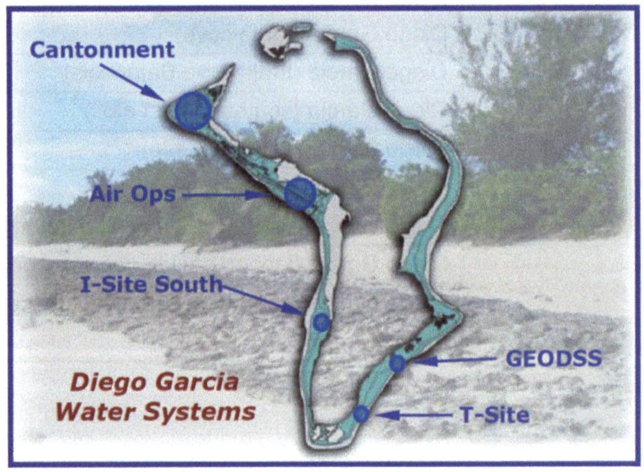

Fig. 21.6 Sites in Diego Garcia where fresh water is extracted from sub-surface water lenses

Fresh Water Supply

The fresh water supply comes almost entirely from five of the water lenses that lie beneath the surface of the island (Fig. 21.6). These accumulate rainwater which percolates into the lenses through the vegetation, during which time it collects organic substances including faecal pathogens from wildlife, especially rodents. Monitoring of freshwater is extensive and includes organic, inorganic and radioactive potential contaminants; in 2011, for example, tens of thousands of tests were conducted on the water supply (see Appendix). The aquifers are shallow and thin, but laterally extensive, and exist because of the relatively high rainfall in the central Indian Ocean. The upper surfaces of the lenses may lie slightly above sea level, while the lower surfaces lie below sea level. They are stable, and surplus input from rain results in freshwater being pushed sideways to seep out into the ocean or lagoon. Indeed, this is the reason for the permanence of the fresh water lenses. The water lenses are not simple underground pools, but each is composed of very many pockets inter-connected with fissures or porous rock; in fact much of the island below surface contains unconsolidated material. These aquifers therefore may be extracted by simple wells, or the wells may be 'horizontal wells' which traverse distances sub-surface though numerous fresh water pockets.

The lenses enable these islands to support lush vegetation. In fact most of the accumulated lens water is used by the vegetation. For example, the trees in a coconut plantation in Diego Garcia transpire over 80 % of the rainfall, hardwood trees about 70 %, while grassland and open spaces transpire about 60 % of total rainfall (Engineering Concepts Inc 2010). Thus only a relatively small proportion of rainfall is available after plant transpiration for extraction and use by a human population. Any over-extraction beyond certain limits would therefore be to the detriment of vegetation and could rapidly become a critical issue (the consequences of over-extraction can be seen, for example, by the highly sulphurous lens beneath the capital of the Maldives to the north, where seawater desalination has become necessary to provide sufficient fresh water). Thus content of e.g. electrolytes is analysed at each point of extraction on a frequent basis (see Appendix), and wells are switched and pumped as necessary.

In Diego Garcia, a water system is installed at or near each aquifer which treats water before distribution. Initially, water from aquifers is aerated to remove hydrogen sulphide gas, is then disinfected with chlorine gas or hypochlorite solution, and is supplied to the distribution systems as non-potable tap water. Thus, in addition to pollutants, monitoring of solutes and electrolytes is needed also to ensure that water is not over-extracted, leading to sea water incursions, sulphur formation and general water lens failure.

Chlorination, used to disinfect the groundwater, can itself create problems as it creates halogenated organic compounds such as trihalomethanes when chlorine reacts with the organic compounds in the water. Because of this, reverse osmosis (nanofiltration) as an additional treatment is applied to some of the extracted water to make it of a potable standard, and this is made available at numerous drinking water outlets. A new project is upgrading the entire system so that all supplied water will be potable in the future.

Sewage Treatment

The sewer system in Diego Garcia consists of separate wastewater collection and treatment systems. The main wastewater treatment systems have gravity sewers, pump stations, force mains, treatment lagoons and ocean outfalls. Sewage and wastewater come from domestic, commercial and industrial sources. These are processed and treated though facultative lagoon systems for biological treatment and are disinfected by chlorine prior to disposal to the ocean.

Remote sites have individual wastewater treatment systems with gravity sewers, septic tanks and leach fields. The majority of septic tanks receive only domestic wastewater. Also, wastewater holding tanks service a few isolated buildings. Wastewater from the holding tanks is pumped on regular basis and are discharged not to the ocean but to the main wastewater treatment systems.

Environmental compliance criteria for wastewater are defined in Chap. 4 of Final Governing Standards. Effluent limitations for biochemical oxygen demand (BOD5), total suspended solids (TSS) and pH are set as primary criteria. However, other characteristics like visual and odour, dissolved oxygen (DO), residual chlorine and volume are also measured to ensure efficiency of wastewater treatment processes.

Outfalls are located on the western ocean side of the atoll, and waste is discharged into an area close to depths of several hundred metres and where currents are moderately strong. Regular inspections invariably have shown negligible green algal growth, and no fouling on the beaches at these locations.

Air Pollution

Normal pollutants resulting from automotive and aircraft emissions are believed to be negligible owing to the low volume of traffic. All motorized equipment and vehicles use only lead-free fuel.

Major sources of air emissions include steam boilers, power plants and incinerators. Operating designs of these facilities fully comply with the Final Governing Standards and air monitoring is not required.

Open burning is prohibited, except for infrequent burning of bulky wood items such as construction debris, furniture or debris from emergency clean-up operations. Permission to operate the open burn pit is required from the BIOT authorities.

Holothurian (Sea Cucumber) Poaching

Besides loss of an important natural resource in Chagos, concern arises over potential harmful ecological effects of poaching holothurians (shark poaching is discussed in

Chap. 19). Being largely detritus feeders, holothurians play an important role in the recycling system of sedimentary habitats (Uthicke et al. 2004), including sandy banks and lagoons of coral reefs; they 'condition' the substratum. Further, as Michio et al. (2003) note: "Commercially fished holothurians have important functions in nutrient recycling, which increases the benthic productivity of coral reef ecosystems. Thus, removal of these animals through fishing may reduce the overall productivity of affected coral reefs." Holothurians likely play a pivotal role in maintaining ecosystem integrity and resilience of coral reef systems.

Holothurians are particularly susceptible to overfishing. Evidence of heavy poaching in Chagos is substantial. Photographs of part of a haul comprising an estimated 5,000–7,000 holothurians on Eagle Island, a Strict Nature Reserve (Spalding 2006) is one example. Price et al. (2010) report significantly higher populations on the populated atoll of Diego Garcia, but where exploitation is absent, than on the uninhabited outer atolls where poachers, particularly from Sri Lanka, visit (Fig. 21.7). Significant reduction in total holothurian abundance (all species) has been observed in Salomon atoll between 2006 (2281 individuals) and 2010 (1661 individuals) within a large transect 18.8 km x 4 m encircling Salomon atoll (Price et al. in preparation). Fishing occurrence showed significant negative association with holothurian abundance for these and other datasets.

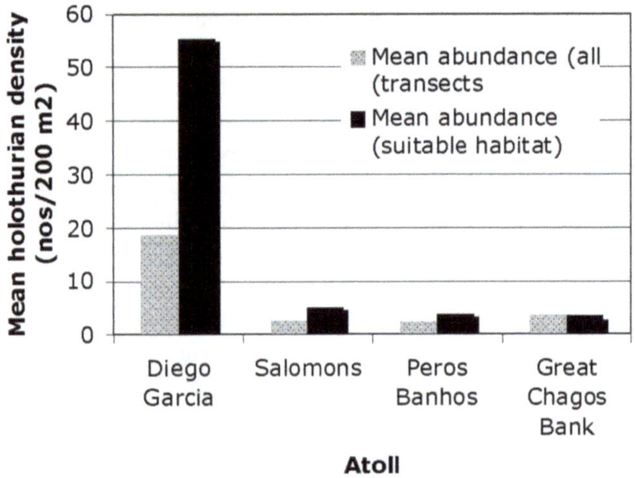

Fig. 21.7 Mean abundance of holothurians on Diego Garcia (21 transects), Salomon (27 transects), Peros Banhos (16 transects) and Great Chagos Bank (8 transects) determined from visual censuses along seventy-two 100 m × 2 m (200 m^2) transects; data shown for all transects and for transects with suitable habitat/ecology, i.e. where holothurians were present (From Price et al. 2010)

Sea Levels, Island Prospects and Shoreline Erosion

Though tiny in total area compared with marine habitats, the islands are central to conservation measures (Chap. 20). Much has been discussed recently about loss of coral islands from climate change; indeed the nearest archipelago, the Maldives, has taken a lead amongst small nations in this matter. There has been debate about the importance of sea level rise in Chagos, and Sheppard et al. (2012) discussed this briefly, but recent data have resulted in a marked predicted increase in global mean rise, such that estimates today are predicted to be within the range 0.5–1.9 m by the end of this century (Vermeer and Rahmstorf 2009; Nichols et al. 2011) which alters previously forecast scenarios. Indeed, many Pacific groups have measured rises of up to 12 mm y^{-1} already (Becker et al. 2012; Meyssignac et al. 2012). Increased predicted rises apply to Chagos also. Healthy reefs provide a breakwater, so an important immediate consideration if predicted mass mortalities re-occur is erosion of shores and inundation of land by the sea, rather than effects of sea level rise per se on coral reefs. Chagos has already suffered considerable overall erosion of many shorelines, most notably in the last decade, and while this has not yet been studied in detail, in Diego Garcia it is of concern because of the tens of millions of dollars required for the sea defence measures currently required there. On several islands, remnants of mature coconut trees are now in the intertidal zone, and pits dug for coconut cultivation are now at or below high tide levels in some places, all indicating net erosion over the past 100–150 years in such places. Island heights and cross-sections are likely

Fig. 21.8 Flooded part of Diego Garcia atoll at high spring tide, March 2012. The shark is a lemon shark, water is approx 50 cm deep

to be important to future events. The highest points of probably all islands occur on the rims, and profiles of all Chagos islands so far measured show a central depression, caused by acidic rain eroding the coral limestone rock (Sheppard 2002), and by accumulation of sand around island rims. Future erosion patterns are unlikely to be smoothly progressive, and may be episodic when a broach occurs. Sea level rise may not even be the most important issue in any case as any future coral mortality events from, for example, warming, may adversely reduce the 'breakwater' effect of coral reefs. There is evidence of episodic broaching in several locations, and a broach could eventually destroy the freshwater lens in that area. In Diego Garcia atoll, for example, several broaches occurred in 2012, such that extensive flooding occurred at several consecutive high spring tides, leading to striking episodes of inundation with lagoon fish and sharks appearing over the island (Fig. 21.8).

Summary

From a chemical contaminant perspective, the marine environment surrounding the Chagos archipelago can be considered as pristine. It is as uncontaminated as sites measured in e.g. the Antarctic or Sargasso Sea. In this respect, therefore, it provides a useful global reference site. There is evidence of holothurian poaching which could have ecological consequences.

Appendix

Pollutants and potential pollutants analysed routinely in Diego Garcia. Left column: substance. Middle column: locations (numbers in brackets indicate number in each type of location). Right column, frequency of analyses.

Monitoring by Labs in Diego Garcia		
1 Water quality monitoring		
Bacteriological: total and fecal coliform	Raw water storage tanks (5), Potable water storage tanks (2), Potable water distribution systems – various locations (13), End of non-potable water distribution systems (10), Entry to non-potable water distribution systems (4) granulated activated carbon at air ops water treatment system -influent and effluent streams (6)	Weekly
Calcium hardness	Potable water storage tanks (2), Raw water for nanofiltration systems (2), Product water from nanofiltration systems (2)	Weekly
Chloride	Groundwater monitoring wells (84)	Monthly
	Raw water supply wells (104)	Weekly
	Entry to potable water distribution systems (2), Entry to non-potable water distribution systems (5)	Daily
Chlorine residual	Potable water distribution systems (65)	Daily
	Product water from nanofiltration systems (2)	Hourly
	Non-potable water distribution systems (20)	Daily
Nitrates and nitrites	Entry to potable water distribution systems (2)	Annually
	Entry to non-potable water distribution system (5)	
pH	Potable water storage tanks (2)	Daily
	Non-potable water distribution systems (5)	Weekly
Turbidity	Product water from nanofiltration systems (2)	Daily
	Raw water storage tanks (5)	Daily
Conductivity	Groundwater monitoring wells (84)	Monthly
	Raw water supply wells – P-1 (69)	Weekly
	Raw water supply wells – others and modules (136)	4x weekly
	Potable water storage tanks (2)	Daily
	Non-potable water storage tanks (3)	Daily
	Entry to potable water distribution systems (2)	Daily
	Entry to non-potable water distribution systems (5)	Daily
2 Wastewater monitoring		
Biochemical oxygen demand, BOD_5	Air-ops wastewater treatment facility – influent and effluent streams (2)	
	R-site/cantonment wastewater treatment facility – influent and effluent streams (2)	Weekly

(continued)

(continued)

Total suspended solids, TSS	Air-ops wastewater treatment facility – influent and effluent streams (2)	Daily
	R-site/cantonment wastewater treatment facility – influent and effluent streams (2)	
pH	Air-ops facility – sewage lagoons (4)	Daily
	R-site/cantonment facility – sewage lagoons (4)	
Dissolved oxygen, DO	Air-ops wastewater treatment facility – sewage lagoons (4)	Daily
Observation: visual and odor	R-site/cantonment wastewater treatment facility – sewage lagoons (4)	
Chlorine residual	Air-ops wastewater treatment facility – contact chamber (1)	Daily
Effluent discharge volume	R-site/cantonment wastewater treatment facility – contact chamber (1)	
3 Diego Garcia Lagoon Seawater Monitoring – Ships in Lagoon (Approximately 10 ships per month)		
Bacteriological: total and fecal coliform	One each from the seawater directly located nearest to the discharge port and 5 m away from the discharge port downwind (2)	Monthly
	Baseline – approximately center of all ships berthed inside DG lagoon (1)	
Physical and chemical: temperature, pH, Ammonia, Nitrites, Nitrates, Phosphates, Alkalinity, Dissolved Oxygen	One each from the seawater directly located nearest to the discharge port and 5 m away from the discharge port downwind (2)	Monthly
	Baseline – approximately center of all ships berthed inside DG lagoon (1)	
4 Solid Waste Management Center – Groundwater and Leachate monitoring		
Physical and chemical: temperature, Conductivity, TSS, pH, BOD_5	SWMC groundwater monitoring wells (6)	Monthly
	SWMC leachate pond (1)	
Inorganic nonmetallic: Ammonia, Nitrate, Chloride, Sulfate	SWMC Groundwater monitoring wells (6)	Semi-annual
	SWMC Leachate pond (1)	
ANALYSES OF SAMPLES SENT OFF-ISLAND, by certified labs of US EPA or US State with primacy of water program		
5 Water quality monitoring		
Dieldrin	Groundwater monitoring wells (8)	Annually
Not a requirement, but monitored to ensure it is < 0.01 ppb, the maximum contaminant level set by U.S. EPA	End of non-potable water distribution system (8)	Annually
	Granulated activated carbon at air ops water treatment system – influent and effluent streams (6)	Quarterly
Total Trihalomethanes, TTHM, Haloacetic Acid, HAA5	Potable water storage tanks (2)	Quarterly
	End of potable water distribution systems (5)	
	End of non-potable water distribution systems (4)	
	Laboratory distilled water – blank sample (1)	
Lead, Copper	End of potable water distribution systems (17)	Annually
	End of non-potable water distribution systems (5)	
Asbestos	Entry to potable water distribution systems (2)	Every 9 years
	Entry to non-potable water distribution system (5)	
Corrosivity	Entry to potable water distribution systems (2)	Once
	Entry to non-potable water distribution system (5)	
Radionuclides: Gross Alpha, Gross Beta, Combined Radium-226 and Radium-228	Representative sampling point for potable water distribution systems (2)	Every 4 years
	Representative sampling point for non-potable water distribution systems (5)	
Antimony, Arsenic, Barium, Beryllium, Cadmium, Chromium, Cyanide, Fluoride, Mercury, Nickel, Selenium, Sodium, Thallium	Point of entry to potable water distribution systems (2)	Annually
	Point of entry to non-potable water distribution system (5)	

(continued)

(continued)

Synthetic Organic Chemicals, Volatile: Benzene, Carbon tetrachloride, o-Dichlorobenzene, cis-1, 2-, Dichloroethylene, trans-1, 2-Dichloroethylene, 1,1-, Dichloroethylene, 1,1,1-Trichloroethane, 1, 2-, Dichloroethane, Dichloromethane, 1,1,2-, Trichloroethane, 1,2,4-Trichlorobenzene, 1, 2-, Dichloropropane, Ethylbezene, Monochlorobenzene, para-Dichlorobenzene, Styrene, Tetrachloroethylene, Trichloroethylene, Toluene, Vinyl chloride, Xylene (total), Acrylamide and Epihydrochlorin	Entry to potable water distribution systems (2)	Every 3 years
	Potable water distribution distribution system (5)	
Synthetic Organic Chemicals, Pesticides/PCBs: Alachlor, Aldicarb, Aldicarb sulfone, Aldicarb sulfoxide, Atrazine, Carbofuran, Chlordane, 2, 4-D, 1, 2-Dibromo-3-Chloropropane (DBCP), Endrin, Ehtylene dibromide(EDB), Heptachlor, Heptchlorepoxide, Hexachlorocyclopentadiene, Lindane, Methoxychlor, PCBs (as decachlorobiphenyls), Pentachlorophenol, Toxaphene, 2, 4, 5-TP (Silvex), Benzo[a]pyrene, Dalapon, Di-(2-ethyhexyl) adipate, Di-(2-ethyhexyl), phthalate, Dinoseb, Diquat, Endothal, Glyphosphate, Hexachlorobenzene, Oxamyl (Vydate), Picloram, Simazine and 2,3,7,8-TCDD (Dioxin)	Entry to potable water distribution systems (2)	Every 3 years
	Entry to non-potable distribution system (5)	
6 Solid Waste: Ash, Groundwater and Leachate quality monitoring		
Volatile Organics: Acetone (2-Propanone), Acrylonitrile, Benzene, Bromochloromethane, Bromodichloromethane, Bromoform, Carbon Disulfide, Carbon tetrachloride, Chlorobenzene, Chloroethane, Chloroform, Dibromochloromethane, DBCP, EDB, o-Dichlorobenzene, p-Dichlorobenzene, trans 1,4-Dichloro-2-butene, 1,1-Dichloroethane, 1,2-Dichloroethane, 1,1-Dichloroethylene; 1,1-Dichloroethene, Cis-1,2-Dichloroethylene, trans-1,2-Dichloroethylene, 1,2-Dichloropropane, Cis-1,3-Dichloropropene, trans-1,3-dichloropropene, Ethylbenzene, 2-Hexanone, Methyl bromide, Methyl chloride, Methylene bromide, Methylene chloride, MEK, Methyl iodide, 4-methyl-2-pentanone, Styrene, 1,1,1,2-Tetrachloroethane, 1,1,2,2-Tetrachloroethane, Tetrachloroethylene, Toluene, 1,1,1-Trichloroethane, 1,1,2-Trichlorethane, Trichloroethylene, CFC-11, 1,2,3-Trichloropropane, Vinyl acetate, Vinyl chloride, Xylenes	Groundwater monitoring wells (6)	Semi annual
	SWMC Leachate pond (1)	
Inorganics: Antimony, Arsenic, Barium, Beryllium, Cadmium, Chromium, Cobalt, Copper, Lead, Nickel, Selenium, Silver, Thallium, Vanadium, Zinc	SWMC Groundwater monitoring wells (6)	Semi annual
	SWMC Leachate pond (1)	
Physical: reactivity (Reactive Cyanide and Sulfide), Ignitability, Corrosivity, Toxicity	SWMC Leachate pond (1)	Semi annual
Inorganics (Toxicity Characteristic Leaching Procedure): Arsenic, Barium, Cadmium, Chromium, Lead, Selenium, Silver, Mercury	SWMC Ash – one each from ash bin and ash as landfill cover material (2)	Semi-annual

Acknowledgements Some of this material was published previously in Sheppard et al. (2012) and is reproduced with permission of Wiley & Sons.

References

Ahrens L (2011) Polyfluoroalkyl compounds in the aquatic environment: a review of their occurrence and fate. J Environ Monit 13(1):20–31

Andrady AL (2011) Microplastics in the marine environment. Mar Pollut Bull 62:1596–1605

Becker M, Meyssignac B, Letetrel C, Llovel W, Cazenave A, Delcroix T (2012) Sea level variations at tropical pacific islands since 1950. Global Planet Change 80–81:85–98

Browne MA, Dissanayake A, Galloway TS, Lowe DM, Thompson RC (2008) Ingested microscopic plastic translocates to the circulatory system of the mussel, *Mytilus edulis* (L.). Environ Sci Technol 42:5026–5031

Browne MA, Galloway TS, Thompson RC (2010) Spatial patterns of plastic debris along Estuarine shorelines. Environ Sci Technol 44:3404–3409

Browne MA, Crump P, Niven SJ, Teuten E, Tonkin A, Galloway T, Thompson R (2011) Accumulation of microplastic on shorelines worldwide: sources and sinks. Environ Sci Technol 45:9175–9179

Cole M, Lindeque P, Halsband C, Galloway TS (2011) Microplastics as contaminants in the marine environment: a review. Mar Pollut Bull 62:2588–2597

Cripps GC (1994) Hydrocarbons in the Antarctic marine environment: monitoring and background. Int J Environ Anal Chem 55:3–13

Everaarts JM, Booij K, Fisher CV, Maas YEM, Nieuwenhuize J (1999) Assessment of the environmental health of the Chagos archipelago. Linn Soc Occas Publ 2:305–326

Fowler SW (1990) PCBs in the environment: the Mediterranean marine ecosystem. In: Waid JS (ed) PCBs and the environment, vol III. CRS Press, Boca Ratón, pp 209–239

Guitart C, Sheppard A, Frickers T, Price ARG, Readman JW (2007) Negligible risks to corals from antifouling booster biocides and triazine herbicides in coastal waters of the Chagos archipelago. Mar Pollut Bull 54:226–232

Hirai H, Takada H, Ogata Y, Yamashita R, Mizukawa K, Saha M, Kwan C, Moore C, Gray H, Laursen D, Zettler ER, Farrington JW, Reddy CM, Peacock EE, Ward MW (2011) Organic micropollutants in marine plastics debris from the open ocean and remote and urban beaches. Mar Pollut Bull 62:1683–1692

Meyssignac B, Becker M, Llovel W, Cazenave A (2012) An assessment of two-dimensional past sea level recontructions over 1950–2009 based on tide-gauge data and different input sea level grids. Surv Geophys 33:945–972. doi:10.1007/s10712-011-9171-x

Michio K, Kengoa K, Yasunoria K, Hitoshia M, Takayukia Y, Hideakia Y, Hiroshib S (2003) Effects of deposit feeder *Stichopus japonicus* on algal bloom and organic matter contents of bottom sediments of the enclosed sea. Mar Pollut Bull 47:118–125

Ministry of Defence (2008) Marine environmental radioactivity surveys at nuclear submarine berths 2006. Ministry of Defence, London, Document Identifier: DSTL/PUB27329, pp 81–85

Nichols RJ, Marinova N, Lowe JA, Brown S, Vellinga P, de Gusmao D, Hinkel J, Tol RS (2011) Sea level rise and its possible impacts given a 'beyond 4oC world' in the twenty first century. Phil Trans R Soc A 369:161–181

Price ARG (1999) Broadscale coastal environmental assessment of the Chagos Archipelago. In: Sheppard CRC, Seaward MRD (eds) Ecology of the Chagos Archipelago. Westbury Publishing/Linnean Society, London, pp 285–296

Price ARG, Harris A (2009) Decadal changes (1996–2006) in coastal ecosystems of the Chagos Archipelago determined from rapid assessment. Aquat Conserv Mar Freshwater Ecosyst 19:637–644

Price ARG, Harris A, McGowan A, Venkatachalam AJ, Sheppard CRC (2010) Chagos feels the pinch: assessment of holothurian (sea cucumber) abundance, illegal harvesting and conservation prospects in British Indian Ocean Territory. Aquat Conserv Mar Freshwater Ecosyst 20:117–126

Readman JW, Tolosa I, Bartocci J, Cattini C, Price ARG, Jolliffe A (1999) Contaminant levels and the use of molecular organic markers to characterize the coastal environment of the Chagos archipelago. Linn Soc Occas Publ 2:297–304

Readman JW, Fillmann G, Tolosa I, Bartocci J, Villeneuve J-P, Cattini C, Mee LD (2002) Petroleum and PAH contamination of the Black Sea. Mar Pollut Bull 44:48–62

Shen L, Reiner EJ, Macpherson KA, Kolic TM, Sverko E, Helm PA, Bhavsar SP, Brindle ID, Marvin CH (2010) Identification and screening analysis of halogenated norhornene flame retardants in the Laurentian Great Lakes: Dechloranes 602, 603, and 604. Environ Sci Technol 44(2):760–766

Sheppard CRC (2002) Island elevations, reef condition and sea level rise in atolls of Chagos, British Indian Ocean Territory. In: Linden O, Souter D, Wilhelmsson D, Obura D (eds) Cordio report 2002. Kalmar University, Kalmar, pp 202–211

Sheppard CRC, Ateweberhan M, Bowen BW, Carr P, Chen CA, Clubbe C, Craig MT, Ebinghaus R, Eble J, Fitzsimmons H, Gaither MR, Gan C-H, Gollock M, Guzman A, Graham NAJ, Harris A, Jones R, Keshavmurthy S, Koldewey H, Lundini CG, Mortimer JA, Obura D, Pfeiffer M, Price ARG, Purkis S, Raines P, Readman JW, Riegl B, Rogers A, Schleyer M, Seaward MRD, Sheppard ALS, Tamelander J, Turner JR, Visram S, Vogler C, Vogt S, Yang JM-C, Yang S-Y, Readman JW, Riegl B, Rogers A, Schleyer M, Seaward MRD, Sheppard ALS, Tamelander J, Turner JR, Visram S, Vogler C, Vogt S, Yang JM-C, Yang S-Y, Yesson C (2012) Reefs and islands of the Chagos Archipelago, Indian Ocean: why it is the world's largest no-take marine protected area. Aquat Conserv Mar Freshwat Ecosyst. doi:DOI: 10.1002/aqc.1248, Published online in Wiley Online Library (wileyonlinelibrary.com)

Engineering Concepts Inc and Shimabukuro, Endo & Yoshizaki Inc (2010) Sustainable yield and maximum safe water production study. US Navy Support Facility, Diego Garcia, BIOT. Prepared for Department of the Navy, Hawaii, 242pp

Spalding MD (2006) Illegal sea cucumber fisheries in the Chagos archipelago. SPC Bêche-de-Mer Inf Bull 23:32–34

Stock NL, Furdui VI, Muir DCG, Mabury SA (2007) Perfluoroalkyl contaminants in the Canadian arctic: evidence of atmospheric transport and local contamination. Environ Sci Technol 41(10):3529–3536

Teuten EL, Rowland SJ, Galloway TS, Thompson RC (2007) Potential for plastics to transport hydrophobic contaminants. Environ Sci Technol 41:7759–7764

Thompson RC, Olsen Y, Mitchell RP, Davis A, Rowland SJ, John AWG, McGonigle D, Russell AE (2004) Lost at Sea: where is all the plastic? Science 304:838

Thompson RC, Moore C, vom Saal FS, Swan SH (2009) Plastics, the environment and human health: current consensus and future trends. Philoso Trans R Soc B 364:2153–2166

Tolosa I, Bayona JM, Albaiges J (1995) Spatial and temporal distribution, fluxes, and budgets of organochlorinated compounds in northwest Mediterranean sediments. Environ Sci Technol 29:2519–2527

Uthicke S, Welch D, Benzie JAH (2004) Slow growth and lack of recovery in overfished holothurians on the great barrier reef: evidence from DNA fingerprints and repeated large-scale surveys. Conserv Biol 18:1395–1404

Vermeer M, Rahmstorf S (2009) Global sea level linked to global temperature. Proc Natl Acad Sci 106:21527–21532

Wolschke H, Sturm R, Ebinghaus R (2011) Analysis of flame retardants and polyfluorinated compounds in sediment samples from the Chagos Archipelago. Report to the University Of Warwick and Save Our Seas Foundation. pp 18

Coral Reefs of the Pitcairn Islands

22

Robert Irving and Terry Dawson

Introduction

The Pitcairn Islands (Fig. 22.1) are located approximately equidistant between New Zealand and Peru and are some of the remotest islands on Earth. They lie at the eastern end of the Tuamotu Archipelago in the South Pacific Ocean and are the most south-easterly islands of the Pacific tectonic plate and Indo-Pacific province. Their nearest neighbours are the Gambier Islands group in French Polynesia, which lie 540 km west-northwest. The islands are a British Overseas Territory and comprise Pitcairn Island (the only inhabited island of the group), Henderson Island (the largest), and the two atolls of Ducie and Oeno. There is little reason to group the islands together other than that of geographical proximity (they happen to be neighbouring islands although the outliers, Oeno and Ducie, lie approximately 560 km apart) and political expediency (they were all claimed by Britain during the nineteenth century).

As a result of their isolation, the nearshore environments of the Pitcairn Islands harbour an intriguing array of habitats and species. However, the biodiversity of their reefs in terms of coral, fish and invertebrate species richness is low, which may be attributed to their isolation and low latitude (between 24° and 25°S). Species endemism is also low (at less than 2%).

A Brief History of the Area

It is believed that the islands of Pitcairn and Henderson were first colonized by Polynesians in about AD 900 and their occupation lasted until about 1450 (Weisler 1995). In 1606, the Portuguese explorer Fernández de Quirós was the first European to discover Henderson Island (by now uninhabited) and Ducie Island, though he did not encounter Pitcairn or Oeno. British interest in the islands (and in Pitcairn in particular) began more than a century and a half later in 1767 when HMS *Swallow*, under the command of Captain Philip Carteret, encountered Pitcairn and mistakenly plotted its position 188 nautical miles west of its actual location. This mistake, however, was to be of great benefit to the island's next inhabitants, the mutineers from HMS *Bounty*, who were looking for a safe island hideaway in 1790, remaining undiscovered there for a further 18 years. Pitcairn Island officially became a British dependency on 29 November 1838. The other three uninhabited islands (Henderson, Oeno and Ducie) were annexed by Great Britain in 1902 and were included in the dependency in 1938. The Pitcairn Islands remain the only UK Overseas Territory in the Pacific.

History of Research

It has not always been possible for visiting research expeditions to visit all four islands, even though they might well have done so had time allowed or if the elements were in their favour. Initial collections of biological material were often undertaken on an *ad hoc* basis. The first truly scientific studies (including the collection of various marine shells) in the Pitcairn Islands were under the guidance of Capt. F. W. Beechey on board HMS *Blossom* in 1825. He was also responsible for the first full description of Ducie atoll (Rehder and Randall 1975). Since that time there have been several individual visits and four major expeditions to the four Pitcairn Islands which included marine studies – the 1970–1971 Westward expedition (Rehder and Randall 1975; Randall 1973, 1978, 1999; Rehder 1974); the 1987 Smithsonian Expedition (Paulay and Spencer 1989; Paulay 1989); the 1991/1992 Sir Peter Scott Commemorative Expedition to the Pitcairn Islands (Benton and Spencer 1995a; Irving 1995; Irving et al. 1995); and, most recently, the 2012 National Geographical Society Pristine Seas Expedition to the Pitcairn Islands.

R. Irving (✉)
Sea-Scope Marine Environmental Consultants, Devon, England (UK)
e-mail: Robert@sea-scope.co.uk

T. Dawson
School of the Environment, University of Dundee,
Dundee, DD1 4HN, Scotland (UK)
e-mail: t.p.dawson@dundee.ac.uk

C.R.C. Sheppard (ed.), *Coral Reefs of the United Kingdom Overseas Territories*, Coral Reefs of the World 4,
DOI 10.1007/978-94-007-5965-7_22,

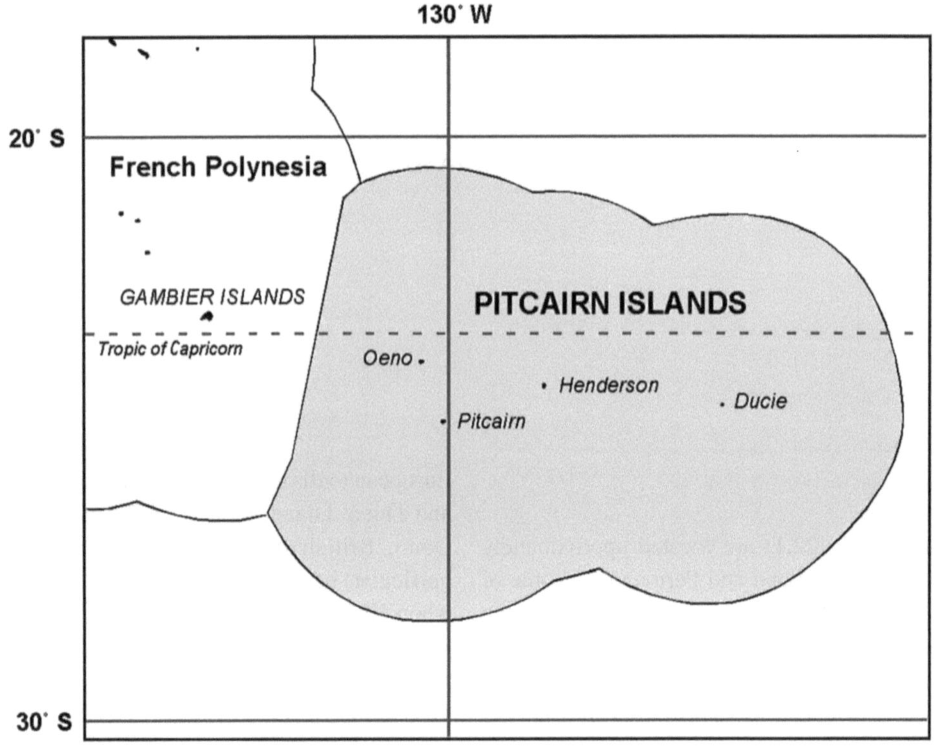

Fig. 22.1 The location of the four islands in the Pitcairn Islands group and their Exclusive Economic Zone (EEZ) which adjoins that of French Polynesia to the west (Illustration adapted from Gillet 2009)

In 2009 a reef monitoring programme off Pitcairn's northwest coast was initiated by the Institut des Récifs Coralliens du Pacifique/Centre de Recherches Insulaires et Observatoire de l'Environnement (Institute for Pacific Coral Reefs/Centre for Island Research and Observatory of the Environment, or IRCP/CRIOBE), based in Moorea, French Polynesia. The programme is part of an ongoing 16-year long-term project monitoring the outer reef slopes of 20 islands spread over the 4 French Polynesian archipelagos (Australs; Marquesas; Society; Tuamotu) and 6 Pacific Island Countries and Territories (PICTs) (Cook Islands; Niue; Kiribati; Tokelau; Kingdom of Tonga; Wallis and Futuna). The project focuses on the effects of natural disturbances on coral reef ecosystems. At Pitcairn, monitoring will be of fish populations (species richness, sizes and abundance), of scleractinian coral populations (photo surveys of species richness and percentage cover) and of a number of physical parameters such as water temperature and swell heights and frequencies (Chancerelle and Lison de Loma 2009).

The Islands Today

Pitcairn remains geographically remote and isolated today, though the installation of a satellite link on the island in 2002 has allowed internet access and the immediacy of e-mail communications. However, getting to and from the island still remains a challenge; there is no airstrip on the island so all visits are by sea. The island supply vessel visits every 3 months but can only carry 12 passengers who can stay for periods of 3 or 10 days. In March 2011, Pitcairn's population stood at 56 resident islanders with 9 non-residents. This compares with a peak in population in the mid-1930s of 250. The island is visited by a number of cruise ships during the austral summer on their way between Easter Island and Tahiti. The three uninhabited islands are rarely visited by the islanders, being largely left to themselves apart from occasional visits by scientific expeditions. Scuba diving is undertaken by a handful of Pitcairners, often linked with catching lobsters and fishes for on-island consumption or for selling on to cruise ships.

Geological Background and Physical Parameters

Much of our knowledge of the deep sea bathymetry of the central South Pacific has come about through interest in tectonics and volcanism during the past 25 years. In December 1989, a number of large volcanic structures were discovered in a virtually uncharted region of the Pacific Ocean some 60–100 km east-southeast of Pitcairn Island (Stoffers et al. 1990). Within an area of about 7,000 km^2, submarine volcanic activity has led to a particularly high density of over 90 volcanic

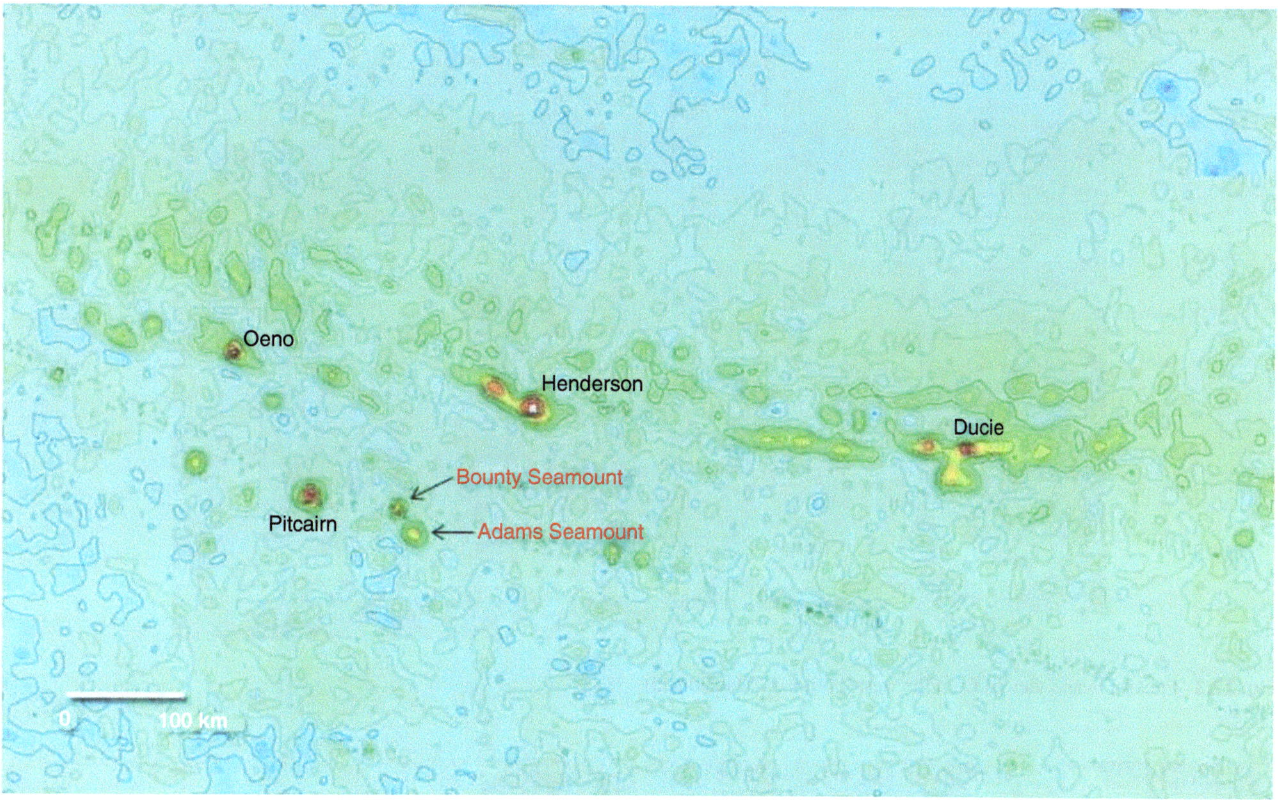

Fig. 22.2 Bathymetric map of the sea floor in the vicinity of the Pitcairn Islands, showing the location of the two seamounts closest to Pitcairn (Adapted from satellite altimetry data [Smith and Sandwell 1997])

cones or seamounts rising from the sea floor at 3,500–3,800 m depth (Fig. 22.2). The largest two edifices, known as Bounty seamount and Adams seamount (lying 90–110 km east-southeast of Pitcairn Island) are still active with steep scarps and fresh lava flows observed on their flanks during submersible diving surveys (Hekinian et al. 2003). The Bounty and Adams seamounts respectively rise to within about 450 and 55 m of the ocean surface. Other smaller edifices in the same locality have been named Christian, Young and McCoy, after some of the original mutineers from HMS *Bounty*.

The seamounts have arisen along two geological 'hotspot' regions, each aligned ESE to WNW. The first, with the youngest features in the south-east, has along it: Ducie (8 Myr); Henderson (13 Myr); and Oeno (16 Myr) [all ages given ± 1 Myr] (Okal and Cazenave 1985). The second active hotspot region has been located in a zone 40–110 km south-east of Pitcairn, and extends through the Gambier Islands to the Duke of Gloucester Islands, running approximately parallel to the first region. This hotspot region was responsible for the creation of Pitcairn Island itself, which is the youngest of the four islands and which was formed as the result of volcanic activity around 0.8–0.9 million years ago (Blake 1995).

When Pitcairn erupted, the weight of the new volcanic island caused the earth's crust beneath it to be depressed, with a consequential uplift approximately 200 km from the load, a see-saw process known as 'lithospheric flexure'. Over thousands of years, this uplift caused Henderson to be raised above sea level and thereby it became an uplifted fossilized reef. The island has been emergent for about 380,000 years. Today, Henderson remains the world's only raised coral atoll with its ecology largely intact.

Annual sea surface temperatures for the area show the monthly mean surface temperature to range from 22.5°C (Aug/Sept) to 26.3°C (Feb/Mar) (Streten and Zillman 1984). This is just above the lower limit for structural coral reef construction which is typically taken as 20°C in the coldest month of the year (Stoddart 1969). At the time of the Oceanic Institute expedition to Ducie in 1970, there was evidence of a relatively recent mass mortality of corals, the cause of which was not identified, although a sudden drop in water temperature was postulated (Rehder and Randall 1975). This may well have come about due to a temporary northwards shift of cooler southern ocean water. All of the islands have regular semi-diurnal tides, with a (measured) spring tidal range of 1.5 m at Henderson (Irving 1995) and 0.4 m at Ducie (Rehder and Randall 1975). The typical water clarity at 20 m depth at Pitcairn is in the region of 50 m, while at Henderson it has been measured at 75 m (R. Irving, pers. obs.). This latter figure probably reflects the fact that there is very little land run-off from Henderson.

Fig. 22.3 Pitcairn Island viewed from the west (Photo: A. MacDonald)

Fig. 22.4 *Left*: View of the landing at Bounty Bay, Pitcairn (Photo: R. A. Irving). *Right*: Tedside, on Pitcairn's NW coast (Photo: R. A. Irving)

Geomorphology and Reef Extent

Each of the four islands which make up the Pitcairn group is different. Not only is this obvious above sea level, but it is also true below sea level. While each of the islands is perched on the tip of its own huge submarine volcano, with steeply sloping sides descending to the abyssal depths, the shallower, nearshore sublittoral environments of each is very different.

Pitcairn

At 25° 04′ S, 130° 06′ W, Pitcairn is just 3.2 km long by 1.6 km wide covering an area of 4.5 km^2 (Fig. 22.3). Its volcanic terrain is rugged and its summit stands 347 m above sea level. Much of its 9.5 km long coastline consists of precipitous cliffs, in places over 120 m high. There are only two landing places – at Bounty Bay and Tedside (Fig. 22.4a, b) and at just a handful of sites where the intertidal zone is accessible from the land. The seabed all the way round the island shelves very gradually from 10 to 30 m in depth for approximately 300–500 m offshore, before plunging to the abyssal depths. In places, a level terrace-like seabed exists, with very little discernable difference in depth for stretches of 150–200 m perpendicular to the shoreline. It is likely that, over tens of thousands of years, the pounding surf affecting the near-shore zone has created a wave-cut platform as the volcanic sand has scoured away the underlying bedrock to a uniform depth.

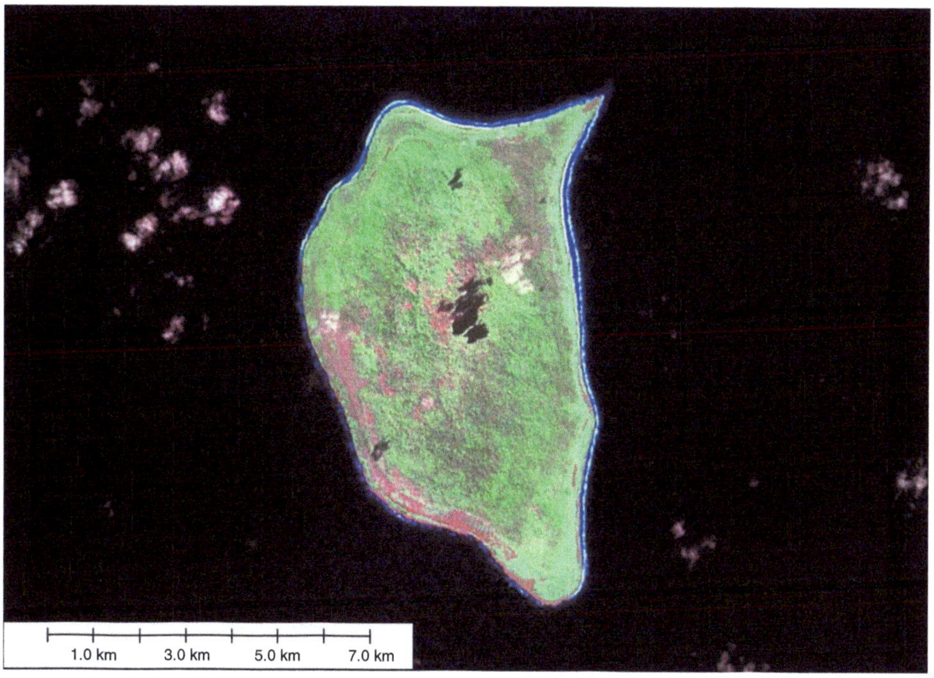

Fig. 22.5 Henderson Island. Colour-enhanced satellite image taken on 20 Sept 2000 (NASA). (Note that the *white patches* are clouds and the *dark areas* are their shadows. *Green areas* indicate dense vegetation/scrub and *pink areas* are unvegetated rock)

Fig. 22.6 Waves pound against the cliffs at the southern end of Henderson Island (Photo: R. A. Irving)

Henderson

Henderson Island, at 24° 21′ S, 128° 19′ W and 200 km east north-east of Pitcairn, is approximately 9.6 km long by 5.1 km wide and covers an area of 43 km^2 (Fig. 22.5). It is a raised coral atoll composed of coralline limestone (makatea) and is surrounded by steep, bare, weathered limestone cliffs (Fig. 22.6), with sandy beaches present off the north, east and north-west coasts (Fig. 22.7). The flat plateau, some 30 m above sea level, is covered by 4–6 m high dense scrub and has a slight depression in the centre where there was once a lagoon.

Fig. 22.7 Henderson's East Beach with its shallow reef platform (Photo: R. A. Irving)

Fig. 22.8 Oeno atoll. Image taken on June 16, 2006 by the Advanced Land Imager on NASA's EO-1 satellite

Henderson has a reef platform adjacent to its North and East Beaches and, to a lesser extent, off the North-West Beach. The near-horizontal platform ranges from 20 to 40 m wide at the North-West beach; 40–75 m wide at the North Beach; and 40–90 m wide at the East Beach (Irving 1995). At low water on spring tides, large areas of smooth horizontal rock are exposed at these beaches. Narrow (<15 cm) channels run perpendicular to the shore every 10–25 m. At the reef front, these channels widen and deepen into larger grooves, allowing water to drain off the flat. The northern end of the East Beach is probably the most biodiverse intertidal area, with small coral heads being exposed at low water on spring tides. There are two narrow channels through the reef on the north and northwest coasts which enable access for small boats, though landing is extremely hazardous. The reef towards the eastern end of the North Beach is formed into a series of spur and groove formations, with live coral dominating the tops and sides of the spurs and extensive areas of coral rubble collecting in the grooves.

Oeno

Oeno atoll lies 120 km to the north-west of Pitcairn at 23° 56′ S, 130° 45′ W (Fig. 22.8). It consists of a central, low-lying island surrounded by a shallow lagoon and a fringing reef (Fig. 22.9). The main island, covered by a mix of trees and other vegetation, is fringed by a narrow beach of sand with bedrock apparent on the north-west side of the island. The lagoon has a shallow entrance/exit connecting with the open sea on its north side. It is about 3 m deep and has an undulating bottom of sand and coral rubble (60%), reef pavement (20%) and patch reefs (20%) (Irving 1995) (Fig. 22.10, left). The patch reefs have near-vertical sides, with some undercut at their bases to form small caves and overhangs. The most striking feature is the large number of 'small' giant clams *Tridacna maxima* embedded within these patch reefs, at an estimated maximum density of 8–10/m^2 (Fig. 22.10,

Fig. 22.9 The north-west shoreline of Oeno and the adjacent lagoon (Photo: R. A. Irving)

Fig. 22.10 *Left*: Assorted corals form a small patch reef within Oeno's lagoon (Photo: R. A. Irving). *Right*: Giant clams *Tridacna maxima* embedded within a patch reef, Oeno lagoon (Photo: R. A. Irving)

right). Beyond the fringing reef, the seabed gradually shelves into deeper water, the angle of steepness increasing beyond 40 m depth.

Ducie

Ducie atoll, at 24° 40′ S, 124° 47′ W, is the easternmost island on the Pacific plate and also the most southerly atoll in the world (UKOTCF 2004) (Fig. 22.11). It comprises a main island (Acadia) and three smaller islets or 'motus' (Edwards, Pandora and Westward) encircling a central lagoon. The islets are composed of coral rubble, echinoid remains and dead shells. Acadia is largely surrounded by reef flats, the reef to the north-west consisting for the most part of an uneven reef pavement flat (Fig. 22.12) and the lagoon contains numerous patch reefs (Fig. 22.13). Most water exchange in to and out of the lagoon takes place via a shallow channel between Westward and Pandora islets.

Biogeographic Background

The Pitcairn Islands are located at the south-eastern extremity of the Indo-West Pacific biogeographic province. This position results in a number of barriers to the establishment of species. Firstly, prevailing winds and currents are dominantly from the east in the central South Pacific, with the

Islands lying upwind and upstream of all Indo-Pacific source areas further to the west. And secondly, the Islands lie just south of the Tropic of Capricorn (23° 26′ S), with relatively cool waters and climates imposing further barriers to the establishment of tropical species.

Studies of groups as diverse as corals, reef fishes, vascular plants and landbirds support the arguments that colonisation of the islands has almost entirely been from the biologically-rich source areas to the west on the margins of South East Asia, taking advantage of the inter-island connectivity of the south west Pacific Ocean (Stoddart 1992). As a consequence of these factors, the diversity of marine species present within the islands' near-shore waters, when compared to island groups further to the west, is impoverished (Benton and Spencer 1995b). It also reflects the lack of some marine and coastal habitats at the four islands, such as mangroves and seagrass beds. Levels of endemism for all four islands are relatively low at around 2% (see also next section). This figure for endemism derives from studies of molluscs (Preece 1995), echinoderms (Paulay 1989) and reef fishes (Randall 1999) in particular. A total of 87 species of sceleractinian corals have been recorded from all four islands (of which 29 are *Acropora* species), together with 19 species of butterflyfish (Family Chaetodontidae) (Irving and Dawson 2012).

Fig. 22.11 Satellite image of Ducie atoll taken on 30 March 2000 (Photo: © GEOEYE/SCIENCE PHOTO LIBRARY)

Biological Characteristics

As already indicated, the nearshore areas around each of the four islands have their own unique characteristics and they are described separately here. For more detail, the marine environment of the Pitcairn Islands has recently been comprehensively reviewed by Irving and Dawson (2012).

Pitcairn

Much of the near-shore seabed around Pitcairn comprises sand-scoured, low-lying rock outcrops surrounded by sand patches. From about 8–15 m depth, the outcrops are colonised by a small number of foliose brown macroalgae, particularly

Fig. 22.12 View south-westwards across the lagoon from Acadia, Ducie atoll (Photo: R. A. Irving)

Fig. 22.13 Patch reef within the lagoon at Ducie (Photo: R. A. Irving)

Sargassum odontocarpum and *Lobophora variegata*. Foliose algae appear to do well at Pitcairn (compared to the other islands), which is probably a reflection of the slightly cooler seawater and the higher nutrient levels in the water column linked with rain run-off from the island.

Coral Cover and Reef Structure

Typically, live coral cover can vary from as little as 5% to as much as 80%, depending on the depth and the location around the island. Coral growth seems to be most prolific in the 12–22 m depth range. However, in 1971 Randall (1999) undertook some of his fish collecting dives in deep water. On one such occasion, when diving a feature known as 'The Bear' off the north-east coast, which rises about 9 m above the seabed at 45 m, he recorded the live coral cover as being almost 100%. One of the most extensive areas of live coral growth is present off Adamstown on the north-east coast at a depth of 18–30 m, covering an area estimated to be about 2 km^2.

A total of 15 coral species have been recorded from Pitcairn (probably an underestimate) which include *Pocillopora damicornis*, *Astreopora myriophthalma*, *Porites* aff. *annae*, *Porites lobata*, *Psammocora haimeana*, *Leptoseris hawaiiensis* and *Favia matthaii*. In general, most reef-building corals reach a height of 1–2 m and where sufficiently dense (>80% cover) these are able to provide a three-dimensional structure for other organisms to utilise, most noticeably fishes (Fig. 22.14, left and right). Occasionally, one may encounter large solitary coral structures ('bommies'), often formed by *Porites lobata*, which may be over 5 m tall.

Other Reef-Associated Taxa at Pitcairn

A total of 270 species of reef fishes have been recorded from Pitcairn. This equates to 77% of the total number of reef

Fig. 22.14 *Left*: Coral formations at 18 m depth off Pitcairn's NE coast. *Right*: Coral formations at 18 m depth off Pitcairn's NE coast. *Pocillopora* sp. in foreground (Photos: R. A. Irving)

Fig. 22.15 Extensive areas of coral rubble are present at Henderson, particularly off the north coast (Photo: R. A. Irving)

fishes (352 species) recorded from all four islands – a relatively modest figure reflecting the Pitcairn Islands' isolation. Thirty three of these species were new to science when first discovered at Pitcairn in 1971 (Randall 1999). The most common fishes observed at diving depths of about 10–20 m at Pitcairn are the wrasses *Thalasoma lutescens* and *Coris* sp., the surgeonfish *Acanthurus leucopareius*, the damselfishes *Chrysiptera galba* and *Stegastes fasicolatus*, and drummer/nanwi *Kyphosus bigibbus* (Randall 1999; R. A. Irving, *pers. obs.*). The two shark species which may be seen at Pitcairn are the Galapagos shark *Carcharhinus galapagensis* and the whitetip reef shark *Triaenodon obesus*.

Reef-associated invertebrate taxa which have been recorded from Pitcairn include 23 species of echinoderms (although almost half of these are brittlestars which have come from dredge samples, Paulay 1989); over 80 molluscan taxa (although the molluscan fauna remains poorly known, Preece 1995); 29 non-ostracod species of crustacean (likely to be an under-representation) and 47 ostracod species (Irving and Dawson 2012). There are very few sessile, filter-feeding organisms present (such as porifera, sessile polychaetes, bivalve molluscs, ascidians etc.), probably a reflection of the poor nutrient levels in the water column and the low reef biodiversity in general.

Henderson

Coral Cover and Reef Structure

An extensive level terrace at 3–6 m depth is present on the shallow fore-reef beyond the seaward edge of the reef platform off the North-West and East Beaches. The smooth bedrock here is pockmarked by small holes containing the urchin *Echinometra* sp. There is a noticeable absence of coral growth on these terraces. Indeed, generally speaking, the fore-reef as a whole is impoverished in terms of coral diversity, with the percentage of live coral cover typically in the order of 10–30% (Irving 1995). Corals of the genus *Pocillopora* are the commonest corals beyond 10 m depths, though the fire coral *Millepora* sp. is also numerous and widespread. A total of 59 species of scleractinian coral have been recorded from Henderson, together with an unspecified number of soft coral species (Family Alcyonacea) (Irving and Dawson 2012). Of the 18 species of acroporid coral known to occur at Henderson, the majority are found off the East Beach.

At several sites on Henderson's fore-reef, bare bedrock is apparent, typically covered by a thin crust of coralline algae. Elsewhere, and particularly in deeper water (>30 m depth), extensive areas of coral rubble are present (Fig. 22.15). Damage to coral formations is likely to have come about as a result of strong wave action during storm events, with coral debris from shallower waters being moved into deeper waters. The richest area of coral diversity appears to be off the northern end of the east coast (Fig. 22.16). Occasional patch reefs or 'bommies' are present within the 25–35 m depth band around the northern half of the island. These massive structures, which can be 7 m or more in height (Fig. 22.17), are typically formed by just one coral (often *Porites lobata*) and may be over 750 years old (Irving 1995).

Fig. 22.16 While most of Henderson's fore-reef has less than 50% coral cover, areas off the East Beach (such as shown here) boast 100% cover (Photo: R. A. Irving)

Fig. 22.17 A large *Porites lobata* 'bommie' (Photo: R. A. Irving)

Other Reef-Associated Taxa in Henderson

A total of 173 species of reef fishes (belonging to 49 families) has been recorded from Henderson, almost half of the total number of species (352) recorded for all four islands (Randall 1999; Irving et al. 1995). The largest family represented is the Labridae (21 species), followed by the Acanthuridae (16 species), Blenniidae (14 species), and the Chaetodontidae, Serranidae and Pomacentridae (all with 12 species) (Irving et al. 1995). Echinoderms are the most conspicuous invertebrates at Henderson, particularly the echinoids (12 spp.) and the holothurians (10 spp.) (Paulay 1989). *Diadema* cf. *savignyi* appears the most numerous sea urchin on the fore-reef, often present in very large aggregations in areas of coral rubble (Fig. 22.18, left). Other conspicuous echinoid species are *Heterocentrotus trigonarius* (common in the shallows of the reef platforms) and *Heterocentrotus mammillatus* (common in crevices in the fore-reef at depths of 8–15 m) (Irving 1995). Asteroids (starfish) and ophiuroids (brittlestars) are rare. A group of about 150 crown-of-thorns starfish *Acanthaster planci* was reported at 32–38 m depth off the island's west coast in December 1991 (Irving 1995) (Fig. 22.18, right).

Oeno

Coral Cover and Reef Structure

The fore-reef at Oeno slopes gradually into deeper water from the reef margin, steepening beyond the 30 m depth contour (Figs. 22.19 and 22.20). The overall cover of live coral is greater than at Henderson though less than at Ducie, ranging from 5% to 70% (Irving 1995). In contrast to Ducie, large areas of sand are present, with a series of sand channels, up to 3 m wide, running perpendicular to the reef between 5 and 20 depth. There are also extensive areas of coral rubble. A total of 17 species of scleractinian coral have been recorded from Oeno (Irving and Dawson 2012).

Fig. 22.18 *Left*: Dense aggregations of the long-spined sea urchin *Diadema* cf. *savignyi* occur on the open fore-reef (Photo: J. Jamieson). *Right*: Crown-of-Thorns starfish *Acanthaster planci*, photographed in December 1991 at Henderson (Photo: R. A. Irving)

Fig. 22.19 Assorted corals on the fore-reef off the west side of Oeno at about 20 m depth (Photo: R. A. Irving)

Other Reef-Associated Taxa

A total of 255 marine molluscs have been recorded from Oeno, including those found within the lagoon. Preece (1995) found the largest number of bivalve species (45) here when compared to collections made at the other three islands, a fact he put down to the soft sediments within the lagoon. Fifteen species of echinoderm have been recorded from Oeno, together with 165 species of reef fishes.

Ducie

Coral Cover and Reef Structure

The greatest seaward extension of the reef at Ducie is off the south-west of the atoll, where the shelf extends 270 m offshore to a depth of 30 m. Beyond this depth, the seabed steepens noticeably. However, coral growth can be seen extending beyond 40 m in places (Irving and Dawson 2012). Cover of live coral was estimated as being 80–100% in the 11–20 m

Fig. 22.20 Assorted corals on the fore-reef off the west side of Oeno at about 16 m depth (Photo: J. Jamieson)

depth range (though note here that soft coral species provide 80% of this cover); 25–100% in the 21–30 m depth range; and 10–85% in the 31–40 m depth range (Irving 1995) (Fig. 22.21a, b). Where cover was greatest (particularly off the south-east of the atoll), at between 80% and 100% cover, a large proportion was provided by one species: *Montipora aequituberculata* (Fig. 22.22). A total of 27 scleractinian coral species have been recorded from Ducie (Irving and Dawson 2012).

Other Reef-Associated Taxa

About 80 species of mollusc have been recorded from Ducie (Preece 1995), though the majority of these remain hidden from view. Twelve species of echinoderms have been recorded, including both the purple and red slate pencil urchins *Heterocentrotus trigonarius* and *H. mammillatus*. The long-spined sea urchin *Diadema* cf *savignyi* has also been found to be abundant on the fore-reef (Irving 1995). A total of 127 species of reef fishes have been recorded from Ducie (Randall 1999) which amounts to 28% of the fish fauna for all four islands, including one species of butterflyfish not found at the other three islands (Fig. 22.23). The most common species observed at Ducie were drummer/nanwi *Kyphosus bigibbus*, sunset wrasse *Thalassoma lutescens*, wrasse *Thalassoma heiseri*, steephead parrotfish *Chlorurus microrhinus*, whitebar surgeonfish *Acanthurus leucopareius*, damselfish *Chrysiptera galba*, Emery's Gregory *Stegastes emeryi* and black jack *Caranx lugubris*. Large numbers of grey reef sharks *Carcharhinus amblyrhynchos* were reported as being present at Ducie in 2012 (Enric Sala, pers. comm.), indicating a healthy reef environment.

Fig. 22.21 *Left*: Assorted corals forming 100% cover off Acadia motu, Ducie at 18 m depth (Photo: R. A. Irving). *Right*: Coral formations at 25 m depth off Acadia motu, Ducie (Photo: R. A. Irving)

Fig. 22.22 One of the dominant coral species on the fore-reef at Ducie is *Monitopora aequituberculata*, rarely seen at the other Pitcairn Islands (Photo: R. A. Irving)

Fig. 22.23 A large shoal of Pacific double-saddle butterflyfish *Chaetodon ulietensis*, a species only found at Ducie within the Pitcairn Islands (Photo: R. A. Irving)

Fig. 22.24 The many-spined butterflyfish *Hemitaurichthys multispinosus*, endemic to Pitcairn (Photo: J. E. Randall)

Endemic Species and Species of Particular Nature Conservation Concern

The number of marine endemic species associated with the Pitcairn Islands is relatively small. It is likely more have yet to be discovered, as not all of the major taxa have been studied to any great degree. The gastropod mollusc *Fusinus galatheae bountyi* is a predatory gastropod which is frequently found in the baited pots set around Pitcairn to catch slipper lobsters. It is found in depths of 40–100 m (Rehder and Wilson 1975) and is still understood to be endemic to Pitcairn. The nudibranch *Bornella irvingi* was first discovered at Ducie in 1991 and remains endemic to that island.

Five species of bony fishes are currently believed to be endemic to the Pitcairn Islands (Randall 1999). These are the Pitcairn sandlance *Ammodytoides leptus*, the many-spined butterflyfish *Hemitaurichthys multispinosus* (both only found at Pitcairn) (Fig. 22.24); the Henderson triplefin *Enneapterygius ornatus* and the squirrelfish *Sargocentron megalops* (both found only at Henderson). There is also an undescribed species of combtooth blenny *Alticus* sp. (Randall 1999), found both at Pitcairn and Henderson. Smith's butterflyfish

Fig. 22.25 Smith's butterflyfish *Chaetodon smithi*. Whilst commonly seen at Pitcairn, this species has a very restricted distribution (Photo: R. A. Irving)

Chaetodon smithi (Fig. 22.25), one of the most frequently seen species of butterflyfish at Pitcairn, was a new species to science when discovered at Pitcairn in 1971. Since then its distribution has been found to extend to SE French Polynesia as well.

A summary of the most endangered conservation status (IUCN 2012) of representatives of each taxonomic group of marine organisms which have been recorded to date from within the Pitcairn Islands Exclusive Economic Zone (EEZ) is given in Table 22.1.

Human Uses and Threats (All Four Islands)

Because of their geographical position and isolation, as well as having a small human population (on Pitcairn only), the coral reefs and their habitats of the Pitcairn Islands do not suffer the usual threats associated with human interference. On Pitcairn itself, the community does not practice any major agriculture, mainly growing vegetables in garden plots and harvesting fruit from natural and semi-natural locations around the island. Consequently, there has been no significant land utilisation or conversion or application of fertilisers resulting in no pollution and minimal sedimentation run-off, save during uncommon extreme rainfall events. The islanders use line-caught and trap methods of fishing only. At the current time, with no reliable trading options, overfishing is also controlled to a degree. Even global warming does not seem to present a significant threat to the Pitcairn corals in the medium term due to their low latitude location where temperatures are at the lower end of the corals' thermal tolerance range. There may be problems from fishing and tourism, particularly associated with the dropping of boat anchors on reefs causing localised damage.

Governance of the Area

The islands' 200 nautical mile EEZ was established in 1980 when the Fisheries Zone Ordinance was constituted under Pitcairn laws to establish a fisheries zone contiguous to the territorial seas of Pitcairn, Henderson, Ducie and Oeno Islands and to regulate fishing practices. Although much of the legislation within this ordinance relates to regulation and licensing of foreign fishing vessels within the Pitcairn EEZ, there are also provisions which allow for the Governor of Pitcairn to limit fishing activities to island residents for the purposes of conservation and management of fisheries resources.

Henderson became a UNESCO World Heritage Site in 1988 on account of its 'unique natural history and ecological intactness'. However, it was not until after the Sir Peter Scott Commemorative Expedition to the Pitcairn Islands in 1991/1992 that the composition of that ecology began to be known. A Management Plan for the period 2004–2009 was subsequently drawn up (Brooke et al. 2004), published by the Foreign and Commonwealth Office, London, in conjunction with the Pitcairn Islands Administration and the Royal Society for the Protection of Birds. This consolidated much of the earlier literature and scientific knowledge of Henderson Island as a consequence of UNESCO designation. The Plan was published to provide a framework for the sustainable management of the island with respect to the Pitcairners, visiting scientists and tourists. Following this, a Pitcairn Islands Environment Management Plan, which focused on the other three islands (Pitcairn, Ducie and Oeno), was published in 2008 (Smyth 2008). This develops ten key objectives for managing the environment of the Pitcairn Islands, based upon the Environmental Charter jointly signed by the Mayor of Pitcairn Island and the UK Government in 2001. Although the UK Government ratified the Convention on Biological Diversity (CBD) in 1994, this has not been extended to the Pitcairn Group. A review of the progress on the CBD in UK Overseas Territories, commissioned by the Worldwide Fund for Nature (WWF), highlighted that the main area of CBD legislation with which the Pitcairn Group was not complying, was in connection with the on-going monitoring of biodiversity (Cross and Pienkowski 1998). A UK Overseas Territories Conservation Forum (UKOTCF) review of existing and potential Ramsar sites in UK Overseas Territories and Crown Dependencies, commissioned by the Department of Environment, Food and Rural Affairs (DEFRA), identified the coastal waters of the Pitcairn Group as potential Ramsar Convention sites of International Importance (Pienkowski 2005), but data remain inadequate to determine designations.

Table 22.1 Number of marine species in major taxonomic groups recorded from within the Pitcairn Islands EEZ (After Irving and Dawson 2012)

Group	No. of species recorded to date (May 2012)	No. of endemic species	Most endangered conservation status represented[a]	Species richness	Level of study
Cetaceans	22	0	Endangered **(EN)**	Low/intermediate	Poor/intermediate
Marine birds (breeding)	13	1	Endangered **(EN)**	Intermediate	Good
Marine birds (non-breeding)	20+	0	Endangered **(EN)**	Low/intermediate	Poor
Turtles	2	0	Critical **(CR)**	Low	Poor
Pelagic fishes	13	0	Near-Threatened (NT)	Low	Poor/intermediate
Reef fishes	352	5	Endangered **(EN)**	Low/intermediate	Good
Echinoderms	>64	3	Not Evaluated (NE)	Intermediate	Intermediate
Molluscs	>502	5+	Not Evaluated (NE)	Low	Intermediate
Crustacea	>42	?	Not Evaluated (NE)	Low	Poor
Ostracods	47	?	Not Evaluated (NE)	Intemediate	Intermediate
Stony corals	87	1	Vulnerable **(VU)**	Intermediate	Intermediate
Hydroids	8	0	Not Evaluated (NE)	Low	Intermediate
Sponges	12	0	Not Evaluated (NE)	Low	Poor
Forams	32	?	Not Evaluated (NE)	Low	Intermediate
Algae	29	0	Not Evaluated (NE)	Low	Poor/intermediate
Blue-green algae	4	0	Not Evaluated (NE)	Low	Poor
Totals:	1,249+	15+			

[a]IUCN (2012)Red list of threatened species status categories

Table 22.2 Ramsar information sheets relating to the proposed sites for designation in the Pitcairn Islands (From www.ukotcf.org)

Ramsar code	Site name	Ramsar Criterion[a]
UK62001	Ducie Island: All islands and marine area to 50 m depth contour; or 1.5 km offshore approximately	1, 3, 4, 6, 7
UK62002	Henderson Island: Island and marine area to 50 m depth contour; or 1.5 km offshore approximately	1, 2, 3, 4, 5, 6, 7
UK62003	Oeno Island: Island and marine area to 50 m depth contour; or 1.5 km offshore from atoll approximately	1, 2, 3, 4, 6, 7
UK62004	Brown's Water, Pitcairn Island: Gully surrounding flowing water feature	1, 2, 3
UK62005	Coastal waters, Pitcairn Island: Marine area to 50 m depth contour; or 1.5 km offshore	7

[a]See Annex II of the *Explanatory Notes and Guidelines* for the Criteria for completing the Information Sheet on Ramsar Wetlands and guidelines for their application (Adopted by Resolution VII.11) available at www.ramsar.org

Protected Areas and Management

At the current time, no marine or terrestrial protected areas have been established on any of the four islands, although, due to their isolation and lack of scheduled transportation links, the uninhabited islands of Henderson, Ducie and Oeno are relatively undisturbed by human interference.

The UKOTCF (2004) review of existing and potential Ramsar Convention sites in UK Overseas Territories and Crown Dependencies has resulted in a drafted set of Ramsar Information Sheets for the proposed Pitcairn sites (Table 22.2).

At the current time, however, the evidence base for justifying the designation under each of the Ramsar Criteria supporting the listing of the Pitcairn sites under the Ramsar Convention remains incomplete. Further research is necessary for establishing a baseline, by describing the ecological functions, products and attributes of the sites that sufficiently identify those benefits and values of international importance.

Under its Global Ocean Legacy programme, the Pew Environment Group is aiming to establish a worldwide system of very large, highly protected marine reserves, where commercial fishing and extractive industries are prohibited (see http://www.pewenvironment.org/campaigns/global-ocean-legacy). Because of its isolation, pristine marine conditions and unique ecological status, the Pitcairn Islands' EEZ, covering an area of approximately 836,100 km^2 (322,823 square miles) of ocean, has been identified as a candidate site for the establishment of the world's largest marine reserve. In collaboration with several UK conservation organisations, which together form the Marine Reserves Coalition (MRC), negotiations are currently underway to progress this aim with the Pitcairn Islands Council, the island community and the UK Government.

Fishing

Historically, the number of commercial foreign fishing vessels operating in the open waters of the Pitcairn EEZ has been very low. Targeting mainly tuna (albacore) species, these fishing activities have been limited to longline fishing by distant water fishing nations, principally Japan, Korea, and Taiwan, under licence from the UK Government. Due to the distances involved and the vessels used, the use of purse-seiners is prohibitive, which reduces the problems of over-extraction and bycatch commonly associated with this method. In recent years, however, no commercial fishing has taken place due to the relatively low catch yields compared to neighbouring French Polynesia. A large number of the small island community regularly fish from the shore or from Pitcairn-based boats for subsistence and for sale to the infrequent passing cruise ships. Virtually all near-shore fishing is conducted using hand-lines catching a number of reef fish, dominated by the drummer *Kyphosus bigibbus* (Fig. 22.26a) although a number of other species are targeted, including the highly-prized coral trout *Variola louti*, various grouper species *Epinephelus* spp. and the sunset wrasse *Thalassoma lutescens* (Fig. 22.26b). Other fishing methods used by some of the islanders include trolling using small outboard powered skiffs mainly for wahoo *Acanthocybium solandri* and the occasional yellowfin tuna *Thunnus albacares*, spear fishing using snorkeling and scuba equipment and using trap pots for catching lobsters. The two lobster species caught locally are the pronghorn or red spiny lobster *Panulirus penicillatus* or crayfish as it is locally known (Fig. 22.26c) and the slipper lobster *Scyllarides haanii* (Fig. 22.26d). The former species is caught by hand using SCUBA equipment at rocky inshore locations during calm weather, whereas the latter species are caught using pots, which are deployed at depths of 30–50 m around the island (Dawson & Christian 2010). The visitation of occasional cruise ships and visiting yachts provides the only opportunity for the Pitcairners to sell or trade their marine resources, mainly in the form of fresh fish (caught in the immediately preceding days and refrigerated), or live lobsters. Anecdotal evidence from local fishermen suggests that the artisanal fisheries and lobster fishing activities are relatively healthy and currently sustainable although in recent months there has been increased fishing effort in order to achieve their previous catch weights and a decline in the number of spiny lobsters caught in trying to meet demand from the cruise ships. Because of a contamination

Fig. 22.26 (**a**) Drummer (local name: nanwi) *Kyphosus bigibbus*, an important food fish for Pitcairners (Photo: T. P. Dawson). (**b**) Sunset wrasse (local name: whistling daughter) *Thalassoma lutescens*, another popular food fish for Pitcairners (Photo: R. A. Irving). (**c**) Pronghorn or red spiny lobster *Panulirus penicillatus* (Photo: R. A. Irving). (**d**) Aesop slipper lobster *Scyllarides haanii* (Photo: T. P. Dawson)

of ciguatoxins in the fish caught on the reefs of French Polynesia, the Secretariat of the Pacific Community (SPC) has proposed that the Pitcairn community develop their reef fisheries commercially for export to neighbouring Mangareva. However, no assessment of fish populations has been undertaken as yet and the ecological sustainability and economics of such a venture have yet to be studied in detail.

Other Concerns

Few other human impacts affect the marine environment of the Pitcairn Islands directly. Successfully eradicated from Oeno and Ducie in 1997, the presence of the Polynesian rat *Rattus exulans* on the Islands of Henderson and Pitcairn continues to have a severe negative impact upon the breeding bird populations. A poor swimmer over long distances, the rat is considered to be a significant marker of human migrations across the Pacific, as the Polynesians accidentally or deliberately introduced it to the islands they visited.

The decline of the Henderson Petrel *Pterodroma atrata* populations, which is listed as endangered on the IUCN Red List and which breeds only on Henderson Island, is thought to be mainly caused by chick predation by rats, although land crabs have also been implicated. A major aerial rat eradication programme was carried out by the Royal Society for the Protection of Birds (RSPB) on Henderson in 2011 using helicopters and poison bait drops. However, a positive rat sighting in March 2012 (and subsequently) suggests that the operation was not successful in eradicating the rats. Climate change may also have a negative impact on the Pitcairn Islands in the long term. Sea-level rise may inundate the low lying areas of some parts of both Oeno and Ducie islands and major storms and other extreme climatic events may cause extensive damage to coral reefs as well as the terrestrial habitats. For example, a species of thintail grass *Lepturus* sp. recorded from earlier expeditions had disappeared from Ducie when storm waves deforested the island some time before the 1975 Smithsonian expedition (Rehder and Randall 1975). Rehder and Randall (1975) also commented on a

Fig. 22.27 Heavy coastal and near-shore sedimentation on Pitcairn Island arising from the storm event in February 2012 (Photo: B. Young)

mass mortality of the lagoon corals at Ducie over all the sites they surveyed and hypothesised that they were possibly killed exceptionally by an unusual influx of low temperature water from more southerly latitudes. More recently, a major rainfall event occurred on Pitcairn Island in February 2012 when 600 mm of rain fell over 2 days, the highest since records began, which caused several landslide events across the island resulting in high sedimentation loads entering the near-shore marine environment (Fig. 22.27). Whilst prolonged suspended sediment in the water column is known to have a deleterious effect on coral reef health due to the reduction in the amount of light reaching coral reefs and other shallow benthic systems, the prevailing strong currents surrounding Pitcairn Island are likely to mitigate against these rare events over longer timescales.

Live specimens of Smith's butterflyfish *Chaetodon smithi* (see Fig. 22.25) have become sought after by aquarists on account of their extremely restricted distribution and their bold colouration. However, their export would require a licence and no such licences have been granted to date.

Conclusions

Situated in the central South Pacific at the eastern edge of the Pacific plate, the Pitcairn Islands remain one of the most pristine marine environments in the world. Their nearshore waters have escaped the ravages of modern fishing methods and the degradation often associated with coastal industries. Their extreme isolation and the low human population are the main reasons for this but these factors also create some of their greatest challenges for sustainable management, including the fragile social structure, limited transport access (only accessible by sea) and small economic base. As the sole remaining UK Overseas Territory in the Pacific, the UK's budgetary aid (2012/2013) to meet the territory's reasonable needs for public services, providing transport (shipping) subsidies and maintain the Pitcairn Island Office in New Zealand is £2.9 million, which accounts for 90–95% of the Island's economy (DFID 2012). Since the Island's revenues from postage stamps declined significantly in the 1990s, the UK Government recognises the current prospects for economic self-sufficiency on Pitcairn Island to be very low. The UK Government is currently, alongside other activities, encouraging private sector initiatives to engage the Pitcairn islanders in developing fish exports to Mangareva (DFID 2012), but the authors consider this course of action to be unwise until a full ecological impact assessment of the fisheries and sustainability appraisal has been undertaken. More positively, the UK Department for International Development identifies tourism as a valuable means of significantly increasing the Pitcairn Islands' revenue base through investments in a new alternative harbour facility (accommodating more frequent and year-round island visits to Pitcairn) and through private enterprise in eco-adventure tourism (DFID 2012). Unlike the scheme to expand commercial fisheries for export, this latter development strategy is conducive with the proposed designation of the Pitcairn EEZ as a fully-protected marine reserve. Whilst increasing tourism numbers is not incompatible with maintaining fragile, pristine, and relatively undisturbed natural areas, tourism development must be handled with care and sensitivity to ensure activities are sustainable and environmentally friendly (ecotourism). In many ways the Pitcairn Islands are a microcosm of the ecological and economic changes occurring in the world. With the health of their natural environment vitally important to their economy and wellbeing, the Pitcairners today stand at a crossroads that could change the course of their islands' future. Careful environmental stewardship of the Pitcairn Islands will ensure their unique biodiversity and natural heritage persists for many generations to come.

References

Benton TG, Spencer T (eds) (1995a) The Pitcairn Islands: biogeography, ecology and prehistory. Biol J Linn Soc 56:1–422

Benton TG, Spencer T (1995b) Biogeographic processes at the limits of the Indo-West Pacific Province. Biol J Linn Soc 56:243–244

Blake SG (1995) Late quaternary history of Henderson Island, Pitcairn Group. Biol J Linn Soc 56:43–62

Brooke M deL, Hepburn I, Trevelyan RJ (2004) Henderson Island world heritage site management plan, 2004–2009. Foreign & Commonwealth Office, London

Chancerelle Y, Lison de Loma T (2009) Polynesia Mana Monitoring Network: Pitcairn site, first campaign, September 2009. Report of fish and coral monitoring studies by the Institute for Pacific Coral Reefs/Centre for Island Research and Observatory of the Environment (IRCP/CRIOBE), Moorea, French Polynesia

Dawson TP, Christian M (2010) Pitcairn Island marine and fisheries preliminary assessment report 2010. Unpublished report to the Darwin Initiative Overseas Territories Challenge Fund

DFID (2012) Pitcairn Budget Aid 2012/12: business case and intervention summary, Department for International Development. http://projects.dfid.gov.uk/project.aspx?Project=203135. Accessed 30 Aug 2012

Gillett R (2009) Fisheries in the Economies of the Pacific Island Countries and Territories. Asia Development Bank, Manila, 1-483 pp

Irving RA (1995) Near-shore bathymetry and reef biotopes of Henderson Island, Pitcairn Group. Biol J Linn Soc 56:309–324

Irving RA, Dawson TP (2012) The marine environment of the Pitcairn Islands. A report to Global Ocean Legacy, a project of the Pew Environment Group, Dundee University Press, ISBN 97818458616122

Irving RA, Jamieson J, Randall JE (1995) Initial checklist of fishes from Henderson Island, Pitcairn Group. Biol J Linn Soc 56:329–338

IUCN (2012) IUCN red list of threatened species, Version 2012.1. http://www.iucnredlist.org. Accessed 16 Feb 2012

Okal EA, Cazenave A (1985) A model for the plate evolution of the East-Central Pacific based on SEASAT investigations. Earth Planet Sci Lett 72:99–116

Paulay G (1989) Marine invertebrates of the Pitcairn Islands: species composition and biogeography of corals, molluscs and echinoderms. Atoll Res Bull 326:1–28

Preece RC (1995) The composition and relationships of the marine molluscan fauna of the Pitcairn Islands. Biol J Linn Soc 56:339–358

Randall JE (1973) Expedition to Pitcairn. Oceans 6:12–21

Randall JE (1978) Marine biological and archaeological expedition to southeast Oceania. National Geographic Society Research Reports 1969 Projects, pp 473–495

Randall JE (1999) Report on fish collections from the Pitcairn Islands. Atoll Res Bull 461:1–36

Rehder HA (1974) Marine biological research in southeastern Polynesia. National Geographic Society Research Reports 1967 Projects, pp 243–254

Rehder HA, Randall JE (1975) Ducie Atoll: its history, physiography and biota. Atoll Res Bull 183:1–40

Rehder HA, Wilson BR (1975) New species of marine mollusks from Pitcairn Island and the Marquesas. Smithson Contrib Zool 203:16

Smith WHF, Sandwell DT (1997) Global sea floor topography from satellite altimetry and ship depth soundings. Science 277:1956–1962

Stoddart DR (1969) Ecology and morphology of recent coral reefs. Biol Rev 44:433–498

Stoddart DR (1992) Biogeography of the tropical Pacific. Pac Sci 46:276–293

Stoffers P, Hekinian R, Ackermand D, Binard N, Botz R, Devey CW, Hansen D, Hodkinson RA, Jeschke G, Lange J, von der Perre E, Scholten J, Schmitt M, Sedwick P, Woodhead JD (1990) Active Pitcairn hotspot found. Mar Geol 95(1):51–55

Streten NA, Zillman JW (1984) Climate of the South Pacific Ocean. In: Van Loon H (ed) Climates of the world, vol 15, World survey of climatology. Elsevier, Amsterdam, pp 263–429

UKOTCF (2004) Ramsar information sheet, 62001 (JNCC). United Kingdom Overseas Territories Conservation Forum, 13 Nov 2004

Weisler MI (1995) Henderson Island prehistory: colonization and extinction on a remote Polynesian island. Biol J Linn Soc 56:377–404

Index

A
Acanthaster (crown-of-thorns), 230, 309, 310
Acidification, 70, 72, 83, 138, 182, 185, 186, 281
Acropora
  A. cervicornis, 64, 66, 101, 113
  A. palmata, 17, 19, 38, 50, 62, 64, 65, 93, 101, 103, 104, 113
  disease, 19
Active fringing reef, 63–65, 104
Aeolianites, 115, 122
Agaricia, 17, 62, 64, 66, 67, 101, 103–105, 123, 136, 156, 160, 189, 191
Ahermatypic, 219
Algae, 17, 19, 20, 38, 41, 52, 53, 62–66, 72, 75, 77, 79, 82, 83, 91, 101, 102, 104, 109, 112, 118, 123, 127, 128, 135, 144, 159, 177, 178, 186, 213, 215–219, 221, 242, 244, 247, 266, 276, 307, 308, 314
Algae-benthic, 17, 19, 20, 102, 109, 123, 215, 266
Algal ridges, 225, 281
Algal-vermetid reefs, 128, 129
Allopatric speciation, 162
Amenity value, 206
Anchor damage, 34, 39, 45, 75, 82, 83, 117
Anegada Limestone Formation, 25, 29, 49–52, 55, 57
Angelfish, 20, 161, 164, 220, 256
Antifouling compounds, 175, 283
Antipatharia, 123, 189–199
Argus/Plantagenet bank, 118, 158
Arthropoda, 166
Artificial reefs, 37, 95
Artisanal fishing, 85
Ascension frigatebird, 215
Ascension Island, 2, 213–221
Azooxanthellate, 156, 157, 191–192, 195, 196

B
Ballast water, 248
Bank patch reefs, 101–103, 110
Barrel sponges, 67, 92, 93, 105
Barren rock pavement, 63–65, 109
Barrier islands, 55, 56
Barrier reef, 2–4, 7, 8, 31, 37, 52, 100–102, 104–105, 107–110, 266
Beach debris, 89, 283, 290, 308
Beach ridges, 50, 79, 80
Beachrock, 28, 50, 51, 57, 58, 72, 74–76
Bermuda, 1, 66, 84, 97, 115, 135, 153, 173, 189–199, 201–221
Bermuda Platform, 8, 115–121, 123, 124, 128, 131, 135–148, 154, 210
Biocides, 283, 289
Biodiversity, 1, 8, 41–43, 59, 69, 70, 79, 82, 91, 92, 99, 113, 153–169, 209, 215, 226, 234, 236, 247, 248, 271, 272, 280, 299, 308, 313, 318
Biological connectivity, 228–229
Biological productivity, 236
Birgus latro (coconut crab), 230–232, 272
Blackfish, 217, 219–221
Bleaching, 18, 33, 45, 61, 69, 70, 72–75, 139, 158, 168, 173, 176–180, 201, 244, 247, 248, 254, 265
Bleaching events, 33, 45, 71–75, 112, 168, 169, 176, 179, 184, 201, 241, 246, 248, 259
Blenheim atoll, 8, 225
Blue-green algae, 217, 219, 314
Boobies, 82, 217, 226
Boulder
  ramparts, 64
  and rock zone-Montserrat, 89–91
  slopes, 216, 217
Briareum, 101, 158–160, 168, 169, 194, 195
British Indian Ocean Territory (BIOT) government, 226, 227, 235–237, 271, 283–297

C
Calcareous green algae, 62, 64, 123, 131
Calcification, 83, 121, 137–138, 185, 186
Carbonate banks, 28, 97, 98, 101
Caribbean, 1, 13, 23, 37, 47, 61, 69, 91, 100, 121, 135, 153, 176, 182, 197, 218, 247, 265
Caridean shrimp, 220
Caryophyllia, 191–193, 197
Cats, 215, 280
Cays-sandy, 28
Chagos Environment Network (CEN), 226
Chagos-Laccadive ridge, 7, 225, 234–235
Challenger Bank, 195
Challenger-HMS Expedition, 190, 192, 193
Channel reefs, 101, 102, 110
Charter boat industry, 41
Chelidonisis, 196, 197
Chelonia mydas, 75, 77, 80, 215, 272
CITES convention, 226
Clades-zooxanthellae, 158–159
Cladocora, 123, 154
Climate change, 2, 19–20, 47, 58, 69, 80, 83, 111, 112, 129, 137, 154–155, 158, 182–186, 201, 210, 215, 226, 235, 247, 248, 253–268, 281, 294, 316
Climate change projections, 235
Climate monitoring, 19–21, 111, 138, 235–236, 268
Coastal protection, 79, 201, 208–209
Coastal protection value, 208–209
Coconut crab (*Birgus latro*), 230, 231, 272
Coconut trees (*Cocus nucifera*), 271, 294
Colpophyllia, 105, 123, 154
Commercial fishing/fisheries, 110, 112, 117, 175, 204, 234, 315, 317–318

C.R.C. Sheppard (ed.), *Coral Reefs of the United Kingdom Overseas Territories*, Coral Reefs of the World 4,
DOI 10.1007/978-94-007-5965-7, © Springer Science+Business Media Dordrecht 2013

Community composition, 102, 104, 109, 257
Conch, 13–15, 45, 46, 70, 75, 84, 92, 102, 110, 166, 217
Conservation, 1, 13, 37, 38, 42–45, 54, 70, 72, 83, 84, 110, 141, 142, 153, 156, 167, 168, 173, 174, 198, 203, 204, 207, 208, 210, 215, 226, 232, 234, 235, 237, 241, 265, 274, 278, 294, 312–315
Conservation planning, 1–2
Cook-Capt James, 213
Cool water upwelling, 27, 216, 236
Copra, 224, 271
Coral-dead, 20, 112, 242, 244
Coralline algae, 20, 62, 64, 65, 123, 128, 177, 178, 186, 213, 216, 217, 221, 308
Corallivores, 266
Corals
  cays, 13, 28, 102, 108, 110, 271
  communities, 3, 17, 66, 123, 127, 128, 136, 155, 156, 169, 189–199, 247, 258, 289
  cover, 17, 19, 33, 38, 45, 71, 72, 83, 112, 123, 124, 127, 137, 145, 173, 201, 241–248, 254–259, 261, 262, 265–268, 307–311
  diseases, 45, 179, 248
  extinction, 248
  growth, 63, 65, 83, 89, 99, 105, 123, 124, 137–138, 155, 246, 307, 308, 310
  juvenile, 138, 244, 246
  mortality, 112, 242, 246, 247, 266, 281, 295
  recovery, 244–247
  reef fisheries, 97, 102, 110, 207, 253–268
Corals-deep sea, 192, 226
Coxl-trnM, 157
Cruise ships, 31, 75, 112, 174, 175, 203, 204
Crustacea, 215, 314
Cultural value, 201, 206–208
Current speeds, 119, 120
Cyanobacteria, 19, 55, 276–277

D
Darwin, 70, 84, 224
Deep terrace and reefs, 63, 66–67, 72
Deep-water coral communities, 189–199
Deltocyathoides, 191, 193
Dendropoma, 128
Development, 23, 31, 32, 34, 37, 41, 45, 47, 59, 61, 63, 65, 69–72, 75, 77, 79–83, 94, 95, 100, 102, 107, 109, 111, 112, 117, 121–122, 127, 131, 136, 140, 142, 154, 159, 174–175, 179, 201, 207, 210, 230, 259, 274, 317, 318
Diadema antillarum ascensionis, 217, 219, 220
Diego Garcia, 8, 223, 241, 254, 271, 283
Diffuse attenuation coefficient, 121
Diplodus ascensionis, 220
Diploria, 17, 18, 62, 64–66, 101, 104, 108, 119, 123, 124, 127, 135, 137, 138, 143, 145, 156, 159, 160, 168, 176, 185, 189–191
Discovery of Chagos, 224, 271
Diseases, 17–20, 28, 33, 38, 45, 52, 53, 61, 69, 75, 83, 93, 95, 112, 113, 167, 173, 178–182, 247–248
Dive operators, 21, 41, 203, 210
Diver recreation, 207
Dome collapse-volcano, 89, 91, 95
Dredging, 70, 75, 77, 82, 112, 127, 137, 174, 192
Drop-off, 61, 64, 66, 72, 97, 100, 103, 104
Dunes, 58, 59, 115, 122, 275

E
Eastern Caribbean, 4, 37, 89, 92
Eastern Windward reef System, 93
Echinodermata, 166–167
Echinoids, 62, 167, 216, 217, 305, 309
Echinometra lucunter polypora, 219
Ecological restoration, 282
Ecoregion, 115, 153, 156, 167, 229, 248
Ecosystem
  approach, 37
  condition, 263
  goods and services, 201
  process, 254, 263
Education and sustainable use, 41
Education value, 201, 209
Egmont, 8, 223, 224, 271, 274, 286, 290, 291
Elkhorn coral, 93–95
El Nino-Southern Oscillation (ENSO), 184, 236, 247
Endemic species, 135, 164, 168, 220, 277, 312–314
Environmental awareness, 207
Environmental governance, 1, 8, 313–315
Environment Charters, 226, 313
Eunicea, 101, 127, 144, 160, 193–197
Extinction, 233, 248, 265

F
Faecal steroids, 283, 286
Favia, 101, 135, 138, 157, 160, 176, 190, 191, 307
  F. gravida, 216, 218
Finance industry, 69–70
Fire coral, 218, 219, 308
Fisheries
  commercial, 110, 112, 117, 168, 175, 176, 204, 234, 315, 317
  commercial value, 205
  longline, 232, 233, 263, 315
  purse seine, 232, 233, 315
  recreational, 16, 76, 168, 186, 204–205, 207, 232, 253, 254, 257, 261, 268
Fishermen, 41, 44, 54, 95, 96, 176, 204, 205, 315
Fishes, 102, 110, 135–148, 161, 162, 164, 168, 176, 204, 217, 219, 220, 228–230, 232, 236, 253, 254, 258, 259, 265–267, 300, 306–312, 314
  abundance, 207, 254
  biomass, 253–263, 265, 268
Fishing, 13, 31, 41, 76, 95, 110, 117, 140, 173, 203, 215, 226, 253, 290, 315–316
Fishing-deep water, 235
Flame-retardants, 283, 286–289
Fore-reef slope, 61, 63, 66, 123, 124, 126, 144, 176
Fore-reef slope reefs, 66–67
Forest, 19, 20, 53, 55, 78, 79, 89, 274–276
Fossil corals, 122
Fouling organisms, 247
Fregata aquila, 215
Fringing reef rubble flat, 64
Fungi, 276–277

G
Gene flow, 157, 158, 165–167, 230, 231
Genetic variation, 157–158, 165–167
Geomorphological map, 5, 9, 10
Georgetown-Ascension, 216
GIS, 1, 2, 39, 41, 100, 142, 143
Global comparison of reef areas, 3, 248, 249

Goods and services, 201, 209
Gorgonia, 101, 127
Grazing, 19, 55, 143, 219, 220
Great Chagos Bank, 223–225, 228, 246, 248, 254, 256, 262, 264, 271, 279, 284–286, 290, 291, 294
Green turtle *Chelonia mydas*, 75, 215
Growth, 17, 19, 23, 27, 61–63, 65–67, 76, 82–85, 89, 93, 99, 104, 105, 109, 111, 122–124, 126, 127, 135, 137–139, 144, 155, 174, 177, 178, 181, 185, 210, 225, 237, 246, 250, 278, 293, 307, 308, 310
Gulf Stream, 119, 153, 154, 201

H
Habitat degradation, 19, 254, 265–267
Halichoeres, 164
Hawksbill turtle *Eretmochelys imbricata*, 272
Herbivory, 253
Hermaphroditism, 162
Historical taxonomy, 198
Holacanthus, 161, 164
Horseshoe Reef, 31, 37, 52, 59
Human impacts, 42, 45, 173, 247, 254, 263, 267, 268, 316
Hurricane effects, 17
Hurricanes, 17–19, 27, 33, 45, 53, 58, 63, 71, 81–83, 85, 99, 109, 113, 119, 120, 208–210
Hybrid species, 164
Hydrocarbons, 283–286

I
Illegal fishing, 176, 233, 253, 263, 290
Important Bird Areas (IBAs), 273, 274, 279
Incident solar radiation, 119
Indian Ocean Dipole (IOD), 236
Insects, 280
Interglacial highstand, 47, 50
Intertidal pools, 220
Intraspecific genetic diversity, 157, 158
Invasive species, 21, 70, 83, 93, 173, 176, 248
Inventory of reef geomorphological diversity, 1
Ironshore, 63, 71, 80–82
Islands, 1, 13, 23, 37, 47, 61, 69, 89, 97, 115, 140, 154, 175, 189, 213, 223, 249, 253, 271, 284, 299
Isophyllia, 102, 127, 159, 160, 189, 190, 198
ITS, 159–161

J
Javania, 191, 193

K
K_d/K_dPAR, 121

L
Lagoonal reefs, 124–127, 140, 143, 173, 174
Lagoons, 1, 8, 28, 50, 52, 65, 71, 75–77, 79, 80, 82, 102, 104, 113, 183, 225, 226, 243, 245, 293, 294, 296
Land based activities, 41
Landsat, 3
Larval duration, 164–166, 228, 230, 232
Larval flows, 165, 167, 228, 230
Leeward islands, 13, 89
Legislation, 15, 31–33, 70, 82, 84, 85, 168, 175, 210, 313
Leptogorgia, 144, 195, 196, 198
Lesser Antilles, 3, 13, 50, 53, 70, 154
Licences, 215, 226, 232, 254, 317
Limestone platform, 7, 13
Linear reef(s), 38, 39, 107, 127, 128
Lionfish, 21, 34, 61, 69, 70, 83, 85–87, 93, 112, 147, 162, 167, 176
Lithothamnia, 217, 221
Lobster, 13–16, 45, 70, 75, 79, 84, 102, 110, 117, 147, 166, 176, 204, 205, 235, 300, 312, 315, 316
Location of UK Dependent territories, 1–10
London representation of territories, 226, 234, 313

M
Macroalgae, 17–20, 72, 75, 128, 130, 247, 306
Madracis, 124, 127, 138, 143
Madrepora, 191, 193
Management, 8, 13, 16, 21, 22, 28, 32, 33, 37–46, 61, 69, 70, 72, 75, 82–84, 97, 110, 112, 113, 121, 139, 153, 167–169, 173, 176, 186, 208–210, 226, 232–235, 237, 248, 250, 253–268, 271, 274, 276, 281, 296, 313, 315, 317
Management plan, 32, 33, 41, 42, 44, 167, 313
Mangroves
  Avicennia marina, 275
  Lumnitzera racemosa, 275
  Pemphis acidula, 275
Mapping, 1–3, 7, 8, 19, 109, 120, 141–142, 228
Marine oxygen isotope stages (MIS), 122
Marine parks and protected areas, 37, 110
Marine Protected Areas (MPAs), 14, 15, 37–46, 70, 72, 82, 95, 144, 147, 168, 175, 210, 223–237, 248, 253, 254, 261, 263, 268, 272, 273, 281
Melichthys niger, 217, 219, 220
Mesophotic zone, 156, 158, 193
Metal pollution, 289
Microplastics, 284, 290, 291
Microplate, 23, 47
Microsatellites, 157, 159, 160, 166, 229
Millennium Coral Reef Mapping Project, 2, 7
Millepora complanata, 62, 64, 102, 104, 218
Milleporidae, 102, 218, 219
MIS 5e, 122
Mitochondrial DNA, 164, 166
Molluscs, 67, 92, 123, 154, 158, 167, 213, 220, 306, 308, 310, 314
Monitoring programme, 19, 41, 83, 235–236, 277, 284, 300
Montastraea, 17, 18, 53, 75, 101, 119, 126, 135, 137, 139, 143, 145, 156, 176, 190
Morphological units, 3, 7, 8
Mosses, 275, 276
MPA network design, 41
Multilateral environmental agreements, 41
Muricea, 143, 160, 194–197
Mycteroperca, 139, 140, 161, 166, 168, 176

N
Near-shore fringing reefs, 101–104
Near-shore patch reefs, 101, 102, 104, 110
New records, 158, 189–199
Noddies, 216
Noddy terns *Anous tenuirostris*, 274
Non-reefal hard bottom communities, 105–106
Northern reef system-Montserrat, 93
Nutrients, 21, 75, 79, 82, 112, 117, 120, 121, 139, 175, 184, 186, 236, 247, 248, 294, 307, 308

O

Ocean acidification, 70, 72, 83, 138, 182, 185, 186, 281
Octocorallia, 156, 158, 167, 189–199, 235
Oculina, 127, 143, 160, 189–191
Oil, 39, 83, 137, 174, 224, 271, 284–286
Oil Islands, 224, 271
Onychoprion fuscatus, 215, 274
Organochlorines, 284, 286
Overfishing, 14, 19, 21, 95, 226, 237, 258, 294, 313
Overseas Territories Environment Programme (OTEP), 13, 41, 45

P

Palythoa, 64, 101, 104, 217, 219, 221
Parantipathes, 197
Patch reefs, 2–4, 7, 8, 17, 31, 34, 37, 38, 52, 61–64, 66, 72, 74, 92, 99–104, 110, 116, 119, 122–124, 127, 137, 140, 141, 143–147, 174, 176, 178, 304, 305, 307, 308
Pelagic larval duration, 165–167, 228
Pelagic MPAs, 234
Peros Banhos atoll, 256, 275, 291
Persistent organic pollutants (POPs), 283, 284, 286
Pew Environment-Ocean Legacy Program, 226, 315
*p*H, 83, 120–121, 185, 293, 295, 296
Photic zones, 227
Phylogenetic clade, 158
Pillar coral, 93, 94, 108
Pinnacle reefs, 127
Piscivores, 253, 265
Plantations, 224, 271, 272, 274, 275, 277–281, 292, 293
Platform margin bank barrier reefs, 101, 104–105, 107, 108
Platform margin deep reefs, 101, 105, 106
Pleistocene, 25, 29, 47–59, 63, 81, 98, 107, 121–123, 154, 155, 162, 198
- reefs, 47–59, 122–123
- sea-levels, 121

Plexaurella, 101, 138, 144, 158–160, 176, 179, 193, 195–197
Pollution, 21, 72, 83, 109, 175, 187, 207, 226, 250, 283–297, 313
Polychlorinated biphenyls (PCBs), 277, 283, 286, 287, 290, 297
Polyfluorinated compounds, 283, 286–289
Polymyces, 191, 193
Populations, 13, 15, 16, 19, 21, 23, 34, 47, 59, 69–71, 75, 76, 79–82, 85, 89, 92, 93, 95, 97, 104, 109–111, 117, 136, 139, 140, 145, 153, 154, 157–159, 162, 164–168, 175–179, 195, 201, 207, 208, 215, 226, 229–232, 234, 235, 237, 250, 253, 254, 259, 263, 265, 266, 268, 271–274, 277–280, 292–294, 300, 313, 316, 317
- connectivity, 157–158, 165–168
- growth, 111, 237, 250

Porifera, 215, 248, 308
Porites, 17, 18, 62, 64, 66, 104, 105, 123, 124, 143, 176, 242, 307
Postage stamps, 317
Predation, 215, 253, 265, 280, 316
Procaris ascensionis, 220
Producer surplus-economic, 203, 204
Protected areas, 13–15, 37–46, 70, 72, 74, 76–79, 81–83, 95, 110–113, 144, 147, 168, 175, 210, 223–237, 253–268, 272–274, 281, 315
Public relations campaign, 42
Pyroclastic volcano flow, 89, 95

R

Rabbits, 215
Rainfall, 27, 28, 55, 79, 119, 236, 293, 313, 317
Raised reefs, 225
Ramsar, 53, 226, 273, 278, 313, 315
Rats, 215, 271, 274, 280, 281, 316
Recreational fisheries value, 204–205
Recreational value, 205
Recruitment, 83, 109, 127, 135–137, 140–141, 154, 162, 166–168, 174, 175, 220, 230, 244, 278
Red-footed Booby Sula sula, 273
Reef
- areas, 1–4, 7, 16–19, 21, 75, 76, 91, 136, 227, 246, 248, 253, 263
- assessments and threats, 110–113
- condition, 241, 248–250, 266
- crest, 28, 38, 50, 52, 64, 71, 72, 74, 104, 106, 107, 109
- definition, 3
- fish, 16, 19, 61, 84, 85, 96, 97, 102, 110, 139–140, 145–148, 162–164, 168, 176, 205, 207, 228–230, 232, 247, 253–268, 306, 307, 309–311, 314–316
- geomorphology, 61, 97, 110
- locations, 209, 263
- management, 61, 69, 97, 112, 121, 167, 210
- mapping, 3, 7, 141–142
- mapping and areas, 1–4, 7, 16–19, 21, 75, 76, 91, 136, 141–142, 227, 246, 248, 253, 263
- morphology, 28, 37
- mortality, 19, 20, 112, 144, 178, 241–244, 247
- typology, 1–3, 8
- zonation, 63–64, 101, 109, 142–148

Remote sensing, 1, 8, 98
Reproduction, 135–136, 139–140
Reptiles, 277, 280, 281
Research value, 201, 209
Resilience, 41, 45, 158, 167–169, 250, 268, 284, 294
Resplendent, 220
Rhizopsammia, 156, 191, 192
Ridge plain, 25, 29, 50–52, 55–59
Ridges and low profile rock shelves-Montserrat, 89, 91–92
Rose Tree *Barringtonia asiatica*, 274
Royal Society for the Protection of Birds (RSPB), 215, 313, 316

S

Salomon atoll, 245, 248, 256, 286, 287, 289, 291, 294
Salt ponds, 21, 28, 47, 53–56, 100
Sand plain, 61, 63, 66, 72, 74, 100, 105, 109
Sargasso Sea, 8, 115, 119–121, 138, 153, 167, 175, 286, 295
Sargassum, 62, 65, 107, 153, 166, 167
Satellite remote sensing, 8
Scaevola toccada shrubs, 276
Scleractinia, 63, 101, 106, 135, 156–157, 160, 167, 169, 178, 189–199, 213, 217, 218, 235, 300, 308, 309, 311
Seabirds, 59, 76, 82, 100, 110, 215, 234, 272–274, 278–280
Sea cucumber, 220, 226, 233, 284, 293–294
Seagrasses, 17, 29, 38, 41, 52, 53, 76, 77, 82, 123, 128, 155, 225
Sea level, 2, 3, 7, 8, 13, 47, 50, 51, 55, 59, 62–64, 80, 81, 97, 102, 104, 115, 121–122, 128, 154, 182, 184, 186, 216, 281, 293–295, 301–303, 316
Sea level rise, 8, 47, 50, 59, 80, 122, 182, 184, 186, 281, 294, 295, 316
Seamounts, 2, 8, 115, 116, 118, 154, 158, 176, 192, 196, 225, 234–235, 301
Seasonality, 139, 153, 155–156, 179
Sea surface salinity, 120
Sea surface temperature, 72, 113, 119, 120, 122, 153, 154, 184, 236, 301
Sea urchins, 62, 64, 77, 219, 309–311. *See also* Echinoids
Sediment, 8, 20, 50, 72, 75, 76, 82, 92, 93, 99, 102, 105, 122, 123, 127, 136, 144, 174, 216, 248, 284–288, 290, 317

Sedimentation, 32, 33, 42, 45, 82, 92, 93, 99, 136, 137, 173, 174, 250, 313, 317
Serranidae, 162, 164, 309
Settlement, 13, 82, 110, 115–117, 136–137, 140, 175, 237, 254, 265, 274
Sewage, 45, 82, 129, 137, 175, 283, 286, 293, 296
Sewage bacteria, 175
Shallow Marine Surveys Group, 215
Shallow terrace, 61, 63, 65–66
Shallow terrace and reefs, 61, 63, 65–66
Shark fin, 263
Sharks, 120, 176, 204, 226, 233–235, 253, 254, 256, 257, 259, 263–265, 268, 293–295, 308, 311
Shellfish, 14
Siderastrea, 17–19, 62, 64, 66, 102, 105, 109, 136, 137, 143, 160, 176, 189–191, 198, 199, 217, 218
Soft coral cover, 20, 71, 72, 178, 202, 242, 245, 246, 311
Sooty Tern *Onychoprion fuscatus*, 274
South Equatorial Current, 216
Southern Equatorial Counter-Current, 216
Southern reef System, 93
Spawning aggregation, 41, 84, 85, 102, 139, 140
Species-island area curve, 275
Spur and grooves, 7, 8, 38, 61, 65, 66, 71, 74, 100, 102–107, 109, 123, 304
SRP54, 157
Stakeholder review process-stakeholder feedback, 41
Standardised comparison of reef areas, 256, 262
Stegastes lubbocki, 220
Stenocyathus, 191, 193
Stepping-stone-genetic, 229, 232
St Helena, 1, 213, 215, 218, 220
Stichopathes, 197
Stresses-human induced, 248
Strict Nature Reserves, 42, 43, 274, 292, 294
Structural complexity, 246, 256, 257, 259, 261
Submarine shelf, 4, 90, 92
Submersible exploration, 192
Sustainable resource use, 37, 43
Swamp-*Typha*, 275
Symbiodinium, 158–161, 168–169, 249

T
Tar balls, 283, 284, 286
Taxonomy, 198
Telestula, 196, 197
Temperature increase, 72
Temperature monitoring, 27, 120, 235, 296
Tethocyathus, 191, 193
Thalassoma ascensionis, 220
Threatened species, 226, 233, 272, 277, 314
Tidal range, 27, 61, 119, 216, 301
Topography, 50, 58, 61–62, 71, 80, 90, 92, 100–101, 104, 216, 235, 275
Topography-island, 100–101
Total economic value, 201–210
Tourism, 13, 15, 16, 23, 31, 32, 38, 43, 61, 69–70, 79, 80, 82, 84, 85, 110, 112, 117, 173–175, 201–204, 209, 210, 253, 313, 317, 318
Tourism values, 201–204
Tourist revenue, 201, 203
Treaties, 111
Triggerfish, 20, 162, 165, 217
Tristan da Cunha, 1, 215
Trophic, 253, 254, 256, 257, 259, 260, 263, 265, 286
Tropical cyclones, 47, 71, 115, 186
Tropical Northwestern Atlantic (TNA), 115, 153, 154, 156, 161, 168
β-Tubulin, 157
Tuna, 205, 226, 232–234, 263, 315
Turtles, 15, 17, 37, 41, 59, 70, 75, 77, 80, 83, 84, 95, 100, 110, 173, 213, 215, 226, 229, 230, 234, 272, 273, 277, 278, 281, 290, 291, 314

U
Upwelling, 27, 99, 112, 216, 236

V
Value, 3, 13, 14, 42, 43, 71, 79, 81, 85, 110, 119–124, 127, 157, 174–176, 201–210, 226–228, 232, 234, 237, 242, 244–246, 249, 253, 254, 256, 257, 259–261, 264, 265, 273, 292, 315
Vegetation assemblages, 274, 275
Video surveys, 245
Volcanic activity, 89, 91–93, 95, 121, 300, 301
Volcanic Flanks, 93
Volcanic islands, 4, 8, 23, 28, 301
Volcanism, 117, 300

W
Waifs, 162
Warming events, 242
Warming water, 184, 242, 249, 250
Wave-cut platform, 216, 302
West Africa, 27, 213, 217, 218, 220
Western Leeward reef System-Montserrat, 92
White band disease, 17, 19, 28, 33, 38, 52, 93, 95, 180
Willingness to pay, 203, 204, 207–208
Wrasse-Ascension, 220

X
Xestospongia, 67, 105

Y
Yacht charter, 31

Z
Zonation, 61, 63–64, 67, 97, 101, 109, 142–148
Zooxanthellae, 75, 139, 158–159, 169, 249

MIX
Papier aus verantwortungsvollen Quellen
Paper from responsible sources
FSC® C105338

If you have any concerns about our products,
you can contact us on
ProductSafety@springernature.com

In case Publisher is established outside the EU,
the EU authorized representative is:
Springer Nature Customer Service Center GmbH
Europaplatz 3, 69115 Heidelberg, Germany

Printed by Libri Plureos GmbH
in Hamburg, Germany